Dynamic Simulation
of Electric Machinery

Dynamic Simulation of Electric Machinery

Using Matlab/Simulink

Chee-Mun Ong

School of Electrical & Computer Engineering
Purdue University
West Lafayette, Indiana
E-mail: ong@ecn.purdue.edu

To join a Prentice Hall PTR
Internet mailing list point to:
http://www.prenhall.com/mail_lists/

PRENTICE HALL PTR
Upper Saddle River, New Jersey 07458

Library of Congress Cataloging-in-Publication Data

Ong, Chee-Mun.
    Dynamic simulation of electric machinery / Chee-Mun Ong.
      p. cm.
    Includes bibliographical references and index.
    ISBN 0-13-723785-5
    1. Electric machinery--Computer simulation.   I. Title.
  TK2391.0533   1997
  621.31'042'011353042--dc21   97-20313
                                             CIP

Editorial Production: bookworks
Acquisitions Editor: Russ Hall
Manufacturing Buyer: Alexis R. Heydt
Cover Designer Director: Jerry Votta
Marketing Manager: Miles Williams
Cover Designer: Design Source

 © 1998 by Prentice Hall PTR
Prentice-Hall, Inc.
A Simon & Schuster Company
Upper Saddle River, NJ 07458

Prentice Hall books are widely used by corporations and government agencies for training, marketing, and resale.

The publisher offers discounts on this book when ordered in bulk quantities.
For more information, contact:

  Corporate Sales Department
  Phone: 800-382-3419
  FAX: 201-236-7141
  E-mail: corpsales@prenhall.com

or write:

  Corporate Sales Department
  Prentice Hall
  One Lake Street
  Upper Saddle River, New Jersey 07458

Printed in the United States of America

10 9 8 7 6 5 4 3 2 1

ISBN 0-13-723785-5

Prentice-Hall International (UK) Limited, *London*
Prentice-Hall of Australia Pty. Limited, *Sydney*
Prentice-Hall Canada, Inc., *Toronto*
Prentice-Hall Hispanoamericana, S. A., *Mexico*
Prentice-Hall of India Private Limited, *New Delhi*
Prentice-Hall of Japan, Inc., *Tokyo*
Simon & Schuster Asia Pte. Ltd., *Singapore*
Editora Prentice-Hall do Brasil, Ltda., *Rio de Janeiro*

Contents

Preface

This book is about the techniques of modeling and simulating electric machinery. It evolved from a graduate-level course in simulation at Purdue that is aimed at introducing students to the modeling of power components and to using computer simulation as a tool for conducting transient and control studies. Simulation can be especially helpful in gaining insights to the dynamic behavior and interactions that are often not readily apparent from reading theory. Next to having an actual system to experiment on, simulation is often chosen by engineers to study transient and control performance or to test conceptual designs. Its use in teaching is growing now that inexpensive and powerful personal computers with simulation capabilities previously available only in specialized facilities are widely accessible to students.

There are many excellent textbooks on electric machine models, operating principles, and steady-state characteristics, but few dwell on the implementation of simulations. While the primary focus of this book is on the techniques applicable to the modeling and simulation of electric machinery, it also illustrates the usefulness of knowing the physical aspects of machines, the assumptions made when modeling machines, and the basic operating characteristics of these machines in verifying the correctness of an implemented simulation and in interpreting simulation results. The ability to correctly interpret and verify results is important not only for simulation, but also in other work where engineers are increasingly relying on software packages developed by others. Exercises to develop the skills to discern are an important aspect of the projects given at the back of each chapter. The chapter projects are of two kinds: one kind reinforces the learning of the operating principles and characteristics, and of the implementation and modeling techniques; the other kind focuses on selected problems of practical interest about which students can explore and learn more about the phenomena first-hand.

Simulation must be taught with a tool with which students can practice. I have chosen MATLAB/SIMULINK[®1] as the platform for this book simply because of the short learning curve that most students require to start using it, its wide distribution, and its general-purpose nature. The platform allows one to conveniently switch back and forth between simulation on SIMULINK and computation on MATLAB, such as when performing control design using the Control Toolbox or when analyzing signals using the Signal Processing Toolbox. For the convenience of the reader, I have provided the SIMULINK implementations for almost all of the chapter projects in a CD-ROM that accompanies the book. One set of implementations is for use on MATLAB version 4.2c and SIMULINK version 1.3, the software on which these simulations were originally developed. The other set of implementations is for those using MATLAB version 5 and SIMULINK version 2. The differences between the two sets are minor, mostly the later versions eliminate warnings about unconnected input or output ports. These implementations primarily facilitate learning, though many of the machine models may find use in other studies. Nevertheless, the modeling and simulation techniques discussed in this book can be extended to other situations and are by no means limited to the chosen platform.

The materials are presented in traditional order. The first chapter contains some general remarks about modeling and dynamic simulations. The second chapter introduces the reader to MATLAB/SIMULINK. The third chapter reviews the properties of ferromagnetic materials and the basics of magnetic circuits that are essential to the modeling of electric machines. Chapter 3 also discusses the origin of circuit parameters in line models that are used to model the connections between machines and sources. The fourth chapter on transformers introduces the techniques of modeling and simulating stationary magnetically coupled circuits and core saturation that are equally applicable to rotating machines, but without the complications of motion. Chapter 5 reviews the fundamentals of excitation, field distribution, winding inductances, and energy conversion common to rotating machines that are covered in later chapters. Chapters 6 and 7 describe two of the most common three-phase ac machines: induction and synchronous machines. Besides deriving nonlinear models of these machines and describing how to implement them in simulations, these chapters also illustrate how to obtain small-signal or linearized models of the same machines from their SIMULINK simulations. Chapter 8 covers modeling, simulation, and control of dc machines. The modes of operation of a four-quadrant drive and the basics of speed and torque control are introduced using the simpler operational behavior of a dc machine. Chapter 9 is about the operation of an induction machine from an adjustable frequency source. It focuses mainly on the volts/hertz control and field-oriented control commonly used in modern induction motor drives. The tenth chapter describes synchronous machine models used in power system studies where the size and complexity of the system often justify the use of models with different levels of detail. It also describes the self-control methods used in small and large synchronous machine drives.

---

[1] MATLAB and SIMULINK are registered trademarks of the The MathWorks, Inc.

I am grateful to my graduate students for their suggestions and criticisms of the course material and to Russell Kerkman, Kraig Olejniezak, Rich Schiferl, Hong Tsai and David Triezenberg for their reviews of the manuscript. It has been a pleasure working with the staff of Prentice Hall. For that I would like to thank Russ Hall for his guidance and encouragement on the preparation, Camie Goffi for her careful editing, and Lisa Garboski for her help in the production of the book. I would like to also thank Todd Birchmann for recording the compact disk. Finally, I am deeply grateful to my wife, Penny, for her encouragement and understanding, and to our children, Yi-Ping, Yi-Ching and Yan, for their patience and understanding.

Chee-Mun Ong
West Lafayette, Indiana

*Dynamic Simulation
of Electric Machinery*

1

Introduction

1.1 ON MODELING IN GENERAL

A theory is often a general statement of principle abstracted from observation and a model is a representation of a theory that can be used for prediction and control. To be useful, a model must be realistic and yet simple to understand and easy to manipulate. These are conflicting requirements, for realistic models are seldom simple and simple models are seldom realistic. Often, the scope of a model is defined by what is considered relevant. Features or behavior that are pertinent must be included in the model and those that are not can be ignored. Modeling here refers to the process of analysis and synthesis to arrive at a suitable mathematical description that encompasses the relevant dynamic characteristics of the component, preferably in terms of parameters that can be easily determined in practice.

In mathematical modeling, we try to establish functional relationships between entities which are important. A model supposedly imitates or reproduces certain essential characteristics or conditions of the actual—often on a different scale. It can take on various forms: physical, as in scale-models and electrical analogs of mechanical systems; mental, as in heuristic or intuitive knowledge; and symbolic, as in mathematical, linguistical, graphical, and schematical representations.

Simulation can be very useful in many scientific studies that proceed as follows:

1. Observing the physical system.

2. Formulating a hypothesis or mathematical model to explain the observation.

3. Predicting the behavior of the system from solutions or properties of the mathematical model.

4. Testing the validity of the hypothesis or mathematical model.

Depending on the nature of the actual physical system and the purpose of the simulation, the definitions of modeling and simulation will vary. Broadly speaking, simulation is a technique that involves setting up a model of a real situation and performing experiments on the model. In this book, we will define simulation to be an experiment with logical and mathematical models, especially mathematical representations of the dynamic kind that are characterized by a mix of differential and algebraic equations. Mathematical models may be classified in many ways, some of which are as follows:

Linear or Nonlinear   Linear models can be described by just linear mathematical relations which obey the principle of superposition. Nonlinear models, on the other hand, possess mathematical relations that are not linear.

Lumped or Distributed Parameter   Lumped parameter models are described by ordinary differential equations with only one independent variable. Distributed parameter models are described by partial differential equations usually with time and one or more spatial coordinates as independent variables.

Static and Dynamic   Static models do not take time variations into account, whereas dynamic models take into account time-varying characteristics and interactions.

Continuous or Discrete   Continuous time models are described by equations in which the dependent variables are continuous in time. Discrete time models are described by difference equations whose dependent variables are defined at distinct instances.

Deterministic or Stochastic   A model is deterministic if chance factors are absent, stochastic if chance factors are taken into account.

The procedure for developing a model is often an iterative one. The cycle begins with identifying the purpose of the model and its constraints, as well as the kinds of simplifying assumptions or omissions that can be made, determining the means of obtaining parameters for the model, and defining the available computational facilities. Both an understanding of and insight into the discipline are essential to making appropriate simplifying assumptions. Simplicity is the hallmark of most good models. Whereas oversimplification and omissions may lead to unacceptable loss of accuracy, a model that is too detailed can be cumbersome to use. There can be more than one model for the same physical system, differing in precision, aspect, and range. For example, a transmission line may be represented by a distributed parameter model, a lumped RLC model, or a lumped RL model. Similarly, the model of an electromagnetic component with an iron core may or may not include the effects of magnetic saturation or hysteresis of the core depending on the purpose of the simulation.

Every model has parameters which are to be estimated. The model must lend itself to methods by which these parameters can be determined experimentally, otherwise the model will be incomplete.

The developed model must be verified and validated. Verification involves checking the consistency of the mathematics, the solution procedure, and the underlying assumptions. Validation is the determination of how adequately the model reflects pertinent aspects of the actual system that are represented by it. When a discrepancy is unacceptably large, the model must be revised and the cycle repeated. Data used for estimating the parameters ought not be the same as those used to validate the model.

Modeling and simulation have appropriate uses. They are especially beneficial in situations where the actual system does not exist or is too expensive, time-consuming, or hazardous to build, or when experimenting with an actual system can cause unacceptable disruptions. Changing the values of parameters, or exploring a new concept or operating strategy, can often be done more quickly in a simulation than by conducting a series of experimental studies on an actual system. Simulation can also be a very useful training aid; it is a technique by which students can learn more and gain greater insight and better understanding about the system they are studying.

A frequent question about simulation is its validity. Do the simulation results reflect those of the actual system for the condition simulated? Even with valid component models, the use of them in a larger simulation must be done carefully with consistency and a well-defined goal in mind; otherwise, the results could be meaningless. Finally, in interpreting the results, we should not overlook the simplifications and assumptions made in devising the model.

1.2 MODELING POWER COMPONENTS

Actual experimentation on bulky power components can be expensive and time-consuming. For many, simulation offers a fast and economical means if not the only means by which they can conduct studies to learn more about these components.

Power components should be designed to withstand expected stresses caused by temporary overvoltages, surges, and faults. Since extreme stresses usually occur during abnormal operation and transient conditions, the design of these components is often dictated by transient considerations. Some examples are: persistent overvoltages which can affect insulation coordination; surges caused by lightning and switching; undesirable interactions such as ferroresonance between nonlinear magnetizing inductance and circuit capacitance; and subsynchronous resonance between the torsional modes of the turbine shaft and natural frequency of the network.

It is beneficial for us to briefly review how others have modeled and simulated power components. Power systems are usually so large and complex that modeling all components in detail is seldom attempted. Aside from the sheer number of components, the spread of their frequency response can also be very broad. Significant reduction in dimension and complexity of the model is often done by making judicious approximations, by limiting fidelity or coverage, or by employing both of these means. Examples of such abstraction are

- Physical partitioning, as in the use of simple equivalents to model distant parts of the system which have minor effects on the behavior that is of interest because of its localized nature, as in the use of network equivalents at remote buses for fast transient studies.
- Frequency domain partitioning, as in the selective use of low-frequency models known to be adequate for portraying the behavior of interest.

Models of various degrees of mathematical complexity have been developed to exploit such abstractions in specific situations. We should therefore be aware of the original purpose and assumptions when using these models. Indiscriminate use of elaborate models for all circumstances will not only result in cumbersome and inefficient simulations, but also in compromised accuracy if the parameters for these models are in question.

1.3 TOOLS FOR DYNAMIC SIMULATION

Except for very simple systems, computing tools are needed to handle the complex models used for transient situations. The three computing tools which have been employed for transient studies are

- Network analyzers.
- Analog and hybrid simulators.
- Digital computers.

On network analyzers, a scaled circuit model of a selected portion of the system is constructed. Usually a limited number of components essential to the situation under study is modeled. Scaling in voltage, current, and frequency are often used to reduce the size and cost of the circuit model. The degree of detail and extent of the model will depend on the situation and the capacity of the analyzer. For example, if surge studies on specific lines are to be studied, the lines directly affected could be represented in greater detail using several more II sections; lines further away could be represented by lesser models. But, if a network analyzer is to be used to study synchronous machine stability, line models can be simpler because the frequency range of the electromechanical interactions involved in such situations need not be as high as the range used for surge studies.

On analog and hybrid simulators, a mathematical rather than circuit model of a selected portion of the system is implemented. On analog simulators, the mathematical model is implemented using electronic operational amplifiers capable of performing the basic mathematical operations of integrating, summing, multiplying(dividing), and generating some simple nonlinear functions—all in parallel. The output level is usually scaled down to ± 10 or ± 20 volts to stay within the amplifier's saturation limits. Also to stay within the frequency response of the amplifiers and auxiliary equipment, such as plotters, time scaling is often used. Time scaling applied to the integrators can slow the simulated response in proportion to the actual response time, thus enabling the same equipment to be used for studying a broad range of transients from the very fast to the very slow. Greater capability and ease of use can be obtained by operating an analog simulator in conjunction with a digital computer. Such an installation is referred to as a hybrid simulator. With hybrid simulators, the digital computer is often used to automate the setting up of the parameters on the analog computer, to control the study sequence in multiple case studies, to simulate long delays, to emulate control loop functions, and to process acquired data—the kinds of tasks at which a digital computer excels.

For large-scale system studies, the burden of abstraction can be avoided when the tool has the capacity to handle large models. The power of the digital computer to process large amounts of information systematically was recognized very early on. Many activities that used the digital computer to perform loadflow and short-circuit calculations were well on their way by the 1950's. Today, there is a wide selection of software available to do loadflows, short-circuit calculations, protection settings, dynamic and transient stability studies, electromagnetic transients, and operational and design optimization studies.

2

Introduction

to

MATLAB/SIMULINK

2.1 INTRODUCTION

Simulation packages may be loosely divided into general-purpose and application-specific packages. Most general-purpose packages are equation-oriented in that they require input in the form of differential or differential-algebraic equations. An application-specific package, on the other hand, may provide convenient, ready-to-use modules or templates of commonly used components for a specific application. A user can also add his/her own modules in equation form. These modules are then incorporated in a network-like fashion by the package. Among the better known simulation packages, ACSL [1], ESL [2], EASY5 [3], and PSCSP [5] are for general systems, whereas SPICE2 [4], EMTP [6], and ATOSEC5 [7] are mainly for simulating electrical or electronic circuits. IESE and SABER [8] are examples of general-purpose electrical network simulation programs that have provisions for handling user-defined modules. Besides these, there are also standard differential and differential-algebraic solvers in callable routines, such as IMSL [9], ODEPAK [26], and DASSL [127].

SIMULINK®[1] is a toolbox extension of the MATLAB[1] program. It is a program for simulating dynamic systems. Student editions of MATLAB 5 and SIMULINK 2 are currently available through Prentice Hall [148]. The SIMULINK simulations given in the CD-ROM accompanying this text were originally developed on MATLAB version 4.2c and

---

[1]MATLAB and SIMULINK are registered trademarks of the The MathWorks, Inc.

6

SIMULINK version 1.3c. To accommodate the newly released MATLAB version 5 and SIMULINK version 2, files for these versions are also provided on the CD-ROM. For the applications in this text, the differences between these new and the older versions are minor.

Briefly, the steps of using SIMULINK involve first defining a model or mathematical representation and the parameters of your system, picking a suitable integration method, and setting up the run conditions, such as the run-time and initial conditions. In SIMULINK, model definition is facilitated by the graphical interface and the library of templates or function blocks that are commonly used in mathematical descriptions of dynamic systems. The objective of this chapter is to guide a new user through some simple setups, bringing to his/her attention some features that are useful for common tasks. For more details, the reader should refer to the Reference and Tutorial sections of the *SIMULINK User's Guide* by The MathWorks, Inc. [13] or to the online Help Desk in MATLAB 5. Those who can program in C or Fortran might be interested to know that SIMULINK has provisions for you to create and mask your own functional block using a MATLAB *.M* file, or a *.MEX* C or Fortran file.

2.2 CD-ROM FILES

The CD-ROM that accompanies this book contains MATLAB and SIMULINK files for the exercises in Chapter 2 and almost all of the projects in Chapters 3 through 10. Further information about the CD-ROM's contents can be found in Appendix B.

Some of these simulations are quite large. For the SIMULINK screen to display the full simulation, you may have to increase the display resolution of your computer monitor.

2.3 GETTING TO SIMULINK

SIMULINK is an extension of the MATLAB program. To get to it, you will need to run MATLAB. From the directory in which you intend to work with your SIMULINK files, *enter* **matlab** to start the MATLAB program. By *enter*, I mean type in the text and press the Enter or Return key on your keyboard. The MATLAB program starts with a display indicating the version of the MATLAB program after which it will display the MATLAB prompt ≫ in a MATLAB *command window*. At this point, the program is waiting for a MATLAB command. To start SIMULINK, enter **simulink** after the MATLAB prompt ≫ in the MATLAB *command window*. The program will display the SIMULINK block library as shown in Fig. 2.1. If you are using SIMULINK most of the time, you may want to insert the **simulink** command in your MATLAB *STARTUP* file, in which case SIMULINK will be automatically activated whenever you start MATLAB.

2.4 CREATING A SIMULINK SIMULATION

Prior to setting up a system model in SIMULINK, you will need to have a mathematical description of the system you want to simulate. A typical mathematical description of a

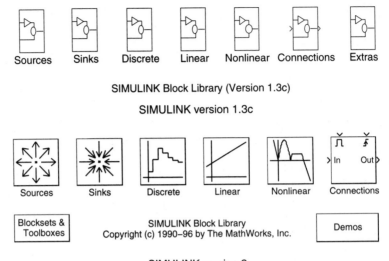

Figure 2.1   SIMULINK Block Library (courtesy of the The MathWorks, Inc.).

dynamic system may consist of a mix of integral and algebraic equations. These equations may have to be manipulated further to eliminate potential algebraic loops. Have an idea as to which variables in your mathematical model are independent and which are dependent. Rewrite the integral equations with the dependent-state variable expressed as some integral of a combination of independent variables and dependent variables, including itself. Construction of the SIMULINK model can then follow the rearranged mathematical description of the model.

Figure 2.1 shows the main SIMULINK block libraries of SIMULINK version 1.3c and SIMULINK version 2. Double-clicking on a menu heading will open up a display of the menu commands under that heading. For example, if you are starting anew and want to create a new SIMULINK model, open the **file** menu heading and select the menu command **new** to produce a blank SIMULINK model screen onto which you can then select and drag components from the block library to assemble your SIMULINK simulation. Note that this screen will initially be named **Untitled** until you give it a name using the **save as** menu command under the **file** menu heading. The assigned filename is automatically appended with an **.mdl** extension in SIMULINK 2 or an **.m** extension in SIMULINK 1.

A variety of function blocks or templates are grouped under the different library blocks. A template can be copied from a library block onto the SIMULINK model screen by first selecting the template and then dragging it to the desired location in the SIMULINK model screen, or by a sequence of copy and paste commands, which are found under the **edit** menu. Selection is accomplished by positioning the mouse pointer on the desired template and clicking the left mouse button once. When a template is selected, its icon becomes highlighted. A selected item can be copied into a buffer using the **copy** menu command under the **edit** menu heading. The copied item can be pasted at some location on a screen by first marking the desired location with a click on the spot and then clicking the **paste** menu

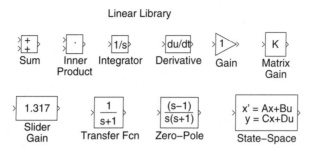

Figure 2.2   Contents of the Linear library block (courtesy of The MathWorks, Inc.).

command under the **edit** menu heading. When starting a new simulation, copying items from the library block is more conveniently done by first selecting the desired template in the library block and dragging it onto the SIMULINK model screen. The combined actions of selecting and dragging are carried out by clicking on the desired template with the left-most mouse button, keeping that button depressed, and moving the pointer to the desired location on the SIMULINK model screen.

Table 2.1 gives a partial listing of function templates in some of the library blocks. The templates can be accessed by double-clicking on the corresponding library block icon to open the library. Then, a copy of the desired template or block can be copied onto your SIMULINK model screen by first selecting the template and then dragging the highlighted template onto the model screen. For example, if you need summers and gains, you first open up the Linear library block by double-clicking on the icon of the Linear library block to display its contents. Figure 2.2 shows the contents of the Linear library block, in which you will find the Sum and Gain templates.

Many templates have internal parameters which you must specify before you can use these templates in a simulation. To view or to set up these parameters, you must double-click on the template, upon which a dialog window will appear with boxes for parameter values to be inserted. You can enter the required information as constants or as named variables. Named variables can be initialized in the MATLAB workspace before you start your simulation. Using named variables may be preferable when you have to perform multiple runs using different parameters values, as in a parameter sensitivity study. Parameters and initial values of named variables and flags may be entered into the MATLAB workspace by typing them in directly, by reading a data file, or by running an M-file that is written to do other things, such as set up the appropriate system condition. Such an M-file can also be started from the SIMULINK screen using a masked block. For a large simulation, the M-file approach is recommended. Creating and debugging the M-file can be conveniently done with the M-file Editor/Debugger from the MATLAB 5 toolbar.

2.5 CHOOSING AN INTEGRATION METHOD

After you have created a SIMULINK model of a system and before you start the simulation, you will need to choose an integration method and specify some run conditions. Under the

TABLE 2.1   SOME TEMPLATES OF SIMULINK 1.3C

Sources	
Clock	Provide and display system time
Constant	Inject a constant value
From File	Read data from a file
From Workspace	Read data from a matrix in Workspace
Signal Generator	Generate various waveforms
Sine Wave	Generate a sine wave
Step Fcn	Generate a step function
Repeating Sequence	Repeat an arbitrary signal
White Noise	Generate random noise

Sinks	
Scope	Display signals during simulation
To File	Write data to a file
To Workspace	Write data to a matrix in Workspace

Connections	
Inport	Input port to a masked block
Outport	Output port of a masked block
Mux	Multiplex several scalar inputs into a vector input
Demux	Demultiplex vector input into scalar component input

Linear	
Derivative	Output a time derivative of the input
Gain	Multiply an input by a constant
Integrator	Integrate an input signal(s)
State-Space	Linear state-space system
Sum	Sum inputs
Transfer Fcn	Linear transfer function in a domain
Zero-Pole	Linear system specified in poles and zeros

Nonlinear	
Abs	Absolute value of input
Backlash	Model hysteresis
Dead Zone	Zero-output dead-zone
Fcn	Any legal C function of input
Look Up Table	Perform piece-wise linear mapping
MATLAB Fcn	Apply a MATLAB function to an input
Product	Multiply inputs together
Rate Limiter	Limit the rate of change of an input
Relay	Switch an output between two values
Saturation	Limit the excursion of a signal
S-function	Make an S-function into a block
Switch	Switch between two inputs
Transport Delay	Delay an input signal by a given amount of time

parameters sub-menu of the main menu heading **simulation**, you can select one of several integration methods and enter the values of simulation parameters, such as tolerance and minimum and maximum step size. In SIMULINK 1.3c, the integration routines available are

- **linsim** A method for solving linear dynamical equations. It computes the derivative and output at each output step.

- **rk23**   The Runge-Kutta third-order method is for solving nonlinear systems with discontinuities. It is recommended for mixed continuous and discontinuous systems, but not for numerically stiff systems. The method takes three internal steps $[0, \frac{1}{2}, 1]$ between output points.

- **rk45**   The Runge-Kutta fifth-order method has characteristics similar to those of *rk23*, except that it takes six internal steps between output points.

- **gear**   The Gear method is a predictor-corrector method for numerically stiff systems. It takes a variable number of internal steps between output points. Since it relies on predictions based on past values, it may not work well when future input changes rapidly.

- **adams**   The Adams method is also a predictor-corrector method that uses a variable number of internal steps between output points. It is particularly efficient when handling systems that have smooth response and are not too numerically stiff.

- **euler**   The Euler method is also a one-step method. It computes the derivative and output at each output point.

SIMULINK 2 has five ODE solvers. The first three are for non-stiff problems and the last two are for stiff problems.

- **ode45**   This method is based on Dormand-Prince(4,5), which is an explicit, one-step Runge-Kutta that is recommended as a *first try* method.

- **ode23**   This method is based on Bogacki-Shampine(2,3), which is also an explicit, one-step Runge-Kutta. It may be more efficient than *ode45* when tolerances are wide.

- **ode113**   This is a multi-step, variable-order Adams-Bashforth-Moulton PECE solver. It is recommended when function evaluation is time-consuming and tolerances are tight.

- **ode15s**   This is a multi-step, variable-order solver based on a backward differentiation formula.

- **ode23s**   This is a one-step solver based on the Rosenbrock formula of order 2. It has the A-stability property.

Depending on your version of SIMULINK and your choice of integration method, you may have to specify some or all of the following integration step size control parameters:

- **tolerance**   Used by the integration routine to control the amount of relative error at each step. The routine tends to take smaller steps when the specified tolerance is small; thus, the run-time will be longer. For the class of problems that we will be handling, the error tolerance may range from $1e^{-3}$ to $1e^{-6}$. If you are not sure initially of what is best for your system, experiment by first starting with something conservative in terms of accuracy and gradually loosen it to reduce run-time until you have a compromise between the accuracy you need and the run-time you are willing to accept. This remark also applies to the initial selection of minimum and maximum step sizes.

- **minimum step size**   This is used to start/restart the integration at the beginning of a run and after a discontinuity. With variable step-size methods, such as the Gear or Adams methods, the specified minimum step size does not affect the accuracy in that the internal step size is varied to give the necessary accuracy, but the specified minimum step size is observed in generating the output. Thus, it is recommended that you specify the minimum step size to be the same as the maximum step size for these methods.

- **maximum step size**   This limits the step length to achieve a smooth appearance in the plot of the output.

Fixed step size simulation of the system model can be obtained with the linsim, rk23, rk45, or Euler integration methods by setting both the minimum and maximum step size to the desired step length.

2.6 STARTING AND RUNNING A SIMULATION

Other than the parameters for the integration method, you will also need to specify the start and stop time of the run before you can start your simulation. The simulation can be started by clicking on the **Start** button under the main menu heading **Simulation** of either the main SIMULINK screen or the model screen. Before you start your simulation, you might want to set up the scope and open up the clock template to monitor the progress of the simulation.

2.6.1 Viewing Variables During a Run

Often in the debugging stage, we need to view certain key variables to check whether the simulation is progressing satisfactorily and working correctly. Progress can be monitored with a display of the simulation time from a *clock* module placed inside the screen. Double-clicking on the *clock* opens up a running display of the simulated time that will give an indication as to how smoothly the integration is proceeding or at which point it got hung up.

SIMULINK provides several types of output devices in the **Sinks** library block for you to monitor variables. The **Scope** provided has a single input which will accept multiplexed signals. The multiplexer template in the **Connections** library block may be used to multiplex two or more signals into one scope input. A scope with input that is left *floating* will have an input that is selected by the mouse. Clicking the mouse pointer on an output port or line will select that as the input to the *floating* scope. In SIMULINK 2, a warning is given regarding unconnected input or output unless you check the box for floating scope on the **Settings** dialog box under the **Properties** menu of the scope's toolbar. Furthermore, in SIMULINK 2, a scope can only display up to six signals. If the multiplexed input is wider than six and you do not want to use multiple scopes, you can use a *selector* switch to select up to six signals of the multiplex input to be fed through to the scope.

2.7 PRINTING

2.7.1 Printing a Figure

A hard-copy printout of a MATLAB plot or a SIMULINK model can be obtained using the print command. The print command for a MATLAB figure or window is

> ≫ **print** [-d*devicetype*] [-*options*] [-f*MATLAB_figure*] [*filename*]

The print command for a SIMULINK window or figure is

> ≫ **print** [-d*devicetype*] [-*options*] [-s*SIMULINK_figure*] [*filename*]

Just typing **print** in the MATLAB command window with no option will print the current MATLAB figure window on the default printer. Adding a *filename* to the print command as in **print** *filename* will save the current MATLAB figure to the named file using the default device format. When the figure is to be used in a report, the figure file may have to be in a device format usable by a text processing program. The device format or type can be specified by adding the **-d***devicetype* option to the print command. Among the many supported device formats are -d*ps2* for Level 2 Postscript and -d*eps* for an encapsulated Postscript (EPSF). Other options for the print command include -P*printer_id* to specify the printer, and -f*figure_name* to specify the MATLAB figure or -s*SIMULINK_figure* to specify the SIMULINK screen. When the title of a figure contains spaces, the option and title ought to be enclosed within single quotes, as in '**-s***Two-word Title*'.

A file created by the *print* command will have a name, such as *filename.devicetype*. To print or preview the file within the MATLAB command window, you need to preface your UNIX command with the escape character, !. For example, to obtain a hard-copy printout of an .eps file, use **!lp -d***printer_id filename.eps*, where *printer_id* is the ID or address of the printer. A figure file in Postscript format may be viewed using Ghostview, Ghostscript, or any graphic program that supports the Postscript format.

2.7.2 Saving Data

Figure 2.3 shows two different ways of monitoring a variable. The output of the signal generator can be viewed directly during the simulation run using a **Scope**. Or, the desired output, along with the run-time from a clock, can be stored as a MATLAB data file using the **To File** template in **Sinks** if the data are to be plotted or processed later. Instead of writing the result directly into a file, the output can be stored temporarily in an array, *yout*, in the MATLAB workspace using the **To Workspace** template given in **Sinks**.

Stored in this manner, the output in the array, *yout*, can also be used by another part of the same SIMULINK simulation. The names of the data file and array, *yout*, associated with the **To File** and **To Workspace** templates can be renamed by the user in the SIMULINK window. When using a **To Workspace** template, an adequate buffer length in **To Workspace** must be specified before starting your simulation. If the buffer length is not sufficient, the stored data may be overwritten with data from the end of the run. In

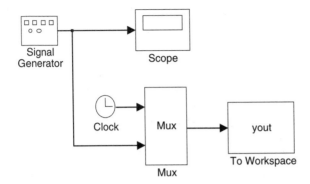

Signal Generator

Scope

Clock

Mux

yout

To Workspace

Mux

Figure 2.3   Monitoring and storing data in the MATLAB workspace.

SIMULINK 1, the second parameter field of the **To Workspace** dialog box allows for entry of the following three parameters: [*nrows, decimation, sample_time*]. The first parameter, *nrows*, is the buffer length. The second, an integer, say **n**, specifies the saving of simulation output data every **n**th integration step. It can be used to space out the data point when your computer does not have sufficient memory storage or there is no need to store the results from every integration time step. The last parameter allows you to specify the sample interval at which to collect data; it is useful when uniformly sampled values are required for later analysis or plotting.

You can also save the array that you have written in the MATLAB workspace for later use, such as plotting or further processing, by using the MATLAB **save** command:

> ≫ **save** *filename yout*

MATLAB will create a binary file, *filename.mat*, in your current directory with the array, *yout*, in it. The data can also be saved in ASCII format by specifying the option in the save command, that is

> ≫ **save** *filename yout -ascii*

Reloading information that has been saved in a previous MATLAB or SIMULINK session back into the workspace can be accomplished using the MATLAB **load** command:

> ≫ **load** *yout*

2.7.3 Plotting

In dynamic simulation studies, plots of variables versus time are usually made to examine the transient response. We will briefly describe how such plots can be made using the MATLAB plot command in the MATLAB window. Let's begin with getting a single plot of some variable versus time on the computer screen for verification purposes. We will assume that the variable has been stored along with the corresponding time in array *yout* of length N, say with time stored in the first column of *yout* and the variable of interest saved in the second column. The following basic MATLAB commands will enable you to customize the plot somewhat:

```
title('insert desired title of plot here')
xlabel('insert x axis label here')
ylabel('insert y axis label here')
grid \% if you want grid lines on plot
plot(yout(:,1),yout(:,2),'linetype')% plot yout(:,1)on x-axis and yout(:,2) on y-axis
```

The optional entry *'linetype'* in the plot command allows you to specify the color and type of line to be used to plot the curve. It is a useful option when there is a need to distinguish one curve from another. For more details, refer to **plot** in the online Help Desk or the *MATLAB User's Guide*. To graph two or more variables versus time on the same plot, say *yout(:,2)* and *yout(:,3)* versus time, you could use the **hold** command or put them in one plot command as shown below:

```
title('insert desired  title of plot here' )
xlabel('insert x axis label here')
ylabel('insert y axis label here')
grid  % if you want grid lines on plot
plot\=(yout(:,1),yout(:,2),'-', yout(:,1),yout(:,3),':')
% plot vector yout(:,2)versus time using solid curve
% plot vector yout(:,3)versus time using dotted curve
```

2.8 EXERCISE 1: VARIABLE-FREQUENCY OSCILLATOR

In this exercise, we will go through the steps to construct a SIMULINK simulation to generate an orthogonal set of sine and cosine time functions of fixed amplitude but adjustable frequency beginning with the following equation of an oscillator:

$$\frac{d^2 y_1}{dt^2} = -\omega^2 y_1 \tag{2.1}$$

where ω is the angular frequency.

The Laplace transform of an ideal differentiator is s; that of an ideal integrator is $1/s$. Letting $s = j\omega$, we can see that the gain of a differentiator is directly proportional to frequency, whereas that of an integrator is inversely proportional to frequency. In other words, a differentiator will be more susceptible to high-frequency noise than an integrator. For this reason, integration is preferred over differentiation.

To implement the oscillator equation, we will first convert the second-order differential equation into two first-order differential equations and then express them in integral form. Let's introduce a new state, $y_2 = (1/\omega)dy_1/dt$. Substituting the time derivative of y_2 into Eq. 2.1, we obtain

$$\frac{dy_2}{dt} = -\omega y_1 \tag{2.2}$$

Rewriting the above equation and that defining y_2 in integral form, we have

$$y_2 = -\omega \int y_1 dt$$

$$y_1 = \omega \int y_2 dt \tag{2.3}$$

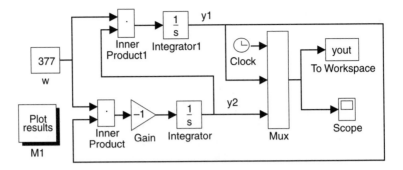

In this simulation, the system parameters and initial conditions have been entered as constants. After running the simulation, double-click on masked block M1 to plot simulated results on the MATLAB figure window.

(a) SIMULINK Simulation sl

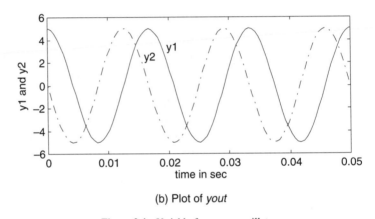

(b) Plot of *yout*

Figure 2.4   Variable-frequency oscillator.

Figure 2.4a shows the SIMULINK implementation of Eq. 2.3 given in the *sl* file. It uses the integrator, multiplier, and gain templates taken from the Linear block library of SIMULINK. Note the use of a multiplexer to stack y_1, y_2, and time into a vector for display and also the use of temporary storage in an array named *yout* in the MATLAB workspace.

The two integrators in Fig. 2.4 ought to be initialized with the proper values, $y_1(0)$ and $-y_2(0)$, of the two states. Their outputs, y_1 and y_2, are orthogonal sinusoids. They can be made to start with any desired initial phase angle by using the appropriate initial conditions for y_1 and y_2. For instance, using the initial values of $y_1(0) = V_{pk}$ and $y_2(0) = 0$ will produce outputs of $y_1(t) = V_{pk}cos\omega t$ and $y_2(t) = -V_{pk}sin\omega t$. The sample result shown in Fig. 2.4b has been obtained with $y_1(0) = 5$, $y_2(0) = 0$, and $\omega = 377$ radians/sec, using the RK5 algorithm with a minimum step size of 0.0001, a maximum step size of 0.01, and an error tolerance of $1e^{-5}$. Labels on the curves are placed with the MATLAB **gtext**('string') command, and the plot is generated by the following MATLAB plotting commands in the *m1* MATLAB file:

```
plot(yout(:,1),yout(:,2),'-', yout(:,1),yout(:,3),'-.')
xlabel('time in sec')
ylabel('y1 and y2')
```

After running the simulation, *s1*, double-click on the masked *plot* block in *s1* to plot the simulation results. The masked block, *plot*, executes the command *eval('m1')* in its *dialog string field*. To see how that block is masked, select it and then select **Create Mask** under **Edit** in SIMULINK 2 or **Mask Block** under **Options** in SIMULINK 1.

2.9 EXERCISE 2: PARALLEL RLC CIRCUIT

The second exercise is to simulate the dynamic response of a parallel RLC circuit. Again, we will go through the steps of rearranging the circuit equations, conducting simulation runs, and storing and plotting the results. This time we will also describe a way to verify the simulation using analytical knowledge of the circuits.

Consider the implementation and verification of the simple parallel RLC circuit shown in Fig. 2.5. Convince yourself that the circuit shown in Fig. 2.5 has the following circuit equations:

KVL of mesh 1:

$$-v_S + i_S R_S + v_C = 0 \tag{2.4}$$

KCL at node 2:

$$-i_S + i_L + i_C = 0 \tag{2.5}$$

Equation for inductor branch:

$$v_C = L\frac{di_L}{dt} \tag{2.6}$$

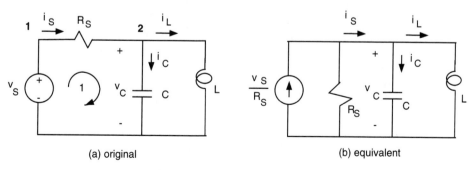

(a) original   (b) equivalent

Figure 2.5   Parallel RLC circuit.

Equation for capacitor branch:

$$v_C = \frac{1}{C} \int i_C dt \tag{2.7}$$

Choosing the inductor current and capacitor voltage as states of the system, the above circuit equations can be rewritten in terms of the two states and the input voltage as follows:

$$i_L = \frac{1}{L} \int v_C dt$$

$$v_C = \frac{1}{C} \int \left(\frac{v_S - v_C}{R_S} - i_L \right) dt \tag{2.8}$$

For the convenience of determining the analytical results that we will use to verify the simulation, we will consider the transient response of the RLC circuit to be a step function voltage. Start up MATLAB and SIMULINK, double-click on the SIMULINK **file** menu and use the **open** file command to select and open the *S2* file containing the model of the parallel RLC circuit. As shown in Fig. 2.5, the step voltage excitation is simulated by a step input generator from the SIMULINK **Sources** block library. Check the connections of the various blocks in *S2* to confirm that the simulation agrees with the mathematical model given in Eqs. 2.8.

If you would like to experiment with implementing the simulation on your own, start by opening a new SIMULINK screen. Open up the **Sources** block library, locate the template for the step input generator, and click on it to select and drag that template onto the new screen. Double-click on the *step input* template to open up its dialog window. Set the *step time* to 0.05 seconds and the *initial* and *final* values of the *step input* voltage to zero and 100 volts, respectively. Similarly, select and drag the integrator and sum templates from the *Linear* block library to synthesize the rest of the equations given in Eqs. 2.8. Fill in the connections between these templates as shown in Fig. 2.5 to set up the SIMULINK simulation of the above two integral equations.

The orientation of the templates on the screen can be changed using the **Flip block** and **Rotate block** commands listed under the **Format** menu in SIMULINK 2. Similar commands are placed under the **Options** menu in SIMULINK 1.

A line connection between any two points can be made by clicking on one of the points, keeping the mouse button depressed while dragging the cursor to the other point, and releasing the mouse button when the cursor is at the second point. The layout of lines drawn between points can be further manipulated manually using a combination of the mouse button and *Shift* key. Line(s) can also be automatically set in SIMULINK 1 by first selecting the line(s) and then selecting the **reroute lines** command under the **Options** menu. Lines can be individually selected by clicking on any part of a line, or they can be selected area-wise by clicking on one corner of a rectangular area enclosing several lines and holding and dragging the cursor to define that area, or by using **Select All** under the **Edit** menu.

Figure 2.6 shows the use of a floating scope, *Scope2*, to display the variable i_C, the selected input to the floating scope being indicated by the little black square on the output line.

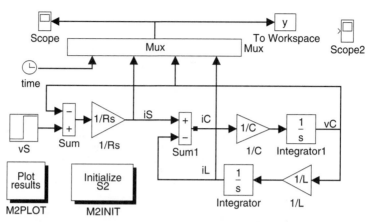

Double-click on masked block M2INIT to intialize before starting
simulation. After simulation, double-click on masked block M2PLOT
to plot simulated results on MATLAB window.

Figure 2.6   SIMULINK Simulation S2 of RLC circuit.

Before we can run the simulation, we must first initialize the circuit parameters to represent a circuit with $R_S = 50\Omega$, $L = 0.1H$, and $C = 1000\mu F$. Next, we must set the desired initial condition of the system, that is, the initial values of the system states. The initial values of the inductor current and capacitor voltage can be set by either directly entering the desired initial values in the dialog boxes of the respective integrators with inductor current and capacitor voltage as output, or by using named variables in these dialog boxes and assigning these variables desired initial values from the MATLAB command window. For this run, set the initial values of the inductor current and capacitor voltage directly to zero.

Finally, we must choose the integration method to use and set the start time, stop time, step size range, and error tolerance. All these are to be specified under parameters of the **Simulation** menu. For this run, we select the Adams/Gear integration method, set the start time to zero, the stop time to 0.5 seconds, the minimum step size to 0.1 msec, the maximum step size to 1 msec, and the error tolerance to $1e^{-5}$.

Before starting the simulation, click on the scope in the SIMULINK screen to open the scope's display screen and set the vertical and horizontal ranges of these scopes. Open the **Simulation** menu again and select **start** to start the simulation. The scope should now display all variables that are multiplexed into its input. The floating scope may be used to display any variable of interest during the simulation run. As shown in Fig. 2.6, the results are stored as the array y in the MATLAB workspace. The variable values in y are stored column-wise in a stacking order corresponding to the order of the multiplexed input. As shown, the clock time will be stored in the first column of y, corresponding values of i_S in the second column, and so on.

Plot the values of i_S, v_C, and i_L at the first 0.5-second interval. A sample output response of the RLC circuit with zero initial capacitor voltage and inductor current set to a step input voltage of 100 V step applied at 0.05 sec. is given in Fig. 2.7.

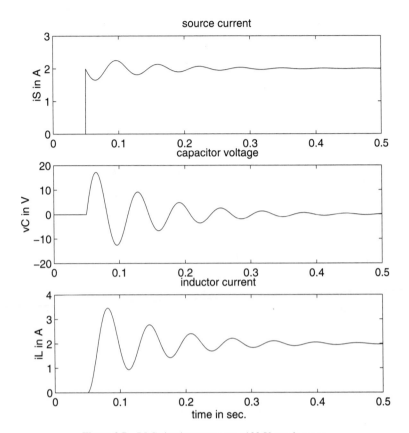

Figure 2.7   RLC circuit response to a 100-V step input, v_S.

The simulation *s2* can be initialized with the desired circuit parameters and initial condition by running the *m2* file shown below in the MATLAB window. The `keyboard` command in *m2* will put the program in *keyboard mode,* in which you can examine or make changes to variables in the MATLAB workspace and conduct your simulation run. *Keyboard mode* is indicated by a *K* prompt in the MATLAB window, during which MATLAB will respond to any valid MATLAB command. In this case, we use the *keyboard mode* to run the SIMULINK simulation. Upon the conclusion of a successful run, enter *return* after the *K* prompt to exit from the *keyboard mode* and continue with the *m2* script.

```
% M-file for Exercise 2 RLC circuit simulation
% input parameters and initial conditions
Rs = 50; %Rs = 50 ohms
L = 0.1; %L = 0.1 Henry
C = 1000e-6;  % C = 1000 uF
VS_mag = 100; % magnitude of step voltage Vs in Volts
tdelay = 0.05; % initial delay of step voltage in sec
vCo = 0; % initial value of capacitor voltage
iLo = 0; % initial value of inductor current
```

```
tstop = 0.5; % stop time for simulation
disp('run simulation, type ''return'' when ready to return')
keyboard
subplot(3,1,1)
plot(y(:,1),y(:,2))
title('source current')
ylabel('iS in A')
subplot(3,1,2)
plot(y(:,1),y(:,3))
title('capacitor voltage')
ylabel('vC in V')
subplot(3,1,3)
plot(y(:,1),y(:,4))
title('inductor current')
xlabel('time in sec.')
ylabel('iL in A')
```

Alternatively, the initialization and plot functions of *m2* can be divided into two M-files, *m2init* and *m2plot*, and their operations can be placed on-command using the separate masked blocks of Fig. 2.6.

Besides checking the correspondence of the simulated system with the integral equations given in Eqs. 2.8 to verify the implementation, we should also cross-check the simulated result against that obtained experimentally or predicted from analysis. In this case, a closed-form response of the transient response to a step input voltage can be obtained as follows:

1. The circuit shown in Fig. 2.5a can be transformed into the equivalent form shown in Fig. 2.5b using a Thevenin to Norton transformation.
2. Applying KCL to the top node of the parallel RLC circuit of Fig. 2.5b, we obtain the following circuit equation in terms of v_C:

$$\frac{v_C}{R_S} + \frac{1}{L}\int_0^t v_C \, dt + C\frac{dv_C}{dt} = \frac{v_S}{R_S}u(t-0.1) \tag{2.9}$$

3. The characteristic equation and its roots are

$$s^2 + \frac{1}{R_S C}s + \frac{1}{LC} = 0 \tag{2.10}$$

$$s_1, s_2 = -\alpha \pm j\omega \tag{2.11}$$

For the given circuit parameters, the attenuation factor, $\alpha = 1/(2R_S C)$, is 20, the natural resonant frequency, $\omega_o = \sqrt{1/LC}$, is 100 rad/sec., the damped oscillation frequency of the response, $\omega = \sqrt{\omega_o^2 - \alpha^2}$, is 97.98 rad/sec., and the period of the damped oscillations given by $2\pi/\omega$, is 0.064 sec.

4. The complete response of the capacitor voltage to a step input voltage may be expressed as the sum of the transient and steady-state components as in

$$v_C(t) = \underbrace{A_1 e^{s_1 t} + A_2 e^{s_2 t}}_{\text{transient}} + \underbrace{v_C(t \to \infty)}_{\text{steady-state}} \tag{2.12}$$

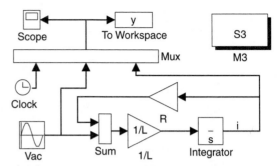

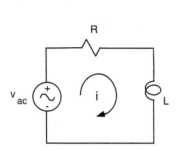

Double-click on the masked block M3 in this screen or run M3.M
in the MATLAB command window to initialize this simulation.
After simulation, type return in the MATLAB window to plot results.

(a) RL circuit with ac excitation   (b) Simulation of an RL circuit with ac excitation

Figure 2.8   AC energization of an RL circuit.

5. With a step input voltage, the steady-state component is zero. Since ω is real, that is $\omega_o^2 \geq \alpha^2$, the under-damped response can expressed in the form:

$$v_C(t) = e^{-\alpha(t-0.05)}(B_1 \cos\omega(t-0.05) + B_2 \sin\omega(t-0.05))u(t-0.05) \qquad (2.13)$$

Constants B_1 and B_2 can be established using the initial conditions of $v_C(0.05) = v_C(0) = 0$ and $i_L(0.05) = i_L(0) = 0$. From $v_C(0.05) = 0$, we obtain $B_1 = 0$; from $i_L(0.05) = 0$, Cdv_C/dt at $t = 0.05^+$ is v_S/R_S or $B_2 = v_S/(\omega C R_S)$. Substituting all these values, Eq. 2.13 for the capacitor voltage reduces to

$$v_C(t) = \frac{v_S}{\omega C R_S} e^{-\alpha(t-0.05)} \sin\omega(t-0.05)u(t-0.05) \qquad (2.14)$$

2.10 EXERCISE 3: AC ENERGIZATION OF AN RL CIRCUIT

In this third exercise, we will examine the transient response of a simple RL circuit to an ac voltage energization; in particular, we will look at the dc offset in the response current from energizing at different points of the ac voltage wave. The circuit and its SIMULINK simulation are shown in Fig. 2.8.

The circuit equation for the RL circuit with an ac source is

$$V_{ac} = iR + L\frac{di}{dt} \qquad (2.15)$$

For simulation purposes, the above equation is rearranged into its integral form:

$$i(t) = \frac{1}{L}\int_0^t (V_{ac} - iR)dt + i(0) \qquad (2.16)$$

The above integral equation is implemented in the SIMULINK file *s3* using a summer and an integrator. The ac voltage source is represented by a sinusoidal signal source from the **Sources** library block. Set the circuit parameter values of $R = 0.4\Omega$ and $L = 0.04H$, and set the initial value of the current to zero, the magnitude of the sinusoidal signal source to 100 volts, its frequency, ω_s, to $314\,rad/sec$, and its phase angle, θ, to zero. Select the *ode45* or *Linsim* numerical integration method and set the start time to zero, the stop time to 0.5 seconds, the minimum step size to 1 msec, the maximum step size to 10 msec, and the error tolerance to $1e^{-5}$. Run the simulation.

Repeat the run using two different values of θ, $\theta = \pi/2$ and $\theta = \tan^{-1}\omega_s L/R$. Examine the simulated response of $i(t)$ to determine whether the initial dc offset, the decay of the dc offset, and the final steady-state magnitude and phase of $i(t)$ agree with the corresponding values predicted from the following analysis:

The complete response of the current to a sinusoidal excitation of $V_{pk}\sin(\omega_s t + \theta)u(t)$ may be expressed as

$$i(t) = \underbrace{Ae^{-t/\tau}}_{\text{transient}} + \underbrace{\frac{V_{pk}}{|Z|}\sin(\omega_s t + \theta - \phi)}_{\text{steady-state}} \qquad (2.17)$$

where the time constant τ is L/R, the impedance, Z, is $\sqrt{R^2 + \omega_s^2 L^2}$, and the power factor angle, ϕ, is $\tan^{-1}(\omega_s L/R)$. Constant A can be determined from the initial condition of $i(0) = 0$, that is

$$i(0) = 0 = A + \frac{V_{pk}}{|Z|}\sin(\theta - \phi) \qquad (2.18)$$

Thus,

$$i(t) = -\frac{V_{pk}}{|Z|}\sin(\theta - \phi)e^{-t/\tau} + \frac{V_{pk}}{|Z|}\sin(\omega_s t + \theta - \phi) \qquad (2.19)$$

It is evident from the above expression that the current, $i(t)$, will have some dc offset when the circuit is being energized at a point of the wave other than at $\theta = \phi$, and that the dc offset decays off at a rate equal to the L/R time constant of the RL circuit.

Figure 2.9 shows the result for the case when the RL circuit with zero initial inductor current is being energized with a sine wave voltage of $100\sin 314t$ V. Determine the value of the power factor angle of the RL circuit ϕ, use it in a run where the sine wave voltage is set equal to $100\sin(314t + \phi)$, and note the initial offset in the response of $i(t)$.

2.11 EXERCISE 4: SERIES RLC RESONANT CIRCUIT

Figure 2.10a shows a series resonant inverter circuit with resistive loading. If the switching transients of the inverter are not of primary interest, the inverter output may be replaced with an equivalent source. Figure 2.10b shows a simple equivalent circuit representation of a system in which the transformer and load are represented by a referred RL impedance and the inverter output is represented by an equivalent square wave source. The system equation is given by the following KVL equation of mesh 1:

$$v_L + v_R + v_C = v_S \qquad (2.20)$$

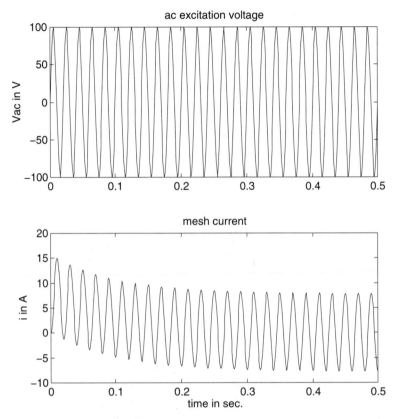

Figure 2.9   RL circuit response to ac excitation.

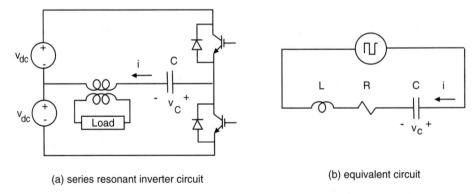

(a) series resonant inverter circuit

(b) equivalent circuit

Figure 2.10   Series resonant circuit.

Using $i_L = C\,dv_C/dt$, $v_L = LC\,d^2v_C/dt^2$, and $v_R = i_L R$, the system equation, expressed in terms of the capacitor voltage, v_C, is

$$\frac{d^2v_C}{dt^2} + \frac{R}{L}\frac{dv_C}{dt} + \frac{v_C}{LC} = \frac{v_S}{LC} \tag{2.21}$$

For a response of v_C that is under-damped, the roots of the characteristic equation of the above system equation are complex. Defining $2\alpha = R/L$ and the series resonant frequency, $\omega_o = 1/\sqrt{LC}$, the pair of complex roots may be written as

$$s_{1,2} = -\alpha \pm \sqrt{\alpha^2 - \omega_o^2} \tag{2.22}$$

from which we see that the damped resonant frequency is $\omega_d = \sqrt{\omega_o^2 - \alpha^2}$.

The admittance of the series RLC branch at the frequency ω is

$$Y(j\omega) = \frac{1}{j\omega L + R + 1/j\omega C} = \frac{1}{R}\frac{2\alpha(j\omega)}{(j\omega)^2 + 2\alpha(j\omega) + \omega_o^2} \tag{2.23}$$

In terms of the quality factor, Q, defined as ω_o/α,

$$Y(j\omega) = \frac{1}{R}\frac{(1/Q)(j\omega/\omega_o)}{(j\omega/\omega_o)^2 + (1/Q)(j\omega/\omega_o) + 1} \tag{2.24}$$

Using circuit elements with a higher Q value will result in a sharper or more selective $Y(j\omega)$ characteristic, but then the peak voltage across the capacitor which is equal to Q times the source voltage at resonance will be correspondingly higher.

Figure 2.11 shows the SIMULINK simulation of the equivalent circuit representation of the system that is given in file *s4*. It makes use of the inductor current and capacitor voltage as system states. The integral equations of these two states are

$$i = \frac{1}{L}\int (v_S - v_C - Ri)dt$$

$$v_C = \frac{1}{C}\int i\,dt \tag{2.25}$$

Assuming that the dc voltage source and the switches of the inverter are ideal, the output voltage of the inverter may be represented as an adjustable frequency square wave voltage of amplitude equal to the dc source voltage magnitude. Feedback of the load power, in this case, is obtained by sensing the instantaneous current and computing the instantaneous power delivered to the load resistor. The filtered output of the load power is then compared with the reference power value and the error is put through a proportional integral (PI) compensator. Since the output power decreases on either side of the resonant value, the control logic is simplified by limiting the operation to only one side, either below or above the resonant frequency.

Let's now consider a circuit where $R = 12\Omega$, $L = 231mH$, $C = 0.1082251\mu F$, and V_{dc} has a value of 100 volts. Since the power delivered to the load resistor, R, is $i^2 R$, the current, i, is controlled by varying the frequency of the square wave inverter output. The fundamental component of the square wave voltage of magnitude V_{dc} has a peak value of $4V_{dc}/\pi$, or an rms value of $4V_{dc}/\pi\sqrt{2}$. For the given values of RLC and V_{dc}, a maximum

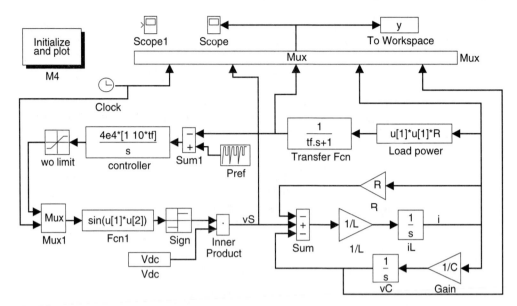

Double-click on the masked block of M4 in this screen or run M4.M in the MATLAB command window to initialize simulation. After simulation, type return after the K>> prompt in the MATLAB command window to obtain a plot of the simulated results.

Figure 2.11   SIMULINK simulation of series resonant circuit.

power of 675.47 W is deliverable to R at the resonant frequency of $200e^3$ rad/sec. Figure 2.12 shows the plots of the admittance of the series RLC circuit and the power deliverable to R with the fixed dc voltage close to its resonant frequency.

To set up the system parameters and run conditions, and to obtain the desired plots, we make use of the *m4* file listed below:

```
% M-file for Exercise 4 series resonant circuit simulation
% input parameters and initial conditions

R = 12; %R in ohms
L = 0.231e-3; %L in H
C = 0.1082251e-6;%C in Farad
wo = sqrt(1/(L*C))  % series resonant frequency in rad/sec
Vdc = 100; % magnitude of ac voltage =  Vdc Volts
iLo = 0; % initial value of inductor current
vCo = 0; % initial voltage of capacitor voltage
tf = 10*(2*pi/wo);  % filter time constant
tstop = 25e-4; % stop time for simulation
% set up time and output arrays of repeating sequence for Pref
Pref_time = [ 0 6e-4 11e-4 11e-4 18e-4 18e-4 tstop ];
Pref_value = [ 0 600 600 300 300 600 600 ];

% determine steady-state characteristics of RLC circuit
we = (0.5*wo: 0.01*wo: 1.5*wo);% set up freq range
wind = 0 % index for w loop
```

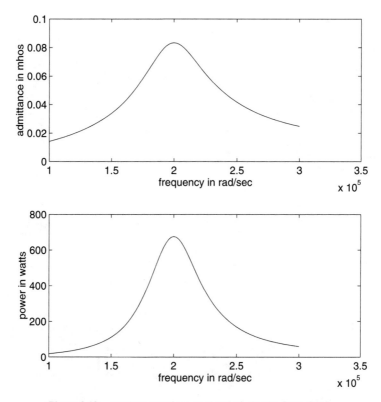

Figure 2.12   Admittance and power absorbed of series RLC circuit.

```
for w = we; % w for loop to compute admittance
wind = wind + 1;
Y(wind) = 1/(R + j*w*L + 1/(j*w*C));
Irms = (4*Vdc/(pi*sqrt(2)))*abs(Y(wind)); % rms value of i
PR(wind)= Irms*Irms*R;
end; % for w

% plot circuit characteristics
clf;
subplot(2,1,1)
plot(we,abs(Y));
xlabel('frequency in rad/sec');
ylabel('admittance in mhos');
subplot(2,1,2)
plot(we,PR);
xlabel('frequency in rad/sec');
ylabel('power in watts');

disp('run simulation, type ''return'' when ready to continue')
keyboard
clf
```

```
subplot(4,1,1)
plot(y(:,1),y(:,2))
title('excitation voltage')
ylabel('Vs in V')
subplot(4,1,2)
plot(y(:,1),y(:,3))
title('load power')
ylabel('PR in W')
subplot(4,1,3)
plot(y(:,1),y(:,4))
title('RLC current')
ylabel('i in A')
subplot(4,1,4)
plot(y(:,1),y(:,5))
xlabel('time in sec')
title('capacitor voltage')
ylabel('VC in V')
```

Figure 2.13 shows the plot of some results from a simulation run using the programmed sequence where the reference power was initially ramped up to 600 W, stepped

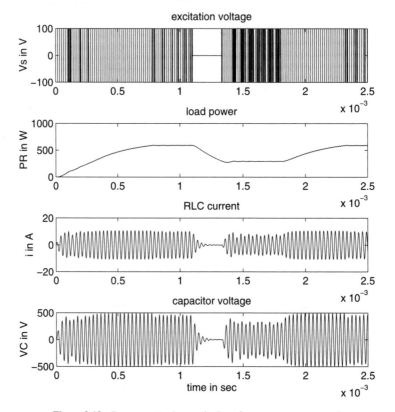

Figure 2.13   Responses to changes in the reference power command.

down to 300 W, and stepped back up to 600 W. Run the simulation as programmed or with a new control sequence. Comment on the magnitude of v_C and v_L relative to that of v_s, and on the change in load power with the change in frequency of the inverter output voltage. Determine the value of Q of the given RLC circuit. Show, by modifying the M-file or by simulation, that at resonant, the peak value of the capacitor voltage is Q times V_{dc}.

$$ 3 $$

Basics of Magnetics

and

Line Modeling

3.1 INTRODUCTION

The machines that we will be discussing in later chapters are constructed with copper or aluminum windings to carry the currents and iron to direct the flux. In many instances, the machines will also be connected by cables or lines to other components or to the supply network. In this chapter, we will briefly review the properties of ferromagnetic material, and the basics of modeling magnetic circuit and electrical lines. For this purpose, line parameter calculations may be limited to simple overhead lines to avoid complex boundary conditions.

3.2 MAGNETIC MATERIALS

The magnetic properties of a material are associated with the atomic magnetic moments produced by the spins and orbiting motion of electrons. Since an electron is negatively charged, its angular momentum, G, is opposite to the magnetic dipole moment, m, that is defined as $m = I \times area\ of\ orbit$. The spin of the electron also produces a magnetic moment m_s of $h/4\pi$, where h is the Planck's constant; m_s is either parallel or anti-parallel to the orbiting magnetic moment m.

In ferromagnetic materials, such as iron, the unpaired 3d electrons are not effectively screened from the influence of an adjacent atom's dipole moment. There is appreciable interaction between neighboring atoms. Unpaired spins are oriented nearly parallel to one another and tend to point in the same direction, giving rise to domains which exhibit spontaneous magnetization, even in the absence of an externally applied field. Such spontaneous magnetization exists only below the Curie temperature. Weiss postulated that in this kind of material are many regions called *domains*, each possessing a net magnetic moment, but the relative orientation of these moments may not be aligned. The structure of the domains are such that the total magnetic energy is a minimum. In equilibrium, the configuration of these domains is determined by a complicated interplay of exchange interaction, crystal anisotropy, magnetostrictive strains, impurities, dislocations, and atomic vacancies in the material.

The external magneto-static field energy is reduced by regions of opposite magnetization separated by an intermediate region, known as the *domain wall*, within which there is a gradual transition of the orientation of magnetization. Within the domain wall, extra exchange energy is involved because the adjacent dipoles are not parallel. There is also extra anisotropic energy, as some dipoles in the wall are not in easy directions of magnetization. The optimum wall thickness will be that for which the sum of the extra exchange and anisotropic energies is a minimum.

A further decrease in external magneto-static energy can be obtained by the presence of closure domains at the ends. Such closure domains involve additional energy due to magnetostrictive and anisotropic effects. Narrower domains will reduce the volume of the closure domains, but decrease the domain wall area. Any increase in exchange interaction energy with domain wall thickness is balanced against that from magnetostrictive and anisotropic effects within the closure domain's volume.

Without an external field, domains are randomly oriented. Under the influence of an external field, however, domains with magnetic moments parallel to the applied field expand while others shrink. The increase in B aided by the internal field is large. The extent of the domain wall's migration at each stage minimizes the total energy. When the domain walls have migrated as far as possible or have vanished, the magnetization direction rotates to align with the applied field. A saturation level is reached when all domains become aligned.

When a material is placed in a magnetic field of H, the interaction between the external and internal atomic magnetic moments increases the resultant, B. If the induced magnetism is denoted by the *magnetic moment density* M Nm. per unit volume, the flux density due to M alone is given by

$$B_{ind} = \mu_o M \tag{3.1}$$

where the magnetic constant, μ_o, has a value of $4\pi \times 10^{-7}$ Wb/Am in the SI system. The total flux density due to M and the external field, H, is

$$B = \mu_o M + \mu_o H \tag{3.2}$$

Let's express M as χH, where χ is the magnetic *susceptibility* of the material. Equation 3.2 for the total flux density becomes

$$B = \mu_o \underbrace{(\chi + 1)}_{\mu_r} H = \mu H \tag{3.3}$$

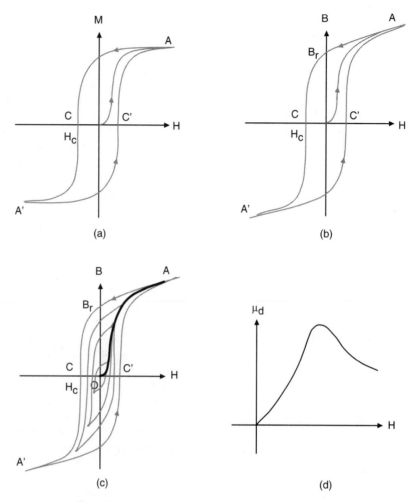

Figure 3.1   Magnetic characteristics of ferromagnetic materials.

where μ_r is the *relative permeability* of the material to free space and μ is referred to simply as the permeability of the material.

3.2.1 BH Characteristics of Ferromagnetic Materials

Figures 3.1a and 3.1b show the typical MH and BH characteristics for soft or high μ ferromagnetic materials. Beginning with an initially unmagnetized specimen, as H increases from zero, B increases along with it—at first with an increasing rate and then with a decreasing rate corresponding to domain rotation as saturation is approached. When H is decreased back to zero, the specimen does not return to the same minimum energy state as before the application of H due to the acquired compatibility of strain in adjacent grains from domain walls that have moved past lattice defects, and from the effects of non-magnetic inclusions. Instead, there is a residual magnetism, B_r. To reduce the magnetism back to zero, it is necessary to apply a reverse magnetizing force, known as the coercive force, H_c. Further increases in H beyond H_c in the negative direction cause a rapid increase in reverse magnetization until saturation at A'. When H is again reversed to the same positive value, the flux density follows the path $A'C'A$. The closed curve, $ACA'C'A$, is known as the *hysteresis loop*. Hysteresis here refers to the change in B lagging that of H. The larger the maximum value of H, the larger the area of the hysteresis loop, until the material is fully saturated whereupon the area of the loop ceases to increase and B_r and H_c reach constant values known as the *remanence* and *coercivity*, respectively. The χ and μ of ferromagnetic materials are very large and from Fig. 3.1 it is evident that they are nonlinear functions of B or H.

With cyclic excitations of lesser amplitudes, the material exhibits *minor hysteresis loops*, such as those shown within the saturation hysteresis loop in Fig. 3.1c. The dark curve, OA, drawn through the tips of the minor hysteresis loops, is referred to as the *normal magnetization curve* of the material. It is often used to represent the magnetization characteristic of ferromagnetic material in engineering calculations. The permeability, given by the ratio of B to H, varies with the point on the BH curve. Figure 3.1d is a plot of the *differential permeability*, μ_d, computed from the ratio of B to H for a small change in H at any point on the normal curve.

If the domain walls of a material are easily moved, its H_c is low and it is easy to magnetize. Such materials are said to be magnetically soft. Magnetically hard materials have a high H_c. Structural irregularities, such as dislocations and non-magnetic inclusions, hinder domain boundary movements. As such, the factors which make materials magnetically hard are also those which generally increase mechanical hardness.

3.2.2 Magnetic Circuits

The behavior of an electromagnetic device whose magnetic field paths can be defined, such as in some confined space, can be modeled by a magnetic circuit. When dealing with machines of not enormous size and very high operating frequencies, the displacement current term is negligible and the integral form of Ampere's circuit law can be written as

$$\oint_C \vec{H} \cdot \vec{dl} = \oint_S \vec{J} \cdot \vec{da} \tag{3.4}$$

where C is a closed contour and S is any surface whose edge is defined by C. In some applications, the closed flux path may be conveniently divided into m separate sections, along which the field intensity can be regarded as uniform. In these applications, the above line integral could be approximated by a summation:

$$\sum_{i=1}^{m} H_i l_i = I \tag{3.5}$$

where l_i is the length of the ith section and I is the net current linked to the flux path.

Similarly, the integral form of the divergence theorem may be expressed as

$$\oint_S \vec{B} \cdot \vec{da} = 0 \tag{3.6}$$

Again, assuming that the cross-sectional area of a flux path can be divided into n sub-sectional areas, wherein the flux density can be regarded as uniform, the above integral can be approximated by

$$\sum_{i=1}^{n} B_i A_i = 0 \tag{3.7}$$

where A_i is the area of the ith subsection.

Magnetic circuit calculations.   Let's consider the derivation of the magnetic and equivalent electric circuits of the magnetic component shown in Fig. 3.2a, whose core's dimensions are symmetrical about the center leg. From its geometry, we can see that the component has more than one flux path. Choosing the two flux paths shown in

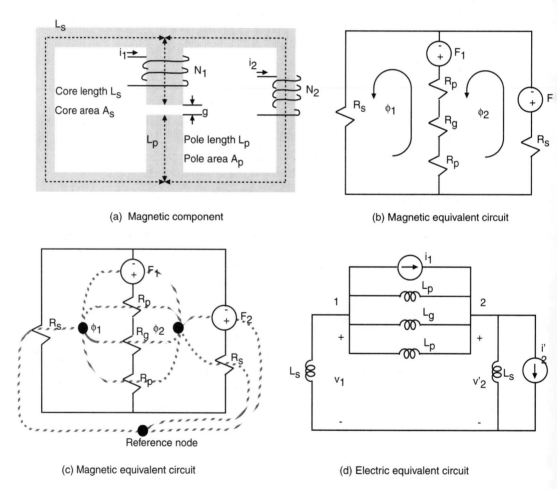

(a) Magnetic component

(b) Magnetic equivalent circuit

(c) Magnetic equivalent circuit

(d) Electric equivalent circuit

Figure 3.2   A simple magnetic system and its equivalent circuits.

Fig. 3.2b, we see that ϕ_1 has in its path the excitation mmf of winding 1, the reluctance of the outer left leg of the core, the reluctance of the gap in the center leg, and the reluctances of the two center leg pole sections. Likewise, going around the path of ϕ_2, we can identify the mmfs and reluctive drops in its path. As drawn in Fig. 3.2b, KVL-like equations of the two flux paths can be systematically written. Going counter-clockwise around the paths of ϕ_1 and ϕ_2, beginning in each case from the top right corner and taking into account the sign of mmf drops, we obtain

$$\phi_1 \mathcal{R}_s + (\phi_1 - \phi_2)(\mathcal{R}_p + \mathcal{R}_g + \mathcal{R}_p) + F_1 = 0$$
$$-F_1 + (\phi_2 - \phi_1)(\mathcal{R}_p + \mathcal{R}_g + \mathcal{R}_p) + \phi_2 \mathcal{R}_s + F_2 = 0 \tag{3.8}$$

The reluctances of the various sections can be determined from the core dimensions of the sections, that is

$$\mathcal{R}_s = \frac{l_s}{\mu A_s}$$

$$\mathcal{R}_p = \frac{l_p}{\mu A_p} \tag{3.9}$$

$$\mathcal{R}_g = \frac{g}{\mu_o A_g}$$

where A_g can be assumed to be the same as A_p if the airgap is so small that fringing of the flux around the airgap can be ignored.

Taking the time derivative of both equations and dividing through by N_1, Eq. 3.8 can be expressed as

$$\frac{dN_1\phi_1}{dt}\frac{\mathcal{R}_s}{N_1^2} + \left(\frac{dN_1\phi_1}{dt} - \frac{dN_1\phi_2}{dt}\right)\left(\frac{\mathcal{R}_p}{N_1^2} + \frac{\mathcal{R}_g}{N_1^2} + \frac{\mathcal{R}_p}{N_1^2}\right) + \frac{di_1}{dt} = 0$$

$$-i_1 + \left(\frac{dN_1\phi_2}{dt} - \frac{dN_1\phi_1}{dt}\right)\left(\frac{\mathcal{R}_p}{N_1^2} + \frac{\mathcal{R}_g}{N_1^2} + \frac{\mathcal{R}_p}{N_1^2}\right) + \frac{dN_1\phi_2}{dt}\frac{\mathcal{R}_s}{N_1^2} + \frac{d\{N_2 i_2/N_1\}}{dt} = 0 \tag{3.10}$$

We can proceed to simplify the form of Eq. 3.11 by introducing the following dual electrical variables:

$$v_1 = \frac{dN_1\phi_1}{dt} \qquad\qquad v_2' = \frac{N_1}{N_2}\frac{dN_2\phi_2}{dt}$$

$$i_2' = \frac{N_2 i_2}{N_1} \qquad\qquad \frac{1}{L_g} = \frac{\mathcal{R}_g}{N_1^2} \tag{3.11}$$

$$\frac{1}{L_s} = \frac{\mathcal{R}_s}{N_1^2} \qquad\qquad \frac{1}{L_p} = \frac{\mathcal{R}_p}{N_1^2}$$

The prime on the variables of winding 2 denotes their referred values to winding 1 or an equivalent winding of the same number of turns as winding 1. In terms of these newly-defined electrical dual variables, Eq. 3.11 simplifies to

$$\frac{1}{L_s}v_1 + (\frac{2}{L_p} + \frac{1}{L_g})(v_1 - v_2') + \frac{di_1}{dt} = 0$$

$$-i_1 + \left(\frac{2}{L_p} + \frac{1}{L_g}\right)(v_2' - v_1) + \frac{1}{L_s}v_2' + \frac{di_2'}{dt} = 0$$

(3.12)

Based on the form of the above equations, we can deduce the electric equivalent circuit shown in Fig. 3.2d. The electric equivalent circuit shown in Fig. 3.2d can also be obtained directly from the equivalent magnetic circuit of Fig. 3.2b using mesh-to-nodal circuit dual techniques.

The diagram in Fig. 3.2c illustrates the technique. A point outside the magnetic circuit is first chosen as the electrical reference node. Then, into each flux mesh we place an electrical node. Next, going around each flux mesh, we insert the electrical dual of the magnetic element encountered between the electrical nodes representing the adjacent flux meshes. The outer space is considered a "mesh" as it translates to the reference node for the nodal voltages. The electrical dual of a flux mesh is a nodal voltage, that of a permeance is an inductance, and that of an mmf source is a current source. The polarity of an mmf source can be translated to a proper direction for its dual current source if we are consistent with the sign convention used in the mmf drops around the flux mesh and in the currents entering or leaving a node. For example, following ϕ_1 counter-clockwise around its loop, we encounter the positive polarity of F_1 first; thus, its dual current, i_1, should also be a positive current in the KCL equation for node 1, which by sign convention would be a current leaving node 1.

We will next illustrate how the inductances of the windings in Fig. 3.2a can be determined by solving the magnetic circuit of Fig. 3.2c. The flux linked by each of the two windings can be expressed as

$$\lambda_1 = L_{11}i_1 + L_{12}i_2 = L_{11}i_1 + L_{12}'i_2'$$

$$\lambda_2 = L_{21}i_1 + L_{22}i_2 = L_{21}i_1 + L_{22}'i_2'$$

(3.13)

The L coefficients are the inductances. The self- and mutual inductances of winding 1 can be determined as follows:

$$L_{11} = \frac{\lambda_1|_{i_2=0}}{i_1} = \frac{N_1 i_1}{i_1(2\mathcal{R}_p + \mathcal{R}_g + \mathcal{R}_s/2)} = \frac{2N_1^2}{\mathcal{R}_s + 4\mathcal{R}_p + 2\mathcal{R}_g}$$

$$L_{12} = \frac{\lambda_1|_{i_1=0}}{i_2} = \frac{N_1}{i_2} \frac{-N_2 i_2}{\mathcal{R}_s + \dfrac{\mathcal{R}_s(2\mathcal{R}_p + \mathcal{R}_g)}{\mathcal{R}_s + 2\mathcal{R}_p + \mathcal{R}_g}} = -\frac{N_1 N_2}{\mathcal{R}_s + 4\mathcal{R}_p + 2\mathcal{R}_g}$$

(3.14)

Those of winding 2 can be determined in the same manner. By reciprocity, the mutual inductance, L_{21}, will be the same as L_{12}.

3.3 ENERGY CONVERSION

In this section we will review the basic relations of energy conversion using a simple system that has the bare essentials for the conversion of electrical energy to mechanical work or

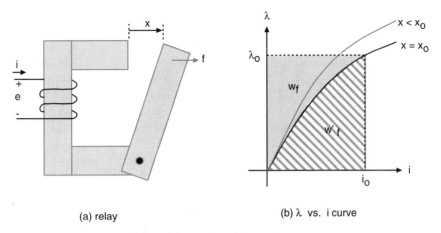

(a) relay

(b) λ vs. i curve

Figure 3.3   A relay and its λ vs. i curve.

vice versa. The relay shown in Fig. 3.3a has an electrical winding, a magnetic circuit, and a movable armature. Movement is essential for mechanical work, as force alone without movement does not amount to mechanical work.

If the copper losses in the winding, the core losses in the iron, and the frictional losses in the armature are to be accounted for externally, the balance of energy flow over an elemental period of time, dt, when the armature moves by dx of the remaining ideal system may be expressed as

$$\underbrace{ei\,dt}_{\substack{\text{electrical}\\\text{input}}} = \underbrace{f\,dx}_{\substack{\text{mechanical}\\\text{output}}} + \underbrace{dW_f}_{\substack{\text{change in}\\\text{field energy}}} \tag{3.15}$$

Replacing the back emf, e, by $d\lambda/dt$ and rearranging Eq. 3.15, the change in magnetic field energy over that same time interval will be given by

$$dW_f = id\lambda - f\,dx \tag{3.16}$$

The above relation indicates that the field energy stored is a function of λ and x. We can also deduce the functional dependence of the magnetic field energy for the system by first identifying the field energy stored in the λ versus i plot. Let's begin by examining the situation where the armature is held fixed at a position, x_o, while the winding current is raised. Flux in the core and armature of the relay will increase along with the excitation current. The rise in flux will not be as rapid as that in a solid iron core because of the high reluctance of the gap between armature and relay core. Figure 3.3b shows the typical flux linkage versus magnetizing current curves at two armature positions. Note that the flux corresponding to a given level of current excitation is higher when the gap reluctance is smaller. If the armature position is held fixed, that is, dx in Eq. 3.16 is zero, the change in field energy is $id\lambda$. From this we can deduce that the electrical energy used to magnetize the relay from an unexcited state to a state of (i_o, λ_o) on the flux linkage curve, with x held fixed at x_o, will be stored as field energy given by the shaded area above the λ vs. i curve for $x = x_o$. Geometrically, we can see that the shaded area is bounded by the $\lambda = \lambda_o$ line and the λ vs. i curve for $x = x_o$. As noted earlier, λ vs. i is in itself a function of x. Thus, the stored energy is a function of λ and x. With $W_f(\lambda, x)$, its total differential with respect to the independent variables is

$$dW_f(\lambda, x) = \frac{\partial W_f}{\partial \lambda} d\lambda + \frac{\partial W_f}{\partial x} dx \qquad (3.17)$$

Equating the coefficients of the $d\lambda$ and dx terms in Eq. 3.16 with those in Eq. 3.17 we obtain

$$i = \frac{\partial W_f}{\partial \lambda}$$
$$f = -\frac{\partial W_f}{\partial x} \qquad (3.18)$$

In practice, it is easier to hold the excitation current, i, rather than the flux linkage, λ, constant. In Fig. 3.3b, the cross-hatched area complementary to the field energy in the rectangle defined by λ and i is referred to as the *coenergy*, that is

$$W_f' = i\lambda - W_f(\lambda, x) \qquad (3.19)$$

When the λ vs. i characteristic is linear, W_f' will be equal in value to W_f.

Taking the total derivative of Eq. 3.19, and replacing the $d(i\lambda)$ term on the right by $id\lambda + \lambda di$ and dW_f by the right side of Eq. 3.16, we obtain

$$dW_f' = \lambda di + f dx \qquad (3.20)$$

from which we can deduce that the coenergy is a function of i and x. This is evident from the area of W_f' being bounded by i and the λ vs. i curve which is dependent on x. Equating the coefficients of the di and dx terms in Eq. 3.20 with those in Eq. 3.17 we now obtain

$$\lambda = \frac{\partial W_f'}{\partial i}$$
$$f = \frac{\partial W_f'}{\partial x} \qquad (3.21)$$

3.3.1 ac Excitations

Soft ferromagnetic materials of high permeability are widely used in ac machines and transformers. Their high permeability and low coercivity enable the required flux to be established with moderate ac excitation losses. With ac excitation, besides resistive losses in the winding, there will be core losses in the ferromagnetic core. The main core losses are those due to hysteresis and eddy current flow in the core. *Hysteresis loss* is the energy expanded in the movement of the domains during cyclic ac magnetization. For example, for each cycle of magnetization of a toroidal specimen of the material of mean length, l_c, and cross-sectional area, A_c, by a winding of N turns, as shown in Fig. 3.4, the energy absorbed is

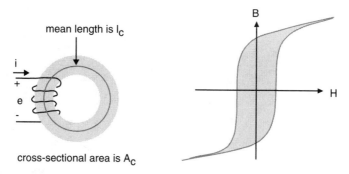

Figure 3.4   The ac magnetization of a specimen and an area of a BH loop.

$$W_h = \int_0^T eidt = \int_0^T N\frac{d\phi}{dt}idt = \oint Nid\phi = \oint Hl_c A_c dB = A_c l_c \underbrace{\oint HdB}_{\text{BH loop area}} \qquad (3.22)$$

When the cyclic excitation is repeated at a rate of f cycles per second, the power loss due to hysteresis may be expressed as

$$P_h \propto f(\text{BH loop area})(\text{volume magnetized}) = k_h f B_{max}^n (\text{volume magnetized}) \qquad (3.23)$$

where n, the Steinmetz constant, depends on B_{max} and the core material with a value typically between 1.5 to 2.5.

When the core is exposed to a time-varying field, the induced voltages from the time-varying field cause currents, known as eddy currents, to flow in the conducting core. Assuming that the core flux density is sinusoidally varying, that is $B(t) = B_{max} \sin \omega t$, the induced emf driving the eddy current encircling an area A_c of the core is

$$e(t) = \frac{d\phi(t)}{dt} = \omega B_{max} A_c \sin(\omega t + \pi/2) \qquad (3.24)$$

The root-mean-square (rms) value of $e(t)$ is

$$E_{rms} = \frac{\omega N B_{max} A_c}{\sqrt{2}} = 4.44 f N A_c B_{max} \qquad (3.25)$$

Figure 3.5 is a simplified model of eddy current flow in a thin lamination whose thickness, t, is much smaller than its width, w. In the thin lamination, the eddy current path may be assumed to be laminar and the short connecting sections at both ends may be neglected. The resistive loss of the eddy current path along the two elemental strips at a distance y from the centerline of the lamination, as shown in Fig. 3.5, may be written as

$$dP_e = \frac{E_i^2}{2R_{strip}} = \frac{(4.44 f B_{max} wy)^2}{2(\rho W/ldy)} \qquad (3.26)$$

where E_i is the induced emf due to the time-varying B field in the elemental loop of area wy, ρ is the resistivity of the core material, and R_{strip} is the resistance of each strip to

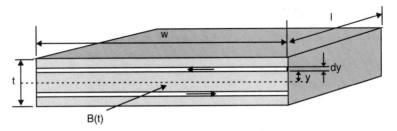

Figure 3.5　Laminar eddy current flow in a thin lamination.

the eddy current flow in an elemental strip of thickness dy. Integrating the right side of Eq. 3.26 from the centerline to $t/2$, we obtain the power loss due to eddy current flow in the lamination:

$$P_e = \frac{(4.44 f B_{max})^2 w l t^3}{12\rho} \tag{3.27}$$

It is evident from the above expression that eddy current loss for a given volume of core material will be reduced by using thinner laminations and material of high resistivity.

If the index, n, of Eq. 3.23 is assumed to be 2, the power losses due to hysteresis and eddy current flow will be proportional to the square of the maximum flux density, B_{max}. Since the rms value of the back emf is proportional to B_{max}, these core losses are proportional to the square of the back emf; as such, they can be accounted for by a resistive loss across the voltage, $e(t)$.

3.4 PERMANENT MAGNETS

Permanent magnetism is produced in a ferromagnetic material by magnetizing the material with a strong dc current or pulses of dc current to insure saturation. The remanence B_r is not truly permanent as it may be reduced by mechanical vibration, thermal agitation, or the application of a reversed field. Many of the alloys of iron, nickel, and cobalt have been known to have more durable remanence than that of iron. Since these alloys are subjected to heat treatment, which results in mechanical hardness of the finished magnet, they are known as hard magnetic materials. But, there is another meaning to magnetically hard and soft. With soft magnetic materials, the early stage of easy magnetization at low values of H is due to domain wall movement. Only at high values of H does domain rotation make an *elastic* contribution to magnetization which disappears as soon as H is lowered. The remanence of soft material is mostly due to domains whose walls have not returned to the position that they occupied before H was applied. The composition of many hard magnetic materials and the heat treatment produce very small elongated crystals, a structure that inhibits domain wall movement. In hard magnetic materials, most of the magnetic magnetization occurs as a result of abrupt inelastic switching of the orientation of entire domains. The persistence of the domains in their switched orientations after H is removed give rise to the high remanence and a coercivity that may be several thousand times greater than that of a soft material.

3.4.1 Permanent Magnet Materials

The Alnicos are alloys of iron with aluminum, nickel, copper, and cobalt. The initial alloy of iron, aluminum, and nickel, developed by Nishima in 1932, has since been modified to include cobalt, copper, and titanium to improve its properties, They can be shaped by casting in dry sand molds or by dry-pressing sintered powder in hydrogen. The formed parts are then heat treated in a magnetic field to produce strong magnetic properties. The advantages of Alnicos are high service temperature, good thermal stability, and high flux density. The disadvantages are their low coercive force coupled with a squarish BH characteristic, which often would require magnetization in situ to achieve higher flux, and that they are mechanically brittle and hard. Since they are hard and brittle, any finishing must be ground.

Barium and strontium ferrites were introduced in 1951 and 1963, respectively. They are widely used today in many kinds of permanent magnets. Hard or ceramic ferrite magnets, made by dry- or wet-pressing a ferrite mixture, are brittle. Bonded ferrites, on the other hand, can be extruded into flexible magnets when bonded with rubber, or molded using die or injection molding when mixed with plastic binders. Although the magnetic properties of bonded or plasto ferrites are not as good as the parent material, they have desirable mechanical properties, such as strength, pliability, elasticity, and resistance to fracture that make them easy to form and machine into desired shapes and dimensions. The important advantages are their low cost, the plentiful supply of raw material, they are easy to produce, and their process is suited for high volume, as well as moderately high service temperature (400°) C. The H_c for hard ferrites is around $240\,kA/m$, that of bonded ferrites, $125\ kA/m$. Their almost linear demagnetization characteristics permit magnetization before assembly with low magnetization forces, about $800kA/m$. A disadvantage is their low B_r, which is about $0.4\ Wb/m^2$ for hard ferrite and $0.2\ Wb/m^2$ for bonded ferrite.

Samarian-cobalt magnets, first introduced in 1963, are made from compounds of iron, nickel, cobalt, and the rare-earth Samarian. They have a remanence of about that of Alnico-type material and coercivities greater than those of ferrites. The advantages of Samarian-cobalt magnets are an exceptionally high BH_{max}, ranging from 40 to $200\,kJ/m^3$, a high H_c of 400 to $720\ kA/m$, and linear demagnetization characteristics. The service temperature can be as high as 300° C, with good temperature stability of about $-0.03°$ C. The disadvantages are that Samarium is not plentiful and is costly, and the magnetizing force required is high, 1300 to $3200\,kA/m$. Also, the magnet may require care in handling since the ground powder and air mixture is pyrophoric and the magnetic force is strong.

Neodymium-iron-boron magnets, first introduced in 1983, are compounds of iron, nickel, and neodynium. They have exceptionally high coercive force, a low service temperature of 150° C, and are susceptible to oxidization unless protected by a coating. They have exceptionally high H_c and $(BH)_{max}$, up to $740\ kA/m$ and $400\ kJ/m^3$, respectively. The corresponding values for the bonded type are lower, $480\ kA/m$ and $120\,kJ/m^3$. The demagnetization characteristics of both types are linear. The disadvantages of neodymium-iron-boron magnets are their low service temperature, under 150° C, and their temperature stability, which exhibits as much as a -0.13% change in B per degree C, which is inferior to Sm-Co magnets. Like Sm-Co, they require very high magnetization

TABLE 3.1   ORDERING BY MEASURES USEFUL IN MOTOR APPLICATIONS

$(BH)_{max}$	B_r	H_c	Energy to magnetize	Max service temperature	Temperature stability	Relative cost to stored energy
Nd-Fe-B	Alnico	Sm-Co	Sm-Co	Alnico	Alnico	Sm-Co
SmCo	Nd-Fe-B	Nd-Fe-B	Nd-Fe-B	Sm-Co	Sm-Co	Nd-Fe-B
Alnico	Sm-Co	Ba,Sr ferrites	Ba,Sr ferrites	Ba,Sr ferrites	Ba,Sr ferrites	Alnico
Ba,Sr ferrites	Ba,Sr ferrites	Alnico	Alnico	Nd-Fe-B	Nd-Fe-B	Ba,Sr ferrites

force, up to $2900 \, kA/m$, and special care in handling since the powder and air mixture is pyrophoric and the magnetic force is strong.

The choice of magnet for a motor is influenced by factors that could affect the performance and costs of the motor. Weight and size considerations will influence the $(BH)_{max}$, B_r, and recoil characteristic needed. The shape of the demagnetization characteristic and the value of H_c might affect the placement of magnets or require further consideration of ways to prevent demagnetization. In addition, tolerances, finishing, assembly, handling, and magnetizing of the magnet can affect the ease of fabrication and overall cost.

Table 3.1 lists some of the commonly used magnet types in decreasing order of measures often considered important in motor applications. It should be borne in mind that within each type there are variations in values from manufacturer to manufacturer and from one compound mix to another. The ordering should therefore be used as a rough guide. For example, hard ferrite has the lowest cost per unit energy listed, bonded ferrite magnets are easily cut and formed to shape, Samarium-Cobalt has high service temperature in combination with high coercivity and energy product, and neodymium-iron-boron has the highest coercivity and energy product.

3.4.2 Equivalent Magnetic Circuits

The demagnetization segment of the BH curve shown in Fig. 3.6 is usually where a permanent magnet is designed to operate. The initial demagnetization from zero to a stabilized value of H_s shifts the operating point initially along the BH curve in the second quadrant. A reversal of the field from H_s to zero will, however, return along a minor hysteresis loop to an effective remanence of B_{r1}. Figure 3.6 shows a complete recoil loop when the applied field cycles between zero and H_s. The dashed center line drawn through a recoil loop is often referred simply as the *recoil line*, and its slope is the recoil permeability, μ_{rc}. For operations where the applied demagnetization field does not fully return to zero, the operating point of the magnet will move along smaller recoil loops within the complete recoil loop shown. If the losses associated with these minor hysteresis loops and the small deviations in recoil permeability from partial demagnetization can be ignored, the behavior of the permanent magnet for $H \geq H_s$ can often be characterized by just the center line of the complete recoil loop from H_s. Should the magnet be demagnetized beyond H_s, say to H_t, subsequent operation will be along a new recoil loop below that shown and its new effective remanence will be lowered. This reduction in effective remanence

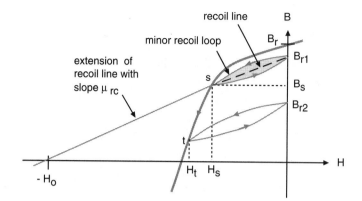

Figure 3.6   Demagnetization curve and recoil loop.

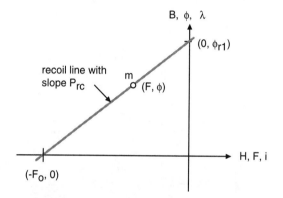

Figure 3.7   Recoil line.

by improper demagnetization can be substantial, especially for magnets with a squarish demagnetization curve, such as for Alnico types. But this is not a concern for magnets with linear demagnetization characteristics, such as Nd-Fe-B, Sm-Co, and hard ferrites.

Figure 3.7 shows the case of a linear recoil characteristic of a magnet through its current operating point m (H_m, B_m), mapped directly onto the mmf-flux (F vs. ϕ) or the current-flux linkage (i vs. λ) plane. For a design of fixed configuration, the axes of the B-H, ϕ-F, and λ-i plots shown in Fig. 3.7 will differ only by scaling constants. It will be shown next that the two simplified circuit models of a permanent magnet in use today are really equivalent, differing only in the choice of variables.

The recoil line can be described by the equation of the line through the point $(-F_o, 0)$ with slope P_{rc}. Using the coordinates (F, ϕ) of point m on the recoil line, the slope, P_{rc}, may be determined from

$$\frac{\phi - 0}{-F - (-F_o)} = P_{rc} \tag{3.28}$$

From Fig. 3.7, we can also deduce that $\phi_r = P_{rc}F_o$. When Eq. 3.28 is rearranged to the form given below, it suggests the equivalent magnetic circuit [16] shown in Fig 3.8a.

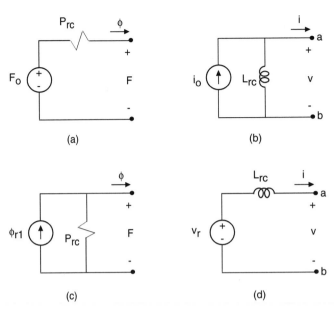

Figure 3.8 Equivalent magnetic circuits.

$$F = F_o - \frac{\phi}{P_{rc}} \tag{3.29}$$

It can also be shown that mmf, flux, and permeance are the mathematical duals of current, voltage, and inductance, respectively [15]. Thus, Fig. 3.8b is an electric dual of the magnetic circuit of Fig. 3.8a.

The line equation of the same recoil line can also be expressed in terms of the effective residual flux at zero field, ϕ_{r1}. Using the relation, $\phi_{r1} = P_{rc}F_o$, to replace F_o with ϕ_{r1}, the line equation in Eq. 3.29 yeilds

$$\phi = \phi_{r1} - P_{rc}F \tag{3.30}$$

The equivalent magnetic circuit and its electric dual corresponding to Eq. 3.30 are shown in Figs. 3.8c and 3.8d, respectively. In circuit analysis, the electrical equivalents of Figs. 3.8b and 3.8d are referred to as the Norton's and Thevenin's equivalents, respectively. If their parameters are not time-varying, they are linear equivalents of the same v-i characteristic as viewed from the terminals, a and b. The current i_o is the current when the output voltage v is zero, just as F_o is the mmf when the output flux is zero. Likewise, v_r is the available voltage when the external current i is zero, and ϕ_{r1} is the available flux when the external mmf is zero.

3.5 LINE PARAMETERS

The RLC circuit parameters of lines are related to the size and spacing of the line conductors and the electrical properties of the insulating medium. In this section we will derive some of the relationships for the R, L, and C of overhead lines. Similar techniques can be applied to cable, albeit less clear-cut because of complications associated with proximity, shielding, and bonding.

3.5.1 Resistance

When a direct current (dc), I, flows through a length of conductor, the voltage drop, V, across the length of conductor obeys Ohm's law, that is

$$V = R_{dc} I \qquad V \tag{3.31}$$

where R_{dc} is the dc resistance in ohms. For a conductor with a uniform cross-sectional area and uniform current distribution, the dc resistance is given by

$$R_{dc} = \frac{l}{\sigma A} \qquad \Omega \tag{3.32}$$

where l is the length of the conductor in the direction of current flow, σ is the conductivity of the material, and A, is the cross-sectional area normal to the current flow. Since conductivity varies with temperature, the value of resistance of a piece of conductor will vary with operating temperature. Given its dc resistance R_{dc1} at T_1, its dc resistance R_{dc2} at temperature T_2 can be computed using

$$\frac{R_{dc2}}{R_{dc1}} = \frac{(T + T_2)^\circ C}{(T + T_1)^\circ C} \tag{3.33}$$

where T is a function of the composition of the conductor material.

The resistance of irregularly-shaped conductors can be determined using numerical field solution methods, such as that which discretizes the region into tubes of flow and slices of equipotential [18].

An alternating current (ac) flowing through the same conductor will encounter an effective resistance higher than the dc value because the current distribution inside the conductor is affected by not only the resistance, but also by the inductance. A time-varying current flowing in a conductor will create a time-varying magnetic field around the conductor. In a circular cross-section conductor, the portion of the ac current nearer to the center of the cylindrical cross-section conductor will link more of the flux that it creates; thus, it will experience a higher self-inductance than the portion of the current flowing near the circumference of the conductor. If we were to examine the variation in field and current distribution inside a semi-infinite block of conductor that is exposed to an alternating magnetic field at the conductor surface, we would find that the field and induced current density within the conductor decay exponentially with depth into the conductor, that is, the current density, J, has the form

$$J = J_s e^{-x/\delta} \qquad A/m^2 \tag{3.34}$$

where J_s is the magnitude at the surface, x is the distance into the conductor from its surface, and δ is the skin depth. The skin depth is defined as the depth at which J decays to $1/e$ of its maximum value. Figure 3.9 shows the current density distribution with depth inside the conductor. At an excitation frequency of ω rad/sec, a homogeneous conductor of conductivity, σ, has a skin depth of $\sqrt{2/\mu\sigma\omega}$. For copper at 60 Hz, assuming a μ_r of 1 and a σ of 0.5×10^8 S/m, the skin depth is

$$\delta = \left(\frac{2}{4\pi \times 10^{-7} \times 0.5 \times 10^8 \times 2\pi \times 60} \right)^{1/2} = 9.2 \; mm \tag{3.35}$$

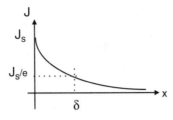

Figure 3.9 Current density distribution vs. depth into conductor.

Where skin effect is pronounced, as much as 86% of the conductor current flows within the outer cylindrical surface of thickness equal to the skin depth, δ.

For a waveform of the current that is nonsinusoidal or complex, the effective ac resistance can be determined from

$$\text{Effective resistance } R_{eff} = \frac{Power\ loss}{I_{rms}^2} \qquad \Omega \qquad (3.36)$$

3.5.2 Inductance

The flow of current in a circuit establishes a magnetic field around the circuit. When the surrounding media is magnetically linear, the flux density produced is proportional to the field strength, which in turn is proportional to the current enclosed. For such a situation, the flux linked by the circuit, λ, due to the current, I, is linear.

$$\lambda = LI \qquad Wb \qquad (3.37)$$

The constant of proportionality, L, is known as the inductance. The unit of inductance is Henry, or Wb/A.

Let's begin by determining the expression for the inductance per meter length of the two-conductor line shown in Fig. 3.10. For now, we will assume that the length of the line and its height above the ground are much greater than the conductor spacing, $D_{aa'}$, so that both end and ground effects can be ignored. Furthermore, $D_{aa'}$ is much larger than the conductor radius, r, that proximity effects on the current density and H field distribution may also be ignored.

Note that the flux within the conductor links only part of the conductor current, whereas that encircling the whole conductor links all of the conductor current. For clarity, we will determine the total flux linkage by the current, I, as the sum of two parts: the internal flux linkage due to flux within the conductor and external flux linkage due to flux outside the conductor.

Flux inside conductor. As shown in Fig. 3.11a, for the flux linkage inside a meter length of the a conductor, consider the flux linked by the portion $x \leq r$ of the current, I_a, flowing inside a cylinder of radius x. The field intensity may be determined from Ampere's law:

$$\int H dl = I_{enclosed} \qquad A.turn \qquad (3.38)$$

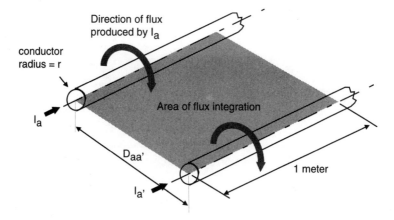

(a) Flux linkage of meter length circuit

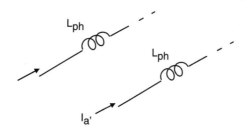

(b) Equivalent circuit of meter length circuit

Figure 3.10   Area of flux integration and equivalent circuit representation.

Ignoring end and proximity effects, the H distribution is concentric about the center of the conductor, with only a tangential component. The tangential component at radius x from the center, H_x, is given by

$$2\pi x H_x = \frac{\pi x^2}{\pi r^2} I_a \qquad A \tag{3.39}$$

and the tangential flux density at radius x is

$$B_x = \mu H_x = \frac{\mu x I_a}{2\pi r^2} \qquad Wb/m^2 \tag{3.40}$$

The elemental flux linkage from the circumscribing B field within the thin cylindrical annulus of internal radius x and external radius $x + dx$ is given by the effective turns times flux linked. Since the current flowing inside the circle of radius x is only a fraction of I_a, the effective turn is equal to the fraction, $\pi x^2/\pi r^2$. The elemental flux linkage per meter

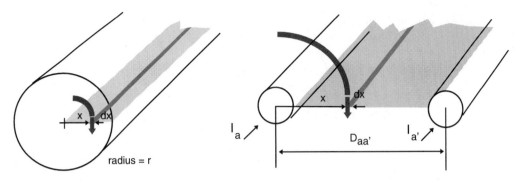

(a) Within conductor   (b) Outside conductor

Figure 3.11   Flux linkages inside and outside of conductor.

length of the conductor is thus given by this fractional turn times the elemental flux due to B_x within the area, $dx \times 1$ meter, that is

$$d\lambda_{int} = \frac{\pi x^2}{\pi r^2}(B_x \times 1 dx) \tag{3.41}$$

Integrating the above over the entire internal cross-section of the conductor, we obtain

$$\lambda_{int} = \int_0^r d\lambda_{int} = \frac{\mu I_a}{8\pi} \quad Wb.\,turn/m \tag{3.42}$$

Flux outside conductor.   The external portion of the flux linkage between the conductors, a and a', due to the field component of I_a only can be determined by integrating the flux linked by the entire current, I_a, from its outer radius outwards. It can be seen from Fig. 3.11b that when the elemental circle of radius x encloses both conductors, the enclosed current will be zero. Although the field is finite, its line integral along circular paths for $x \geq D_{aa'} + r$ will be zero. Hence, the integration need not be made from r to infinity, but rather to the point when the total enclosed current becomes zero. With a finite conductor cross-section, the total current linked actually tapers off to zero from $x = D_{aa'} - r$ to $x = D_{aa'} + r$, but for simplicity, we will approximate that by an abrupt drop to zero at $x = D_{aa'}$,

$$B_x = \frac{\mu_0 I}{2\pi x} \quad D_{aa'} \geq x \geq r \tag{3.43}$$

$$\lambda_{ext} = \int_r^{D_{aa'}} d\lambda_{ext} = \int_r^{D_{aa'}} B_x dx = \frac{\mu_0 I_a}{2\pi} \ln \frac{D_{aa'}}{r} \quad Wb.\,turn/m \tag{3.44}$$

Likewise, the flux established by the flow of $I_{a'}$ in the other conductor of the line links the same one-meter length circuit, albeit in the other direction because the circuit is on the left side of the current. It can be shown that the resultant flux linkage of the one-meter length circuit due to fluxes created by I_a and $I_{a'}$ is

$$\lambda_{aa'} = \frac{\mu}{8\pi}(I_a - I_{a'}) + \frac{\mu_0(I_a - I_{a'})}{2\pi}\ln\frac{D_{aa'}}{r} \qquad Wb.\,turn/m \qquad (3.45)$$

With no other current path but just the aerial conduction, the sum, $I_a + I_{a'}$, must be zero. Replacing $I_{a'}$ by $-I_a$ and μ by $\mu_0\mu_r$, the above expression for the total flux linkage simplifies to

$$\lambda_{aa'} = \frac{\mu}{4\pi}I_a + \frac{\mu_0 I_a}{\pi}\ln\frac{D_{aa'}}{r} \qquad Wb.\,turn/m$$

$$= \frac{\mu_0 I_a}{\pi}\ln\frac{D_{aa'}}{re^{\frac{-\mu_r}{4}}} \qquad Wb.\,turn/m \qquad (3.46)$$

The term $re^{\frac{-\mu_r}{4}}$ is known as the geometric mean radius(GMR) of the solid conductor. With physical radius r, it will be denoted by r'.

From the circuit modeling viewpoint, the per meter length circuit may be considered to have two equal per unit length series inductances, L_{ph} in the path of I_a and $I_{a'}$, and $\lambda_{aa'} = 2L_{ph}I_a$. The inductance per meter length, L_{ph}, of the line may be obtained from Eq. 3.46, that is

$$L_{ph} = \frac{\mu_0}{2\pi}\ln\left(\frac{D_{aa'}}{re^{-\frac{\mu_r}{4}}}\right) \qquad H/m \qquad (3.47)$$

3.5.3 Self- and Mutual Inductances of Circuits with Ground

The presence of a conducting ground plane close to the conductors of a line will affect the magnetic field distribution established by the currents of the line. Analytically, the condition can be handled by the method of images in which the field distribution with ground plane is being replicated by additional image currents. When the ground is approximated by a perfectly conducting plane, the image currents are of the same magnitude as their real currents, but of opposite sign, and are placed in mirror image positions with respect to the ground plane as shown in Fig. 3.12.

Let's begin by considering the flux linkage of a circuit formed by a one-meter length of the conductor 1 and its image $1'$. With $I_{j'} = -I_j$ $j = 1, 2, 3$, the flux linkage of such a circuit can be expressed as

$$\lambda_1 = \underbrace{L_{11}I_1}_{\lambda_{11}} + \underbrace{L_{12}I_2}_{\lambda_{12}} + \underbrace{L_{13}I_3}_{\lambda_{13}} \qquad (3.48)$$

where L_{11} is referred to as the self-inductance of the circuit since it is associated with the flux created by the current flowing in that same circuit, and L_{12} and L_{13} are the mutual inductances between circuit 1 and circuits 2 and 3, respectively, since they are associated with fluxes created by current flow in those two circuits. The three inductances can be separately determined by considering the corresponding flux linkage components. Thus for L_{11}, we will consider the situation when there is only current flow in the circuit formed by

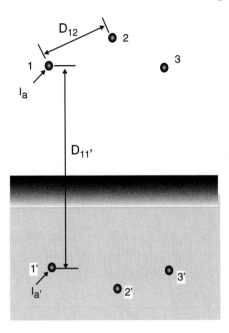

Figure 3.12 Image currents of a three-conductor line.

conductors 1 and $1'$. From the analysis of the two-conductor line in the previous section, in particular Eq. 3.47, we can deduce that

$$\lambda_{11} = L_{11}I_1 = \frac{\mu_0 I_1}{\pi} \ln\left(\frac{D_{11'}}{r'}\right) \tag{3.49}$$

where $D_{11'}$ is the center-to-center distance between conductor 1 and its image conductor $1'$, and r' is the geometric mean radius of the conductors 1 and $1'$.

The mutual inductance between circuits 1 and 2, L_{12}, can be determined by considering the flux linkage of circuit 1 due to only current I_2 in circuit 2, or that of circuit 2 due to current flow in circuit 1. The flux linkage of circuit 1 due to only current I_2 in circuit 2 may be found from

$$\lambda_{12} = L_{12}I_2 = \int_{D_{12}}^{D_{1'2}} \frac{\mu_0 I_2}{2\pi x} dx + \int_{D_{12'}}^{D_{1'2'}} \frac{\mu_0 I_2'}{2\pi x'} dx' \tag{3.50}$$

where x is measured from the center of conductor 2 and x' is measured from the center of $2'$.

Making use of the relations, $I_2' = -I_2$ and $D_{1'2} = D_{12'}$, the above expression for λ_{12} reduces to

$$\begin{aligned}
L_{12}I_2 &= \frac{\mu_0 I_2}{2\pi}\left(\ln\frac{D_{1'2}}{D_{12}} - \ln\frac{D_{1'2'}}{D_{12'}}\right) \\
&= \frac{\mu_0 I_2}{2\pi}\ln\left(\frac{D_{12'}D_{1'2}}{D_{12}D_{1'2'}}\right)
\end{aligned} \tag{3.51}$$

from which we can obtain the expression for the mutual inductance between circuits 1 and 2, that is

$$L_{12} = \frac{\mu_0}{2\pi} \ln\left(\frac{D_{12'}D_{1'2}}{D_{12}D_{1'2'}}\right) \tag{3.52}$$

Using the same approach, we can derive similar expressions for λ_{13} and L_{13}. In general, for a line with n conductors carrying currents $I_1, I_2, \ldots, I_n$, the flux linkages of the meter length circuits formed by the real conductors and their corresponding images can be written in matrix form as

$$\begin{bmatrix} \lambda_1 \\ \lambda_2 \\ \vdots \\ \lambda_n \end{bmatrix} = \begin{bmatrix} L_{11} & L_{12} & \cdots & L_{1n} \\ L_{21} & L_{22} & & L_{2n} \\ \vdots & \vdots & \ddots & \vdots \\ L_{n1} & L_{n2} & \cdots & L_{nn} \end{bmatrix} \begin{bmatrix} I_1 \\ I_2 \\ \vdots \\ I_n \end{bmatrix} \tag{3.53}$$

Both the self- and mutual inductances may in general be computed using a general form of Eqs. 3.49 and 3.52, which is given below:

$$L_{ij} = \frac{\mu_0}{2\pi} \ln\left(\frac{D_{ij'}D_{i'j}}{D_{ij}D_{i'j'}}\right) \qquad i, j = 1, 2, \ldots, n \tag{3.54}$$

where $D_{ii} = D_{i'i'} = r_i'$ is the geometric mean radius of conductor i.

3.5.4 Geometric Mean Radius and Distance

In this subsection, we will derive the general expressions for the geometric mean radius and distance by considering the case where a group of line conductors are divided into two clusters: Cluster Ω has n conductors carrying a total current of I in one direction, and cluster Γ has m conductors carrying the same total current in the opposite direction. For simplicity, we will assume that the total current of each cluster is equally divided amongst its conductors and that the conductors of a cluster are of the same size.

Using Eq. 3.54, we can express the flux linkage of the circuit formed by the ith conductor in cluster Ω and its image as

$$\lambda_i = \frac{\mu_0}{2\pi} \frac{I}{n} \ln\left\{ \prod_{j \in \Omega} \frac{D_{ij'}D_{i'j}}{D_{ij}D_{i'j'}} \right\} - \frac{\mu_0}{2\pi} \frac{I}{m} \ln\left\{ \prod_{k \in \Gamma} \frac{D_{ik'}D_{i'k}}{D_{ik}D_{i'k'}} \right\} \tag{3.55}$$

The n conductors of the cluster Ω contribute to the n product terms for $j \in \Omega$. Likewise, the m conductors of the cluster Γ contribute to the m product terms for $k \in \Gamma$. When the distances between all conductors and their images are much larger than those separating the conductors, we can make the reasonable approximation that $D_{ij'} \approx D_{ik'}$. Using this approximation and that of $D_{ij'} = D_{i'j}$ in Eq. 3.55, we can rewrite it as

$$\lambda_i = \frac{\mu_0 I}{\pi} \ln\left\{ \frac{\sqrt[m]{\prod_{k \in \Gamma} D_{ik}}}{\sqrt[n]{\prod_{j \in \Omega} D_{ij}}} \right\} \tag{3.56}$$

The average value of the flux linkages of the circuits formed by conductors of the cluster Ω is given by

$$\lambda_\Omega^{ave} = \frac{1}{n} \sum_{i \in \Omega} \lambda_i \qquad (3.57)$$

The average loop inductance of the entire Ω cluster of conductors carrying a total current of I may be expressed as

$$L_{loop} = 2L_{ph} = \frac{\lambda_\Omega^{ave}}{I}$$

$$L_{ph} = \frac{\mu_0}{2\pi} \ln \left\{ \frac{\sqrt[mn]{\prod_{i \in \Omega}(\prod_{k \in \Gamma} D_{ik})}}{\sqrt[n^2]{\prod_{i \in \Omega}(\prod_{j \in \Omega} D_{ij})}} \right\} \qquad (3.58)$$

The numerator of the argument of the log term is the geometric mean distance between the two clusters. It is equal to the mnth root of the products of the distances between conductors of the two clusters. The denominator of the same argument is the geometric mean radius, or the self-GMD, of cluster Ω. Likewise, similar expressions of loop and self-inductances may be written for the Γ cluster of conductors.

3.5.5 Transposition of Line Conductors

In practice, conductor spacing is usually not symmetric. With unsymmetric spacing, the position of each phase conductor relative to that of another phase is not the same. As a result, the inductance of one phase will not be the same as that of another. This difference in circuit inductances among the phases can be minimized by rotating the position that each phase conductor occupies over the entire length. The changing over of position, or transposing, is usually done at convenient points along the line. Transposing the line conductors will reduce the imbalance of circuit inductance and capacitance. Figure 3.13 shows the three sets of conductor positions of a transposition cycle. The conductor positions are labeled by numbers, whereas the phases are labeled by alphabetic characters.

Consider a uniformly transposed line in which the positions of the phase conductors are fully rotated. Over the three segments of the transposition cycle denoted by the superscripts, X, Y, and Z, the expressions of the flux linkage for a meter length circuit formed by the conductors, a and a', are

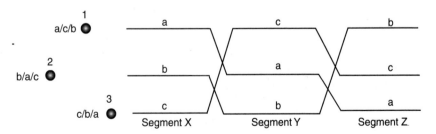

Figure 3.13   The transposition of three-phase line conductors.

$$\lambda_a^X = \frac{\mu_0}{2\pi}\left\{I_a \ln\left(\frac{D_{1'1}D_{11'}}{D_{11}D_{1'1'}}\right) + I_b \ln\left(\frac{D_{1'2}D_{12'}}{D_{12}D_{1'2'}}\right) + I_c \ln\left(\frac{D_{1'3}D_{13'}}{D_{13}D_{1'3'}}\right)\right\}$$

$$\lambda_a^Y = \frac{\mu_0}{2\pi}\left\{I_a \ln\left(\frac{D_{2'2}D_{22'}}{D_{22}D_{2'2'}}\right) + I_b \ln\left(\frac{D_{2'3}D_{23'}}{D_{23}D_{2'3'}}\right) + I_c \ln\left(\frac{D_{2'1}D_{21'}}{D_{21}D_{2'1'}}\right)\right\} \qquad (3.59)$$

$$\lambda_a^Z = \frac{\mu_0}{2\pi}\left\{I_a \ln\left(\frac{D_{3'3}D_{33'}}{D_{33}D_{3'3'}}\right) + I_b \ln\left(\frac{D_{3'1}D_{31'}}{D_{31}D_{3'1'}}\right) + I_c \ln\left(\frac{D_{3'2}D_{32'}}{D_{32}D_{3'2'}}\right)\right\}$$

If the phase conductors are of the same size, we can replace D_{11}, D_{22}, D_{33}, $D_{1'1'}$, $D_{2'2'}$, and $D_{3'3'}$ by r'. Taking the average value of the above three λ_a's, that is,

$$
\begin{aligned}
\lambda_a^{ave} &= \frac{\lambda_a^X + \lambda_a^Y + \lambda_a^Z}{3} \\[2mm]
&= \frac{\mu_0}{2\pi}\left\{ \frac{I_a}{3}\ln\left(\frac{D_{1'1}D_{11'}}{D_{11}D_{1'1'}}\frac{D_{2'2}D_{22'}}{D_{22}D_{2'2'}}\frac{D_{3'3}D_{33'}}{D_{33}D_{3'3'}}\right) \right. \\[2mm]
&\quad + \frac{I_b}{3}\ln\left(\frac{D_{1'2}D_{12'}}{D_{12}D_{1'2'}}\frac{D_{2'3}D_{23'}}{D_{23}D_{2'3'}}\frac{D_{3'1}D_{31'}}{D_{31}D_{3'1'}}\right) \\[2mm]
&\quad \left. + \frac{I_c}{3}\ln\left(\frac{D_{1'3}D_{13'}}{D_{13}D_{1'3'}}\frac{D_{2'1}D_{21'}}{D_{21}D_{2'1'}}\frac{D_{3'2}D_{32'}}{D_{32}D_{3'2'}}\right)\right\}
\end{aligned}
\qquad (3.60)
$$

Note that the numerators and denominators of the logarithm coefficient to I_b and I_c are equal in value, the two subscripts of the corresponding D terms are just permutated, as such Eq. 3.61 can be rewritten as

$$
\begin{aligned}
\lambda_a^{ave} &= \frac{\mu_0}{2\pi}\left\{ \frac{I_a}{3}\ln\left(\frac{D_{1'1}D_{11'}}{D_{11}D_{1'1'}}\frac{D_{2'2}D_{22'}}{D_{22}D_{2'2'}}\frac{D_{3'3}D_{33'}}{D_{33}D_{3'3'}}\right) \right. \\[2mm]
&\quad \left. + \frac{(I_b + I_c)}{3}\ln\left(\frac{D_{1'2}D_{12'}}{D_{12}D_{1'2'}}\frac{D_{2'3}D_{23'}}{D_{23}D_{2'3'}}\frac{D_{3'1}D_{31'}}{D_{31}D_{3'1'}}\right)\right\}
\end{aligned}
\qquad (3.61)
$$

When the phase currents are balanced, that is $I_a + I_b + I_c = 0$, there will not be any ground conduction. When the spacings between conductors are much smaller than their heights above ground, or if the effect of ground is to be ignored by assuming that it is far away from the conductors, all conductor-to-image distances may be approximated by a common value, D_{IR}, in which case the numerator of the two log terms in Eq. 3.61 will be equal. Replacing $I_b + I_c$ by $-I_a$, the D_{ii} and $D_{i'i'}$ terms by r', and $D_{i'j'}$ by D_{ij}, Eq. 3.61 will simplify to

$$\lambda_a^{ave} = \underbrace{\left(\frac{\mu_0}{\pi}\ln\frac{\sqrt[3]{D_{12}D_{23}D_{31}}}{r'}\right)I_a}_{2L_{ph}} \qquad (3.62)$$

where L_{ph} is the average per meter inductance to the current, I_a, flowing in the conductor, a, over a full transposition cycle. A comparison of Eq. 3.62 with Eq. 3.58 shows that the GMD is $\sqrt[3]{D_{12}D_{23}D_{31}}$. The GMD may also be interpreted as the equilateral spacing of an equivalent three-phase line having the same phase inductance. In both of these cases, the expression of the average phase inductance is of the form:

$$L_{ph} = \frac{\mu_0}{2\pi} \ln \frac{GMD}{GMR} \qquad (3.63)$$

Zero-sequence inductance.   With unbalanced currents, that is $\sum I_j \neq 0$, the effects of ground may have to be considered to account for the ground path taken by the so-called zero-sequence current component, I_0. For convenience, we will assume that the heights of the conductors above ground are large relative to the spacings between conductors so that we can approximate the distances between real and image conductors by a common mean distance, D_{IR}. For a non-perfectly conducting ground, D_{IR} will be a function of the frequency and earth's resistivity.

$$D_{IR} \approx k\sqrt{\sigma/f} \qquad (3.64)$$

Assuming full transposition and ground height much larger than conductor spacings, and using the result given in Eq. 3.60 for the flux linkages of a meter length of the circuit formed by conductor a and a' for the other two phases, we can obtain

$$\lambda_a^{ave} = \frac{\mu_0}{2\pi} \left(2I_a \ln \frac{D_{IR}}{r'} + 2I_b \ln \frac{D_{IR}}{GMD} + 2I_c \ln \frac{D_{IR}}{GMD} \right)$$

$$\lambda_b^{ave} = \frac{\mu_0}{2\pi} \left(2I_a \ln \frac{D_{IR}}{GMD} + 2I_b \ln \frac{D_{IR}}{r'} + 2I_c \ln \frac{D_{IR}}{GMD} \right) \qquad (3.65)$$

$$\lambda_c^{ave} = \frac{\mu_0}{2\pi} \left(2I_a \ln \frac{D_{IR}}{GMD} + 2I_b \ln \frac{D_{IR}}{GMD} + 2I_c \ln \frac{D_{IR}}{r'} \right)$$

Equating the zero-sequence flux linkage λ_0 to $2L_0 I_0$, where $\lambda_0 = (\lambda_a^{ave} + \lambda_b^{ave} + \lambda_c^{ave})/3$, $I_0 = (I_a + I_b + I_c)/3$, and L_0 is the per meter zero-sequence inductance to the zero-sequence current of the real abc currents, we have

$$\lambda_0 = \frac{1}{3}(\lambda_a^{ave} + \lambda_b^{ave} + \lambda_c^{ave})$$

$$2L_0 I_0 = 2I_0 \frac{\mu}{2\pi} \ln \left(\frac{GMD^3}{r' D_{IR}^2} \right) \qquad (3.66)$$

Thus,

$$L_0 = \frac{\mu_0}{2\pi} \ln \left(\frac{GMD^3}{r' D_{IR}^2} \right) \qquad (3.67)$$

whose value may be several times the value of the per meter phase inductance, L_{ph}.

Double-circuit line.   Double-circuits are often used when the availability of right-of-way is limited. The phase conductors of a double-circuit may be arranged in more than one way. We will consider the conductor transposition shown in Fig. 3.14, in which the geometric mean distances of the corresponding phases are maximized to reduce phase inductance. In Fig. 3.14, the positions occupied by the transposed conductors are denoted by numbers. The letters by the side of each conductor position indicate the order in which

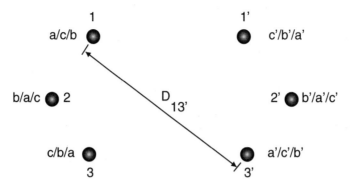

Figure 3.14   Conductor positions of a transposed double-circuit.

the phases occupy that conductor position over the three segments, X, Y, and Z, of a transposition cycle.

The GMDs of the phases for the three parts of a transposition cycle are

$$GMD_{ab}^X = GMD_{ac}^Y = GMD_{bc}^Z = \sqrt[4]{D_{12}D_{12'}D_{3'2'}D_{3'2}} \qquad (3.68)$$

$$GMD_{bc}^X = GMD_{ab}^Y = GMD_{ac}^Z = \sqrt[4]{D_{23}D_{21'}D_{2'1'}D_{2'3}} \qquad (3.69)$$

$$GMD_{ca}^X = GMD_{bc}^Y = GMD_{ab}^Z = \sqrt[4]{D_{31}D_{33'}D_{1'3'}D_{1'1}} \qquad (3.70)$$

The GMD between phases for transposition phase X is given by

$$GMD^X = \sqrt[3]{GMD_{ab}^X GMD_{bc}^X GMD_{ca}^X} \qquad (3.71)$$

Along with similar expressions for the GMDs for the transposition phases, Y and Z, we can determine the geometric mean of these GMDs over the entire transposition cycle from

$$GMD = \sqrt[3]{GMD^X GMD^Y GMD^Z} = GMD^X \qquad (3.72)$$

Similarly, the GMRs of phase a for the three parts of the transposition cycle are

$$GMR_a^X = \sqrt[4]{(r'D_{13'})^2}$$

$$GMR_a^Y = \sqrt[4]{(r'D_{22'})^2} \qquad (3.73)$$

$$GMR_a^Z = \sqrt[4]{(r'D_{31'})^2}$$

The geometric mean of the GMRs of phase a over the entire transposition cycle may be expressed as

$$GMR = \sqrt[3]{GMR_a^X GMR_a^Y GMR_a^Z} = \sqrt[3]{(D_{13'}D_{22'}D_{31'})^{1/2}(r')^{3/2}} \qquad (3.74)$$

With the values of GMD and GMR computed above, we can determine the average phase inductance for each phase of the double-circuit line using

$$L_{ph} = \frac{\mu_0}{2\pi} \ln\left(\frac{GMD}{GMR}\right) \qquad H/m \qquad (3.75)$$

3.5.6 Capacitance

Assuming a uniformly distributed surface charge of q Coulombs per meter on the cylindrical conductor, Gauss' law gives

$$\int_s D\,ds = q_{enclosed} \qquad C \tag{3.76}$$

For a one-meter length of conductor, the electric flux density at a radial distance x from the center is

$$D = \frac{q}{2\pi x} \qquad C/m^2 \tag{3.77}$$

The electric field intensity or force per Coulomb at that distance is

$$E = \frac{q}{2\pi \epsilon x} \qquad V/m \tag{3.78}$$

Since force can be expressed in units of energy or Volts Coulomb over meter, the unit of the electric field intensity is Volts over meter.

The electrical potential at a point is usually defined as the work done in moving a unit of electrical charge from infinity to the point, that is,

$$V = \int_\infty^r -E\,dx \qquad V \tag{3.79}$$

or

$$E = -\text{grad } V = -\frac{\partial V}{\partial n}\hat{\mathbf{n}} \tag{3.80}$$

where $\hat{\mathbf{n}}$ is the normal to the equipotential surface.

The potential drop in going from the equipotential surface at the radial distance, r_1, to that at r_2 in Fig. 3.15 is given by

$$V_{12} = V_{1\infty} - V_{2\infty} = -\int_\infty^{r_1} E\,dx + \int_\infty^{r_2} E\,dx$$

$$= \int_{r_1}^{r_2} E\,dx = \frac{q}{2\pi \epsilon}\ln\left(\frac{r_2}{r_1}\right) \qquad V \tag{3.81}$$

For positive q, V_{12} will be positive when $r_2 > r_1$.

Capacitance of a two-conductor line.   Let's derive the expression of the capacitance for the two-conductor line shown in Fig. 3.16. Using superposition, the voltage drop from 1 to 2 due to the fields established by q_1 and q_2 can be expressed as

$$V_{12} = \int_{r_1}^D E_1\,dx_1 + \int_D^{r_2} E_2\,dx_2 \tag{3.82}$$

where x_1 and x_2 are distances measured from the respective centers of conductors 1 and 2.

$$V_{12} = \frac{q_1}{2\pi \epsilon}\ln\frac{D}{r_1} + \frac{q_2}{2\pi \epsilon}\ln\frac{r_2}{D} \tag{3.83}$$

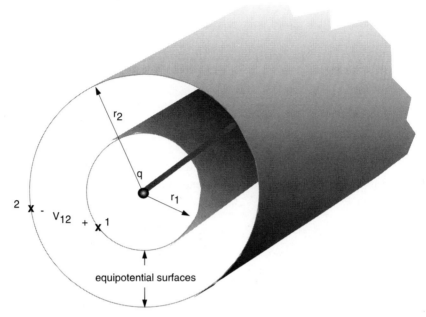

Figure 3.15   Potential between points 1 and 2 from E field.

Figure 3.16   Line charge on two-conductor line.

Imposing the condition of conservation of charge, that is $q_1 + q_2 = 0$, the above equation reduces to

$$V_{12} = \frac{q_1}{2\pi\epsilon} \ln \frac{D^2}{r_1 r_2} \qquad (3.84)$$

Defining capacitance as charge per unit volt, the capacitance between conductors 1 and 2 is given by

$$C_{12} = \frac{q_1}{V_{12}} = \frac{2\pi\epsilon}{\ln\left(\dfrac{D^2}{r_1 r_2}\right)} \qquad F/m \qquad (3.85)$$

where r_1 and r_2 are the radii of the surface envelopes of the two conductors. The above equation has the form:

$$C_{12} = \frac{\pi\epsilon}{\ln\left(\dfrac{GMD}{GMR}\right)} \qquad (3.86)$$

Unlike the GMR in the inductance formula, which also accounts for the flux linkage inside the conductor, the GMR in the capacitance expression uses the envelope radius.

3.5.7 Self- and Mutual Capacitance of a Conductor Group

Using Eq. 3.81 the voltage drop or potential difference between conductors i and n of the conductor group shown in Fig. 3.17 can be written as

$$V_{in} = \int_{D_{1i}}^{D_{1n}} E_1 dx_1 + \int_{D_{2i}}^{D_{2n}} E_2 dx_2 + \ldots + \int_{D_{ni}}^{D_{nn}} E_n dx_n \qquad (3.87)$$

where D_{ii} is the radius of conductor i and D_{ij} is the physical distance between the centers of conductors i and j.

$$V_{in} = \frac{1}{2\pi\epsilon} \sum_{j}^{n} q_j \ln\left(\frac{D_{jn}}{D_{ji}}\right) \qquad (3.88)$$

Imposing the condition that $\sum_j q_j = 0$, and using it to eliminate q_n from the above expression, we obtain

$$V_{in} = \frac{1}{2\pi\epsilon} \sum_{j}^{n-1} q_j \left(\ln\frac{D_{jn}}{D_{ji}} - \ln\frac{D_{nn}}{D_{ni}}\right) \qquad i \ and \ j \neq n \qquad (3.89)$$

$$V_{in} = \frac{1}{2\pi\epsilon} \sum_{j}^{n-1} q_j \left(\ln\frac{D_{jn}D_{ni}}{D_{ji}D_{nn}}\right) \qquad (3.90)$$

Similar expressions for the potential difference between other conductors and conductor n can also be written. These expressions can all be assembled in matrix form as

$$\begin{bmatrix} V_{1n} \\ V_{2n} \\ \vdots \\ V_{(n-1)n} \end{bmatrix} = \begin{bmatrix} P_{11} & P_{12} & \cdots & P_{1(n-1)} \\ P_{21} & P_{22} & \cdots & P_{2(n-1)} \\ \vdots & \vdots & \ddots & \vdots \\ P_{(n-1)1} & P_{(n-1)2} & \cdots & P_{(n-1)(n-1)} \end{bmatrix} \begin{bmatrix} q_1 \\ q_2 \\ \vdots \\ q_{(n-1)} \end{bmatrix} \qquad (3.91)$$

The coefficient matrix of Ps is a symmetric matrix; it contains the self- and mutual potential coefficients of the conductors. Using the basic definition of capacitance as charge per unit

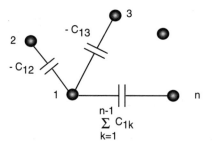

Figure 3.17   Self- and mutual capacitances of two groups of conductors.

volt, the inverse of the $(n-1) \times (n-1)$ potential coefficient matrix yields a symmetric capacitance matrix.

$$
\begin{bmatrix} q_1 \\ q_2 \\ \vdots \\ q_{(n-1)} \end{bmatrix} = \begin{bmatrix} C_{11} & C_{12} & \cdots & C_{1(n-1)} \\ C_{21} & C_{22} & \cdots & C_{1(n-2)} \\ \vdots & \vdots & \ddots & \vdots \\ C_{(n-1)1} & C_{(n-1)2} & \cdots & C_{(n-1)(n-1)} \end{bmatrix} \begin{bmatrix} V_{1n} \\ V_{2n} \\ \vdots \\ V_{(n-1)n} \end{bmatrix} \tag{3.92}
$$

Note that the voltages used in the above expression are those measured with respect to conductor n. To obtain the circuit capacitances between conductors 1, say, and the other conductors, we will have to rewrite the above equation for q_1 in terms of the voltage differences between conductor 1 and the other conductors. That is

$$
\begin{aligned}
q_1 &= (C_{11}+C_{12}+\ldots+C_{1(n-1)})V_{1n}+C_{12}(V_{2n}-V_{1n})+\ldots+C_{1(n-1)}(V_{(n-1)n}-V_{1n}) \\
&= (C_{11}+C_{12}+\ldots+C_{1(n-1)})V_{1n}-C_{12}V_{12}+\ldots+C_{1(n-1)}V_{1(n-1)} \tag{3.93}
\end{aligned}
$$

Written in this manner, we can see that the capacitance between conductors 1 and n is $(C_{11}+C_{12}+\ldots+C_{1(n-1)})$, and the capacitance between conductors 1 and j $(j \neq n)$ is $-C_{1j}$.

Capacitance of a double-circuit line.   Consider again the transposed, double-circuit, three-phase line of Fig. 3.14. If the charge on each phase conductor in the three transposed positions were to remain the same, that is $q_a^X = q_a^Y = q_a^Z$, the voltage between any two phases would vary as the phase conductors occupied different positions of the transposition cycle. For instance, the voltages between phases a and b for the three phases of a transposition cycle are

$$
\begin{aligned}
V_{ab}^X &= \frac{1}{2\pi\epsilon}\left(q_a\ln\frac{D_{12}}{r}+q_b\ln\frac{r}{D_{12}}+q_c\ln\frac{D_{23}}{D_{31}}\right) \\
V_{ab}^Y &= \frac{1}{2\pi\epsilon}\left(q_a\ln\frac{D_{23}}{r}+q_b\ln\frac{r}{D_{23}}+q_c\ln\frac{D_{31}}{D_{21}}\right) \tag{3.94} \\
V_{ab}^Z &= \frac{1}{2\pi\epsilon}\left(q_a\ln\frac{D_{31}}{r}+q_b\ln\frac{r}{D_{31}}+q_c\ln\frac{D_{12}}{D_{23}}\right)
\end{aligned}
$$

The average value of the voltage between the a and b phases over an entire transposition cycle is

$$
\begin{aligned}
V_{ab} &= \frac{V_{ab}^X+V_{ab}^Y+V_{ab}^Z}{3} \\
&= \frac{1}{2\pi\epsilon}\left(q_a\ln\frac{GMD}{r}+q_b\ln\frac{r}{GMD}\right) \tag{3.95}
\end{aligned}
$$

where $GMD = \sqrt[3]{D_{12}\,D_{23}\,D_{31}}$. Using a similar approach, we can derive the expressions for the average voltage between the other phases. That between the a and c phases over a transposition cycle can be written down from the previous equation of V_{ab} by simply replacing subscript b with c, that is

$$V_{ac} = \frac{1}{2\pi\epsilon}\left(q_a \ln\frac{GMD}{r} + q_c \ln\frac{r}{GMD}\right) \tag{3.96}$$

When the three-phase voltages are balanced, that is $V_{an} + V_{bn} + V_{cn} = 0$, we can show from the phasor diagram of these voltages that the following relationship between the line-to-line voltages and the phase-to-neutral voltage holds:

$$|V_{ab} + V_{ac}| = 3|V_{an}| \tag{3.97}$$

Using the above three expressions and $q_a + q_b + q_c = 0$, we can obtain

$$3V_{an} = \frac{3}{2\pi\epsilon}q_a \ln\left(\frac{GMD}{r}\right) \tag{3.98}$$

from which we can deduce that the phase-to-neutral capacitance of the uniformly transposed three-phase line is

$$C_{an} = \frac{2\pi\epsilon}{\ln\left(\dfrac{GMD}{r}\right)} \; F/m \tag{3.99}$$

3.5.8 The Presence of Ground

The presence of ground can be handled analytically using the method of images. For a perfectly conducting ground plane, the depth below ground of the image charges will be the same as the height above ground of the corresponding conductor charges. For the two-conductor line of Fig. 3.18, the expression for the voltage to ground of conductor a with the presence of ground taken care of by image charges can be written as

$$V_{ag} = \frac{1}{2\pi\epsilon}\left\{q_a \ln\frac{h_a}{r} + q_{a'} \ln\left(\frac{h_{a'}}{h_{a'}+h_a}\right) + q_b \ln\left(\frac{h_b}{D_{ab}}\right) + q_{b'} \ln\left(\frac{h_{b'}}{D_{ab'}}\right)\right\} \tag{3.100}$$

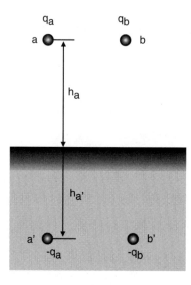

Figure 3.18 Image charges of a two-conductor line.

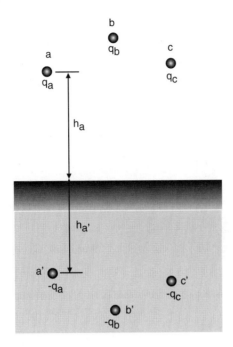

Figure 3.19 Image charges of a three-conductor line.

Substituting $q'_i = -q_i$ and $h'_i = h_i$, $i \in a, b$ we obtain

$$V_{ag} = \frac{1}{2\pi\epsilon} \left(q_a \ln \frac{2h_a}{r} + q_b \ln \frac{D_{ab'}}{D_{ab}} \right) \tag{3.101}$$

Using Eq. 3.101, we can write the following phase-to-ground voltage equations for the three phases of the three-conductor line shown in Fig. 3.19:

$$\begin{bmatrix} V_{ag} \\ V_{bg} \\ V_{cg} \end{bmatrix} = \frac{\ln}{2\pi\epsilon} \begin{bmatrix} \dfrac{2h_a}{r} & \dfrac{D_{ab'}}{D_{ab}} & \dfrac{D_{ac'}}{D_{ac}} \\[2mm] \dfrac{D_{a'b}}{D_{ab}} & \dfrac{2h_b}{r} & \dfrac{D_{bc'}}{D_{bc}} \\[2mm] \dfrac{D_{a'c}}{D_{ac}} & \dfrac{D_{b'c}}{D_{bc}} & \dfrac{2h_c}{r} \end{bmatrix} \begin{bmatrix} q_a \\ q_b \\ q_c \end{bmatrix} \tag{3.102}$$

Inverting Eq. 3.102, we have

$$\begin{bmatrix} q_a \\ q_b \\ q_c \end{bmatrix} = [P]^{-1} \begin{bmatrix} V_{ag} \\ V_{bg} \\ V_{cg} \end{bmatrix} = [C'] \begin{bmatrix} V_{ag} \\ V_{bg} \\ V_{cg} \end{bmatrix} \tag{3.103}$$

where the capacitance matrix is of the form

$$[C'] = \begin{bmatrix} C'_{aa} & C'_{ab} & C'_{ac} \\ C'_{ba} & C'_{bb} & C'_{bc} \\ C'_{ca} & C'_{cb} & C'_{cc} \end{bmatrix} \tag{3.104}$$

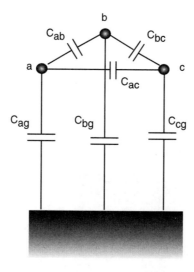

Figure 3.20 Circuit capacitances of a three-phase line.

The elements of the C' matrix can be translated to the capacitances in the circuit representation shown in Fig. 3.20. Rewriting the expression for q_a given by $q_a = C'_{aa}V_{ag} + C'_{ab}V_{bg} + C'_{ac}V_{cg}$, we get

$$q_a = (C'_{aa} + C'_{ab} + C'_{ac})V_{ag} + C'_{ab}(V_{bg} - V_{ag}) + C'_{ac}(V_{cg} - V_{ag}) \qquad (3.105)$$

In accordance with the desired circuit representation shown in Fig. 3.20, the expression of q_a should be

$$q_a = C_{ag}V_{ag} + C_{ab}V_{ab} + C_{ac}V_{ac} \qquad (3.106)$$

Thus,

$$C_{ag} = C'_{aa} + C'_{ab} + C'_{ac} \qquad (3.107)$$

$$C_{ab} = -C'_{ab} \qquad (3.108)$$

$$C_{ac} = -C'_{ac} \qquad (3.109)$$

Double-circuit lines. The phase-to-neutral capacitance of the transposed, double-circuit, three-phase line shown in Fig. 3.14 may be determined from

$$C_{an} = \frac{2\pi\epsilon}{\ln\left(\dfrac{GMD}{GMR}\right)} \qquad F/m$$

where the GMD may be determined from $\sqrt[3]{GMD^X \, GMD^Y \, GMD^Z}$ as shown earlier in Eq. 3.72.

The value of the GMR will, however, be different from that used in the inductance calculation. For the capacitance, the GMR of the phases is given by

$$GMR = \sqrt[3]{GMR_a^X \, GMR_a^Y \, GMR_a^Z}$$

where

$$GMR_a^X = \sqrt[4]{(r_a D_{13'})^2}$$

$$GMR_a^Y = \sqrt[4]{(r_a D_{22'})^2} \qquad (3.110)$$

$$GMR_a^Z = \sqrt[4]{(r_a D_{31'})^2}$$

3.6 LUMPED PARAMETER CIRCUIT MODELS

Several circuit models of transmission lines are used in simulation studies. Usually, lumped parameter circuit models, like the series RL, nominal and equivalent π, are used in steady-state calculations and in studies limited to low-frequency electromechanical behavior, such as transient stability of machines. A distributed parameter line model may be more appropriate for higher-frequency electromagnetic behavior, such as the propagation of lightning and switching surges.

3.6.1 Series RL and Nominal π Circuit Models

Figure 3.21 shows two per phase equivalent circuit models for short and medium length lines. The series RL model is applicable when the shunt capacitance of the line can be ignored. It is usually adequate for lines that are shorter than 50 miles. Where the shunt capacitance of the line is to be included, as in longer lines up to 120 miles in length, a nominal π model is often used. The nominal π equivalent circuit has the structure of a π network. Its series branch is the same as that used in the series RL circuit model, that is

$$\mathbf{Z} = R_{ph} + j\omega L_{ph} \qquad (3.111)$$

In the nominal π, half of the total shunt admittance of the line is lumped at each end of the line, that is

$$\frac{\mathbf{Y}}{2} = j\omega \frac{C_{ph}}{2} \qquad (3.112)$$

The parameters of the circuit models shown in Fig. 3.21 can easily be determined from the per unit length RLC parameters of the line as shown below:

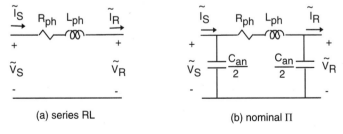

(a) series RL   (b) nominal Π

Figure 3.21   Series RL and nominal π circuit models.

- R_{ph} = (phase resistance/unit length) times line length.
- L_{ph} = (phase inductance/unit length) times line length.
- C_{ph} = (phase-to-neutral capacitance/unit length) times line length.

More than one nominal π circuit may be required when the line is long, when higher fidelity is desired, or when there is a need to introduce a fault or connection at some intermediate point between the busses joined by the lines.

3.6.2 ABCD Transmission Matrix

The sending-end phasor voltage and current of the nominal π circuit model in Fig. 3.21 can be expressed in terms of the receiving-end phasors as follows:

$$\tilde{\mathbf{V}}_S = \left(\tilde{\mathbf{V}}_R \frac{\mathbf{Y}}{2} + \tilde{\mathbf{I}}_R \right) \mathbf{Z} + \tilde{\mathbf{V}}_R \tag{3.113}$$

$$\tilde{\mathbf{I}}_S = \tilde{\mathbf{V}}_S \frac{\mathbf{Y}}{2} + \tilde{\mathbf{V}}_R \frac{\mathbf{Y}}{2} + \tilde{\mathbf{I}}_R \tag{3.114}$$

Arranging these two equations in matrix form, we have

$$\begin{bmatrix} \tilde{\mathbf{V}}_S \\ \tilde{\mathbf{I}}_S \end{bmatrix} = \begin{bmatrix} \left(1 + \dfrac{\mathbf{ZY}}{2}\right) & \mathbf{Z} \\ \mathbf{Y}\left(1 + \dfrac{\mathbf{ZY}}{4}\right) & \left(1 + \dfrac{\mathbf{ZY}}{2}\right) \end{bmatrix} \begin{bmatrix} \tilde{\mathbf{V}}_R \\ \tilde{\mathbf{I}}_R \end{bmatrix} \tag{3.115}$$

The shunt admittance, $\mathbf{Y}$, in these expressions will be zero for a series RL model. The above set of equations enable us to identify the relation of the ABCD parameters for the nominal π line model in what is generally known as the ABCD transmission matrix of the two-port representation. This is done by equating the coefficients of Eq. 3.115 with those of the transmission equations:

$$\begin{bmatrix} \tilde{\mathbf{V}}_S \\ \tilde{\mathbf{I}}_S \end{bmatrix} = \begin{bmatrix} \mathbf{A} & \mathbf{B} \\ \mathbf{C} & \mathbf{D} \end{bmatrix} \begin{bmatrix} \tilde{\mathbf{V}}_R \\ \tilde{\mathbf{I}}_R \end{bmatrix} \tag{3.116}$$

The ABCD transmission representation is especially handy when cascading a number of components. For example, the overall transmission equation for the three segments shown in Fig. 3.22 may be determined from

$$\begin{bmatrix} \tilde{\mathbf{V}}_{S1} \\ \tilde{\mathbf{I}}_{S1} \end{bmatrix} = \begin{bmatrix} \mathbf{A}_1 & \mathbf{B}_1 \\ \mathbf{C}_1 & \mathbf{D}_1 \end{bmatrix} \begin{bmatrix} \mathbf{A}_2 & \mathbf{B}_2 \\ \mathbf{C}_2 & \mathbf{D}_2 \end{bmatrix} \begin{bmatrix} \mathbf{A}_3 & \mathbf{B}_3 \\ \mathbf{C}_3 & \mathbf{D}_3 \end{bmatrix} \begin{bmatrix} \tilde{\mathbf{V}}_{R3} \\ \tilde{\mathbf{I}}_{R3} \end{bmatrix} \tag{3.117}$$

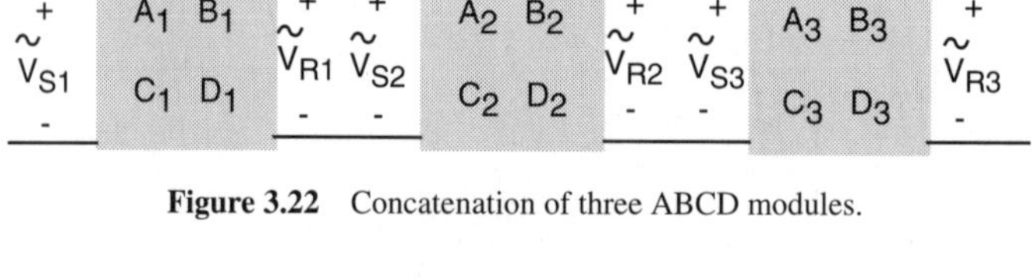

Figure 3.22   Concatenation of three ABCD modules.

3.7 DISTRIBUTED PARAMETER MODEL

If the objective of the study requires higher accuracy or higher fidelity to portray the transient components than what the lumped parameter circuit models are capable of, we can use a distributed parameter circuit model of the line. Figure 3.23 shows the circuit representation of a distributed parameter model which applies the RLC parameter of the line to an elemental section, instead of to the whole length of the line.

Let's determine the voltage and current relationship of both ends of the elemental section shown in Fig. 3.23. Applying KCL to the node at $(x + \Delta x)$, we obtain

$$i(x) - i(x + \Delta x) = (Gv + C\frac{\partial v}{\partial t})\Delta x \tag{3.118}$$

Applying KVL around the elemental loop, we obtain

$$v(x) - v(x + \Delta x) = (Ri + L\frac{\partial i}{\partial t})\,\Delta x \tag{3.119}$$

Letting $\Delta x \to 0$, the above equations can be expressed as

$$-\frac{\partial i}{\partial x} = Gv + C\frac{\partial v}{\partial t}$$
$$-\frac{\partial v}{\partial x} = Ri + L\frac{\partial i}{\partial t} \tag{3.120}$$

When the line is lossless, that is, R and G are zero, the ratio of the two sides of Eq. 3.120 yields

$$\frac{\partial v}{\partial i} = \sqrt{\frac{L}{C}} \tag{3.121}$$

The above ratio is known as the surge impedance of the line. Eliminating the ∂v and ∂i terms between the two equations of Eq. 3.120, we obtain

$$\frac{\partial x}{\partial t} = \frac{1}{\sqrt{CL}} \tag{3.122}$$

Later we will see that $\partial x / \partial t$ corresponds to the propagation speed of the voltage and current wave components propagating along the line.

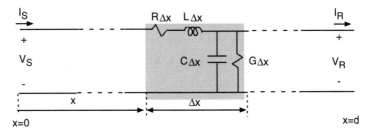

Figure 3.23   Distributed parameter model.

Note that both v and i in Eq. 3.120 are functions of time and position x along the line. Ignoring initial conditions, $v(x, 0^+)$ and $i(x, 0^+)$, and applying the Laplace transform in time to Eq. 3.120, we obtain

$$\frac{d\mathbf{I}}{dx} = -(G + sC)\mathbf{V} = -Y\mathbf{V}$$
$$\frac{d\mathbf{V}}{dx} = -(R + sL)\mathbf{I} = -Z\mathbf{I}$$

$$(3.123)$$

Taking the second derivative with respect to x, and substituting to separate the variables, we obtain two ordinary differential equations:

$$\frac{d^2\mathbf{I}}{dx^2} = (G + sC)(R + sL)\mathbf{I}$$
$$\frac{d^2\mathbf{V}}{dx^2} = (R + sL)(G + sC)\mathbf{V}$$

$$(3.124)$$

Let's now define a propagation constant, γ, as

$$\gamma^2 = (R + sL)(G + sC) \tag{3.125}$$

and a characteristic impedance as

$$Z_c = \sqrt{\frac{(R + sL)}{(G + sC)}} \tag{3.126}$$

The second-order ordinary differential equations of Eq. 3.124 have solutions of $\mathbf{I}$ and $\mathbf{V}$ that are of the form:

$$\mathbf{I}(x, s) = \underbrace{A_1 e^{\gamma x}}_{\mathbf{I}_b} + \underbrace{A_2 e^{-\gamma x}}_{\mathbf{I}_f} \tag{3.127}$$

$$\mathbf{V}(x, s) = \underbrace{Z_c A_2 e^{-\gamma x}}_{\mathbf{V}_f} - \underbrace{Z_c A_1 e^{\gamma x}}_{\mathbf{V}_b} \tag{3.128}$$

The terms in Eqs. 3.128 and 3.128 with a negative exponent in x are often referred to as the forward components, and those with a positive exponent in x are the backward components. The terminology of these terms will become evident later. For now, we see that

$$\mathbf{V}_f = \mathbf{I}_f Z_c \qquad \mathbf{V}_b = -\mathbf{I}_b Z_c \tag{3.129}$$

and

$$\mathbf{V}(x, s) = \mathbf{V}_f + \mathbf{V}_b = Z_c(\mathbf{I}_f - \mathbf{I}_b) \tag{3.130}$$

If the line is terminated by some impedance, $Z_d(s)$, at the receiving end where $x = d$, then

$$\mathbf{Z}_d(s) = \frac{\mathbf{V}(d, s)}{\mathbf{I}(d, s)} = \frac{Z_c(\mathbf{I}_f - \mathbf{I}_b)}{\mathbf{I}_f + \mathbf{I}_b} \tag{3.131}$$

Defining the current reflection coefficient, ρ_i, at that point by the ratio

$$\rho_i = \frac{\mathbf{I}_b}{\mathbf{I}_f},\tag{3.132}$$

and substituting it into Eq. 3.131, we obtain the following relationships:

$$\frac{Z_d}{Z_c} = \frac{1-\rho_i}{1+\rho_i} \quad or \quad \rho_i = \frac{Z_c - Z_d}{Z_c + Z_d}\tag{3.133}$$

Likewise, we can define the voltage reflection coefficient at that same point to be

$$\rho_v = \frac{\mathbf{V}_b}{\mathbf{V}_f} = \frac{Z_d - Z_c}{Z_d + Z_c} = -\rho_i\tag{3.134}$$

The reflection coefficients of some common terminations are given below:

Open-circuit $Z_d = \infty$ $\rho_v = 1$ $\rho_i = -1$
Short-circuit $Z_d = 0$ $\rho_v = -1$ $\rho_i = 1$
Matched termination $Z_d = Z_c$ $\rho_v = 0$ $\rho_i = 0$

The integration constants, A_1 and A_2, are functions of the boundary conditions. Setting $x = d$ in Eq. 3.128 yields

$$\rho_i = \frac{\mathbf{I}_b(d)}{\mathbf{I}_f(d)} = \frac{A_1}{A_2} e^{2\gamma d}$$

Likewise, setting $x = 0$ yields

$$\mathbf{I}(0, s) = A_1 + A_2 = A_1 \left(1 + \frac{e^{2\gamma d}}{\rho_i} \right)\tag{3.135}$$

Between Eqs. 3.124 and 3.135, we can obtain the following expressions for A_1 and A_2:

$$A_1 = \frac{\mathbf{I}(0, s)\rho_i e^{-2\gamma d}}{1 + \rho_i e^{-2\gamma d}}\tag{3.136}$$

$$A_2 = \frac{\mathbf{I}(0, s)}{1 + \rho_i e^{-2\gamma d}}\tag{3.137}$$

Wave propagation characteristics.   We can get some insight into the nature of what we have termed forward and backward components of the solution by examining the solution for the hypothetical case of an infinitely long line. Letting $d \to \infty$, we have from Eq. 3.136 $A_1 \to 0$, and from Eq. 3.137 $A_2 = \mathbf{I}(0, s)$. Thus, for an infinitely long line, both ρ_v and ρ_i are zero; in other words, there are no backward components. And the solution simplifies to

$$\mathbf{I}(x, s) = \mathbf{I}(0, s)e^{-\gamma x} = \mathbf{I}_f\tag{3.138}$$

In practice, the shunt conductance, G, is very small and may often be neglected. When G is set to zero, the truncated Taylor's expansion of γ in Eq. 3.125 simplifies to

$$\gamma = \frac{R}{2} \sqrt{\frac{C}{L}} + s\sqrt{LC}\tag{3.139}$$

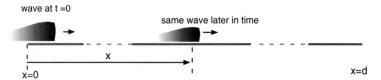

Figure 3.24   Forward component at position x and time t.

Substituting this γ into Eq. 3.138 and taking the inverse transform, we will obtain the following time domain solution of the current at position x and time t:

$$i(x,t) = \underbrace{e^{-\left(\frac{R}{2}\sqrt{\frac{C}{L}}\right)x}}_{\substack{\text{attenuation} \\ \text{factor}}} i(0, t - \sqrt{LC}x) \underbrace{u(t - \sqrt{LC}x)}_{\substack{\text{unit step} \\ \text{function}}} \tag{3.140}$$

Likewise, with no reflected components, the time-domain response of the voltage is

$$v(x,t) = Z_c i(x,t) \tag{3.141}$$

The above voltage and current expressions indicate that the forward component at position x and time t is an attenuated version of the same component that originated at the sending end of the line at time $t = \sqrt{LC}x$. In other words, these components are traveling at a speed of $1/\sqrt{LC}$. Figure 3.24 depicts the attenuated wave traveling down the line.

Likewise, the backward components of the voltage and current may be regarded as traveling waves. The forward component propagates in the direction of increasing x, but the backward component propagates in the opposite direction. These components propagate with the same velocity, $1/\sqrt{LC}$. The voltage components are related to the corresponding current components geometrically by the characteristic impedance of the line, Z_c.

3.7.1 The Bewley Lattice Diagram

The Bewley lattice diagram provides a quick graphical solution for waves traveling in a radial network. The hand-drawn diagram can be used to show the position and direction of the incident, reflected, and transmitted waves at any instant.

We will use this diagram to illustrate what happens when an excitation voltage, V, is applied at time $t = 0$ to a line of length d and characteristic impedance Z_c that is terminated with an impedance, Z_d, at the other end. Upon closing the switch at the sending end, a forward voltage wave component of magnitude V at the sending end travels down the line towards the receiving end. The amplitude of the wave is attenuated by the factor $e^{-\alpha d}$ from one end to the other. The first forward component arrives at the receiving end at $t = \tau$; upon encountering the change in impedance, the attenuated forward wave is reflected as a backward component of magnitude $\rho_{vr} V e^{-\alpha d}$. The voltage reflection coefficient at the receiving end, ρ_{vr}, is determined by $(Z_d - Z_c)/(Z_d + Z_c)$. This backward component propagates toward the sending end, arriving there another τ seconds later. By then, it is further attenuated by $e^{-\alpha d}$. Assuming that the source impedance at the sending end is zero, that is ρ_{vs} is equal to -1, the reflected wave at the sending end will also be inverted. It then

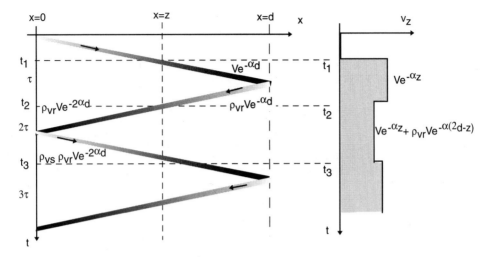

Figure 3.25   Bewley lattice diagram.

propagates down the line as the second forward component. Thus, reflections at both ends of the line result in subsequent forward and backward components in the line.

The lattice diagram of Fig. 3.25 displays the events in two axes: distance along the line in the x-axis, and time transpired in the negative y-axis. Shown on the right side of Fig. 3.25 is the voltage vs. time waveform at a point, $x = z$, that is obtained by noting the magnitude and time of arrival of the forward and backward wave components at that location. The plot assumes ρ_{vr} to be negative, or $Z_d < Z_c$.

3.7.2 Equivalent π Model

The more accurate distributed parameter line representation yields a circuit model that may be used to represent very long lines in low-frequency studies. For steady-state condition, the Laplace operator, s in Eqs. 3.127 and 3.128, may be replaced by $j\omega$, and these equations may be expressed in terms of phasor quantities as

$$\tilde{I}(x) = A_1 e^{\gamma x} + A_2 e^{-\gamma x}$$
$$\tilde{V}(x) = Z_c A_2 e^{-\gamma x} - Z_c A_1 e^{\gamma x} \tag{3.142}$$

where

$$\gamma = \sqrt{(R + j\omega L)(G + j\omega C)} = \alpha + j\beta \tag{3.143}$$

and the characteristic impedance is given by

$$Z_c = \sqrt{\frac{(R + j\omega L)}{(G + j\omega C)}} \tag{3.144}$$

The real and imaginary components of the propagation constant γ, α and β, are known as the attenuation and phase constants of the line, respectively. Since a wavelength corresponds to 2π change in phase, the wavelength, λ_w, of the line may be determined from the relation, $\lambda_w = 2\pi/\beta$.

The two integration constants, $\mathbf{A}_1$ and $\mathbf{A}_2$, can be evaluated using available boundary conditions at either $x = 0$, the sending end, or at $x = d$, the receiving end. For example, using conditions at the receiving end, we obtain

$$\tilde{\mathbf{I}}_R = \mathbf{A}_1 e^{\gamma d} + \mathbf{A}_2 e^{-\gamma d}$$

$$\tilde{\mathbf{V}}_R = \mathbf{Z}_c \mathbf{A}_2 e^{-\gamma d} - \mathbf{Z}_c \mathbf{A}_1 e^{\gamma d} \tag{3.145}$$

The above equations can then be used to determine the constants, $\mathbf{A}_1$ and $\mathbf{A}_2$, in terms of the receiving end quantities, that is

$$\mathbf{A}_1 = \left(\frac{\tilde{\mathbf{I}}_R - \tilde{\mathbf{V}}_R/\mathbf{Z}_c}{2} \right) e^{-\gamma d}$$

$$\mathbf{A}_2 = \left(\frac{\tilde{\mathbf{I}}_R + \tilde{\mathbf{V}}_R/\mathbf{Z}_c}{2} \right) e^{\gamma d} \tag{3.146}$$

Back-substituting these expressions of $\mathbf{A}_1$ and $\mathbf{A}_2$ into Eq. 3.142, we obtain

$$\tilde{\mathbf{I}}(x) = \left(\frac{\tilde{\mathbf{I}}_R - \tilde{\mathbf{V}}_R/\mathbf{Z}_c}{2} \right) e^{-\gamma(d-x)} + \left(\frac{\tilde{\mathbf{I}}_R + \tilde{\mathbf{V}}_R/\mathbf{Z}_c}{2} \right) e^{\gamma(d-x)}$$

$$\tilde{\mathbf{V}}(x) = \mathbf{Z}_c \left(\frac{\tilde{\mathbf{I}}_R + \tilde{\mathbf{V}}_R/\mathbf{Z}_c}{2} \right) e^{\gamma(d-x)} - \mathbf{Z}_c \left(\frac{\tilde{\mathbf{I}}_R - \tilde{\mathbf{V}}_R/\mathbf{Z}_c}{2} \right) e^{-\gamma(d-x)} \tag{3.147}$$

Since the value of x increases from the sending end to the receiving end, the $e^{\gamma(d-x)}$ term of the forward components decreases in magnitude and lags further in phase with x. Thus, its magnitude is diminished and its phase retarded with distance from the source. On the other hand, both magnitude and phase of the $e^{-\gamma(d-x)}$ term of the reflected components will increase with the value of x. When the load at the receiving end is terminated with a load of impedance equal to the characteristic impedance of the line, $\tilde{\mathbf{V}}_R = \tilde{\mathbf{I}}_R \mathbf{Z}_c$, the coefficients of the reflected current and voltage components will be zero.

The characteristic impedance, $\mathbf{Z}_c$, is also known as the surge impedance when the losses of the line are neglected. The surge impedance of the lossless line is given by $\mathbf{Z}_c = \sqrt{L/C}$. The surge impedance loading (SIL) of a line, defined as the power delivered to a purely resistive load equal to $\mathbf{Z}_c$, is

$$SIL = \frac{\tilde{\mathbf{V}}_{rated}^2}{\mathbf{Z}_c} \tag{3.148}$$

The velocity of propagation of the wave may be determined from the number of cycles per second (frequency) times the wavelength, that is

$$\text{velocity of propagation} = f\lambda_w = \left(\frac{\omega}{2\pi} \right) \left(\frac{2\pi}{\beta} \right) = \frac{\omega}{\beta} \tag{3.149}$$

Using the hyperbolic function identities,

$$\cosh\gamma(d-x) = \frac{e^{\gamma(d-x)} + e^{-\gamma(d-x)}}{2} \qquad \sinh\gamma(d-x) = \frac{e^{\gamma(d-x)} - e^{-\gamma(d-x)}}{2} \tag{3.150}$$

Eq. 3.147 can be rewritten as

$$\tilde{\mathbf{I}}(x) = \tilde{\mathbf{I}}_R \cosh\gamma(d-x) + \frac{\tilde{\mathbf{V}}_R}{\mathbf{Z}_c}\sinh\gamma(d-x)$$
$$\tilde{\mathbf{V}}(x) = \tilde{\mathbf{I}}_R \mathbf{Z}_c \sinh\gamma(d-x) + \tilde{\mathbf{V}}_R \cosh\gamma(d-x) \tag{3.151}$$

At the sending end, that is at $x = 0$, the above equation becomes

$$\begin{bmatrix} \tilde{\mathbf{V}}_S \\ \tilde{\mathbf{I}}_S \end{bmatrix} = \begin{bmatrix} \cosh\gamma d & \mathbf{Z}_c \sinh\gamma d \\ (1/\mathbf{Z}_c)\sinh\gamma d & \cosh\gamma d \end{bmatrix} \begin{bmatrix} \tilde{\mathbf{V}}_R \\ \tilde{\mathbf{I}}_R \end{bmatrix} \tag{3.152}$$

Equation 3.152 is the transmission equation from the distributed parameter line model. An equivalent π circuit model for the above transmission equation can be obtained by equating its ABCD matrix with that of the nominal π model in Eq. 3.115. The series and shunt circuit elements of the equivalent π circuit model will be distinguished from those of the nominal π circuit model by a subscript, π. Equating the **B** elements of the two ABCD matrices and making use of the relationship, $\mathbf{Z}_c = \sqrt{\mathbf{Z}/\mathbf{Y}}$, we obtain

$$\mathbf{B} = \mathbf{Z}_\pi = \mathbf{Z}_c \sinh\gamma d = \sqrt{\mathbf{Z}/\mathbf{Y}}\sinh\gamma d = \mathbf{Z}d\frac{\sinh\gamma d}{\sqrt{\mathbf{Z}\mathbf{Y}}d} = \mathbf{Z}\frac{\sinh\gamma d}{\gamma d} \tag{3.153}$$

Similarly, equating the **A** and **D** elements of the two ABCD matrices, we have

$$\mathbf{A} = \mathbf{D} = \cosh\gamma d = 1 + \frac{\mathbf{Z}_\pi \mathbf{Y}_\pi}{2} \tag{3.154}$$

$$\frac{\mathbf{Y}_\pi}{2} = \frac{\cosh\gamma d - 1}{\mathbf{Z}_\pi} = \mathbf{Z}_c\frac{\cosh\gamma d - 1}{\sinh\gamma d}$$

$$= \frac{1}{\mathbf{Z}_c}\tanh\left(\frac{\gamma d}{2}\right) = \sqrt{\frac{\mathbf{Y}}{\mathbf{Z}}}\tanh\left(\frac{\gamma d}{2}\right) \tag{3.155}$$

$$\frac{\mathbf{Y}_\pi}{2} = \mathbf{Y}d\frac{\tanh(\gamma d/2)}{\sqrt{\mathbf{Z}\mathbf{Y}}d} = \frac{\mathbf{Y}}{2}\frac{\tanh(\gamma d/2)}{(\gamma d/2)} \tag{3.156}$$

The circuit representation of the equivalent π model is shown in Fig. 3.26.

3.7.3 Terminations and Junctions

All along the line, including at terminations and junctions, the total voltage and current are related by the effective impedance at that point, that is

$$\mathbf{Z} = \frac{\mathbf{V}}{\mathbf{I}} = \frac{\mathbf{V}_f + \mathbf{V}_b}{\mathbf{I}_f + \mathbf{I}_b} \tag{3.157}$$

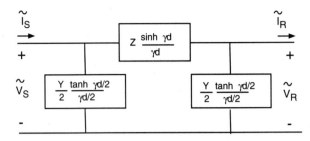

Figure 3.26 Equivalent π circuit model.

Capacitive termination. Let's first consider the response to a step voltage excitation of a line of characteristic impedance Z_c that is terminated by a capacitor, C_t. In the Laplace domain, the reflected component at the termination produced by a forward step component of $\mathbf{V}_f(s) = V_o/s$ is

$$\mathbf{V}_b(s) = \left(\frac{1/C_t s - Z_c}{1/C_t s + Z_c}\right)\mathbf{V}_f(s) = \left(1 - \frac{2s}{s + 1/C_t Z_c}\right)\frac{V_o}{s} \tag{3.158}$$

Taking the inverse Laplace transform, the time-domain expression of the reflected voltage component at the end of the line is

$$v_b(t) = V_o(1 - 2e^{-t/C_t Z_c}) \tag{3.159}$$

The reflected current component is

$$i_b(t) = -\frac{v_b(t)}{Z_c} = \frac{V_o}{Z_c}(2e^{-t/C_t Z_c} - 1) \tag{3.160}$$

The terminal voltage of the capacitor is given by

$$v(t) = v_b(t) + v_f(t) = 2V_o(1 - e^{-t/C_t Z_c}) \tag{3.161}$$

Note that as $t \to \infty$, the terminal voltage approaches the value of $2V_o$. As the capacitor charges up, it behaves like an open-circuit. Figure 3.27 shows the voltage and current waveforms at two instances after the reflection.

Inductive termination. For the case of an inductive termination of $Z_d(s) = Ls$, with an incident component of $\mathbf{V}_f(s) = V_o/s$, the s-domain expression of the reflected component is

$$\mathbf{V}_b(s) = \left(\frac{Ls - Z_c}{Ls + Z_c}\right)\mathbf{V}_f(s) = \left(\frac{2s}{s + Z_c/L} - 1\right)\frac{V_o}{s} \tag{3.162}$$

The time-domain expression of the reflected voltage component is

$$v_b(t) = V_o\left(2e^{-Z_c t/L} - 1\right) \tag{3.163}$$

The corresponding reflected current component is

$$i_b(t) = -\frac{v_b(t)}{Z_c} = \frac{V_o}{Z_c}\left(1 - 2e^{-Z_c t/L}\right) \tag{3.164}$$

And the terminal voltage is given by

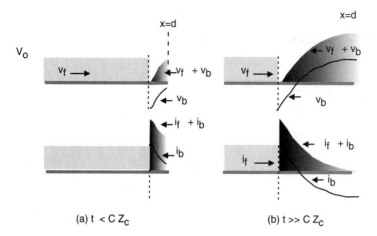

Figure 3.27   Waveforms of capacitive termination.

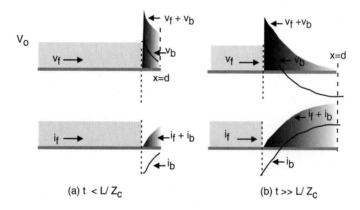

Figure 3.28   Waveforms of inductive termination.

$$v(t) = v_f(t) + v_b(t) = V_o + V_o \left(2e^{-Z_c t/L} - 1\right) \qquad (3.165)$$

In this case, as $t \rightarrow \infty$, the terminal voltage approaches zero as the inductor behaves like a short-circuit in steady-state. Figure 3.28 shows the voltage and current waveforms at two instances after the reflection.

Junctions.   In the case of a transition from one kind of line to another, besides the reflected component, we also have to consider the voltage or current transmitted through the connected line. Since the voltage at the junction must be continuous, the voltage a little to the left of the junction must be the same as the voltage a little to the right, that is $v(x_j^+, t) = v(x_j^-, t)$. Thus, the total transmitted voltage at the junction, v_t, is the same as the total voltage on the other side, that is $v_f + v_b$. Defining the voltage transmission coefficient

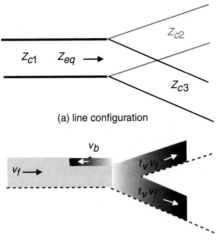

(a) line configuration

(b) reflected and transmitted components

Figure 3.29 Reflected and transmitted components from a junction.

as the ratio of the total transmitted voltage to the incident component, we can express it in terms of the ρ_v:

$$t_v = \frac{v_t}{v_f} = 1 + \frac{v_b}{v_f} = 1 + \rho_v \tag{3.166}$$

Similarly, with no lumped shunt components at the junction, the current at the junction will also be continuous, or $i(x_j^+, t) = i(x_j^-, t)$. Defining the current transmission coefficient in the same manner as for the voltage, we have

$$t_i = \frac{i_t}{i_f} = 1 + \frac{i_b}{i_f} = 1 + \rho_i = 1 - \rho_v \tag{3.167}$$

Knowing either ρ_v or ρ_i in the above expressions will enable us to determine the transmitted voltage and current. As an example, a wave in the line of characteristic impedance Z_{c1} in Fig. 3.29 traveling towards the junction will see an equivalent impedance, Z_{c2}, in parallel with Z_{c3} at the junction. Thus, the voltage reflection coefficient to that wave at the junction is

$$\rho_{v1} = \frac{Z_{eq} - Z_{c1}}{Z_{eq} + Z_{c1}} \quad \text{where} \quad Z_{eq} = \frac{Z_{c2} Z_{c3}}{Z_{c2} + Z_{c3}} \tag{3.168}$$

The above value of ρ_{v1} can then be used to compute the transmission coefficients in Eqs. 3.166 and 3.167, which in turn can be used to determine the transmitted voltage and current components.

3.8 SIMULATION OF A SINGLE-PHASE LINE

In this section we will discuss a technique for implementing a distributed line simulation using the simple single-phase circuit shown in Fig. 3.30. It is assumed that the parameters

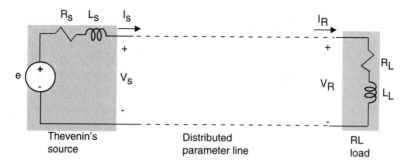

Figure 3.30   A single-phase line connecting a source to an RL load.

of the circuit, including those of the Thevenin's source and load, are known.

As shown earlier, with a distributed parameter line representation, the voltage and current at any position x and time t consist of forward and backward wave components, that is

$$v(x,t) = v_f + v_b \tag{3.169}$$

$$i(x,t) = i_f + i_b \tag{3.170}$$

where $v_f = Z_c i_f$ and $v_b = -Z_c i_b$. For a line of length d meters, the time taken for the waves to travel from one end to the other is given by

$$\tau = d\sqrt{LC} \tag{3.171}$$

where L and C are series inductance and shunt capacitance per meter of the line, respectively.

If the delay, τ, is to be obtained using a circular buffer, at a fixed sampling time of τ_s, the required buffer length, N, is the next higher integer value of the ratio of the total delay, τ, to the sampling time, τ_s.

$$N = Integer\left(\frac{\tau}{\tau_s}\right) \tag{3.172}$$

Two delay buffers will be needed: one for the transport delay of the forward component and the other for the backward component. A common pointer steps through the elements of these two buffers at the rate of one location per sampling instant. With $N = $ integer (τ/τ_s), it will take the pointer τ seconds to cycle once through these buffers. As shown in Fig. 3.31, fresh values of v_{bR} and v_{fS} computed from Eqs. 3.181 and 3.181 and stored in the arrays at a certain time will be retrieved when the pointer returns to the same location the next time around, or τ seconds later.

From the circuit connection of Fig. 3.30, we see that given the voltage, e, and the electrical parameters of the line and series RL load at the receiving end, we can use the voltages across the inductors, L_s and L_L, to determine the sending and receiving end currents, i_S and i_R, respectively. From the source end, we have

$$e - v_S = R_s i_S + L_s \frac{di_S}{dt} \tag{3.173}$$

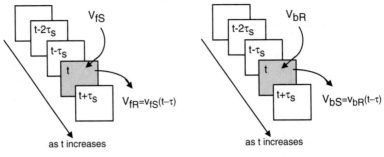

(a) forward component array   (b) backward component array

Figure 3.31   Ring buffers of delay elements.

Expressed in integral form below, we can integrate for the value of i_S:

$$i_S = \frac{1}{L_s} \int (e - v_S - R_s i_S)\, dt \tag{3.174}$$

At the load end, the receiving-end current, i_R, can be determined from the integral form of the voltage equation for the RL load, that is

$$i_R = \frac{1}{L_L} \int (v_R - R_{Li_R})\, dt \tag{3.175}$$

The value of i_S is in turn being used to determine v_S from

$$v_S = v_{fS} + v_{bS} = Z_{ci} f_S + v_{bS} = Z_c(i_S - i_{bS}) + v_{bS} \tag{3.176}$$

$$v_S = Z_c i_S + 2v_{bS} \tag{3.177}$$

The value of v_{bS}, the backward voltage component at the sending end, is derived from the output of a transport delay with the following delay and attenuation applied to v_{bR}:

$$attenuation\ factor = e^{-\frac{R}{2}\sqrt{\frac{C}{L}}d} \qquad delay = d\sqrt{LC} \tag{3.178}$$

where d is the length of the line, and R, L, and C are the per unit length resistance, inductance, and capacitance of the line, respectively.

Likewise, i_R is used to determine v_R from

$$v_R = 2v_{fR} - Z_c i_R \tag{3.179}$$

The value of v_{fR}, the forward voltage component at the receiving end, is derived from the output of another transport delay applied to v_{fS}.

Finally, to complete the simulation, the inputs to the two transport delays, v_{bR} and v_{fS}, are computed from

$$v_{bR} = v_R - v_{fR} \tag{3.180}$$

$$v_{fS} = v_S - v_{bS} \tag{3.181}$$

Figure 3.32 shows the flow of variables in the single-phase line simulation.

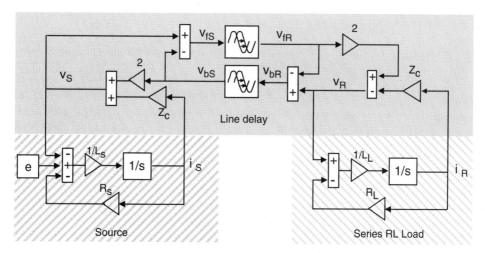

Figure 3.32   Flow of variables in single-phase line simulation.

3.9 PROJECTS

3.9.1 Project 1: Line Parameters and Circuit Models

In this project, we will first determine the per unit length parameters of a 60 Hz, 345 kV, three-phase ac line with conductor arrangement as shown in Fig. 3.33. Next, we will compare the sending end's condition as determined using a series RL and nominal π with that of an equivalent π circuit model of the entire line, 160 km long, delivering 120 MW at 0.9 power factor lagging to a fixed receiving end voltage, $\tilde{\mathbf{V}}_R$, of $345/\sqrt{(3)}\underline{/0°}$ kV. Finally, we will examine the behavior of real and reactive power transfer through the same length of line when the angle and magnitude of the voltage across the line are varied.

Relative accuracy of circuit models for different line lengths.   The line has conductor spacings of $D_{12'} = 7.772$ m and $D_{1'2} = 6.858$ m. Each phase is a bundle of two identically sized conductors, each of diameter 4.4755 cm, an ac resistance of 0.02896 ohms/km, and a GMR of 1.6276 cm.

The GMDs between any two of the bundled phase conductors are

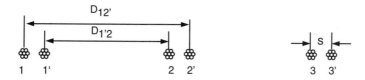

Figure 3.33   Three-phase bundled conductor line.

TABLE 3.2   SENDING END CONDITION

Circuit model	Per phase power $P_s + jQ_s$ in MW	Input power factor	Input voltage magnitude in kV	Input current magnitude in A
Series RL	$40 + j19.401$	0.8998	199.24	223.13
Nominal π	$40 + j19.372$	0.9000	199.24	223.067
Equiv π	$40 + j19.372$	0.9000	199.24	223.067

$$GMD_{12} = \sqrt[4]{D_{12}D_{12'}D_{1'2}D_{1'2'}}$$

$$GMD_{23} = \sqrt[4]{D_{23}D_{23'}D_{2'3}D_{2'3'}} \quad (3.182)$$

$$GMD_{31} = \sqrt[4]{D_{31}D_{31'}D_{3'1}D_{3'1'}}$$

Thus, with full transposition, the GMD over a transposition cycle is

$$GMD = \sqrt[3]{GMD_{12}GMD_{23}GMD_{31}} \quad (3.183)$$

Run the *m1* m-file to determine the RLC parameters of the line, the ABCD transmission matrices of the series RL, nominal π, and equivalent π circuit models of the whole line. Table 3.2 shows sample results for the case where the line length is 160 km and the total power delivered to the receiving end is 120 MW at 0.9 power factor lagging.

Modify the data in the M-file for the case where the line length is 240 instead of 160 km. Compare the results from the equivalent π model with those from the other two circuit models. Modify the M-file to use two nominal π sections for each half of the line and again compare the results obtained with those from the equivalent π model. Note the improvement in accuracy of the results as compared to that from just one nominal π section.

Real and reactive power transfer.   In this part of the project, we will examine the behavior of real and reactive power delivered to the receiving end's load when the voltage magnitudes at both ends are held constant but the relative phase angle between the sending and receiving ends' voltages is varied. We will begin with the base case condition of the 160 km length line delivering 120 MW at 0.9 power factor lagging to a receiving end when the magnitudes of the sending and receiving ends' voltages are held constant at their previous values, that is 199.24 and $345/\sqrt{(3)}$ kV, respectively. Keep the phase angle of $\tilde{\mathbf{V}}_R$ at zero, but vary that of $\tilde{\mathbf{V}}_S$ from $-\pi$ to $+\pi$ radians in one-degree increments.

The complex power delivered to the receiving end can be computed from

$$\mathbf{S}_R = \mathbf{P}_R + j\mathbf{Q}_R = \tilde{\mathbf{V}}_R\tilde{\mathbf{I}}_R^*$$
$$= \tilde{\mathbf{V}}_R\left(\frac{\tilde{\mathbf{V}}_S - \mathbf{A}\tilde{\mathbf{V}}_R}{\mathbf{B}}\right) \quad (3.184)$$

Figure 3.34 shows the loci of the receiving end's complex power for the 160-km line based on the ABCD parameters of its equivalent π circuit model. These plots are obtained by

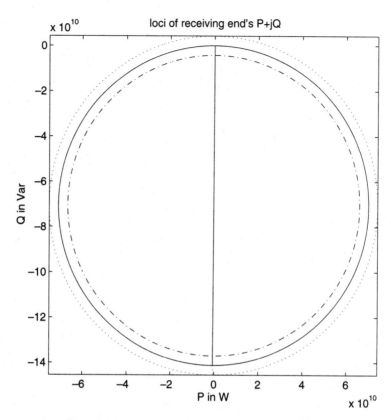

Figure 3.34   Loci of receiving end's real and reactive power.

varying the angle of $\tilde{\mathbf{V}}_S$ over the range of $\pm\pi$. The solid-line plot is the **PQ** locus for a $|V_s|$ of 199.24 kV. The inner dash-dot plot is for six percent less in the sending end's voltage, and the outer dotted-line plot is for six percent more in the sending end's voltage.

From transient stability consideration, the phase difference between the sending end's and receiving end's voltages of lines is to be kept well within 90°. Examine the data of these plots and see if you can identify the portions of these loci that are of practical use from the point of view of lower line current and losses for some given level of power transfer.

3.9.2 Project 2: Switching Transients in a Single-phase Line

The objectives of this project are as follows:

- To implement a transient simulation of one phase circuit of a 500-kV, three-phase line shown in Fig. 3.35, given in the Simulink simulation *s2*.

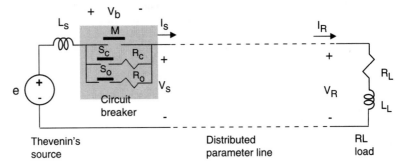

Figure 3.35   Single-phase circuit.

- Use the simulation to study the impact of inserted circuit breaker resistance on the breaker and line transient voltages during and following the opening and closing of the breaker.
- Examine the impact of line loading on the breaker and line transient voltages.

In the single-phase circuit of Fig. 3.35, the supply source at the sending end of the single-phase line is represented by a Thevenin's equivalent consisting of a 60-Hz voltage source, $e = 500\sqrt{\frac{2}{3}}\cos\omega_e t$ kV, and a source inductance, $L_s = 0.1$ H. The single-phase line is 100 miles long and is to be represented by a distributed parameter model. The electrical parameters of the line are

$$R = 0.15 \ \Omega/\text{mile}$$

$$L = 2.96 \ \text{mH/mile}$$

$$C = 0.017 \ \mu\text{F/mile}$$

The circuit breaker shown has one main contact, M, and two auxiliary contacts represented by switches S_c and S_o. Upon receiving a signal to close, the circuit breaker closes its auxiliary contact, S_c, first, inserting the closing resistor, R_c for t_c, seconds before its main contact, M, closes. The closing of the main contact, M, shorts out R_c immediately, following which S_c reopens and remains open until another closing operation.

When a circuit breaker is closed and it receives a signal to open, it closes its auxiliary contact, S_o, first before it opens its main contact, M, at the next breaker's current zero. After the opening of the main contact, the auxiliary contact, S_o, remains closed with R_o inserted for a further t_o seconds. After t_o seconds, S_o opens to interrupt the diminished breaker current at the next current zero.

Figure 3.36a shows the overall diagram of simulation $s2$ of the circuit of Fig. 3.35. The details inside the main blocks of the simulation are given in Figs. 3.36b through 3.36e. As simulated, the breaker can be closed at any point of the wave of the source voltage. It is, however, simpler to obtain the point of the wave of closing onto the unenergized line by arranging for the breaker to close at $t = 0$, using an initial phase of the source voltage, e, that corresponds to the desired point of the wave of closing.

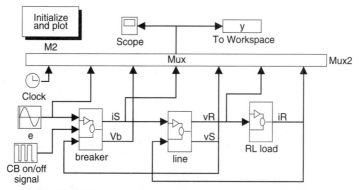

Double-click on masked block M2 in this screen, or run M2.M in
the MATLAB command window to initialize this simulation.
After the simulation, type return after the K>> prompt in the
MATLAB command window to plot results.

(a) Overall diagram of *s2* simulation

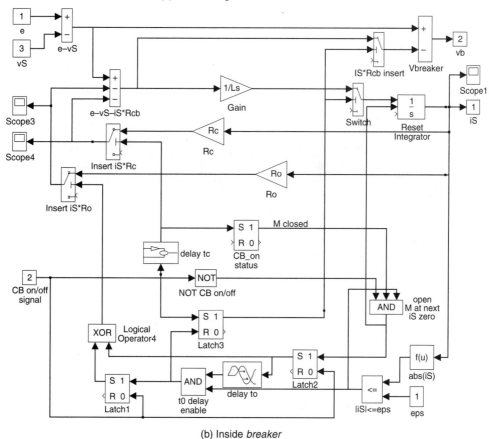

(b) Inside *breaker*

Figure 3.36   Simulation *s2* of distributed line and breaker.

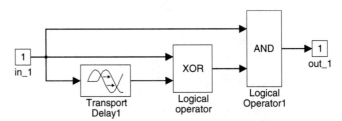

AND used to suppress
spurious operation when
input goes from high to low.

(c) Inside *delay tc* of *breaker*

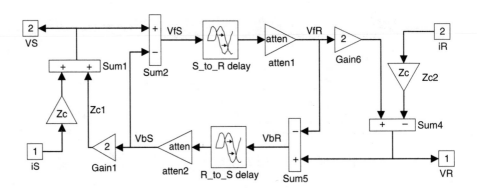

(d) Inside the *line block*

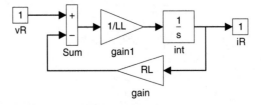

(e) Inside the *RL load block*

Figure 3.36 *(cont.)*   Simulation *s2* of distributed line and breaker.

In Fig. 3.36a, the Thevenin's source inductance and operational logics of the breakers with insertion resistors are simulated in the *source* block, the delay and attenuation of the line in the *line* block, and the series RL load at the receiving end of the line in the *RL load* block.

In Fig. 3.36b, the switches, *Insert iS*Rc* and *Insert iS*Rc*, insert the appropriate amount of breaker insertion resistive drops. These two switches are controlled by logic signals having pulse widths corresponding to t_c and t_o, respectively. As required by the condition that M and S_c open on a current zero, the rising and falling edges of the opening pulse signal are synchronized to the current zeros of the breaker current, i_S. Since the *switch* block operates with a default transition condition of $input(2) \geq 0$, a bias of -0.5 must be added to its logic signal, with $true = 1$ and $false = 0$, to operate the switch properly. As shown, current zero is approximately located by testing for the condition when $abs(i_S) \leq \epsilon$. Care must be exercised in selecting ϵ, as too small an ϵ can result in a current zero being stepped over when the integration step size is not small.

Set up the simulation *s2*. Also given is the MATLAB script file *m2* that can be used to set up the desired parameters of the system and the operating times of the breaker. Modify *m2* to establish the condition for each case, and use *s2* to investigate the following switching operations:

1. Simulate the following cases of the breaker closing on an unenergized line with no load at the receiving end:

    a. $R_c = 0$, $t_c = 1$ cycle.
    b. $R_c = 1\ k\Omega$, $t_c = 1/4$, $1/2$, and 1 cycle.
    c. Use a few golden section searches to locate the approximate value of R_c within zero and $200\ \Omega$ that will minimize the over-voltage transients on v_S and v_R for $t_c = 1$ cycle. The MATLAB functions, *max* and *min*, may be used to determine the maximum and minimum of the values of v_S and v_R that are stored in the MATLAB workspace after each simulation run.

    Determine from the waveform of v_R the time it takes the switching transients to travel from the receiving end to the sending end and back. Check this value against that use for the transport delay.

2. Simulate the following cases of the breaker opening when the energized line has no load at the receiving end:

    a. $R_o = 0$, $t_o = 1$ cycle.
    b. $R_o = 50\ k\Omega$, $t_o = 1$ cycle.
    c. Use a few golden section searches to locate the approximate value of R_o within 3 and $10\ k\Omega$ that will minimize the over-voltage transients in v_b for $t_o = 1$ cycle.

3. Repeat cases (1.c) and (2.c) with the line connected to a single-phase RL load of 30 MVA and 0.8 power factor lagging at its receiving end. The single-phase load is to be represented as a lumped parameter RL impedance, whose value corresponds to the given loading at the rated voltage of the line.

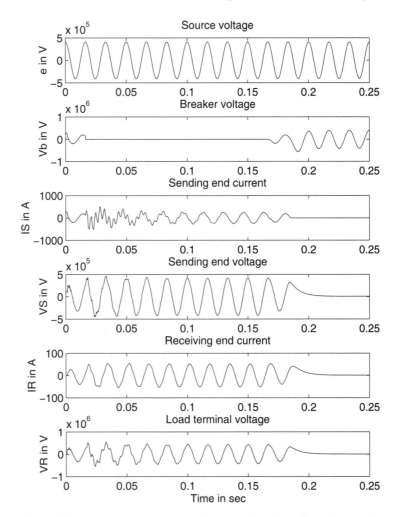

Figure 3.37   Plot of v_b(trace 1), v_S(trace 2), v_R(trace 3), i_S(trace 4), and i_R(trace 5).

In each case, plot the variables e, v_b, v_S, i_S, and v_R.

Given in Fig. 3.37 are sample results for the case where $e = 500\sqrt{(2/3)} \sin(\omega_e t + \pi/2)$ kV and the receiving end of the line is terminated with a fixed impedance corresponding to a three-phase loading of 30 MVA, 0.8 power factor lagging. The circuit breaker was closed at $t = 0$, or at the peak value of the source voltage with $t_c = 1$ cycle and $R_c = 1\ k\Omega$, and reopened 10 cycles later with $t_o = 1$ cycle and $R_o = 1k\ \Omega$. The closing and opening sequence is carried out using a *repeating sequence* signal source for the circuit breaker ON/OFF signal. For the results shown, the time values array of the ON/OFF signal source was set to [0 10*Te 10*Te 15*Te], Te being the period of one 60 Hz cycle, and the corresponding output values array being [1 1 0 0].

4

Transformers

4.1 INTRODUCTION

The main uses of electrical transformers are for changing the magnitude of an ac voltage, providing electrical isolation, and matching the load impedance to the source. They are formed by two or more sets of stationary windings which are magnetically coupled, often but not necessarily, with a high permeability core to maximize the coupling. By convention, the input ac winding is usually referred to as the primary winding, and the other windings from which output is drawn are referred to as secondary windings.

Power transformers working at lower frequencies between 25 to 400 Hz usually have iron cores to concentrate the flux path linking the windings; those for high-frequency applications may have powder ferrite cores or air cores to avoid excessive core losses. Eddy current losses in an iron core can be reduced by using a laminated construction. For 60-Hz transformers, the laminations of the core are typically about 0.014 inch (0.35 mm) thick.

4.2 IDEAL TRANSFORMER

We will begin by considering the relationships between the primary and secondary voltages and currents of an ideal transformer that has no winding copper losses, core losses, leakage fluxes, or core reluctance. Consider the magnetic coupling between the primary and

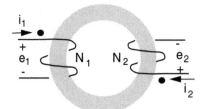

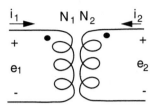

(a) Windings and core of a two-winding transformer   (b) Circuit symbol of a two-winding transformer

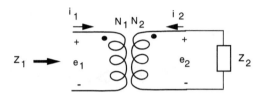

(c) Referred value of Z_2

Figure 4.1   Ideal transformer.

secondary windings of such an ideal transformer shown in Fig. 4.1. The flow of currents in the two windings produces magneto-motive forces (mmf) which in turn set up the fluxes. As shown in Fig. 4.1a, the sense of these windings is such that their mmfs are directed around the core in a counter-clockwise direction for the assumed direction of winding currents shown. With core reluctance neglected, the resultant mmf needed to magnetize the core is zero, that is

$$N_1 i_1 + N_2 i_2 = 0 \ \text{ or } \ \frac{i_1}{i_2} = -\frac{N_2}{N_1} \tag{4.1}$$

The dots in Fig. 4.1a mark the end of each winding that wraps around the core in the same sense. Consequently, the induced voltages, e_1 and e_2, with polarity so marked are in time phase and their ratio yields the following relation between the winding voltages:

$$\frac{e_1}{e_2} = \frac{N_1 \, (d\phi_m/dt)}{N_2 \, (d\phi_m/dt)} = \frac{N_1}{N_2} \tag{4.2}$$

With all losses neglected, the ideal transformer is lossless; in other words, the *net* power flowing into the ideal transformer from both of its windings will be zero, that is

$$e_1 i_1 = \left(\frac{N_1 e_2}{N_2}\right) \left(\frac{-N_2 i_2}{N_1}\right) = -e_2 i_2 \tag{4.3}$$

Figure 4.1b shows the circuit symbol of the ideal transformer with current and voltage relations as given by Eqs. 4.1 and 4.2. Note that the positive polarity of the voltages is on the dotted end of the winding and the direction of the currents is into the dotted end of each winding. Since power flow is usually into one winding and out of another, the actual direction of the currents could be opposite to that shown in Fig. 4.1. When the assumed direction of i_2 is opposite to that shown, the mmfs of the windings will be subtractive, and $-i_2$ ought to take the place of i_2 in all the expressions given above.

In simulation and analysis, circuit variables and elements on one side of the ideal transformer in Fig. 4.1c can be referred to the other side to facilitate the computation or to obtain a simpler equivalent circuit. For example, using p to denote the time derivative operator, the operational impedance, $Z_2(p) = r_2 + pL_2$, on winding 2's side in Fig. 4.1c may be *referred* to winding 1's side using the voltage and current relationships of the ideal transformer, that is

$$Z_1(p) = \frac{e_1}{i_1} = \left(\frac{N_1 e_2}{N_2}\right)\left(\frac{-N_1}{N_2 i_2}\right) = \frac{N_1^2}{N_2^2}\left(\frac{-e_2}{i_2}\right) = \left(\frac{N_1}{N_2}\right)^2 Z_2(p) \qquad (4.4)$$

4.3 MODEL OF A TWO-WINDING TRANSFORMER

In this section we will begin by deriving the flux linkage and terminal voltage equations of a two-winding transformer, taking into account the resistance and leakage fluxes of the windings, and the reluctance of the core. Towards the end of this section, we will use these equations to develop an equivalent circuit representation for the transformer.

4.3.1 Flux Linkage Equations

When leakage fluxes are included, as illustrated in Fig. 4.2, the total flux linked by each winding may be divided into two components: a mutual component, ϕ_m, that is common to both windings, and a leakage flux component that links only the winding itself. In terms of these flux components, the total flux linked by each of the windings can be expressed as

$$\phi_1 = \phi_{l1} + \phi_m \qquad (4.5)$$

$$\phi_2 = \phi_{l2} + \phi_m \qquad (4.6)$$

where ϕ_{l1} and ϕ_{l2} are the leakage flux components of windings 1 and 2, respectively. As in an ideal transformer, the mutual flux, ϕ_m, is established by the *resultant* mmf of the two windings acting around the same path of the core. Assuming that N_1 turns of winding 1

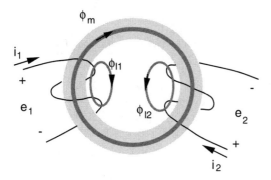

Figure 4.2 Magnetic coupling of a two-winding transformer.

effectively link both ϕ_m and the leakage flux, ϕ_{l1}, the flux linkage of winding 1, defined as the turn times the total flux linked, is

$$\lambda_1 = N_1\phi_1 = N_1(\phi_{l1} + \phi_m) \tag{4.7}$$

The right side of Eq. 4.7 can be expressed in terms of the winding currents by replacing the leakage and mutual fluxes by their respective mmfs and permeances. The leakage flux, ϕ_{l1} is created by the mmf of winding 1, $N_1 i_1$, over an effective path permeance of P_{l1}, say. And the mutual flux, ϕ_m, is created by the combined mmf, $N_1 i_1 + N_2 i_2$, in the mutual path of permeance P_m. Substituting for the leakage and mutual fluxes, Eq. 4.7 can be rewritten as

$$\lambda_1 = N_1(\underbrace{N_1 i_1 P_{l1}}_{\phi_{l1}} + \underbrace{(N_1 i_1 + N_2 i_2) P_m}_{\phi_m}) = \underbrace{(N_1^2 P_{l1} + N_1^2 P_m)}_{L_{11}} i_1 + \underbrace{N_1 N_2 P_m}_{L_{12}} i_2 \tag{4.8}$$

Similarly, the flux linkage of winding 2 can be expressed as

$$\lambda_2 = N_2(\phi_{l2} + \phi_m) = N_2(N_2 i_2 P_{l2} + (N_1 i_1 + N_2 i_2) P_m)$$

$$= \underbrace{(N_2^2 P_{l2} + N_2^2 P_m)}_{L_{22}} i_2 + \underbrace{N_1 N_2 P_m}_{L_{21}} i_1 \tag{4.9}$$

The resulting flux linkage equations for the two magnetically coupled windings, expressed in terms of the winding inductances, are

$$\lambda_1 = L_{11} i_1 + L_{12} i_2 \tag{4.10}$$

$$\lambda_2 = L_{21} i_1 + L_{22} i_2 \tag{4.11}$$

where L_{11} and L_{22} are the self-inductances of the windings, and L_{12} and L_{21} are the mutual inductances between them.

The self-inductance of winding 1 may be considered as the sum of a leakage, L_{l1}, and a magnetizing component, L_{m1}, from its own current. Thus for winding 1, with $i_2 = 0$,

$$L_{11} = \frac{\lambda_{1 i_2=0}}{i_1} = \frac{N_1(\phi_{l1} + \phi_{m1})}{i_1} = \underbrace{N_1^2 P_{l1}}_{L_{l1}} + \underbrace{N_1^2 P_m}_{L_{m1}} \tag{4.12}$$

where $\phi_{m1} = N_1 i_1 P_m$ is the portion of the mutual flux magnetized by i_1. Likewise, for winding 2

$$L_{22} = \frac{\lambda_{2 i_1=0}}{i_2} = \frac{N_2(\phi_{l2} + \phi_{m2})}{i_2} = \underbrace{N_2^2 P_{l2}}_{L_{l2}} + \underbrace{N_2^2 P_m}_{L_{m2}} \tag{4.13}$$

where $\phi_{m2} = N_2 i_2 P_m$ is the portion of the mutual flux magnetized by i_2. In general, core saturation will affect the values of the winding inductances.

Taking the ratio of L_{m2} to L_{m1}, we obtain the following relationship between the magnetizing inductances of the two windings:

$$L_{m2} = \frac{N_2 \phi_{m2}}{i_2} = \frac{N_2 L_{12}}{N_1} = N_2^2 P_m = \left(\frac{N_2}{N_1}\right)^2 L_{m1} \tag{4.14}$$

The total mutual flux linked by each winding can be expressed in terms of its own winding magnetizing inductance times the corresponding magnetizing current. For example, the total flux linked by winding 1, expressed in terms of its own magnetizing inductance, is

$$N_1\phi_m = N_1(\phi_{m1} + \phi_{m2}) = L_{m1}\left(i_1 + \underbrace{\frac{N_2}{N_1}i_2}_{i_2\ referred}\right) \tag{4.15}$$

It is evident from the above expression that the equivalent magnetizing current, as viewed on winding 1's side, is the sum of winding 1's current and the referred current of winding 2.

4.3.2 Voltage Equations

The induced voltage in each winding is equal to the time rate of change of the winding's flux linkage. Thus, using the flux linkage expression given in Eq. 4.10, the induced voltage in winding 1 is given by

$$e_1 = \frac{d\,\lambda_1}{dt} = L_{11}\frac{di_1}{dt} + L_{12}\frac{di_2}{dt} \tag{4.16}$$

Replacing L_{11} by $L_{l1} + L_{m1}$ and $L_{12}i_2$ by $N_2 L_{m1} i_2 / N_1$, the voltage induced in winding 1 can also be expressed as

$$e_1 = L_{l1}\frac{di_1}{dt} + L_{m1}\frac{d(i_1 + (N_2/N_1)i_2)}{dt} \tag{4.17}$$

Whether it is just for convenience of computation or out of necessity, as when the parameters are only measurable from one winding, often quantities of the other winding are *referred* to the side which has the information available directly. This process of referring is equivalent to scaling the number of turns of one winding to be the same as that of the winding whose variables are to be retained explicitly. For instance, the current $N_2 i_2 / N_1$ is the equivalent value of winding 2's current that has been referred to a winding of N_1 turns, chosen to be the same as that of winding 1. Denoting the referred value of i_2 by i_2', Eq. 4.17 becomes

$$e_1 = L_{l1}\frac{di_1}{dt} + L_{m1}\frac{d}{dt}(i_1 + i_2') \tag{4.18}$$

Similarly, the induced voltage of winding 2 may be written as

$$e_2 = L_{l2}\frac{di_2}{dt} + L_{m2}\frac{d}{dt}\left(\frac{N_1}{N_2}i_1 + i_2\right) \tag{4.19}$$

The voltage e_2 can also be referred to winding 1, or scaled to a fictitious winding of N_1 turns, using the relation given in Eq. 4.2. Thus, multiplying Eq. 4.19 through by N_1/N_2, denoting $N_1 e_2 / N_2$ by e_2', and replacing $N_1^2 L_{m2} / N_2^2$ by L_{m1}, Eq. 4.19 can be rewritten into the form:

$$e_2' = L_{l2}'\frac{di_2'}{dt} + L_{m1}\frac{d}{dt}(i_1 + i_2') \tag{4.20}$$

The terminal voltage of a winding is the sum of the induced voltage and the resistive drop in the winding. That for winding 1 is given by

$$v_1 = i_1 r_1 + e_1 = i_1 r_1 + L_{l1}\frac{di_1}{dt} + L_{m1}\frac{d}{dt}(i_1 + i_2') \tag{4.21}$$

Instead of writing a similar equation for the terminal voltage of winding 2, we will write it in terms of quantities referred to winding 1's side, that is

$$v_2' = \frac{N_1}{N_2}v_2 = \underbrace{\left(\frac{N_2}{N_1}i_2\right)}_{i_2'}\underbrace{\left(\frac{N_1}{N_2}\right)^2 r_2}_{r_2'} + e_2' \tag{4.22}$$

$$= i_2' r_2' + L_{l2}'\frac{di_2'}{dt} + L_{m1}\frac{d}{dt}(i_1 + i_2')$$

Equations 4.21 and 4.22 are written with the current and voltage of winding 2 referred to the side of winding 1. The voltage equations with the current and voltage of winding 1 referred to the side of winding 2 can be obtained by simply interchanging the subscripts of 1 and 2 for all the variables and parameters in Eqs. 4.21 and 4.22.

4.3.3 Equivalent Circuit Representation

The form of the voltage equations in Eqs. 4.21 and 4.22 with the common L_{m1} term suggests the equivalent T-circuit shown in Fig. 4.3 for the two-winding transformer. In Fig. 4.3, the prime denotes referred quantities of winding 2 to winding 1. For example, i_2' will be the equivalent current flowing in a winding having the same number of turns as winding 1. (Equivalent in the sense that it will produce the same mmf, $N_2 i_2$, in the common magnetic circuit shared with winding 1, that is $N_1 i_2' = N_2 i_2$.) This is apparent from the ideal transformer part of the equivalent circuit. Similarly, the referred voltage, v_2', satisfies the ideal transformer relationship, $v_2'/v_2 = N_1/N_2$. In a practical transformer, unlike that in an ideal transformer, the core permeance or the mutual inductance is finite. To establish the mutual flux, a finite magnetizing current, $i_1 + i_2'$, flows in the equivalent magnetizing inductance on the winding 1 side, L_{m1}.

The values of circuit parameters of winding 2 referred to winding 1 are determined by the relation given in Eq. 4.4, that is

$$r_2' = \left(\frac{N_1}{N_2}\right)^2 r_2 \tag{4.23}$$

$$L_{l2}' = \left(\frac{N_1}{N_2}\right)^2 L_{l2} \tag{4.24}$$

If we were to include core losses by approximating them as losses proportional to the square of the flux density in the core, or the square of the internal voltage e_m shown in Fig. 4.3, an appropriate core loss resistance could be connected across e_m, in parallel with the magnetizing inductance, L_{m1}. The resultant equivalent circuit would then be the same as that derived from steady-state considerations in a standard electric machinery textbook [26].

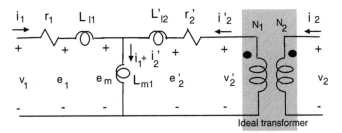

Figure 4.3   Equivalent circuit of a two-winding transformer.

4.4 SIMULATION OF A TWO-WINDING TRANSFORMER

In this section, we will describe an arrangement by which the voltage and flux linkage equations of a two-winding transformer can be implemented in a computer simulation. There is, of course, more than one way to implement a simulation of the transformer even when we are using the same mathematical model. For example, when using the simple model described in the earlier section, we could implement a simulation using fluxes or currents as state variables. Note that the equivalent circuit representation of Fig. 4.3 has a cut set of three inductors. Since their currents obey Kirchhoff's current law at the common node, all three inductor currents cannot be independent. The magnetizing branch current may be expressed in terms of the winding currents, i_1 and i_2', as shown.

In our case, we will pick the total flux linkages of the two windings as the state variables. In terms of these two state variables, the voltage equations can be written as

$$v_1 = i_1 r_1 + \frac{1}{\omega_b} \frac{d\psi_1}{dt} \tag{4.25}$$

$$v_2' = i_2' r_2' + \frac{1}{\omega_b} \frac{d\psi_2'}{dt} \tag{4.26}$$

where $\psi_1 = \omega_b \lambda_1$, $\psi_2 = \omega_b \lambda_2$, and ω_b is the base frequency at which the reactances are computed. The flux linkage per second of the windings can be expressed as

$$\psi_1 = \omega_b \lambda_1 = x_{l1} i_1 + \psi_m \tag{4.27}$$

$$\psi_2' = \omega_b \lambda_2' = x_{l2}' i_2' + \psi_m \tag{4.28}$$

and

$$\psi_m = \omega_b L_{m1}(i_1 + i_2') = x_{m1}(i_1 + i_2') \tag{4.29}$$

Note that ψ_m is associated with the magnetizing inductance referred to winding 1.

The current i_1 can be expressed in terms of ψ_1 and ψ_m using Eq. 4.27. Similarly, i_2' can be expressed in terms of ψ_2' and ψ_m using Eq. 4.28.

$$i_1 = \frac{\psi_1 - \psi_m}{x_{l1}} \tag{4.30}$$

$$i_2' = \frac{\psi_2' - \psi_m}{x_{l2}'} \tag{4.31}$$

Substituting the above expressions of i_1 and i_2 into Eq. 4.29, we obtain

$$\frac{\psi_m}{x_{m1}} = \frac{\psi_1 - \psi_m}{x_{l1}} + \frac{\psi_2' - \psi_m}{x_{l2}'} \tag{4.32}$$

Collecting the ψ_m terms to the right, we obtain the desired expression of ψ_m in terms of the two desired states, that is

$$\psi_m \left(\frac{1}{x_{m1}} + \frac{1}{x_{l1}} + \frac{1}{x_{l2}'} \right) = \frac{\psi_1}{x_{l1}} + \frac{\psi_2'}{x_{l2}'} \tag{4.33}$$

Letting

$$\frac{1}{x_M} = \frac{1}{x_{m1}} + \frac{1}{x_{l1}} + \frac{1}{x_{l2}'} \tag{4.34}$$

Eq. 4.33 can be written more compactly as

$$\psi_m = x_M \left(\frac{\psi_1}{x_{l1}} + \frac{\psi_2'}{x_{l2}'} \right) \tag{4.35}$$

Using Eqs. 4.30 and 4.31 to replace the currents, Eqs. 4.25 and 4.26 can be expressed as integral equations of the two total flux linkages, that is

$$\psi_1 = \int \left\{ \omega_b v_1 - \omega_b r_1 \left(\frac{\psi_1 - \psi_m}{x_{l1}} \right) \right\} dt \tag{4.36}$$

$$\psi_2' = \int \left\{ \omega_b v_2' - \omega_b r_2' \left(\frac{\psi_2' - \psi_m}{x_{l2}'} \right) \right\} dt \tag{4.37}$$

Collectively, Eqs. 4.30, 4.31, 4.35, 4.36, and 4.37 form a basic dynamic model of a two-winding transformer to which magnetic nonlinearity and iron losses may be added if necessary. In this model, the flux linkages are the internal variables, the terminal voltages are the required inputs, and the winding currents are the main outputs. Figure 4.4 shows the flow diagram of a simulation of a two-winding transformer that requires the instantaneous value of the terminal voltages of both windings as input to the simulation. Figure 4.5 shows the SIMULINK simulation given in *s1a* that is in accordance with the flow diagram given in Fig. 4.4. It shows that internal variables, ψ_1, ψ_2', and ψ_m, are also available directly from this simulation.

4.5 TERMINAL CONDITIONS

As shown in Fig. 4.5, the developed simulation of a two-winding transformer uses terminal voltages of the two windings as inputs to the simulation and produces the winding currents as outputs. The input-output set of a subsystem in a simulation need not be the same as that of the physical subsystem as long as the input-output requirements of all subsystems in the simulation can be properly matched by those of connected subsystems, including the actual inputs and outputs of the entire system. For example, if a load is connected to the secondary winding of the transformer and the nature of the load is such that its describing equation can

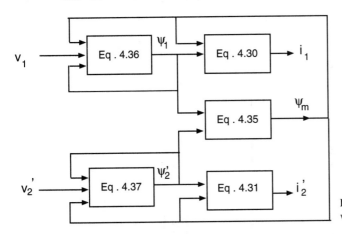

Figure 4.4   Flow of variables in a two-winding transformer simulation—linear case.

be implemented in the simulation to use the load current as its input and the load voltage as its output, the load module input-output requirement is said to match those of the secondary terminal of the two-winding transformer simulation given above.

In manipulating equations to be solved either on the digital or analog computer, the forms to avoid are those with derivatives or algebraic loops. While the derivative form is obvious, the presence of an algebraic loop is less so. Algebraic loops occur when implementing simultaneous algebraic equations where there is direct feed-through of their inputs from one block to another around a loop not involving an integrator.

Let's return to the problem of completing the input requirement for the simulation of the two-winding transformer, specifically that of determining v_2' from the load simulation. The techniques to be presented are useful, not only for combining the above transformer simulation with those of other network elements, but they are also applicable to other electromagnetic devices such as the rotating machines simulations that will be described in later chapters. The input voltage to the primary winding, v_1, is either a known fixed ac voltage or obtainable from the simulation of other components to which the primary winding is connected. That the primary input voltage can also be derived from another part of the simulation will become clear after the following discussion on how to obtain the secondary terminal input voltage.

The condition of a short circuit at the secondary terminal is easily simulated by using $v_2' = 0$ in the simulation. Simulating an open-circuit condition at the secondary terminal is not as straightforward as that for a short-circuit condition. The open-circuit condition at the secondary terminal has an i_2' of zero, which when substituted into Eqs. 4.26 and 4.28, yields $v_{2oc}' = d\psi_m/dt$. To avoid taking the time derivative of ψ_m in the simulation, this value of secondary input voltage can be derived from the value of $d\psi_1/dt$ just before the integrator that yields ψ_1. The relationships used are those between ψ_m and ψ_1 in Eqs. 4.26 and 4.28 and that of Eq. 4.29 for the condition $i_2' = 0$.

$$v_{2oc}' = \frac{d\psi_m}{dt} = \frac{x_{m1}}{x_{l1} + x_{m1}} \frac{d\psi_1}{dt} = \frac{x_{m1}}{x_{l1} + x_{m1}} (v_1 - i_1 r_1)\omega_b \qquad (4.38)$$

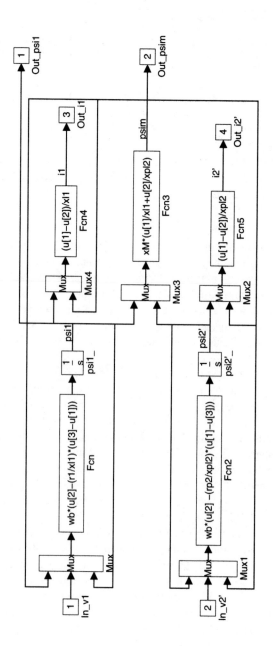

Figure 4.5 SIMULINK simulation *s1a* of a two-winding transformer—linear case.

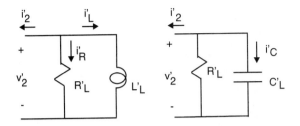

Figure 4.6   Equivalent circuits of load.

The case of finite loading on the secondary terminal is easiest when the load can be represented by an equivalent impedance or admittance. For example, a specified loading of S_L at the rated voltage of V_{2rated} can be translated into an equivalent load admittance referred to the primary side.

$$(G_L' + j B_L')^{-1} = \left(\frac{N_1}{N_2}\right)^2 \frac{V_{2rated}^2}{S_L^*} \tag{4.39}$$

The conductance and susceptance of Eq. 4.39 may be modeled by either one of the two equivalent circuit representations shown in Fig. 4.6. The equations of these parallel circuit equivalent loads can be expressed in integral form with voltage as output and current as input to complement those of winding 2's terminal. For lagging power factor loads, the parallel RL combination of Fig. 4.6 is used. The required v_2' output from the load simulation is obtained from the v vs. i relationship of the resistive element, that is

$$v_2' = i_R' R_L' = - \left(i_2' + i_L'\right) R_L' \tag{4.40}$$

where i_2' is the output current from the simulation of winding 2 of the transformer and i_L' is to be obtained from integrating the voltage across the equivalent load inductance, L_L'.

$$i_L' = \frac{1}{L_L'} \int v_2' dt = \omega_b B_L' \int v_2' dt \tag{4.41}$$

The parallel RC combination of Fig. 4.6 is to be used for leading power factor loads. However, the required v_2' output from the RC load simulation is obtained from the v vs. i relationship of the capacitor, that is

$$v_2' = \frac{1}{C_L'} \int i_C' dt = \omega_b B_L' \int \left\{ -i_2' - \frac{v_2'}{R_L'} \right\} dt \tag{4.42}$$

Note how the common junction voltage between the load and the winding 2 terminal of the transformer can be developed in these two different situations, that is the use of the load resistor in Eq. 4.40 and the use of the capacitor in Eq. 4.42. Clearly, the common junction voltage needs to be defined by only one of the components when there are two or more components connected to that junction.

Sometimes the nature of the interconnected components may not lend themselves to such matching of their input/output requirements at all junctions. This situation is not uncommon in a complex system simulation which is constructed using standard modules of components or templates. There is an advantage in being able to use well-established templates, especially large and complicated modules. The convenience of not having to

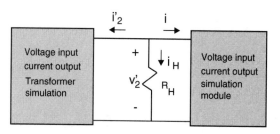

Figure 4.7 Use of a fictitious resistor to develop v_2'.

reformulate may, in some situations, justify the expense of little inaccuracy caused by the introduction of either a very small capacitor or a very large resistor across a junction to develop the common junction voltage required by connected modules whose input/output requirements are all of the voltage input current output kind. While choosing an extremely small capacitor or an extremely large resistor does minimize the inaccuracies caused by the introduction of such non-physically-related elements, too extreme a value may result in system equations that are overly stiff for the solution procedure used. In practice, the smallest capacitor or largest resistor that one can use for such an approximation will be limited by numerical instability.

As a simple illustration of the use of such approximations, consider the sub-circuit shown in Fig. 4.7, in which the voltage, v_2', is the desired input to both the transformer simulation and the other module. Adding a fictitious resistor, R_H, that is much larger than the impedance of the actual circuit elements, would allow us to develop v_2' without introducing too much inaccuracy to the simulated solution. The required voltage input to the connected modules is given by the voltage equation of the fictitious resistor, R_H, that is

$$v_2' = i_H R_H = -(i_2' + i)R_H \qquad (4.43)$$

Alternatively, a small capacitor, C_L, can be used in place of the resistor, R_H, to develop the junction voltage, that is

$$v_2' = -\frac{1}{C_L} \int (i_2' + i)dt \qquad (4.44)$$

Although using a capacitor will introduce an extra state, its voltage does not amplify the noises in the currents as with R_H. In some situations, it may result in a more tolerable approximation than R_H at an acceptable simulation speed.

4.6 INCORPORATING CORE SATURATION INTO SIMULATION

Core saturation mainly affects the value of the mutual inductance and, to a much lesser extent, the leakage inductances. Though small, the effects of saturation on the leakage reactances are rather complex and would require construction details of the transformer that are not generally available. In many dynamic simulations, the effect of core saturation may be assumed to be confined to the mutual flux path. Core saturation behavior can be determined from just the open-circuit magnetization curve of the transformer. The open-circuit curve, such as that shown in Fig. 4.8, is usually obtained by plotting the measured

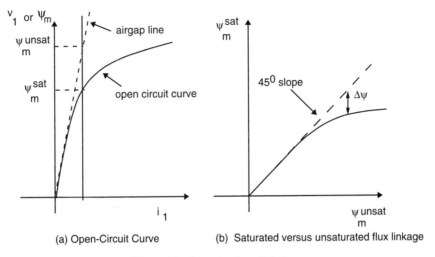

(a) Open-Circuit Curve   (b) Saturated versus unsaturated flux linkage

Figure 4.8   Saturation characteristics.

rms value of the terminal voltage against the no-load current drawn on winding 1 when the secondary terminal is open-circuited. With core losses ignored, the no-load current is just the magnetizing current, I_{mrms}. Furthermore, with only no-load current flowing into winding 1, the voltage drop across the series impedance, $r_1 + j\omega L_{l1}$, is usually negligible compared to that across the large magnetizing reactance, $x_{m1} = \omega L_{m1}$. Since the secondary is open-circuited, i_2' will be zero, thus $V_{1rms} \approx I_{mrms} x_{m1}$. In the unsaturated region, the ratio of V_{1rms}/I_{mrms} is constant, but as the voltage level rises above the knee of the open-circuit curve, that ratio becomes smaller and smaller.

Some of the methods that have been used for incorporating the effects of core saturation in a dynamic simulation are:

(i) Using the appropriate saturated value of the mutual reactance at each time step of the simulation.

(ii) Approximating the magnetizing current by some analytic function of the saturated flux linkage.

(iii) Using the relationship between saturated and unsaturated values of the mutual flux linkage.

In method (i), the saturated value of the magnetizing inductance, x_{m1}^{sat}, can be updated in the simulation using the product of the unsaturated value of the magnetizing inductance, x_{m1}^{unsat}, times a saturation factor, k_s, both of which can be determined from the open-circuit test data. Under open-circuit conditions, the small voltage drop across the $r_1 + jx_{l1}$ can be neglected, thus $V_{1rms} \approx E_{mrms}$. When the excitation flux is sinusoidal, it can be seen from Eqs. 4.25 and 4.29 that $E_{mrms} = \psi_{mrms}$. Thus, the voltage axis of the open-circuit curve in Fig. 4.8a may also be read as that for the flux linkage, ψ_{mrms}. The slope of the airgap line, shown as a dashed line in Fig. 4.8a, is equal to the unsaturated value of the magnetizing reactance, x_{m1}^{unsat}. The saturated magnetizing reactance, x_{m1}^{sat}, at any voltage on the open-circuit curve is equal to the slope of a line joining that point on the open-circuit curve to the origin. The degree of saturation can be expressed by a saturation factor that is defined as

$$k_s = \frac{\psi_{mrms}^{sat}}{\psi_{mrms}^{unsat}} \text{ or } \frac{I_{mrms}^{unsat}}{I_{mrms}^{sat}} \qquad k_s \le 1 \tag{4.45}$$

If the effective saturated magnetizing reactance, x_{m1}^{sat}, is defined as the ratio of $\psi_{mrms}^{sat} / I_{mrms}^{sat}$,

$$k_s = \frac{\psi_{mrms}^{sat}}{I_{mrms}^{sat}} \frac{I_{mrms}^{sat}}{\psi_{mrms}^{unsat}} = \frac{x_{m1}^{sat}}{x_{m1}^{unsat}} \tag{4.46}$$

With certain methods of simulation, such as in analog simulation, it is generally easier to handle a constant rather than a variable magnetizing reactance in Eq. 4.35 to account for the effect of core saturation. Usually in such simulation, the current value of ψ_m^{sat} will be determined from the unsaturated value of mutual flux, ψ_m^{unsat}, that is computed using the value of x_{m1}^{unsat}.

Let's denote the difference between the unsaturated and saturated values by $\Delta \psi$, that is

$$\psi_m^{unsat} = \psi_m^{sat} + \Delta \psi \tag{4.47}$$

The value of $\Delta \psi$ will be positive in the first quadrant but negative in the third quadrant. The relation between $\Delta \psi_m$ and ψ_m^{unsat} or ψ_m^{sat} can be obtained from the open-circuit magnetization curve of the transformer. As shown in Fig. 4.8a, for a given no-load current, i_1, we can determine the corresponding values of ψ_m^{unsat} and ψ_m^{sat}. Repeating the procedure over the desired range for ψ_m^{sat}, we can obtain a sufficient number of paired values of ψ_m^{unsat} and ψ_m^{sat} to plot the curve shown in Fig. 4.8b.

With method (ii), the relation between the peak value of the flux linkage and the peak of the magnetizing current must be established. Since the open-circuit test is usually performed by applying a sinusoidal input voltage, ignoring the series voltage drop, the core flux may be assumed to be also sinusoidally varying in time. But the magnetizing current for a sinusoidal flux excitation into the saturation region will not be sinusoidal. The conversion of the measured rms values of the applied voltage and magnetizing current to instantaneous values when the magnetizing current is non-sinusoidal is not straightforward. A procedure given in [30] for deriving the instantaneous value saturation curve from measured data of the open-circuit test will be described in the next section.

Method (iii) uses the relationship between the saturated and unsaturated values of the flux linkage. Unlike method (ii), it does not require an explicit relationship between the flux linkage and magnetizing current. When flux linkages are chosen as the state variables, as in our case, method (iii) is preferred.

We will next describe the modifications needed to include core saturation using method (iii). For clarity, additional superscripts are introduced in this section to distinguish between unsaturated and saturated values of the mutual flux linkage. Unsaturated and saturated values of the winding currents and total flux linkages are implied by their relationships with the mutual flux linkage. We will begin here by first rewriting Eq. 4.29 as

$$\psi_m^{unsat} = \omega_b L_{m1}^{unsat}(i_1 + i_2') = x_{m1}^{unsat}(i_1 + i_2') \tag{4.48}$$

Similarly, in terms of the saturated flux linkages, the saturated value of the winding currents can be expressed as

$$i_1 = \frac{\psi_1 - \psi_m^{sat}}{x_{l1}} \tag{4.49}$$

$$i_2' = \frac{\psi_2' - \psi_m^{sat}}{x_{l2}'} \tag{4.50}$$

Substituting the above expressions of i_1 and i_2' in Eq. 4.48, we obtain

$$\frac{\psi_m^{unsat}}{x_{m1}^{unsat}} = \frac{\psi_1 - \psi_m^{sat}}{x_{l1}} + \frac{\psi_2' - \psi_m^{sat}}{x_{l2}'} \tag{4.51}$$

Note that the values of ψ_1 and ψ_2' in Eqs. 4.50 to 4.51 are saturated values. Replacing ψ_m^{unsat} by $\psi_m^{sat} + \Delta\psi$ and collecting the ψ_m^{sat} terms, we obtain

$$\psi_m^{sat} = x_M \left(\frac{\psi_1}{x_{l1}} + \frac{\psi_2'}{x_{l2}'} - \frac{\Delta\psi}{x_{m1}^{unsat}} \right) \tag{4.52}$$

where the value of x_M is as in Eq. 4.34 for the unsaturated case, that is

$$\frac{1}{x_M} = \frac{1}{x_{m1}^{unsat}} + \frac{1}{x_{l1}} + \frac{1}{x_{l2}'} \tag{4.53}$$

A comparison of the above set of expressions for the saturated condition, along with the corresponding expressions for the unsaturated or linear case, will show that to account for saturation in the mutual flux, we need to be able to determine the value of $\Delta\psi$ on the right side of Eq. 4.52. This can be done with the help of the functional relation between $\Delta\psi$ and ψ_m^{sat}. The above formulation in which the parameters, in particular x_{m1}^{unsat} and x_M, remain constant is relatively simple to implement. The modified implementation which will account for mutual flux saturation is shown in the flow diagram of Fig. 4.9. Compared to that for the linear case given earlier in Fig. 4.4, the changes are in the last term of Eq. 4.52 and an additional module which is needed to compute $\Delta\psi$ from ψ_m^{sat}. Note, however, that the $\Delta\psi$ loop so introduced is an algebraic one.

In digital simulation, the current value of $\Delta\psi$ can be determined by a look-up table using interpolation, or by simple analytic functions approximating some range of the $\Delta\psi$ vs. ψ_m^{sat} curve. In SIMULINK, a table relating $\Delta\psi$ to ψ_m^{sat} can be implemented using the **Look-up Table** module given in the **Nonlinear** block library. The input-output relation of the **Look-up Table** module is defined by input and output array variables of the same length.

Alternatively, the relation between $\Delta\psi$ and ψ_m^{sat} shown in Fig. 4.10 can be approximated by a simple analytic function. Some analytical approximations studied [27] include linear, power series, exponential, and hyperbolic functions. The choice of representation depends on the application and the range for which the approximation is to be valid. In some studies, the reluctivity or local gradient of the magnetization is important, for others it may be the general fit over a larger range, as in system-behavior-type simulation.

Figure 4.10 shows two examples of three-segment approximations of the $\Delta\psi$ vs. ψ_m^{sat} curve in the first quadrant. The mathematical description of the three segments in Fig. 4.10a are

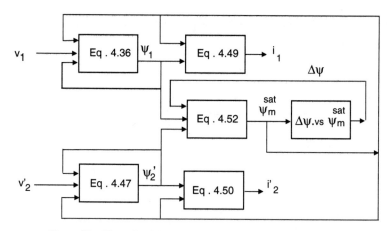

Figure 4.9   Flow of variable in simulation with mutual flux saturation.

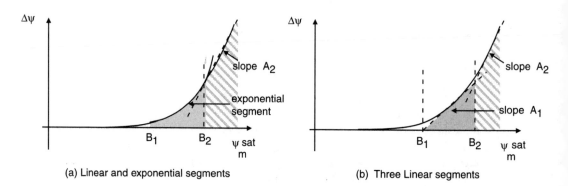

(a) Linear and exponential segments   (b) Three Linear segments

Figure 4.10   Piece-wise approximations of $\Delta\psi$ vs. ψ_m^{sat}.

Linear Region ($\psi_m^{sat} < B_1$)
In the unsaturated portion of the $\Delta\psi$ vs. ψ_m^{sat} curve,

$$\Delta\psi = 0 \tag{4.54}$$

Knee Region ($B_1 < \psi_m^{sat} < B_2$)
The knee region is highly nonlinear. It can be approximated quite well by a suitable exponential function of the form:

$$\Delta\psi = ae^{b(\psi_m^{sat} - B_1)} \tag{4.55}$$

where the value of a is equal to the small step discontinuity at $\psi_m^{sat} = B_1$ and the constant b is to be determined by equating the expression to the value of $\Delta\psi$ at the point $\psi_m^{sat} = B_2$, that is

$$\Delta\psi(B_2) = ae^{b(B_2 - B_1)} \tag{4.56}$$

Fully Saturated Region $(\psi_m^{sat} > B_2)$

In the fully saturated region, the $\Delta\psi$ vs. ψ_m^{sat} curve is approximately linear. Thus for $\psi_m^{sat} > B_2$,

$$\Delta\psi = A_2(\psi_m^{sat} - B_2) + \Delta\psi(B2) \tag{4.57}$$

The mathematical description of the piece-wise linear approximation shown in Fig. 4.10b can be expressed as

$$\Delta\psi = A_1\left(\psi_m^{sat} - B_1\right) + A_2\left(\psi_m^{sat} - B_2\right) \tag{4.58}$$

where

$$A_1 = \begin{cases} slope1 & \text{if } \psi_m^{sat} > B_1 \\ 0 & \text{otherwise} \end{cases}$$

$$A_2 = \begin{cases} slope2 - slope1 & \text{if } \psi_m^{sat} > B_2 \\ 0 & \text{otherwise} \end{cases}$$

The slopes and breakpoints of the second and third linear segments are *slope1* and B_1, and *slope2* and B_2, respectively.

Since ψ_m is alternating, saturation for negative ψ_m must be taken care of by a similar $\Delta\psi$ vs. ψ_m^{sat} approximation in the third quadrant. For negative ψ_m^{sat}, the slope of the linear approximation remains the same, hence the A values remain the same, but the sign of the breakpoint values, that is B, changes with that of ψ_m^{sat}.

4.6.1 Instantaneous Value Saturation Curve

The open-circuit magnetization curve of the transformer can be obtained from the results of an open-circuit test. With $I_2' = 0$, the applied sinusoidal voltage, V_1, to winding 1's terminals is gradually raised from zero to slightly above its rated value. Usually, the measured rms value of winding 2's output voltage and the measured rms value of winding 1's excitation current, that is V_2' vs. I_1, are plotted Since all variables in our simulation model are referred to primary winding 1 and are of instantaneous rather than rms value, the correction for saturation, $\Delta\psi$, should be expressed in terms of the instantaneous variables referred to the primary winding. The measured open-circuit secondary rms voltage can easily be referred to the primary side using the turns ratio, that is

$$V_{rms1}^{oc} = \frac{N_1}{N_2} V_{rms2}^{oc} \tag{4.59}$$

The relation between the peak value of the primary winding flux linkage and the peak value of its magnetizing current can be obtained following the procedure described in [30, 31]. Beginning with the open-circuit curve with rms values referred to the side of the winding that we will use in the simulation and distinguishing the rms values by upper-case letters and the instantaneous values by lower-case letters, mark on the rms open-circuit curve, as shown in Fig. 4.11a, the points 0, 1, 2, 3, ..., N. Point 0 is at the origin and point 1 is at the end of the linear region of the magnetization curves. The other points can be approximately distributed over the saturated portion of the curve. Figure 4.11c shows

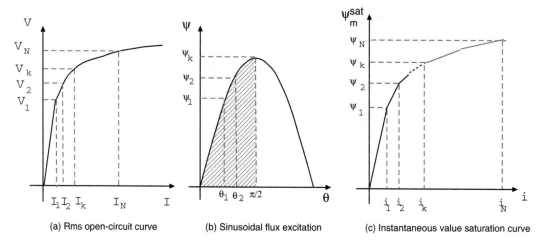

(a) Rms open-circuit curve   (b) Sinusoidal flux excitation   (c) Instantaneous value saturation curve

Figure 4.11   rms value curve to instantaneous value curve.

the corresponding points on the instantaneous ψ_m^{sat} vs. i curve that is to be determined successively, one point at a time. Corresponding to each point on the rms open-circuit curve, consider the open-circuit rms current drawn by the transformer when an input sinusoidal voltage having a peak value of $\sqrt{2}$ times that of the rms voltage is applied, such as that shown in Fig. 4.11b for the kth point. For a sinusoidal input voltage of frequency ω, the corresponding flux linkage will be sinusoidal with a peak value given by

$$\psi_k = \sqrt{2}V_k \qquad k = 0, 1, \ldots, N \tag{4.60}$$

Thus, the values of $\psi_0, \psi_1, \ldots, \psi_N$ in Fig. 4.11b can be determined from the above relations.

Except for the initial value, that is $i_0 = 0$, the peak value of their corresponding magnetizing currents, $i_1, i_2, \ldots, i_N$, however are still unknown at this stage. They are to be determined by equating an expression of the rms value of the instantaneous current in Fig. 4.11c to the measured rms value for the corresponding points in Fig. 4.11b in the manner described below.

When the number of points used is sufficiently large and their level judiciously distributed, the rms value of the instantaneous current corresponding to a sinusoidal voltage excitation can be determined with reasonable accuracy using a simple piece-wise linear analytical representation between adjacent points, as shown in Fig. 4.11c. Let K_j be the slope of the line joining points $(j-1)$ and j, measured from the vertical, that is

$$K_j = \frac{i_j - i_{j-1}}{\psi_j - \psi_{j-1}} \qquad j = 1, \ldots, N \tag{4.61}$$

The value of i_k can then be expressed as

$$i_k = \sum_{j=1}^{k} K_j(\psi_j - \psi_{j-1}) \qquad \text{for } k = 1, \ldots, N \tag{4.62}$$

The algorithm described in [30] determines the values of $K'_j s$ for $j = 1, \ldots, N$ successively by equating the expression of the rms value of the instantaneous current to the corresponding measured rms value for an input sinusoidal voltage having a peak that is $\sqrt{2}$ times the corresponding rms voltage.

Beginning with $j = 1$, the peak of the flux linkage wave corresponding to point 1 on the open-circuit curve of Fig. 4.11a is $\psi_1 = \sqrt{2}V_1$. With the first ψ_m^{sat} vs. i_1 linear segment starting from the origin, the analytical expression of the instantaneous current corresponding to the above excitation is

$$i = K_1 \psi_1 \sin \theta \qquad (4.63)$$

Assuming sinusoidal voltage excitation, the flux linkage wave will also be sinusoidal. When hysteresis is ignored, the magnetizing current with an applied sinusoidal input voltage is quarter-wave symmetric; as such, we need only to consider a quarter-wave excitation when computing its rms value. For example, for the kth point, we need only to consider the rms value of the current due to the hatched area of the applied voltage shown in Fig. 4.11b. For $k = 1$, we equate the expression for the square of the rms value of i with a peak of i_1 from the curve of Fig. 4.11c to the square of the rms current, that is I_1 in Fig. 4.11a.

$$I_1^2 = \frac{2}{\pi} \int_0^{\pi/2} (K_1 \psi_1 \sin \theta)^2 \, d\theta = \frac{K_1^2 \psi_1^2}{2} \qquad (4.64)$$

or

$$K_1 = \frac{\sqrt{2}I_1}{\psi_1} \qquad (4.65)$$

Similarly, for point 2 of Fig. 4.11a, we use an excitation flux wave with a peak value of ψ_2 equal to $\sqrt{2}V_2/\omega$ (note that $\theta_2 = \pi/2$ in this instance). Equating the square of the corresponding measured rms current to the expression for the square of the rms current produced by a sinusoidal voltage of this magnitude, we have

$$I_2^2 = \frac{2}{\pi} \left\{ \int_0^{\theta_1} (K_1 \psi_2 \sin \theta)^2 \, d\theta + \int_{\theta_1}^{\pi/2} (K_1 \psi_1 + K_2(\psi_2 \sin \theta - \psi_1))^2 \, d\theta \right\} \qquad (4.66)$$

where $\theta_1 = \sin^{-1}(\psi_1/\psi_2)$. Using $i_1 = K_1 \psi_1$, Eq. 4.66 can be rearranged into a quadratic equation in K_2 of the form

$$A_2 K_2^2 + B_2 K_2 + C_2 = 0 \qquad (4.67)$$

where

$$A_2 = \int_{\theta_1}^{\pi/2} (\psi_2 \sin \theta - \psi_1)^2 d\theta, \qquad A_2 > 0 \qquad (4.68)$$

$$B_2 = 2K_1 \psi_1 \int_{\theta_1}^{\pi/2} (\psi_2 \sin \theta - \psi_1) d\theta, \qquad B_2 > 0 \qquad (4.69)$$

$$C_2 = i_1^2 (\frac{\pi}{2} - \theta_1) + \int_0^{\theta_1} (K_1 \psi_2 \sin \theta)^2 d\theta - \frac{\pi}{2} I_2^2, \qquad C_2 < 0 \qquad (4.70)$$

With only a positive value of K_2 admissible,

$$K_2 = \frac{-B_2 + \sqrt{B_2^2 - 4A_2C_2}}{2A_2} \tag{4.71}$$

Likewise, it can be shown that the linear segment with slope, K_k say, satisfies a relation similar to that given in Eq. 4.67 obtained for point 2, that is

$$A_k K_k^2 + B_k K_k + C_k = 0 \tag{4.72}$$

where by induction it has been shown in [30] that

$$C_k = d_k + \sum_{j=1}^{k-1}(K_j^2 A_j + K_j B_j + d_j) - \frac{\pi}{2} I_k^2$$

$$d_j = i_{j-1}^2 t_j$$

$$t_j = \theta_j - \theta_{j-1}$$

$$\theta_j = \sin^{-1}\left(\frac{\psi_j}{\psi_k}\right)$$

$$s_j = \frac{1}{2}(\sin 2\theta_j - \sin 2\theta_{j-1}) \tag{4.73}$$

$$g_j = \cos 2\theta_j - \cos 2\theta_{j-1}$$

$$A_j = \frac{\psi_k^2}{2}(t_j - s_j) + 2\psi_k \psi_{j-1} g_j + \psi_{j-1}^2 t_j$$

$$B_j = -2i_{j-1}(\psi_k g_j + \psi_{j-1} t_j)$$

$$\text{for } j = 1, \ldots, k \; ; 1 \le k \le N$$

Beginning with the point at the origin, that is $k = 0$, where $\psi_0 = 0$, $i_0 = 0$, and $\theta_0 = 0$, the values of K_k for $k = 1, \ldots, N$ can be obtained successively using Eqs. 4.72 and 4.73. As shown earlier for the case of $k = 2$, the solution is of the form given by Eq. 4.71.

An M-file based on the above algorithm to obtain the instantaneous value ψ_m^{sat} vs. i curve from the standard open-circuit test data in rms values is given in *mginit*. Figure 4.12 shows the results of the sample case given in *mginit*. The subfigure on the top left of Fig. 4.12 is a plot of the input open-circuit curve data in rms values; the plot of the corresponding instantaneous value ψ_m^{sat} vs. i curve is on the top right. The bottom two subfigures are plots of errors in comparison to the open-circuit rms current when the two magnetization curves are subjected to a sinusoidal voltage excitation whose amplitude is slowly varied to cycle over the full range of the magnetization curves. The one on the bottom left is obtained using the open-circuit curve, but it is scaled by a factor of $\sqrt{2}$, assuming that both voltage and current waveforms are sinusoidal. That on the bottom right is obtained using the instantaneous value ψ_m^{sat} vs. i curve. These results, obtained using the SIMULINK program, *smg*, shown in Fig. 4.13, indicate that the error is negligible when the amplitude of the excitation voltage is below the knee of the saturation curve, but grows rapidly as the

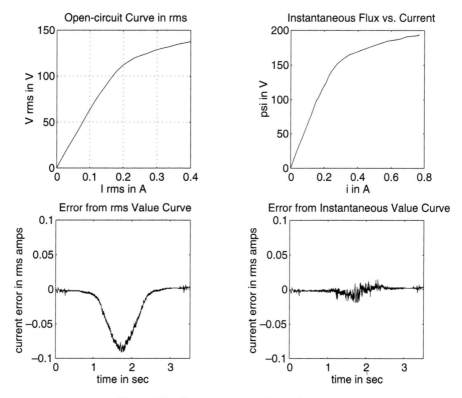

Figure 4.12   Saturation curves and errors in current.

amplitude of the excitation voltage reaches further into the saturated region. As shown in this exercise with the algorithm, the open-circuit curve should be monotonic and the usual scattered points from measured data should be curve-fitted first to avoid excessive kinks in the ψ_m^{sat} vs. i curve.

4.7 THREE-PHASE CONNECTIONS

The generation and distribution of ac electric power are mostly in three-phase. When the three-phase system is operating under balanced conditions, it can be represented for study purposes by an equivalent single-phase representation, thus experiencing savings in computational effort. For unbalanced operating conditions, the three-phase system will have to be represented as-is or by a combination of sequence networks. In this section, we will examine methods that can be used to simulate some of the common three-phase transformer connections, as in electromechanical transient studies.

In general, the operating characteristic of a three-phase transformer will depend not only on its winding connections, but also on the magnetic circuit of its core. When the windings of the different phases share a common core that provides a magnetic path for mutual flux, there will be mutual coupling between them.

Demonstration compares error from using the rms open-circuit curve
to that of using the instantaneous *psi*_sat vs. *i* curve.

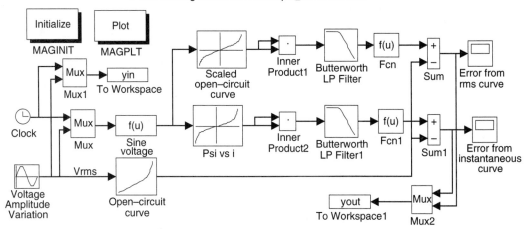

Figure 4.13 Simulation *smg* to compare the accuracy of the two curves.

The two most common winding connections are the Wye and Delta connections. A
Wye connection is favored over a Delta connection at higher voltage levels, but a Delta
connection can be advantageous when the winding current level is high. Other factors
normally considered include those of grounding from the point of safety and protection,
and paths for harmonic currents and fluxes to minimize distortion of the voltage waveform.
For greater detail, the reader is referred to reference [24].

4.7.1 Wye-Wye Connection

We will begin by considering the case where the primary and secondary windings of a
three-phase transformer are Wye-Wye connected as shown in Fig. 4.14a. For illustration
purposes, the neutral of the primary windings is grounded through a resistor, R_N, and those
of the supply sources and secondary windings are solidly grounded. Since the neutral of
the primary windings is not solidly grounded, its potential will be floating with respect to
the system ground, G. Depending on whether the operating condition is balanced or not,
the phase-to-neutral voltages across the transformer primary windings may not be the same
as the corresponding phase-to-neutral voltages of the source, since

$$v_{AN} = v_{AG} - v_{NG}$$

$$v_{BN} = v_{BG} - v_{NG} \qquad (4.74)$$

$$v_{CN} = v_{CG} - v_{NG}$$

where $v_{AG} = v_{AO}$, $v_{BG} = v_{BO}$, and $v_{CG} = v_{CO}$. Figure 4.14b shows the equivalent circuit
representation of only one of the phases of the three-phase system, in which the transformer
has been represented by the equivalent circuit representation given in Fig. 4.3. Recall that
the two-winding transformer simulation developed earlier requires primary and secondary

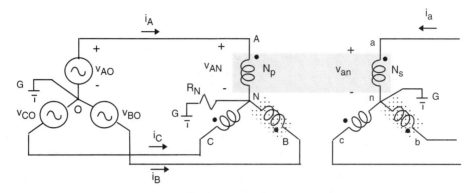

(a) Source and transformer connection

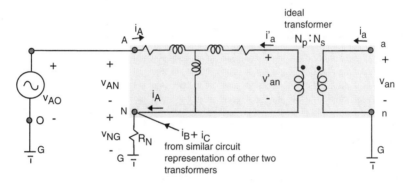

(b) Equivalent circuit representation of one transformer in detail

Figure 4.14   Wye-Wye connected transformer.

voltages as input and provides primary and secondary currents as output. The voltage, v_{NG}, can be determined from the relationship of the voltage drop across R_N, that is

$$v_{NG} = (i_A + i_B + i_C)R_N \tag{4.75}$$

R_N could be a large fictitious resistor in the case where the neutral point, N, is ungrounded.

Figure 4.15 shows a SIMULINK simulation of the three-phase transformer connection given in Fig. 4.14a for the case where three separate two-winding transformers are used. The assumption of three separate two-winding units here is used so that we can use three two-winding transformer modules, as shown in Fig. 4.16 of the inside of one module, and the connection relationships expressed in Eqs. 4.74 and 4.75.

4.7.2 Delta-Wye Connection

Figure 4.17 shows a Delta-Wye connected transformer, where the secondary neutral point, n, is grounded through resistor R_n. When the windings are connected as shown, the effective secondary-to-primary turns ratio of the three-phase connection is $\sqrt{3}N_s/N_p$, and the

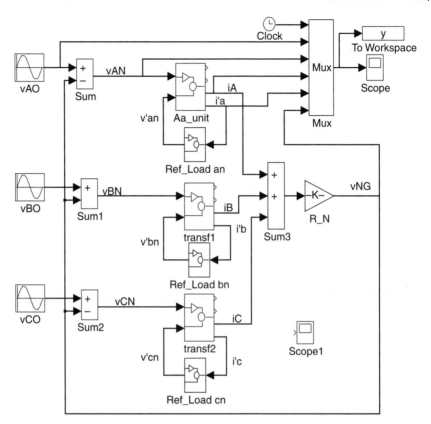

Figure 4.15 Simulation of Wye-Wye connected transformer.

secondary phasor voltages are phase-shifted by 30° counter-clockwise to the corresponding primary phasor voltages.

With the primary windings of the three transformers connected as shown in Fig. 4.17, the input voltages to the primary windings can be computed directly from the ac source phase voltages using

$$v_{AB} = v_{AO} - v_{BO}$$
$$v_{BC} = v_{BO} - v_{CO} \qquad (4.76)$$
$$v_{CA} = v_{CO} - v_{AO}$$

The input voltages to the secondary windings of the transformers, on the other hand, are a function of the secondary neutral point voltage to system ground, that is

$$v_{an} = v_{an} - v_{nG}$$
$$v_{bn} = v_{bn} - v_{nG} \qquad (4.77)$$
$$v_{cn} = v_{cn} - v_{nG}$$

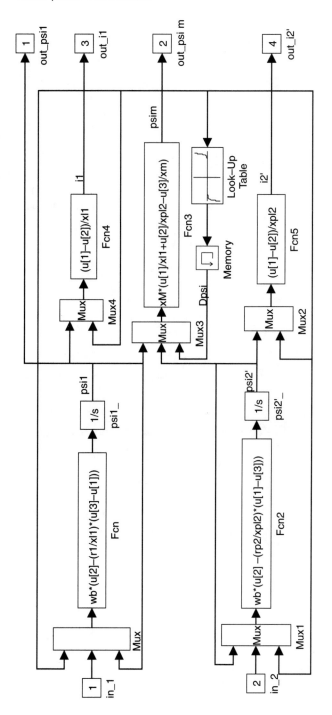

Figure 4.16 Inside the *Aa* unit.

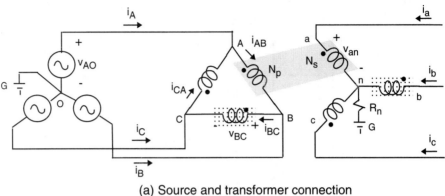

(a) Source and transformer connection

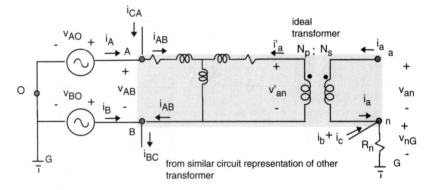

(b) Equivalent circuit representation of one transformer in detail

Figure 4.17   Delta-Wye connected transformer.

where $v_{nG} = (i_a + i_b + i_c)R_n$. In the simulation, the line currents on the primary side can be computed from the primary winding output currents using

$$i_A = i_{AB} - i_{CA}$$

$$i_B = i_{BC} - i_{AB} \qquad (4.78)$$

$$i_C = i_{CA} - i_{BC}$$

A SIMULINK simulation of the system shown in Fig. 4.17a is given in Fig. 4.18. Again, for simplicity, we have assumed that the Delta-Wye transformer is formed by three separate single-phase units. The simulation shown uses three two-winding transformer modules like the one given in Fig. 4.16, and the relations given in Eqs. 4.76, 4.77, and 4.78.

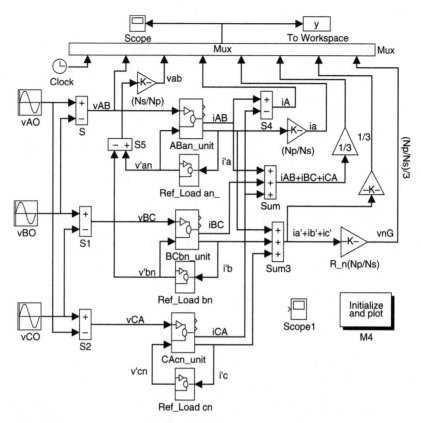

Figure 4.18   Simulation *s4* of Delta-Wye connected transformer.

4.8 PROJECTS

The circuit parameters and magnetization curve of the 120/240, 1.5 kVA, 60-Hz, pole-type, two-winding distribution transformer for Projects 1 and 3 are as follows:

$$r_1 = 0.25\Omega \qquad\qquad r_2' = 0.134\Omega$$

$$x_{l1} = 0.056\Omega \qquad\qquad x_{l2}' = 0.056\Omega$$

$$x_{m1} = 708.8\Omega$$

All parameters given above are referred to the 120V, winding 1 side. Figure 4.19 is a plot of the magnetization curve of the transformer in terms of the instantaneous values of the flux linkage and magnetizing current on the 120V, winding 1 side.

4.8.1 Project 1: Short-circuit and RL Load Terminations

The objectives of this project are to implement a SIMULINK simulation of a single-phase two-winding transformer, use it to examine two operating conditions: (1) the secondary

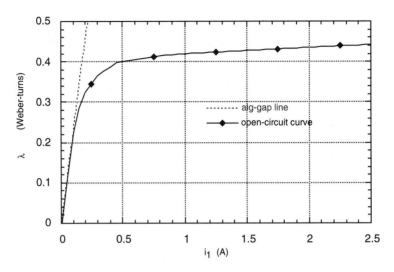

(a) Instantaneous flux linkage vs. magnetizing current

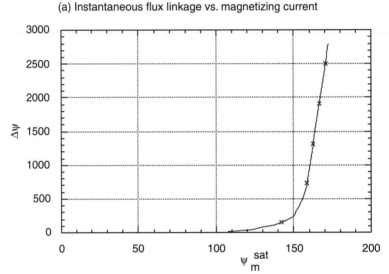

(b) Plot of instantaneous $\Delta\psi$ vs. ψ_m^{sat}

Figure 4.19   Magnetization curve of distribution transformer.

terminals short-circuited and (2) the secondary terminals connected to a non-unity power factor load, and to verify the results obtained with those predicted from an analysis using the equivalent circuit given in Fig. 4.3.

The SIMULINK simulation of a two-winding transformer can be set up using the voltage-input, current-output model described earlier in this chapter. Core saturation can be handled using either a piece-wise linear analytic approximation of the saturation curve,

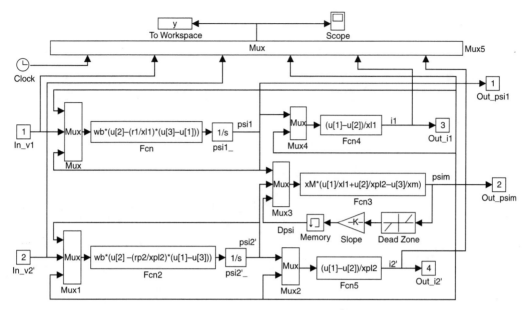

Figure 4.20 Simulation *s1b* of a two-winding transformer with a piece-wise approximation of the saturation curve.

such as that of Fig. 4.20 and given in *s1b*, or a look-up table between ψ_m^{sat} and $\Delta\psi$, such as that of Fig. 4.21 and given in *s1c*. The **Look-up Table** and **Memory** modules to break the algebraic loop are found in SIMULINK's **Nonlinear** block library.

Use the MATLAB script file *m1* to set up the parameters of the SIMULINK simulation *s1c*. Check the parameter settings in *m1* to make sure that they correspond to the parameters of the 120/240, 1.5 kVA, 60-Hz, pole-type distribution transformer given above. Begin by trying the ode15s or the Adams/Gear numerical method with a minimum step size of 0.1 msec, a maximum step size of 1 msec, and an error tolerance of $1e^{-7}$. Conduct the following runs on the simulation using a sinusoidal input voltage of $v_1 = 120\sqrt{2}\sin(120\pi t + \theta)$ V:

1. With the 240V side terminals short-circuited, that is $v_2' = 0$, and no initial core flux, energize the transformer at the following point of the wave of the voltage, v_1:

   a. The peak of the sinusoidal supply voltage, using a θ of $\pi/2$.

   b. The zero of the sinusoidal supply voltage, using a θ of zero.

   Plot the values of v_1, i_1, ψ_m^{sat}, v_2', and i_2'.

2. Replace the short-circuit termination on the secondary terminal with a fixed impedance representing 1.5 kVA, 0.8 lagging power factor of loading at rated voltage and repeat the above energization studies with the transformer on the load.

Discuss the results obtained, checking some of them against analytical results based on the same equivalent circuit. For example, check the decay times of the dc offset in the

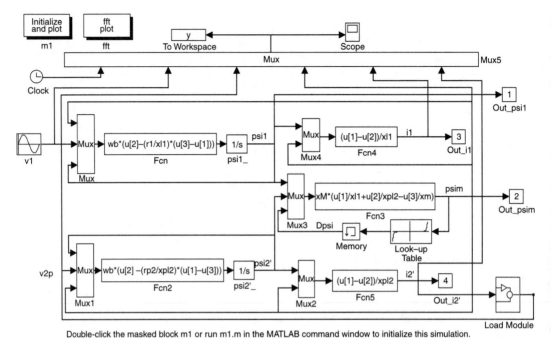

Figure 4.21   Simulation *s1c* of a two-winding transformer using a look-up table for the saturation curve.

input current or flux against the values of the time constant for the corresponding terminal condition on the secondary side, with saturation and without saturation. Note the impact that saturation has and explain it. Compare the magnitude of the currents obtained from the simulation when it reaches steady-state with those computed from steady-state calculations.

4.8.2 Project 2: Open-circuit Termination, In-rush Current, and dc Bias Core Saturation

The objectives of this project are

- To implement an open-circuit termination.
- To examine the phenomena of in-rush currents during the energization of a transformer.
- To study the core saturation caused by a dc excitation component, as in the case of a geomagnetic-induced current.

Open-circuit termination.   It is assumed that you have implemented the two-winding transformer simulation called for in the previous project. Approximate the open-circuit condition on the secondary terminals in your simulation by using a large fictitious resistor across the open-circuit terminals. In implementing the volt-current relationship for the fictitious resistor, take heed of the current direction and voltage polarity of the resistor shown in Fig. 4.7. The larger the fictitious resistor value that you can use in your simulation to approximate the open-circuit condition, the closer the approximation. Typically, a resistor value about 100 times the base impedance, defined as rated voltage to rated current, will be acceptable if it does not present a numerical problem. The simulation can have difficulty converging when the resulting system equation is numerically ill-conditioned, as in the case of tightly-coupled windings with relatively small leakages. A good example of this is the situation at hand, where we are attempting to use a large resistor to simulate the open-circuit operation of the given two-winding transformer that has tightly coupled windings and a saturation curve with sharp knee. To avoid numerical instability, we will have to pick an integration method that is suitable for stiff systems, keep the step size and/or the error tolerance small, and if necessary, use a smaller fictitious resistor at the expense of less accuracy.

Experiment with a few values of the fictitious resistor in the range between 50 to 200 times base impedance in combination with some reasonable step size to see whether your simulation of the transformer with saturation will yield satisfactory results, that is, no noticeable glitches riding on the secondary voltage waveform. Note the value of the secondary current in each case.

In-rush current.   Large magnetizing in-rush current can occur with a certain combination of point of wave energization and residual core flux. Here we will examine the in-rush of magnetizing current in an unloaded transformer that is energized at the instant when the supply voltage and residual core flux are both zero.

Use the **To Workspace** block to store the values of t, v_1, i_1, ψ_m, and v_2' over the initial 20 cycles for plotting and analysis to be performed later. To perform a fast Fourier analysis, values of the function at a uniformly spaced interval are required. Unless you set the maximum and minimum step size to the same value, the result from a variable step size integration algorithm is unlikely to be uniformly spaced. Uniform samples can be obtained by using the save feature of **To Workspace**, or by interpolation of the simulated output. The save feature of **To Workspace** allows you to specify the sample time and also the decimation, that is the number of sample times to skip between successive stores. However, with a simulation that is highly nonlinear or with many discrete changes, using a uniform time step may not give satisfactory results unless the specified time step is very small.

Given in the CD-ROM that accompanies this book is an M-file, *FFTPLOT*, for obtaining the discrete Fourier transform plot of a variable that is stored in column array form in the MATLAB workspace, as in the **To Workspace** block. The values of the variable need not be taken at uniformly spaced time intervals. The M-file, *FFTPLOT*, uses interpolation to first obtain uniformly sampled values of the variable required by the MATLAB fast Fourier transform function, *fft*, and then plots the normalized value of the discrete transform sequence. Double-clicking on the masked block, *fftplot* in *s1c* shown in Fig. 4.21, will execute the given *FFTPLOT* M-file, if it has been properly placed in MATLAB's search path beforehand.

For purposes of comparison, we will first collect data from a transformer energization with no core saturation by leaving open the *Dpsi* feedback loop as shown in Fig. 4.21. With zero initial core flux, energize the transformer with a sinusoidal input voltage of $v_1 = 120\sqrt{2}\sin(120\pi t)$ V and plot the values of v_1, i_1, and ψ_m vs. t.

Close the *Dpsi* loop in the transformer simulation and repeat the energization. Plot the values of v_1, i_1, and ψ_m vs. t. Comment on the differences between the i_1 and ψ_m of the two sets.

Use *FFTPLOT* to obtain the discrete Fourier transform of the magnetizing in-rush current, i_1, for the case of the transformer with an open-circuit secondary being energized with a sine input voltage. Figure 4.22 shows the sample plot, i_1, and its discrete transform for the above condition with a stop time of 0.2 second. Note the dc component and even harmonic components in the transform sequence of the in-rush current. The prominent second harmonic current component of the in-rush current is employed in some circulating-current protection schemes to restrain the overcurrent relay from operating during the in-rush period.

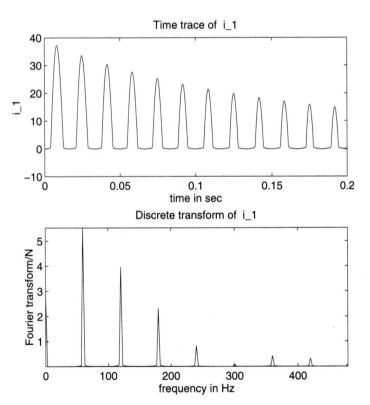

Figure 4.22   Discrete transform of an in-rush current with sine voltage excitation.

dc bias core saturation.   Half-cycle saturation and its attendant problems can be caused by inadvertent excitation of an asymmetric nature, as in rectifier circuits or during a geomagnetic storm. We can observe this behavior in simulation by introducing a small dc offset component to the primary voltage, v_1.

Modify your simulation to allow you to switch in an additional dc component voltage to v_1 besides the main ac component of $120\sqrt{2}\sin(120\pi t)$ V. Connect the 1.5 kVA, 0.8 power factor lagging load to the secondary terminals. Start the simulation with the transformer with a v_1 of $120\sqrt{2}\sin(120\pi t)$ V and the load on the secondary. When steady-state is reached, slowly ramp up the dc voltage component in the primary voltage, v_1, from zero to 10 V and record the behavior of ψ_m, i_1, and v_2'. Comment on the potential adverse effects in practice from the point of view of harmonic distortion and additional var requirements.

Sample results for the case of no load on the secondary terminal are given in Figs. 4.23 and 4.24. These results are obtained using the simulation shown in Fig. 4.21 with an Adams/Gear numerical method, a minimum step size of 0.1 msec, a maximum step size of 1

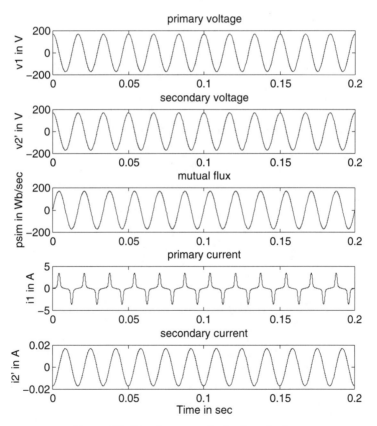

Figure 4.23   Transformer energized at the peak of v_1.

msec, and an error tolerance of $1e^{-7}$. The open-circuit terminal condition on the secondary terminal was approximated with a large resistor of $1\ k\Omega$.

4.8.3 Project 3: Autotransformer Connection

Show how you can use the transformer simulation given in Fig. 4.21 to simulate the autotransformer winding connection shown in Fig. 4.25, where the voltage on the high-voltage side is the sum of the winding voltages, v_1 and v_2, and the voltage on the low-voltage side of the autotransformer is v_2.

4.8.4 Project 4: Delta-Wye Transformer

The objectives of this project are to determine the voltage and current ratios of the Delta-Wye connected transformer shown in Fig. 4.17, and to examine the waveforms of the zero-sequence current components on both sides of the transformer.

Set up a SIMULINK simulation *s4* of the Delta-connected transformer and ac source system shown in Fig. 4.17a. Figure 4.18 shows the overall diagram of the SIMULINK simulation, *s4*. Each of the transformer modules in *s4* contains a full representation of the two-winding transformer used in Projects 1 and 2. Let's consider the situation when the load across each secondary winding is 1.5 KVA, unity power factor. Note that the turns ratio, N_p/N_s, is 0.5 and that the primed secondary voltages and currents are referred to the primary; as such, the load value used within the three load modules should also be the load impedance referred to the primary side. The parameters and magnetization characteristic of the three single-phase transformers forming the Delta-Wye arrangement are identical. The voltages of the three ac sources on the primary side are

$$v_{AO} = 120\sqrt{\frac{2}{3}}\cos(120\pi t)\qquad V$$

$$v_{BO} = 120\sqrt{\frac{2}{3}}\cos\left(120\pi t - \frac{2\pi}{3}\right)\qquad V\qquad(4.79)$$

$$v_{CO} = 120\sqrt{\frac{2}{3}}\cos\left(120\pi t - \frac{4\pi}{3}\right)\qquad V$$

Voltage and current ratios.   For this part of the project, we will ignore core saturation. The core saturation in each transformer module can be deactivated by opening the *Dpsi* loop. Use the MATLAB script file *m4* to set up the parameters and initial conditions in the MATLAB workspace for simulating *s4*. The duration of the simulation run, *tstop*, is set inside *m4* to be 1.5 seconds to allow time for the variables to settle to steady-state condition; however, the length of the **To Workspace** array *y* is purposely kept short so that only the last few cycles of the waveforms will be plotted. Using ode15s or the Adams/Gear numerical method, *s4* will run satisfactorily with a minimum step size of 0.2 msec, a maximum step size of 10 msec, and an error tolerance of $2e^{-6}$.

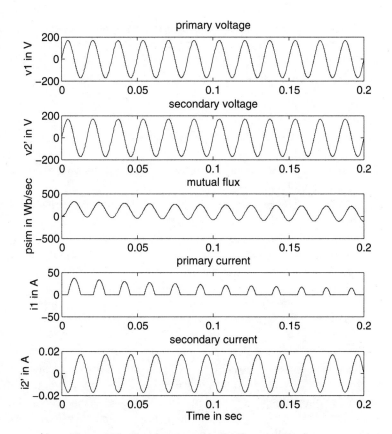

Figure 4.24   Transformer energized at the zero point of v_1.

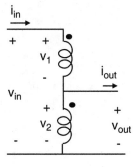

Figure 4.25   Autotransformer winding connection.

Run *m4* from the MATLAB command window and respond to its prompt to input the ohmic value of the secondary neutral resistor to ground, R_n, to 10Ω. Run the simulation with the transformer operating with balanced supply and load, and obtain a plot of the line voltages and currents, v_{AB}, v_{ab}, i_A, and i_a, the zero-sequence component of the winding and line currents, $(i_{AB}+i_{BC}+i_{CA})/3$ and $(i_a+i_b+i_c)/3$, and the secondary neutral to ground voltage v_{nG}. From the circuit diagram in Fig. 4.17b, the voltage, v_{nG}, may be expressed as

$$v_{nG} = (i_a+i_b+i_c)R_n = \frac{N_p}{N_s}(i'_a+i'_b+i'_c)R_n \tag{4.80}$$

From the plotted waveforms, determine the following:

1. The ratio of the primary-to-secondary line-to-line voltages. Check it against the nominal ratio of $N_p/\sqrt{3}N_s$.
2. The ratio of the primary-to-secondary line currents. Check it against the nominal ratio of $\sqrt{3}N_s/N_p$.
3. The phase shift between the primary and secondary line-to-line voltages, v_{AB} and v_{ab}. Compare it with the phase shift between the primary and secondary line currents, i_A and i_a. Sketch the phasor diagram to illustrate and cross-check these phase shifts in the voltages and currents for the unity power factor load condition. For the primary winding connection used, show that the sum of the primary line current, that is $(i_A+i_B+i_C)$, is zero.

Zero-sequence current.   Repeat the above study, but with core saturation by closing the *Dpsi* loops in all three transformer modules. Set R_n to $10\ \Omega$ and run the simulation to obtain plots of the same set of variables. Comment on the difference in the wave shape of the zero-sequence component of the primary and secondary windings, as well as the primary line currents.

Next, unbalance the secondary loading by halving only the resistive component of RL load on phase *a*. With R_n still at $10\ \Omega$, repeat the run and examine the plots of the zero-sequence components of the winding and line currents, $(i_{AB}+i_{BC}+i_{CA})/3$ and $(i_a+i_b+i_c)/3$. Is there a change in the zero-sequence component on the secondary winding due to the single-phase unbalance in loading? Will there be a similar change in the zero-sequence component of the primary line currents, $(i_A+i_B+i_C)/3$? Why not?

Finally, increase the value of R_n to some large value, say 100 times the base impedance, Z_b, of $120^2/1500\ \Omega$, to approximate an ungrounded neutral of the Wye winding. With the secondary loading still unbalanced, repeat the run to obtain a plot of the same three zero-sequence currents. Explain the presence or absence of the zero-sequence current component in the primary and secondary winding currents and in the primary line currents. Figure 4.26 shows the voltages and currents of the Delta-Wye transformer with the ungrounded neutral connection of its Wye secondary windings approximated by $100Z_b$ when the load on each secondary phase is 1.5 kW.

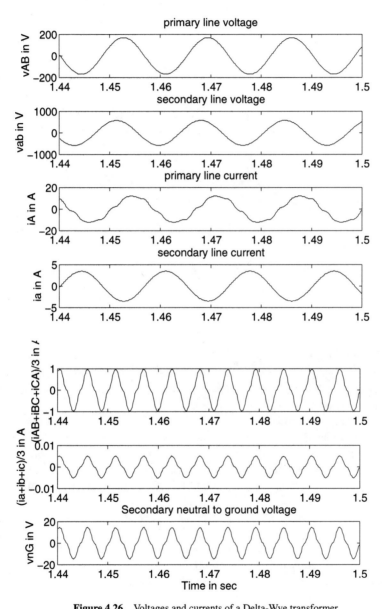

Figure 4.26   Voltages and currents of a Delta-Wye transformer.

5

Basics

of Electric Machines

and Transformations

5.1 INTRODUCTION

Most of today's electric machinery uses iron as a medium to concentrate or to direct the useful flux, air as a medium to store energy and to provide relative motion, and copper or aluminum to conduct the current and to collect the induced voltage. The windings in an electric machine usually consist of many turns of copper or aluminum conductor. A voltage is induced in a winding when the flux it links changes with time. In transformers, the windings are stationary, and the voltages induced in the windings are due to the time-varying flux linking them. But in rotating electric machines, changes in flux linkage with time are usually created by either relative motion between a winding and a spatially distributed magnetic field, or by a variation of reluctance in the magnetic circuit linking the winding as a result of rotor motion.

In electric machines and transformers, windings are usually assembled from identically shaped coils formed by one or more turns of copper or aluminum conductors. Concentric windings are formed by coiling roughly similar sized turns concentrically about a common axis. They are suitable for transformers, and are also used in small electric machines. Although less expensive to construct, they are nevertheless impractical in large machines where the large cross-sectional area of the coil sides would require an unacceptably large slot. Widening the slots to accommodate the coil sides will reduce the iron area available to the useful flux, and deepening the slots can result in excessive

122

localized saturation at the narrower tooth base. The alternative is a distributed winding that distributes the coils of the winding over several smaller slots which are adjacent to one another.

Windings are also referred to by their principal functions. A winding whose primary function is to handle the converted electrical power is referred to as an armature winding, whereas that which is used to provide the excitation flux is known as a field winding. Sometimes the same winding performs both functions, power carrying and excitation, as in the windings of an ac induction machine.

Let's briefly review the fundamentals which can help us understand not only how most electric machines operate, but also to know why they are built that way. From the fundamentals of electrostatics, we learned that electric charges exert a force on one another. When the charges are stationary, the force may be attributed to the presence of an electric field created by other stationary charges. But when the charges are moving, there is an additional force, besides that exerted by the electrostatic field. This additional force on the moving charge is from the magnetic field of the other moving charges. The movement of charges or current creates a magnetic field. The force on the moving charge can be expressed in terms of the magnetic field. Figure 5.1a shows the force acting on a charge, q, moving in a magnetic field of flux density, $\vec{\mathbf{B}}$. In MKS units, with the unit of charge expressed in Coulomb, its velocity in meters per second, and the field $\vec{\mathbf{B}}$ in Webers per square meter, the force, in Newton, is given by

$$\vec{\mathbf{F}} = q\vec{\mathbf{v}} \times \vec{\mathbf{B}} \qquad N \tag{5.1}$$

The magnitude of the force, F, is $qBv\sin\theta$, where $v\sin\theta$ is the component of the velocity, $\vec{\mathbf{v}}$, that is perpendicular to the field, $\vec{\mathbf{B}}$. A steady current flow, as from a continuous stream of moving charges in a conductor of length L meters exposed to a uniform magnetic field $\vec{\mathbf{B}}$, will result in a force on the conductor given by

$$\vec{\mathbf{F}} = L\vec{\mathbf{i}} \times \vec{\mathbf{B}} \qquad N \tag{5.2}$$

Such a developed force can be harnessed to perform useful work, as in the situation shown in Fig. 5.1c, where a counter-clockwise torque is produced by forces acting on the conductor segments, ab and cd, which turn the rotor holding the conductors. Assuming that the two conductor segments through which the common current flows are of identical length, the forces on them will be equal and opposite when they are placed under equal but opposite field polarity, that is with $\vec{\mathbf{B}}$ radially inwards on ab and radially outwards on cd. With a symmetrical configuration, the two forces combine to produce a torque with no net translational component. This simple setup has many of the basic features of a working motor: a winding to carry the current to interact with the magnetic field to produce a torque and a movable element so that the developed torque can produce useful work, as force alone without movement produces no work.

As the movable element rotates under the influence of the torque produced, an induced voltage is produced in the moving conductors. If the velocity of the conductors is $\vec{\mathbf{v}}$ meters per second relative to the magnetic field, $\vec{\mathbf{B}}$, the induced voltage in each coil side is given by

$$\vec{\mathbf{e}} = L\vec{\mathbf{v}} \times \vec{\mathbf{B}} \qquad V \tag{5.3}$$

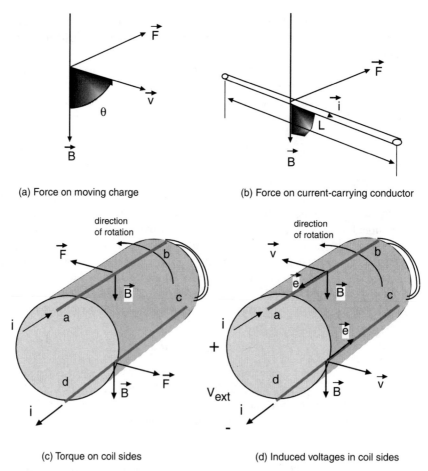

(a) Force on moving charge   (b) Force on current-carrying conductor

(c) Torque on coil sides   (d) Induced voltages in coil sides

Figure 5.1   Forces and torque in motoring.

With $\vec{B}$ radially outward and $\vec{v}$ as shown, Fig 5.1d shows that the induced voltages on the two conductor segments are of a polarity that will oppose the current, $\vec{i}$, shown in Fig. 5.1c. In other words, if torque is to be developed continuously in the direction shown in Fig. 5.1c, to keep the conductors rotating counter-clockwise, the external voltage source, V_{ext}, driving the current, $\vec{i}$, will have to force the current flow against the induced voltages in ab and cd.

The situation that we have just described is commonly referred to as the motoring mode of operation, where an external voltage source must overcome the internal induced voltages to circulate the current that interacts with the magnetic field to produce a useful torque. In motoring, energy from the external voltage source is being converted by the machine to mechanical work.

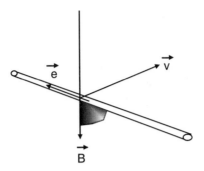

(a) Induced voltage in a moving conductor

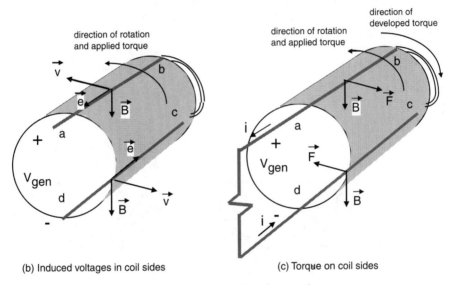

(b) Induced voltages in coil sides   (c) Torque on coil sides

Figure 5.2   Induced voltages in generating.

In the so-called generating mode of operation, an external torque is applied to the rotor carrying the conductors, rotating them relative to the magnetic field to produce an induced voltage. As placed in Fig. 5.2, the conductors are under opposite field polarity so that their induced voltages will reinforce one another, or become additive, around the loop. When no current flows in these conductors, there will not be a force acting on them from the magnetic field. The externally applied torque need only overcome whatever friction and windage losses to maintain motion at a certain speed. However, if the induced voltages were to be used to supply electrical power to some electrical circuit connected externally, the flow of current which accompanied the supply of electrical power would again interact with the magnetic field. This time, the forces on the conductors will produce a torque that opposes the externally applied torque. Thus, for the machine to rotate at the same speed and continue to generate electrical power, the externally applied torque would have to overcome the developed torque and also the friction and windage torques.

The above description, of course, is a rather simple viewpoint. When it comes to designing a practical machine, we will have to handle numerous engineering details and address fundamental questions such as how do we produce the **B** field so that the machine can generate steady power or produce a steady torque.

5.2 ELECTRICAL RADIANS AND SYNCHRONOUS SPEED

Around the airgaps of most ac machines, the magnetic field alternates in polarity. Figure 5.3 shows the field distributions of two- and four-pole ac machines with round and salient rotors. For a P-pole machine, the spatial flux density may be viewed as having $P/2$ pairs of north-south poles around the airgap. A winding linked to such a spatial flux density and moving at a uniform relative speed with respect to the flux distribution will in one rotation of the rotor have $P/2$ cycles of induced voltage. Thus for a P-pole machine, the relation between the angular measures of the electrical and mechanical variables is

$$\theta_e = \frac{P}{2}\theta_m \qquad elect.\, rad \qquad\qquad (5.4)$$

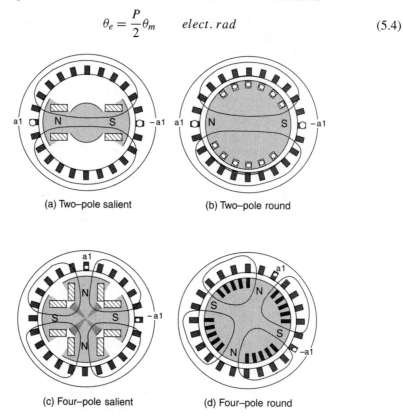

(a) Two–pole salient   (b) Two–pole round

(c) Four–pole salient   (d) Four–pole round

Figure 5.3   Flux distribution and induced voltage.

Differentiating both sides of Eq. 5.4 with respect to time, we obtain

$$\omega_e (electrical\ rad/sec) = \frac{P}{2}\omega_m (mechanical\ rad/sec) \qquad (5.5)$$

The subscripts, e and m, have been added to distinguish between the electrical and mechanical measures.

The relation between the frequency, f, of the induced voltage in cycles per second and the mechanical speed, N, in revolutions per minute, can be shown to be

$$f = \frac{P}{2}\frac{N}{60} \qquad Hz \qquad (5.6)$$

where f is $(\omega_e/2\pi)$ cycles per second and N is $(\omega_m/2\pi) \times 60$ revolutions per minute. The value of ω_m or N, which satisfy the relationships given in Eqs. 5.5 and 5.6, is known as the synchronous speed of the machine.

5.3 FLUX PER POLE AND INDUCED VOLTAGE

The induced voltage and developed torque of an electric machine are dependent on the amount of flux under each pole. When the flux density distribution in a pole is known, the flux under the pole can be determined by integrating the flux density distribution over the pole area. For example, Fig. 5.4 shows the developed view of a sinusoidally distributed flux density over two poles with $B(\theta_e) = B_{pk}\cos\theta_e$. The flux per pole for such a field distribution is given by

$$\phi_{pole} = \int_{-\pi_e/2}^{\pi_e/2} B_{pk}\cos\theta_e LRd\theta_m \qquad Wb \qquad (5.7)$$

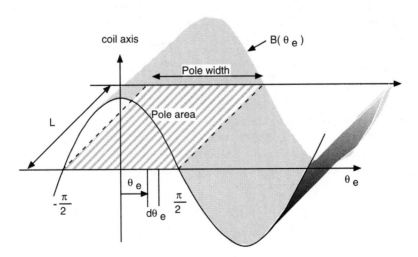

Figure 5.4   Flux per pole.

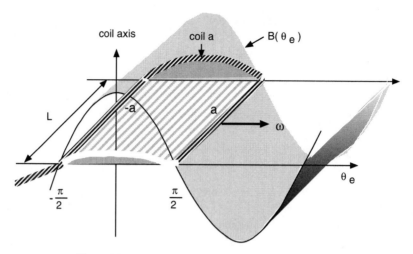

Figure 5.5   Full-pitch coil moving relative to flux density, B.

where L is the effective length of the pole area, R the mean radius of the air-gap annulus, and $-\pi_e/2$ to $\pi_e/2$ is the angular span of a full pole width in electrical radians. Substituting for $d\theta_m$ using

$$d\theta_m = \frac{2}{P}d\theta_e \qquad (5.8)$$

we obtain

$$\phi_{pole} = \frac{2}{P}B_{pk}LR\int_{-\pi_e/2}^{\pi_e/2}\cos\theta_e\,d\theta_e = \frac{4}{P}B_{pk}LR \qquad (5.9)$$

For a sinusoidal flux distribution, the average value of the flux density, B_{ave}, is $(2/\pi)B_{pk}$. The flux per pole can also be obtained from the average flux density under a pole area times the pole area, that is

$$\phi_{pole} = B_{ave}\underbrace{\left(\frac{2\pi LR}{P}\right)}_{(pole\ area)} = \frac{4}{P}B_{pk}LR \qquad (5.10)$$

Figure 5.5 shows a full-pitched coil with N turns that is moving laterally with respect to the sinusoidally distributed flux density at a constant relative velocity of $\omega_e = d\theta_e/dt$. The flux linked by the coil varies with its position with respect to the sinusoidal flux density. The maximum flux linked by the full-pitched coil of N turns is $N\phi_{pole}$. If we let time t be zero when the coil's axis coincides with the peak of wave B, the flux linked by the coil at any time t may be expressed as

$$\lambda(t) = N\phi_{pole}\cos\omega_e t \qquad Wb.\ turn \qquad (5.11)$$

The induced voltage in the full-pitch coil is given by

$$e = \frac{d\lambda(t)}{dt} = N\frac{d\phi_{pole}}{dt}\cos\omega_e t - \omega_e N\phi_{pole}\sin\omega_e t \qquad V \qquad (5.12)$$

The first term of Eq. 5.12 is known as the transformer voltage term and the second term is the speed voltage.

In most rotating ac machines, the flux density distribution, $B(\theta_e)$, and the flux per pole, ϕ_{pole}, are usually constant under steady-state operation. Thus in steady-state, the induced voltage is from the speed voltage, or $e = -\omega_e N \phi_{pole} \sin \omega_e t$.

5.3.1 RMS Value of the Induced Voltage

The root-mean-square (rms) value of the above sinusoidally varying speed voltage component can be expressed as

$$E_{rms} = \frac{\omega_e N \phi_{pole}}{\sqrt{2}} = 4.44 f N \phi_{pole} \qquad V \qquad (5.13)$$

In ac machines, especially those of higher ratings, a phase winding is formed by interconnecting identical coils whose coil sides are positioned in two or more adjacent slots. These distributed windings may also have coils that are short pitched, that is, coils which span less than $180°$ of the flux wave. The effects of both distribution and short pitching of the coils can be separately accounted for by a distribution and pitch factor, k_d and k_p, respectively. The combined effect of both distribution and short pitching on the induced voltage is usually represented by the winding factor, k_w. Thus, the rms value of the induced voltage of a winding with a distribution factor, k_d, and pitch factor, k_p, can be written as

$$E_{rms} = 4.44 f k_w N \phi_{pole} \qquad (5.14)$$

where

$$k_w = k_p k_d \qquad (5.15)$$

5.3.2 Distribution Factor

The phase windings of an ac machine may consist of series and/or parallel combinations of more than one coil under a different pole-pair. Within each pole-pair region, the coils of a distributed winding are distributed over many pairs of slots. As shown in Fig. 5.6 the voltages induced in the component coils forming the phase winding occupying adjacent slots will be phase displaced from one another by the slot angle separating them, α_s^e, which is the electrical angle subtended by the arc between two adjacent slots.

Defining the distribution factor as the ratio of the resultant voltage with coils distributed to the resultant voltage if those same coils were concentrated in one location, we have

$$k_d \triangleq \frac{Resultant\ voltage\ of\ coils\ under\ a\ pole\text{-}pair\ |E_{pole}|}{Arithmetic\ sum\ of\ the\ coil\ voltages\ \sum_i |E_{ci}|} \qquad (5.16)$$

For the case of a phase winding having q coils/phase/pole, $|E_{pole}| = 2R_E \sin(q\alpha_s^e/2)$ and $|E_{ci}| = 2R_E \sin(\alpha_s^e/2)$, its distribution factor is

$$k_d = \frac{\sin(q\alpha_s^e/2)}{q \sin(\alpha_s^e/2)} \qquad (5.17)$$

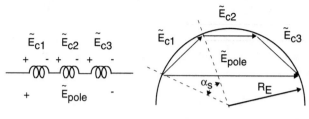

Coils of a pole   Result of coil's phasor voltages

Figure 5.6 Resultant voltage of series full-pitch coils under each pole.

5.3.3 Pitch Factor

Short-pitching, or chording, refers to using coils with width less than one pole-pitch such as that shown in Fig. 5.7. Chording is often used in machines with fractional-slot windings (non-integral slots per pole or slots per pole per phase) in a double-layer winding arrangement. It allows the use of a total number of armature slots that is not necessarily a multiple of the number of poles, thus enabling a finite set of stampings with a certain number of slots to be used for a range of machines running at different speeds. Chording may also be used to reduce or suppress certain harmonics in the phase emfs. Chording results in shorter

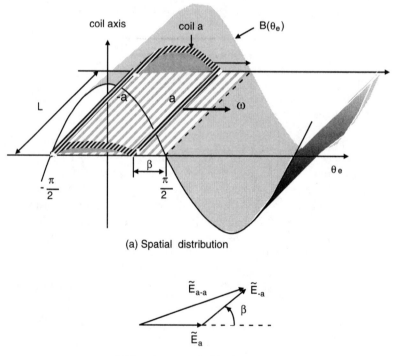

(a) Spatial distribution

(b) Resultant of coil-side phasor voltages

Figure 5.7 Resultant of coil-side voltages of a fractional-pitch coil.

end connections, but the resultant fundamental phase emf is reduced with chording. As such, coil span is rarely made less than 2/3 pole-pitch because the additional turns necessary will offset whatever savings gained in the end connections.

The effect of chording can be accounted for by the pitch factor, k_p, defined as the ratio of the induced voltage in a short-pitch coil to that in a full-pitch coil:

$$k_p = \frac{\text{Resultant voltage in short-pitch coil}}{\text{Arithmetic sum of voltages induced in coil sides}} \qquad (5.18)$$

With sinusoidal induced voltages, each coil voltage is the phasor sum of its two coil-side voltages, thus for coil a, $\tilde{\mathbf{E}}_{ca} = \tilde{\mathbf{E}}_a + \tilde{\mathbf{E}}_{-a}$, where $|\tilde{\mathbf{E}}_a| = |\tilde{\mathbf{E}}_{-a}|$,

$$k_p = \frac{\tilde{\mathbf{E}}_{ca}}{2|\tilde{\mathbf{E}}_a|} = \cos\frac{\theta_e}{2} \qquad (5.19)$$

5.4 SPATIAL MMF DISTRIBUTION OF A WINDING

Let's examine the mmf that is established by a current, i, flowing in a single concentric coil of n_c turns as shown in Fig. 5.8. Ignoring the reluctive drop in iron and applying Ampere's law to a closed path drawn symmetrically around the bottom coil side, we obtain

$$F_1 - F_2 = n_c i \qquad A.turn \qquad (5.20)$$

By symmetry of the chosen path, the mmf across the two gaps in a closed path are equal and opposite, that is, $F_2 = -F_1$. Thus

$$F_1 = \frac{n_c i}{2} \qquad (5.21)$$

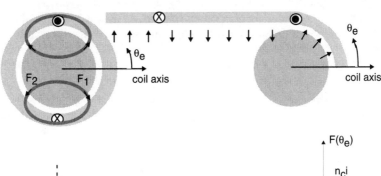

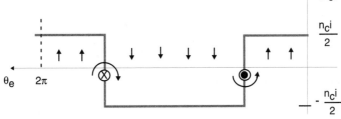

Figure 5.8   mmf distribution of concentrated coil.

A developed view of the airgap mmf distribution can be obtained by slitting a side of the stator and unrolling it as shown in Fig. 5.8. The direction and magnitude of the airgap mmf on either side of the two coil sides can be deduced from the figure and the above relationships between F_1 and F_2. For the single coil of n_c turns, the airgap mmf distribution is a quasi-square of amplitude $n_c i/2$.

The fundamental component of the quasi-square mmf distribution of magnitude $n_c i/2$ is given by

$$F_{a1} = \frac{4}{\pi}\frac{n_c}{2}i\cos\theta_e \tag{5.22}$$

The expression for the corresponding fundamental component of the airgap flux density in a uniform airgap machine is

$$B_1 = \mu_0\frac{F_{a1}}{g} = \frac{4}{\pi}\frac{\mu_0}{g}\frac{n_c}{2}i\cos\theta_e \qquad Wb/m^2 \tag{5.23}$$

where the airgap length, g, is assumed to be the same all around.

5.4.1 The Effect of Distributing the Phase Coils

In practice, the coils of the winding are distributed over several slots on the inner bore of the stator stampings or on the outer circumference of the rotor stampings. Because of the discrete location of the coil sides, the spatial mmf distribution of such a distributed winding consists of steps at every location of the coils sides. For instance, Fig. 5.9 shows the distribution of coil sides in a two-layer arrangement of a set of three-phase windings in a two-pole stator. As shown, each of the abc phase windings are formed by connecting four coils identified in the figure by numbers 1 through 4. The two coil sides of a coil are numbered with the same phase and coil number, distinguished only by an extra minus sign for one of the coil sides. For a full-pitch winding in a two-pole machine, coils 1 and 2 of the a-phase winding have coil sides on the top and bottom slots vertically opposite one another, that is, their coil sides are one pole-pitch apart. Assuming that each of the four coils has n_c turns, the sum of the fundamental mmf components produced by coils $a1$ and $a2$ is given by

$$F_{a1-a1} + F_{a2-a2} = \frac{4}{\pi}n_c i\cos\left(\theta_e + \frac{\alpha_s}{2}\right) \tag{5.24}$$

where θ_e is the angle measured from the a-phase winding axis and α_s is the angle between the center lines of adjacent slots, both measured in electrical radians. Similarly, the sum of the fundamental mmf components produced by coils $a3$ and $a4$ is

$$F_{a3-a3} + F_{a4-a4} = \frac{4}{\pi}n_c i\cos\left(\theta_e - \frac{\alpha_s}{2}\right) \tag{5.25}$$

The distribution factor is defined as the ratio of the phasor resultant to the arithmetic sum of their magnitudes. Using the cosine rule for the resultant, we have

$$Distribution\ factor\ k_d = \frac{1}{2}\sqrt{\cos^2\left(\theta_e + \frac{\alpha_s}{2}\right) + \cos^2\left(\theta_e - \frac{\alpha_s}{2}\right) + 2\cos\left(\theta_e + \frac{\alpha_s}{2}\right)\cos\left(\theta_e - \frac{\alpha_s}{2}\right)}$$

$$= \cos\frac{\alpha_s}{2} \tag{5.26}$$

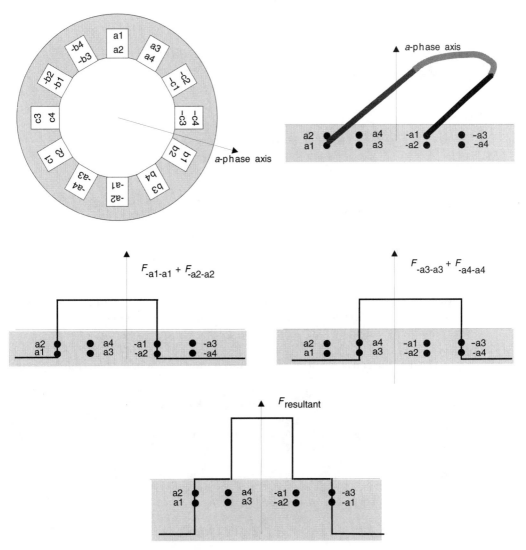

Figure 5.9   The resultant mmf of a distributed winding.

5.4.2 The Effect of Short-pitching

The span of a full-pitch coil is 180 electrical degrees; that of a short-pitch coil is less. For example, Fig. 5.10 shows the layout of windings which are short-pitch by one slot angle, the stator has the same number of slots as in the full-pitch winding case shown in Fig. 5.9. If we were to compare the position of the inner-layer coil sides of the full- to the short-pitch arrangement, we would see that the short-pitch arrangement could be obtained from the full-pitch case by rotating the inner layer of coil sides by one slot angle in the direction which shortens the span of each coil, in this case counter-clockwise. The effect of short-

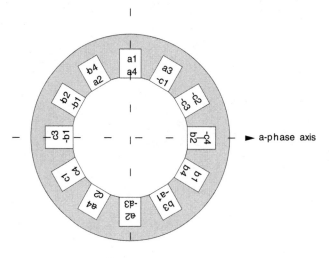

Figure 5.10 A fractional pitch winding arrangement.

pitching the phase coils can be conveniently accounted for by considering the space mmf of these fictitious full-pitch coils. For example, in Fig. 5.10, we can imagine the equivalent space mmf to be set up by four fictitious full-pitch coils: $(a_1, -a_3)$, $(a_4, -a_2)$, $(a_2, -a_1)$, and $(a_3, -a_4)$.

The fundamental component of the resultant from these four fictitious full-pitch coils is

$$
\begin{aligned}
F_a &= F_{a_1,-a_3} + F_{a_4,-a_2} + F_{a_3,-a_4} + F_{a_2,-a_1} \\
&= \frac{4}{\pi} n_c i \left\{ \cos\theta_e + \frac{1}{2}\cos(\theta_e - \alpha_s) + \frac{1}{2}\cos(\theta_e + \alpha_s) \right\} \\
&= \frac{4}{\pi} n_c i \cos\theta_e (1 + \cos\alpha_s) = \frac{4}{\pi} 2 n_c i \cos\theta_e \cos^2 \frac{\alpha_s}{2}
\end{aligned}
\tag{5.27}
$$

By comparing the above expression of F_a to the form

$$
F_a = \frac{4}{\pi} 2 n_c i k_d k_p \cos\theta_e
\tag{5.28}
$$

and allowing for the distribution of the four full-pitch coils by a k_d of $\cos\alpha_s/2$ found in Eq. 5.26, the factor attributable to short-pitching by one slot angle may be identified as

$$
k_p = \cos\frac{\alpha_s}{2}
\tag{5.29}
$$

In general, the expression for the fundamental mmf of a distributed winding with a winding factor of k_w and a total of n_{pole} turns over a two-pole region may be expressed as

$$
F_{a1} = \frac{4}{\pi} \frac{n_{pole}}{2} k_w i \cos\theta_e
\tag{5.30}
$$

Assuming that the total phase turns of N_{ph} in the P-pole machine are divided equally among $P/2$ pole-pair regions, the number of turns per pole-pair is

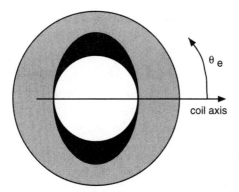

Figure 5.11 Sinusoidal current sheet.

$$n_{pole} = \frac{N_{ph}}{(P/2)} \quad or \quad \frac{n_{pole}}{2} = \frac{N_{ph}}{P} \tag{5.31}$$

In terms of N_{ph}, the fundamental mmf is

$$F_{a1} = \frac{4}{\pi}\left(\frac{N_{ph}k_w}{P}\right)i\cos\theta_e \tag{5.32}$$

By comparing the above expression with Eq. 5.22, which gives the fundamental mmf component produced by a concentric winding, we can deduce that the same fundamental mmf component can be produced by a winding with N_{eff} turns of full-pitch concentric coils per pole-pair, where

$$N_{eff} = \frac{2N_{ph}k_w}{P} \tag{5.33}$$

For analytical purposes, the same fundamental mmf can be produced by an equivalent continuous current sheet, shown in Fig. 5.11, that is produced by the same current i flowing in a hypothetical winding of sinusoidally varying turn density,

$$n(\theta_e) = \frac{N_{sine}}{2}\sin\theta_e \tag{5.34}$$

Such a current sheet has a total number of turns per pole pair of N_{sine}. It produces a sinusoidally distributed spatial mmf that is given by

$$F_a = \frac{1}{2}\int_{\theta_e}^{\theta_e+\pi} n(\theta_e)i\,d\theta_e = \frac{N_{sine}i}{2}\cos\theta_e \tag{5.35}$$

By comparing the above mmf expression with that of Eq. 5.32, and using the relationship given in Eq. 5.33, we can obtain the following relationships between the different turns mentioned so far:

$$\frac{N_{sine}}{2} = \frac{4}{\pi}\frac{N_{ph}k_w}{P} = \frac{4}{\pi}\frac{N_{eff}}{2} \tag{5.36}$$

The expression of the fundamental flux density, assuming a uniform airgap of g, is

$$B_{a1} = \frac{\mu_0 F_{a1}}{g} = \frac{4}{\pi}\mu_0\frac{N_{ph}k_w}{Pg}i\cos\theta_e \tag{5.37}$$

The expression for the flux per pole due to the fundamental flux density component is

$$\phi_{pole} = L \int B_{a1} R d\theta_m = LR \int_{-\pi/2}^{\pi/2} B_{a1}(\theta_e) d\theta_e \left(\frac{2}{P}\right)$$

$$= \frac{4}{\pi} \frac{\mu_0 LR}{g} \underbrace{\frac{2N_{ph}k_w}{P}}_{N_{eff}} i \left(\frac{2}{P}\right) = \frac{4}{\pi} \frac{\mu_0 LR}{g} \left(\frac{2}{P}\right) N_{eff} i \tag{5.38}$$

5.5 WINDING INDUCTANCES

In this section, we shall derive the expressions of the self- and mutual winding inductances for the elementary machine shown in Fig. 5.12. The self-inductance of the winding, s, on the stator of the elementary machine with N_{effs} turns per pole pair linking ϕ_{pole}, not including leakages fluxes, is given by

$$L_{ss} = \frac{(P/2)N_{effs}\, \phi_{pole}}{i} = \frac{4}{\pi} \frac{\mu_0}{g} N_{effs}^2 LR \tag{5.39}$$

The flux linked by the rotor winding, r of N_{effr} effective turns, due to the field established by the stator winding, s of N_{effs} effective turns, is

$$\lambda_{rs} = \left(\frac{P}{2}\right) N_{effr} \left\{ L \int_{-\pi/2+\alpha}^{\pi/2+\alpha} \frac{4}{\pi} \frac{\mu_0}{g} \frac{N_{effs}}{2} i_s \cos\theta\; R\; d\theta_e \left(\frac{2}{P}\right) \right\}$$

$$\tag{5.40}$$

$$= \frac{4}{\pi} \frac{\mu_0}{g} N_{effr} N_{effs} LRi_s \cos\alpha$$

$$L_{rs} = \frac{\lambda_{rs}}{i_s} = \frac{4}{\pi} \frac{\mu_0}{g} N_{effr} N_{effs}\; LR\cos\alpha \tag{5.41}$$

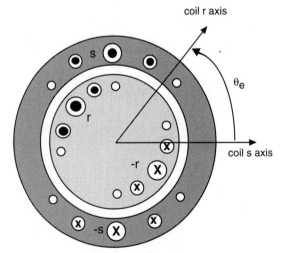

Figure 5.12 Mutual coupling between stator and rotor windings.

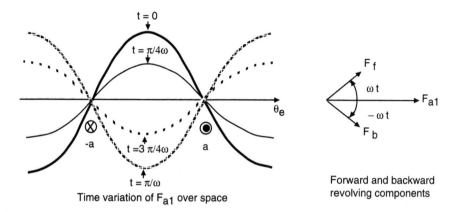

Figure 5.13   mmf of a single-phase coil.

It can be readily shown that $L_{sr} = L_{rs}$ and that

$$L_{rr} = \frac{4}{\pi} \frac{\mu_0}{g} N_{effr}^2 \, LR \tag{5.42}$$

5.6 ROTATING FIELDS

Using Eq. 5.32, the fundamental component of the space mmf that is established by a single-phase winding with a current $i = I_a \cos \omega t$ may be written as

$$F_{a1} = F_{m1} \cos \omega t \cos \theta_a \tag{5.43}$$

where the peak value of the fundamental mmf, F_{m1}, is equal to $(4/\pi)k_w(N_{ph}/P)I_a$, and θ_a is the electrical angle measured in the counter-clockwise direction from the winding axis. Replacing the product of the two cosines with a sum and difference, we obtain

$$F_{a1} = \underbrace{\frac{1}{2}F_{m1}\cos(\theta_a - \omega t)}_{F_f} + \underbrace{\frac{1}{2}F_{m1}\cos(\theta_a + \omega t)}_{F_b} \tag{5.44}$$

The fundamental component of the space mmf may be interpreted as a pulsating standing wave by the form of Eq. 5.43, or as two counter-revolving mmf waves, each of half the amplitude of the resultant's by Eq. 5.44. Both interpretations are illustrated in Fig. 5.13. The forward component, F_f, rotates counter-clockwise in the direction of increasing θ_a at a speed ω, whereas the backward component, F_b, rotates in the opposite direction at the same speed.

In a three-phase machine, the axes of the three-phase windings are spaced $2\pi/3$ electrical radians from one another, that is $\theta_b = \theta_a - 120°$ and $\theta_c = \theta_a + 120°$. Assuming balanced operation, that is the phase currents are of the same magnitude and 120° phase apart,

$$i_a = I_m \cos \omega t$$

$$i_b = I_m \cos(\omega t - 120°)$$ $\qquad$ (5.45)

$$i_c = I_m \cos(\omega t + 120°)$$

Using Eq. 5.44, the fundamental airgap mmfs of the three phases, expressed in terms of θ_a, are

$$F_{a1} = \frac{1}{2} F_{m1} \{\cos(\theta_a - \omega t) + \cos(\theta_a + \omega t)\}$$

$$F_{b1} = \frac{1}{2} F_{m1} \{\cos(\theta_a - \omega t) + \cos(\theta_a + \omega t - 240°)\} \qquad (5.46)$$

$$F_{c1} = \frac{1}{2} F_{m1} \{\cos(\theta_a - \omega t) + \cos(\theta_a + \omega t + 240°)\}$$

The second terms of these mmfs form a balanced set of a zero resultant. The resultant from a superposition of these three winding mmfs is yielded by the first terms, that is

$$F_{a1} + F_{b1} + F_{c1} = \underbrace{\frac{3}{2} F_{m1}}_{F_s} \cos(\theta_a - \omega t) \qquad (5.47)$$

Thus, the resultant airgap mmf is a constant amplitude, sinusoidal-shaped, revolving wave whose peak coincides with the magnetic axis of the a-phase winding at $t = 0$ and rotates with a speed of ω in a direction corresponding to the sequence of the peaking of the phase currents.

5.7 DEVELOPED TORQUE OF A UNIFORM AIRGAP MACHINE

Using the basic principle of energy conversion, the torque developed by a machine can be obtained by considering the change in coenergy, W'_{fld}, of the system produced by a small change in rotor position, θ_m, when the currents are held constant.

$$T = \frac{\partial W'_{fld}}{\partial \theta_m} \Big|_{i \ held \ constant} \qquad (5.48)$$

The coenergy of the magnetic system is defined as the complement of the field energy, W'_{fld}, that is

$$W'_{fld} = \lambda i - W_{fld} \qquad (5.49)$$

Depending on the problem at hand, one form of the torque expression may be preferred over another. Here, we shall examine the expressions of the developed torque for the elementary two-pole machine with one stator and one rotor winding shown in Fig. 5.12. We will first use the coupled-circuit, and then we will use the field approach.

   Coupled-circuit approach.   If we ignore magnetic nonlinearity, the coenergy, W'_{fld} is equal to the stored energy in the field. In the circuit approach, the stored energy of the two coupled windings in the idealized machine of Fig. 5.14a is given by

$$W_{fld} = \frac{1}{2}L_{ss}i_s^2 + \frac{1}{2}L_{rr}i_r^2 + L_{sr}i_s i_r \cos\delta_{sr} \qquad J \qquad (5.50)$$

where L_{ss} and L_{rr} are the self-inductances of the stator and rotor windings, respectively, L_{sr} is the maximum value of mutual inductance between the stator and rotor windings, and δ_{sr} is the angle of inclination between the two windings in electrical radians. Using Eq. 5.48, the torque developed is given by

$$T = \frac{\partial W'_{fld}}{\partial\delta_{srm}} = \frac{\partial' W_{fld}(\delta_{sr}, i_s, i_r)}{\partial\delta_{sr}} \frac{\partial\delta_{sr}}{\partial\delta_{srm}}; \qquad \delta_{sr} = \frac{P}{2}\delta_{srm}$$

$$= -\frac{P}{2}L_{sr}i_s i_r \sin\delta_{sr} \qquad N.m \qquad (5.51)$$

The negative sign in Eq. 5.51 indicates that the developed torque is in the direction opposite to that of increasing δ_{sr}.

   Magnetic field approach.   In the magnetic field approach, the stored energy can be obtained by multiplying the volume of the space containing the field by the energy

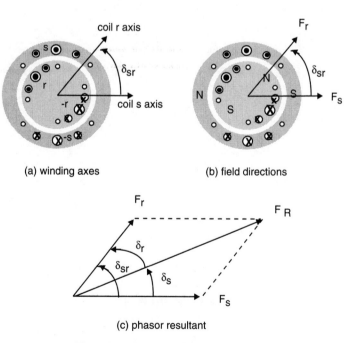

(a) winding axes   (b) field directions

(c) phasor resultant

Figure 5.14   Field components and resultant.

density. If F_s and F_r are the peak values of the fundamental mmf components in the airgap that are produced by currents of the stator and rotor windings, respectively, and if the electrical phase angle between them is δ_{sr}, the peak value of the resultant mmf F_R as given by the cosine rule is

$$F_R^2 = F_s^2 + F_r^2 + 2F_s F_r \cos\delta_{sr} \tag{5.52}$$

The peak and rms values of the resultant field intensity in the airgap are

$$H_{peak} = \frac{F_R}{g}, \quad H_{rms} = \frac{H_{peak}}{\sqrt{2}} \qquad A/m \tag{5.53}$$

The average energy density is equal to $(\mu_0 H_{rms}^2)/2$ or $\mu_0 F_R^2/4g^2$. Ignoring the reluctive drops in the iron, the stored energy is mainly from the airgap annulus. For an annulus of length L, mean diameter D, and radial thickness g, the stored energy is

$$W_{fld} = \frac{\mu_0}{4}\frac{F_R^2}{g^2}(\pi DLg) \qquad J$$
$$= \frac{\mu_0 \pi DL}{4g}(F_s^2 + F_r^2 + 2F_s F_r \cos\delta_{sr}) \tag{5.54}$$

With the nonlinearity of the magnetic material disregarded, W_{fld}' will be equal in value to W_{fld}. Since the above expression for W_{fld} is already in terms of the mmf and rotor angle, the independent variables of W_{fld}', the torque expression may again be obtained using Eq. 5.48 with W_{fld}' equal to the right side of Eq. 5.54, that is

$$T = \frac{\partial W_{fld}}{\partial \delta_{srm}} = -\frac{P}{2}\frac{\mu_0}{2}\frac{\pi DL}{g}F_s F_r \sin\delta_{sr} \qquad N.m \tag{5.55}$$

The torque can also be expressed in other forms using the following relationships:

$$F_s \sin\delta_{sr} = F_R \sin\delta_r$$
$$F_r \sin\delta_{sr} = F_R \sin\delta_s$$
$$B_{pk} = \frac{\mu_0 F_R}{g}, \tag{5.56}$$
$$\phi = B_{ave}(pole\ area) = \left(\frac{2}{\pi}B_{pk}\right)\left(\frac{\pi DL}{P}\right)$$

For example, the developed torque can be expressed in terms of the peak value of the flux density in the airgap, that is

$$T = -\frac{P}{2}\frac{\pi DL}{2}B_{pk}F_r \sin\delta_r \qquad N.m \tag{5.57}$$

or, it can be written in terms of the flux per pole:

$$T = -\frac{\pi}{2}\left(\frac{P}{2}\right)^2 \phi_{pole}F_r \sin\delta_r \qquad N.m \tag{5.58}$$

5.8 THREE-PHASE TRANSFORMATIONS

In the study of power systems, mathematical transformations are often used to decouple variables, to facilitate the solution of difficult equations with time-varying coefficients, or to refer all variables to a common reference frame. For example, the method of symmetrical components developed by Fortescue uses a complex transformation to decouple the *abc* phase variables:

$$[f_{012}] = [T_{012}][f_{abc}] \tag{5.59}$$

Variable f in Eq. 5.59 may be the currents, voltages, or fluxes, and the transformation $[T_{012}]$ is given by

$$[\mathbf{T_{012}}] = \frac{1}{3} \begin{bmatrix} 1 & 1 & 1 \\ 1 & \mathbf{a} & \mathbf{a}^2 \\ 1 & \mathbf{a}^2 & \mathbf{a} \end{bmatrix} \tag{5.60}$$

where $\mathbf{a} = e^{j\frac{2\pi}{3}}$. Its inverse is given by

$$[\mathbf{T_{012}}]^{-1} = \begin{bmatrix} 1 & 1 & 1 \\ 1 & \mathbf{a}^2 & \mathbf{a} \\ 1 & \mathbf{a} & \mathbf{a}^2 \end{bmatrix} \tag{5.61}$$

The symmetrical component transformation is equally applicable to steady-state vectors or instantaneous quantities.

Another commonly-used transformation is the polyphase to orthogonal two-phase (or two-axis) transformation. For the *n*-phase to two-phase case, it can be expressed in the form:

$$[f_{xy}] = [T(\theta)][f_{1\,2\,3\ldots n}] \tag{5.62}$$

where

$$[\mathbf{T}(\theta)] = \sqrt{\frac{2}{n}} \begin{bmatrix} \cos\frac{P}{2}\theta & \cos\left(\frac{P}{2}\theta - \alpha\right) \ldots \cos\left(\frac{P}{2}\theta - (n-1)\alpha\right) \\ \sin\frac{P}{2}\theta & \sin\left(\frac{P}{2}\theta - \alpha\right) \ldots \sin\left(\frac{P}{2}\theta - (n-1)\alpha\right) \end{bmatrix} \tag{5.63}$$

and α is the electrical angle between adjacent magnetic axes of the uniformly distributed *n*-phase windings. The coefficient, $\sqrt{2/n}$, is introduced to make the transformation power invariant.

Important subsets of the general *n*-phase to two-phase transformation, though not necessarily power-invariant, are briefly discussed in the following subsections.

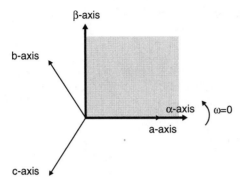

Figure 5.15 Relationship between the $\alpha\beta$ and abc quantities.

5.8.1 Clarke'sTransformation

The stationary two-phase variables of the Clarke's transformation are denoted as α and β. As shown in Fig. 5.15, the α-axis coincides with the phase-a axis and the β-axis lags the α-axis by $\pi/2$. Just so that the transformation is bidirectional, a third variable known as the zero-sequence component is added:

$$[\mathbf{f}_{\alpha\beta 0}] = [\mathbf{T}_{\alpha\beta 0}][\mathbf{f}_{abc}] \tag{5.64}$$

where the transformation matrix, $[\mathbf{T}_{\alpha\beta 0}]$, is given by

$$[\mathbf{T}_{\alpha\beta 0}] = \frac{2}{3}\begin{bmatrix} 1 & -\frac{1}{2} & -\frac{1}{2} \\ 0 & \sqrt{\frac{3}{2}} & -\sqrt{\frac{3}{2}} \\ \frac{1}{2} & \frac{1}{2} & \frac{1}{2} \end{bmatrix} \tag{5.65}$$

The inverse transformation is

$$[\mathbf{T}_{\alpha\beta 0}]^{-1} = \begin{bmatrix} 1 & 0 & 1 \\ -\frac{1}{2} & \sqrt{\frac{3}{2}} & 1 \\ -\frac{1}{2} & -\sqrt{\frac{3}{2}} & 1 \end{bmatrix} \tag{5.66}$$

5.8.2 Park'sTransformation

The Park's transformation is a well-known three-phase to two-phase transformation in synchronous machine analysis. The transformation equation is of the form:

$$[\mathbf{f}_{dq0}] = [\mathbf{T}_{dq0}(\theta_d)][\mathbf{f}_{abc}] \tag{5.67}$$

where the $dq0$ transformation matrix is defined as

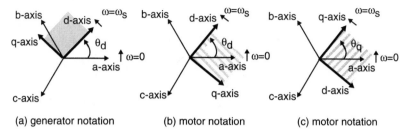

Figure 5.16   Relationship between the dq and abc quantities.

$$[\mathbf{T}_{dq0}(\theta_d)] = \frac{2}{3} \begin{bmatrix} \cos\theta_d & \cos\left(\theta_d - \dfrac{2\pi}{3}\right) & \cos\left(\theta_d + \dfrac{2\pi}{3}\right) \\ -\sin\theta_d & -\sin\left(\theta_d - \dfrac{2\pi}{3}\right) & -\sin\left(\theta_d + \dfrac{2\pi}{3}\right) \\ \dfrac{1}{2} & \dfrac{1}{2} & \dfrac{1}{2} \end{bmatrix} \tag{5.68}$$

and its inverse is given by

$$[\mathbf{T}_{dq0}(\theta_d)]^{-1} = \begin{bmatrix} \cos\theta_d & -\sin\theta_d & 1 \\ \cos\left(\theta_d - \dfrac{2\pi}{3}\right) & -\sin\left(\theta_d - \dfrac{2\pi}{3}\right) & 1 \\ \cos\left(\theta_d + \dfrac{2\pi}{3}\right) & -\sin\left(\theta_d + \dfrac{2\pi}{3}\right) & 1 \end{bmatrix} \tag{5.69}$$

Park's transformation is used to transform the stator quantities of a synchronous machine onto a dq reference frame that is fixed to the rotor, with the positive d-axis aligned with the magnetic axis of the field winding. The positive q-axis is defined as leading the positive d-axis by $\pi/2$ in the original Park's transformation. The internal voltage, $\omega L_{af} i_f$, is along the positive q-axis.

Some authors define the q-axis as lagging the d-axis by $\pi/2$. Defined in this manner, the q-axis coincides with the induced voltage, which is the negative of the internal voltage. The transformation with the q-axis lagging the d-axis is given by

$$[\mathbf{T}_{dq0}(\theta_d)] = \frac{2}{3} \begin{bmatrix} \cos\theta_d & \cos\left(\theta_d - \dfrac{2\pi}{3}\right) & \cos\left(\theta_d + \dfrac{2\pi}{3}\right) \\ \sin\theta_d & \sin\left(\theta_d - \dfrac{2\pi}{3}\right) & \sin\left(\theta_d + \dfrac{2\pi}{3}\right) \\ \dfrac{1}{2} & \dfrac{1}{2} & \dfrac{1}{2} \end{bmatrix} \tag{5.70}$$

Others use a $qd0$ transformation in which the q-axis leads the d-axis and the transformation is usually expressed in terms of the angle, θ_q, between the q-axis and the a-axis, as shown in Fig. 5.16c:

$$[\mathbf{f}_{qd0}] = [\mathbf{T}_{qd0}(\theta_q)][\mathbf{f}_{abc}] \tag{5.71}$$

where

$$[\mathbf{T}_{qd0}(\theta_q)] = \frac{2}{3}\begin{bmatrix} \cos\theta_q & \cos\left(\theta_q - \dfrac{2\pi}{3}\right) & \cos\left(\theta_q + \dfrac{2\pi}{3}\right) \\ \sin\theta_q & \sin\left(\theta_q - \dfrac{2\pi}{3}\right) & \sin\left(\theta_q + \dfrac{2\pi}{3}\right) \\ \dfrac{1}{2} & \dfrac{1}{2} & \dfrac{1}{2} \end{bmatrix} \tag{5.72}$$

with an inverse given by

$$[\mathbf{T}_{qd0}(\theta_q)]^{-1} = \begin{bmatrix} \cos\theta_q & \sin\theta_q & 1 \\ \cos\left(\theta_q - \dfrac{2\pi}{3}\right) & \sin\left(\theta_q - \dfrac{2\pi}{3}\right) & 1 \\ \cos\left(\theta_q + \dfrac{2\pi}{3}\right) & \sin\left(\theta_q + \dfrac{2\pi}{3}\right) & 1 \end{bmatrix} \tag{5.73}$$

The relation between θ_q and θ_d, which is defined in the original Park's transformation, is

$$\theta_q = \theta_d + \frac{\pi}{2} \tag{5.74}$$

Substituting the above relation into $[\mathbf{T}_{qd0}(\theta_q)]$ and making use of the trigonometric reduction formulas:

$$\cos(\theta_d + \frac{\pi}{2}) = -\sin\theta_d \tag{5.75}$$

$$\sin(\theta_d + \frac{\pi}{2}) = \cos\theta_d \tag{5.76}$$

we can show that the two transformations, $[\mathbf{T}_{qd0}(\theta_q)]$ and $[\mathbf{T}_{qd0}(\theta_d)]$, are basically the same, except for the ordering of the d and q variables.

5.9 $qd0$ TRANSFORMATION APPLIED TO LINE ELEMENTS

In this section, we shall illustrate the use of the $qd0$ transformation by deriving the $qd0$ circuit models of the series RL and shunt capacitance circuits given in Figs. 5.17 and 5.20. These $qd0$ circuit models may be used in a variety of situations. For example, they can be combined to form the equivalent circuit models of three-phase lines represented by π circuit models, or of three-phase RL and RC loads.

5.9.1 $qd0$ Transformation to Series RL

We shall begin by deriving the $qd0$ equations for the three-phase series RL line with a ground return given in Fig. 5.17 in a $qd0$ reference frame that is rotating at an arbitrary speed of ω. The angle, θ_q, is determined by

$$\theta_q(t) = \int_0^t \omega(t)dt + \theta_q(0) \qquad rad \tag{5.77}$$

The sending end voltage with respect to local ground is given by the equation:

$$v_{asgs} = i_a r_a + L_{aa}\frac{di_a}{dt} + L_{ab}\frac{di_b}{dt} + L_{ac}\frac{di_c}{dt} + L_{ag}\frac{di_g}{dt} + v_{argr} + v_{grgs} \tag{5.78}$$

Using the relation $i_g = -(i_a + i_b + i_c)$, the voltage drops across the three phases of the line can be expressed in matrix form as

$$[v_S] - [v_R] = [\mathbf{R}][\mathbf{i}] + p[\mathbf{L}][\mathbf{i}] \tag{5.79}$$

where

$$[v_S] = \begin{bmatrix} v_{asgs} \\ v_{bsgs} \\ v_{csgs} \end{bmatrix} \qquad [v_R] = \begin{bmatrix} v_{argr} \\ v_{brgr} \\ v_{crgr} \end{bmatrix}$$

$$[\mathbf{R}] = \begin{bmatrix} r_a + r_g & r_g & r_g \\ r_g & r_b + r_g & r_g \\ r_g & r_g & r_c + r_g \end{bmatrix}$$

$$[\mathbf{L}] = \begin{bmatrix} L_{aa} + L_{gg} - 2L_{ag} & L_{ab} + L_{gg} - L_{bg} - L_{ag} & L_{ac} + L_{gg} - L_{cg} - L_{ag} \\ L_{ab} + L_{gg} - L_{ag} - L_{bg} & L_{bb} + L_{gg} - 2L_{bg} & L_{bc} + L_{gg} - L_{cg} - L_{bg} \\ L_{ac} + L_{gg} - L_{ag} - L_{cg} & L_{bc} + L_{gg} - L_{bg} - L_{cg} & L_{cc} + L_{gg} - 2L_{cg} \end{bmatrix}$$

The equation of the voltage drop across the ground path is

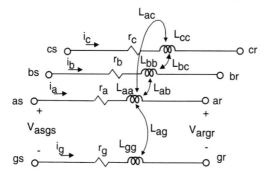

Figure 5.17 Three-phase RL line with ground return.

$$v_{grgs} = -v_{gsgr} = -i_g r_g - L_{gg}\frac{di_g}{dt} - L_{ag}\frac{di_a}{dt} - L_{bg}\frac{di_b}{dt} - L_{cg}\frac{di_c}{dt}$$

$$= r_g(i_a + i_b + i_c) + (L_{gg} - L_{ag})\frac{di_a}{dt} \tag{5.80}$$

$$+ (L_{gg} - L_{bg})\frac{di_b}{dt} + (L_{gg} - L_{cg})\frac{di_c}{dt}$$

for a uniformly transposed line, $r_a = r_b = r_c$, $L_{ab} = L_{bc} = L_{ca}$, and $L_{ag} = L_{bg} = L_{cg}$. Letting $L_s = L_{aa} + L_{gg} - 2L_{ag}$, $L_m = L_{ab} + L_{gg} - 2L_{ag} = L_s - L_{aa} + L_{ab}$, $r_s = r_a + r_g$, and $r_m = r_g$, the resistance and inductance matrices simplify to

$$[\mathbf{R}] = \begin{bmatrix} r_s & r_m & r_m \\ r_m & r_s & r_m \\ r_m & r_m & r_s \end{bmatrix} \quad \text{and} \quad [\mathbf{L}] = \begin{bmatrix} L_s & L_m & L_m \\ L_m & L_s & L_m \\ L_m & L_m & L_s \end{bmatrix}$$

The $qd0$ equations for the uniformly transposed line can be obtained separately by considering the resistive and inductive drops of the a-phase equation. Let's first consider the resistive drop in the a-phase given by

$$r_s i_a + r_m(i_b + i_c) \tag{5.81}$$

Substituting $i_0 = (i_a + i_b + i_c)/3$ to eliminate i_b and i_c, we obtain

$$(r_s - r_m)i_a + 3r_m i_0 \tag{5.82}$$

Expressing i_a in terms of the $qd0$ currents, the resistive drop in the a-phase becomes

$$(r_s - r_m)(i_q \cos\theta_q + i_d \sin\theta_q + i_0) + 3r_m i_0. \tag{5.83}$$

Similarly, for the inductive drop term in the a-phase, we have

$$L_s\frac{di_a}{dt} + L_m\frac{d(i_b + i_c)}{dt} \tag{5.84}$$

Eliminating i_b and i_c,

$$(L_s - L_m)\frac{di_a}{dt} + 3L_m\frac{di_0}{dt} \tag{5.85}$$

Using the $qd0$ transformation of Eq. 5.71 to express i_a in terms of the $qd0$ currents, the inductive drop in the a-phase becomes

$$(L_s - L_m)p(i_q \cos\theta_q + i_d \sin\theta_q + i_0) + 3L_m p i_0 \tag{5.86}$$

Similarly, applying the same $qd0$ transformation to the voltage across the sending and receiving ends of the a-phase on the right-hand side of Eq. 5.79 and equating the coefficients of the $\cos\theta_q$, $\sin\theta_q$, and constant terms, we obtain

$$\Delta v_q = (r_s - r_m)i_q + (L_s - L_m)\frac{di_q}{dt} + (L_s - L_m)i_d\frac{d\theta_q}{dt} \tag{5.87}$$

$$\Delta v_d = (r_s - r_m)i_d + (L_s - L_m)\frac{di_d}{dt} - (L_s - L_m)i_q\frac{d\theta_q}{dt} \tag{5.88}$$

$$\Delta v_0 = (r_s + 2r_m)i_0 + (L_s + 2L_m)\frac{di_0}{dt} \tag{5.89}$$

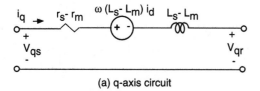

(a) q-axis circuit

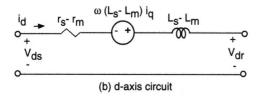

(b) d-axis circuit

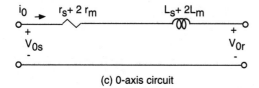

(c) 0-axis circuit

Figure 5.18 Equivalent $qd0$ circuits of series RL of a transposed line.

It is worthwhile to note that the corresponding voltage drop equation of the same line in symmetrical components is

$$\Delta \begin{bmatrix} v_0 \\ v_1 \\ v_2 \end{bmatrix} = \begin{bmatrix} z_s + 2z_m & & \\ & z_s - z_m & \\ & & z_s - z_m \end{bmatrix} \Delta \begin{bmatrix} i_0 \\ i_1 \\ i_2 \end{bmatrix} \tag{5.90}$$

The $qd0$ equations of the voltage drops across the line suggest the equivalent circuit representation shown in Fig. 5.18. In terms of the original phase parameters of the line, we have

$$r_s - r_m = r_a \tag{5.91}$$

$$r_s + 2r_m = r_a + 3r_g \tag{5.92}$$

$$L_s - L_m = L_{aa} - L_{ab} \tag{5.93}$$

$$L_s + 2L_m = L_{aa} + 2L_{ab} + 3(L_{gg} - 2L_{ag}) \tag{5.94}$$

When the mutual inductances between phases and between phase to ground are zero, that is $L_{ab} = L_{ac} = L_{bc} = 0$ and $L_{ag} = L_{bg} = L_{cg} = 0$, then $L_s = L_{aa} + L_{gg}$ and $L_m = L_{gg}$. The $qd0$ equivalent circuit representation of the line simplifies to that given in Fig. 5.19. The equivalent circuits of Fig. 5.19 are often used for shunt RL loads where the mutual inductances are zero. Assuming voltage input, the $qd0$ currents can be obtained by solving the following integral equations on the computer:

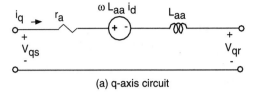

(a) q-axis circuit

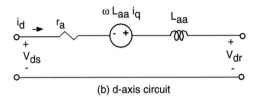

(b) d-axis circuit

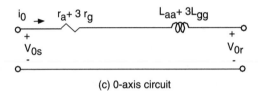

(c) 0-axis circuit

Figure 5.19 Equivalent $qd0$ circuits of series RL line with no mutual coupling between phases.

$$i_q = \frac{1}{L_{aa}} \int (v_{qs} - v_{qr} - \omega L_{aa} i_d - i_q r_a) dt \tag{5.95}$$

$$i_d = \frac{1}{L_{aa}} \int (v_{ds} - v_{dr} + \omega L_{aa} i_q - i_d r_a) dt \tag{5.96}$$

$$i_0 = \frac{1}{L_{aa} + 3 L_{gg}} \int (v_{0s} - v_{0r} - i_0 r_a + 3 r_g) dt \tag{5.97}$$

where $\omega = d\theta_q / dt$.

5.9.2 $qd0$ Transformation to Shunt Capacitances

In this subsection, we shall derive the $qd0$ equations for the voltage drops across the shunt capacitances of the three-phase line shown in Fig. 5.20. Besides the phase to neutral capacitance of the phases, we have also included the mutual capacitances between phases.

Let $C_{ab} = C_{bc} = C_{ca} = C_m$, $C_{an} = C_{bn} = C_{cn}$, and $C_s = C_{an} + 2C_{ab}$. The equation of the a-phase current in Fig. 5.20 may be expressed as

$$i_a = C_{an} \frac{d}{dt} v_{an} + C_{ab} \frac{d}{dt} (v_{an} - v_{bn}) + C_{ac} \frac{d}{dt} (v_{an} - v_{cn}) \tag{5.98}$$

$$i_a = (C_{an} + C_{ab} + C_{ac}) \frac{dv_{an}}{dt} - C_m \frac{dv_{bn}}{dt} - C_m \frac{dv_{cn}}{dt} \tag{5.99}$$

Exchanging the b and c phase voltages with $v_0 = (v_{an} + v_{bn} + v_{cn})/3$ gives

$$i_a = (C_s + C_m) \frac{dv_{an}}{dt} - 3C_m \frac{dv_0}{dt} \tag{5.100}$$

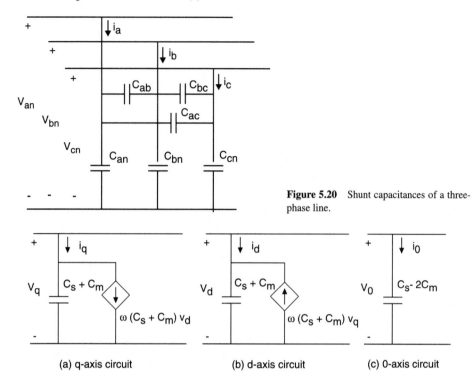

Figure 5.20   Shunt capacitances of a three-phase line.

(a) q-axis circuit   (b) d-axis circuit   (c) 0-axis circuit

Figure 5.21   Equivalent $qd0$ circuit of shunt capacitances of a transposed three-phase line.

Applying the $qd0$ transformation to the current and voltage of the a-phase, we obtain

$$i_q \cos\theta_q + i_d \sin\theta_q + i_0 = (C_s + C_m)\frac{d}{dt}(v_q \cos\theta_q + v_d \sin\theta_q + v_0) - 3C_m \frac{dv_0}{dt} \qquad (5.101)$$

Equating the coefficients of the $\cos\theta_q$, $\sin\theta_q$, and constant terms, respectively, we obtain the following equations for the $qd0$ currents:

$$i_q = (C_s + C_m)\frac{dv_q}{dt} + (C_s + C_m)v_d\omega \qquad (5.102)$$

$$i_d = (C_s + C_m)\frac{dv_d}{dt} - (C_s + C_m)v_q\omega \qquad (5.103)$$

$$i_0 = (C_s - 2C_m)\frac{dv_0}{dt} \qquad (5.104)$$

where $\omega = d\theta_q/dt$. In terms of the original phase parameters,

$$C_s + C_m = C_{an} + 3C_{ab} \qquad (5.105)$$

$$C_s - 2C_m = C_{an} \qquad (5.106)$$

The above set of equations for the $qd0$ currents suggests the equivalent circuits given in Fig. 5.21. The equations in integral form are

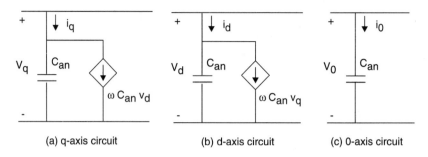

(a) q-axis circuit   (b) d-axis circuit   (c) 0-axis circuit

Figure 5.22   Equivalent $qd0$ circuits of a set of three-phase shunt capacitances.

$$v_q = \frac{1}{C_s + C_m} \int \left\{ i_q - (C_s + C_m)v_d \frac{d\theta_q}{dt} \right\} dt \qquad (5.107)$$

$$v_d = \frac{1}{C_s + C_m} \int \left\{ i_d + (C_s + C_m)v_q \frac{d\theta_q}{dt} \right\} dt \qquad (5.108)$$

$$v_0 = \frac{1}{C_s - 2C_m} \int i_0 dt \qquad (5.109)$$

When $C_m = 0$ and $C_{an} = C_{bn} = C_{cn} = C_s$, the $qd0$ equivalent circuits simplify to those shown in Fig. 5.22.

5.10 SPACE VECTORS AND TRANSFORMATIONS

This section describes the mathematical basis of the commonly used $qd0$ transformations in electric machine analysis and also the idea of space vectors which has become popular in vector control schemes of induction and synchronous machines.

Using Eqs. 5.32 and 5.35, we can express the airgap mmf due to $i_a(t)$ flowing in the a-phase winding alone as

$$F_{a1} = \frac{4}{\pi} \frac{k_w N_{ph}}{P} i_a(t) \cos\theta_a = \frac{N_{sine}}{2} i_a(t) \cos\theta_a \qquad (5.110)$$

F_{a1} has a cosine spatial distribution around the airgap centered about the a-phase winding axis. Its magnitude along the a-phase winding axis is $N_{sine}i_a(t)/2$. As $i_a(t)$ varies in time, the magnitude of the whole distribution of F_{a1} fluctuates along with it. Essentially, F_{a1} is a standing wave having nodes at $\theta_a = \pm\pi/2$. Equation 5.110 may be written in space vector notation as

$$\vec{\mathbf{F}}_{a1} = \frac{N_{sine}}{2} \vec{\mathbf{i}}_a \qquad (5.111)$$

where $\vec{\mathbf{i}}_a$ denotes a current space vector whose magnitude may be time-varying, that is $i_a(t)$. The space vector, $\vec{\mathbf{i}}_a$, has a cosine spatial distribution about the a-phase winding axis or along the direction of $\theta_a = 0$. It therefore may be interpreted as a vector of magnitude proportional to that of $i_a(t)$ in the direction of $\theta_a = 0$. The mmf space vector, $\vec{\mathbf{F}}_{a1}$, may be interpreted likewise.

Assuming linearity, the resultant airgap mmf established by the currents flowing in all three-phase windings whose axes are along the directions, $\theta_a = 0$, $\theta_b = 0$, and $\theta_c = 0$, is

$$\vec{\mathbf{F}}_s = \vec{\mathbf{F}}_{a1} + \vec{\mathbf{F}}_{b1} + \vec{\mathbf{F}}_{c1} \tag{5.112}$$

Using Eqs. 5.110 and 5.111, the above expression can be written as

$$\vec{\mathbf{F}}_s = \frac{N_{sine}}{2}(\vec{\mathbf{i}}_a + \vec{\mathbf{i}}_b + \vec{\mathbf{i}}_c) = \frac{N_{sine}}{2}(i_a \cos\theta_a + i_b \cos\theta_b + i_c \cos\theta_c) \tag{5.113}$$

where θ_b and θ_c are angles to the same position given by θ_a, but measured from the b-phase and c-phase axes, respectively; that is $\theta_b = \theta_a - \frac{2\pi}{3}$ and $\theta_c = \theta_a - \frac{4\pi}{3}$. Using Euler's identity, we can rewrite $\vec{\mathbf{F}}_s$ as

$$\vec{\mathbf{F}}_s = \frac{N_{sine}}{4}\left\{ e^{j\theta_a}(i_a + i_b e^{-j\frac{2\pi}{3}} + i_c e^{-j\frac{4\pi}{3}}) + e^{-j\theta_a}(i_a + i_b e^{j\frac{2\pi}{3}} + i_c e^{j\frac{4\pi}{3}}) \right\} \tag{5.114}$$

Adopting the notation used for symmetrical components, that is $\mathbf{a} = e^{j2\pi/3}$, and $\mathbf{a}^2 = e^{j4\pi/3} = e^{-j2\pi/3}$, the above equation becomes

$$\vec{\mathbf{F}}_s = \frac{N_{sine}}{4}\left\{ e^{j\theta_a}\underbrace{(i_a + \mathbf{a}^2 i_b + \mathbf{a} i_c)}_{\vec{\mathbf{i}}_2} + e^{-j\theta_a}\underbrace{(i_a + \mathbf{a} i_b + \mathbf{a}^2 i_c)}_{\vec{\mathbf{i}}_1} \right\}$$

$$= \frac{N_{sine}}{4}(\vec{\mathbf{i}}_2 e^{j\theta_a} + \vec{\mathbf{i}}_1 e^{-j\theta_a}) \tag{5.115}$$

where $\vec{\mathbf{i}}_1$ and $\vec{\mathbf{i}}_2$ are the positive-sequence current and negative-sequence current space vectors of the three-phase winding currents.

Let's pause to examine the components of sequence space vectors $\vec{\mathbf{i}}_1$ and $\vec{\mathbf{i}}_2$. Their components, i_a, $\mathbf{a} i_b$, and so on, may also be regarded as space vectors along the a, b and c-axes, respectively. Written explicitly in terms of real and imaginary components, we have

$$\vec{\mathbf{i}}_1 = i_a + \left(-\frac{1}{2} + j\frac{\sqrt{3}}{2}\right)i_b + \left(-\frac{1}{2} - j\frac{\sqrt{3}}{2}\right)i_c$$

$$\hspace{6cm} \tag{5.116}$$

$$= \frac{3}{2}i_a + j\frac{\sqrt{3}}{2}(i_b - i_c) - \frac{1}{2}(i_a + i_b + i_c)$$

$$\vec{\mathbf{i}}_2 = i_a + \left(-\frac{1}{2} - j\frac{\sqrt{3}}{2}\right)i_b + \left(-\frac{1}{2} + j\frac{\sqrt{3}}{2}\right)i_c$$

$$\hspace{6cm} \tag{5.117}$$

$$= \frac{3}{2}i_a - j\frac{\sqrt{3}}{2}(i_b - i_c) - \frac{1}{2}(i_a + i_b + i_c)$$

It is evident from Eqs. 5.116 and 5.118 that space vectors $\vec{\mathbf{i}}_1$ and $\vec{\mathbf{i}}_2$ are complex conjugates of each other, that is $\vec{\mathbf{i}}_2 = (\vec{\mathbf{i}}_1)^*$.

Returning now to the right-hand side of the expression of the resultant mmf given in Eq. 5.115, since $\vec{\mathbf{i}}_2$ is the conjugate of $\vec{\mathbf{i}}_1$, $\vec{\mathbf{i}}_1 e^{-j\theta_a}$, $\vec{\mathbf{i}}_2 e^{j\theta_a}$ in Eq. 5.115 are a conjugate pair and their resultant will be real. In other words, $\vec{\mathbf{F}}_s$ of Eq. 5.115 is a real quantity.

For balanced three-phase current excitation, that is

$$i_a = I_m \cos(\omega_e t)$$

$$i_b = I_m \cos\left(\omega_e t - \frac{2\pi}{3}\right) \tag{5.118}$$

$$i_c = I_m \cos\left(\omega_e t - \frac{4\pi}{3}\right)$$

the zero-sequence current space vector is zero and Eq. 5.116 reduces to

$$
\begin{aligned}
\vec{i}_1 &= \frac{3}{2} I_m \cos(\omega_e t) + j\frac{\sqrt{3}}{2} I_m \left\{ \cos\left(\omega_e t - \frac{2\pi}{3}\right) - \cos\left(\omega_e t - \frac{4\pi}{3}\right) \right\} \\
&= \frac{3}{2} I_m \cos(\omega_e t) + j\frac{\sqrt{3}}{2} I_m \left\{ -2\sin(\omega_e t)\sin\left(\frac{-2\pi}{3}\right) \right\} \\
&= \frac{3}{2} I_m (\cos(\omega_e t) + j\sin(\omega_e t)) = \frac{3}{2} I_m e^{j\omega_e t}
\end{aligned}
\tag{5.119}
$$

The above expression shows that the positive-sequence current space vector has a peak value of 3/2 times the peak value of the balanced phase currents. It may be represented by a sinusoidal spatially distributed current sheet of magnitude $3I_m/2$ that rotates clockwise in the positive direction at an angular velocity of ω_e. The negative-sequence current space vector given by the conjugate:

$$\vec{i}_2 = (\vec{i}_1)^* = \frac{3}{2} I_m e^{-j\omega_e t} \tag{5.120}$$

has the same peak value but rotates in the opposite direction at the same angular speed.

Substituting Eqs. 5.119 and 5.120 into Eq. 5.115 for $\vec{F}_s$, we obtain

$$
\begin{aligned}
\vec{F}_s &= \frac{N_{sine}}{4} \left(\frac{3}{2} I_m\right) (e^{j(\theta_a - \omega_e t)} + e^{-j(\theta_a - \omega_e t)}) \\
&= \frac{N_{sine}}{2} \left(\frac{3}{2} I_m\right) \cos(\theta_a - \omega_e t)
\end{aligned}
\tag{5.121}
$$

Equation 5.121 indicates that the resultant airgap mmf, $\vec{F}_s$, may be considered as a rotating mmf space vector. $\vec{F}_s$ has a sinusoidal spatial distribution around the airgap and rotates at a speed of ω_e in the positive direction of θ_a. Its peak value is 3/2 times the peak value of the mmf space vector of each of the phase windings.

For ease of visualizing the transformation, some people prefer to introduce a scaling factor that makes the magnitude of the current space vector the same as the peak value of the phase currents. This is accomplished by defining

$$\vec{i} \triangleq \frac{2}{3}\vec{i}_1 = i_a + j\frac{(i_b - i_c)}{\sqrt{3}} - i_0 \tag{5.122}$$

where i_0, the corresponding zero-sequence current space vector that is equal to $(i_a + i_b + i_c)/3$, is a real quantity. From the above relationship, we can express the actual a-phase current in terms of $\vec{i}$ as

$$i_a - i_0 = \Re(\vec{\mathbf{i}}) \tag{5.123}$$

We can also show that

$$\mathbf{a}^2\vec{\mathbf{i}} = \frac{2}{3}(\mathbf{a}^2 i_a + \mathbf{a}^3 i_b + \mathbf{a}^4 i_c)$$

$$= i_b + j\left(\frac{i_c - i_a}{\sqrt{3}}\right) - \frac{1}{3}(i_a + i_b + i_c) \tag{5.124}$$

or

$$i_b - i_0 = \Re(\mathbf{a}^2\vec{\mathbf{i}}) \tag{5.125}$$

Likewise, we can also show that

$$i_c - i_0 = \Re(\mathbf{a}\vec{\mathbf{i}}) \tag{5.126}$$

As intended, the scaled positive-sequence current space vector, $\vec{\mathbf{i}}$ defined in Eq. 5.122, is a sinusoidal spatially distributed current sheet of the same peak value as the phase current; it too rotates in the clockwise direction at an angular speed of ω_e.

5.10.1 Transformation between *abc* and Stationary *qd*0

The defined relationships of the three current space vectors, $\vec{\mathbf{i}}_1$, $\vec{\mathbf{i}}_2$, and $\vec{\mathbf{i}}_0$ with i_a, i_b and i_c, can be expressed in the same form as the classical symmetrical transformation, that is

$$\begin{bmatrix} \vec{\mathbf{i}}_1 \\ \vec{\mathbf{i}}_2 \\ \vec{\mathbf{i}}_0 \end{bmatrix} = \begin{bmatrix} 1 & \mathbf{a} & \mathbf{a}^2 \\ 1 & \mathbf{a}^2 & \mathbf{a} \\ \frac{1}{3} & \frac{1}{3} & \frac{1}{3} \end{bmatrix} \begin{bmatrix} i_a \\ i_b \\ i_c \end{bmatrix} \tag{5.127}$$

From Eqs. 5.120 and 5.122, $\vec{\mathbf{i}}_2 = (\vec{\mathbf{i}}_1)^* = 3/2(\vec{\mathbf{i}})^*$, the above matrix equation can therefore be written as

$$\begin{bmatrix} \vec{\mathbf{i}} \\ (\vec{\mathbf{i}})^* \\ \vec{\mathbf{i}}_0 \end{bmatrix} = \frac{2}{3}\begin{bmatrix} 1 & \mathbf{a} & \mathbf{a}^2 \\ 1 & \mathbf{a}^2 & \mathbf{a} \\ \frac{1}{3} & \frac{1}{3} & \frac{1}{3} \end{bmatrix} \begin{bmatrix} i_a \\ i_b \\ i_c \end{bmatrix} \tag{5.128}$$

From the latter form we can see that row two can be eliminated without any loss of information. Letting $\vec{\mathbf{i}} = i_q^s - ji_d^s$, and writing the real and imaginary components in two separate rows, we obtain the following real transformation:

$$\begin{bmatrix} i_q^s \\ i_d^s \\ i_0 \end{bmatrix} = \frac{2}{3}\begin{bmatrix} 1 & \Re(\mathbf{a}) & \Re(\mathbf{a}^2) \\ 0 & -\Im(\mathbf{a}) & -\Im(\mathbf{a}^2) \\ \frac{1}{2} & \frac{1}{2} & \frac{1}{2} \end{bmatrix} \begin{bmatrix} i_a \\ i_b \\ i_c \end{bmatrix} \tag{5.129}$$

$$\begin{bmatrix} i_q^s \\ i_d^s \\ i_0 \end{bmatrix} = \frac{2}{3}\begin{bmatrix} 1 & -\frac{1}{2} & -\frac{1}{2} \\ 0 & -\frac{\sqrt{3}}{2} & \frac{\sqrt{3}}{2} \\ \frac{1}{2} & \frac{1}{2} & \frac{1}{2} \end{bmatrix} \begin{bmatrix} i_a \\ i_b \\ i_c \end{bmatrix} \tag{5.130}$$

The above matrix equation can be written more compactly as

$$[i^s_{qd0}] = [\mathbf{T}^s_{qd0}][i_{abc}] \tag{5.131}$$

where $[i^s_{qd0}]$ and $[i_{abc}]$ are column vectors of the $qd0$ current components and the phase currents, respectively. The matrix $[\mathbf{T}^s_{qd0}]$ is the coefficient matrix in Eq. 5.130; it, in effect, transforms the abc phase currents to the $qd0$ currents. The above transformation is usually referred to as the abc to stationary $qd0$ transformation. The superscript, s, is to denote the stationary reference frame. The inverse transformation, that is from stationary $qd0$ back to abc, can be accomplished using $[i^s_{abc}] = [\mathbf{T}^s_{qd0}]^{-1}[i_{qd0}]$. The matrix inverse of $[\mathbf{T}^s_{qd0}]$, $[\mathbf{T}^s_{qd0}]^{-1}$, can be shown to be

$$[\mathbf{T}^s_{qd0}]^{-1} = \begin{bmatrix} 1 & 0 & 1 \\ -\frac{1}{2} & -\frac{\sqrt{3}}{2} & 1 \\ -\frac{1}{2} & \frac{\sqrt{3}}{2} & 1 \end{bmatrix} \tag{5.132}$$

For a balanced set of three-phase currents given by

$$\begin{aligned} i_a &= I_m \cos(\omega_e t + \phi) \\ i_b &= I_m \cos\left(\omega_e t - \frac{2\pi}{3} + \phi\right) \\ i_c &= I_m \cos\left(\omega_e t - \frac{4\pi}{3} + \phi\right) \end{aligned} \tag{5.133}$$

the transformation of Eq. 5.130 yields

$$\begin{aligned} i^s_q &= I_m \cos(\omega_e t + \phi) \\ i^s_d &= -I_m \sin(\omega_e t + \phi) = I_m \cos\left(\omega_e t + \phi + \frac{\pi}{2}\right) \\ i_0 &= 0 \end{aligned} \tag{5.134}$$

Thus, the scaled current space vector for the balanced currents is

$$\begin{aligned} \vec{i} = i^s_q - ji^s_d &= I_m\{\cos(\omega_e t + \phi) + j\sin(\omega_e t + \phi)\} = I_m e^{j(\omega_e t + \phi)} \\ &= I_m e^{j\phi} e^{j\omega_e t} = \sqrt{2}\tilde{I}_a e^{j\omega_e t} \end{aligned} \tag{5.135}$$

where $\tilde{I}_a$ is the rms time phasor of the phase-a current.

Clearly, for balanced three-phase currents, the qd currents, i^s_q and i^s_d, are orthogonal and they have the same peak value as the abc phase currents. From the above expressions, we can also see that i^s_d peaks $\frac{\pi}{2}$ radians ahead of i^s_q, and that the resultant current, $\vec{i}$, rotates counter-clockwise at a speed of ω_e from an initial position of ϕ to the a-phase axis at $t = 0$. Equation 5.135 also indicates the relationship between space vector and traditional time phasor.

5.10.2 Transformation between *abc* and Rotating *qd*0

For certain studies or applications, we may see a need or an advantage to transform the stationary qd variables onto yet another qd reference frame that is rotating. Equation 5.135 shows that the resultant $\vec{\mathbf{i}}$ rotates at ω_e. We can therefore deduce that an observer moving along at that same speed will see the current space vector, $\vec{\mathbf{i}}$, as a constant spatial distribution, unlike the time-varying qd components on the stationary qd axes given in Eq. 5.134.

To see through the eyes of the so-called moving observer is equivalent mathematically to resolving whatever variables that we want to see onto a rotating reference moving at the same speed as the observer. Since we are dealing with two-dimensional variables, the rotating reference can be any two independent basis vectors, which for convenience, we will use another pair of orthogonal qd axes. The zero-sequence component remains the same as before. Figure 5.23 shows the geometrical relationship of the rotating qd axes with respect to the earlier stationary qd axes. Consider, for example, the resolution of the space current vector $\vec{\mathbf{i}}$ of the balanced set of abc currents, given in Eqs. 5.133–5.134. In terms of its stationary qd components, the resolved components of $\vec{\mathbf{i}}$ on the new rotating qd axes can, by geometry, be shown to be given by

$$\begin{bmatrix} i_q \\ i_d \end{bmatrix} = \begin{bmatrix} \cos\theta & -\sin\theta \\ \sin\theta & \cos\theta \end{bmatrix} \begin{bmatrix} i_q^s \\ i_d^s \end{bmatrix} \tag{5.136}$$

The angle, θ, is the angle between the q-axes of the rotating and stationary qd axes; it is a function of the angular speed, $\omega(t)$, of the rotating qd axes and the initial value, that is

$$\theta(t) = \int_0^t \omega(t)dt + \theta(0) \tag{5.137}$$

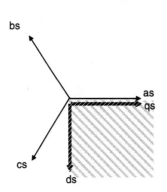

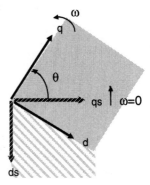

a) three-phase and stationary qd   (b) stationary and rotating qd

Figure 5.23   Relationship between stationary and rotating qd axes.

where $\theta(0)$ is the initial value of θ at time $t = 0$. When the qd components are combined into the space vector form, we have

$$i_q - ji_d = i_q^s \cos\theta - i_d^s \sin\theta - j(i_q^s \sin\theta + i_d^s \cos\theta)$$
$$= (i_q^s - ji_d^s)e^{-j\theta} \tag{5.138}$$

The inverse transformation can be shown to be

$$\begin{bmatrix} i_q^s \\ i_d^s \end{bmatrix} = \begin{bmatrix} \cos\theta & \sin\theta \\ -\sin\theta & \cos\theta \end{bmatrix} \begin{bmatrix} i_q \\ i_d \end{bmatrix} \tag{5.139}$$

Alternatively, the reverse transformation can be expressed as

$$i_q^s - ji_d^s = (i_q - ji_d)e^{j\theta} \tag{5.140}$$

The factor, $e^{j\theta}$, may be interpreted as a rotational operator that rotates whatever vector components it multiplies by the angle, θ. Thus, Eq. 5.138 indicates that the resolution of stationary qd variables to a set of synchronously rotating qd axes that is θ ahead is equivalent to rotating the stationary qd components backward by the same angular displacement of θ.

The choice of rotational speed and θ_0 will depend on the kinds of simplification or formulation best suited to the application at hand. Besides the stationary qd, where $\omega = 0$, the other frequent choices are the synchronously rotating qd, when ω is equal to the excitation frequency, ω_e, and the rotating qd at rotor speed, with the new d-axis aligned to the physical field winding axis.

Let's examine the nature of the qd components on synchronously rotating qd axes. This is equivalent to what will be seen by an observer rotating along with $\vec{\mathbf{i}}$. Denoting the variables in the new synchronously rotating qd axes by the superscript e to distinguish them from those with the superscript s on the stationary qd axes, and noting that the synchronous speed is constant, we have

$$\theta_e(t) = \int_0^t \omega_e dt + \theta_e(0) = \omega_e t + \theta_e(0) \tag{5.141}$$

The space vector, $\vec{\mathbf{i}}$ in the new qd coordinates, is

$$(i_q^e - ji_d^e) = (i_q^s - ji_d^s)e^{-j(\omega_e t + \theta_e(0))}$$
$$= I_m e^{j(\omega_e t + \phi)} e^{-j(\omega_e t + \theta_e(0))} = I_m e^{j(\phi - \theta_e(0))} \tag{5.142}$$
$$= I_m \cos(\phi - \theta_e(0)) + jI_m \sin(\phi - \theta_e(0))$$

Since ϕ and $\theta_e(0)$ are both constant, the values of i_q^e and i_d^e in the synchronously rotating qd axes are steady, as anticipated earlier. The $\phi - \theta_e(0)$ term indicates a relative angular measure, with the q-axis of the synchronously rotating qd axis as a reference. We can simplify the angular measure by starting the rotating qd axes with their q-axis aligned with the a-phase winding axis at time $t = 0$, that is making $\theta_e(0) = 0$. In which case, Eqs. 5.135 and 5.142 can be expressed in the following way to show the relationships between the different forms of representation for the set of balanced three-phase currents:

$$\vec{i} = i_q^s - ji_d^s = \sqrt{2}\tilde{I}_a e^{-j\omega_e t} = (i_q^e - ji_d^e)e^{-j\omega_e t} \tag{5.143}$$

or

$$(i_q^e - ji_d^e) = \sqrt{2}\tilde{I}_a \tag{5.144}$$

Equation 5.144 shows that the q and d components in the synchronous reference frame are the same as the real and imaginary components of the peak-value phasor of the a-phase current.

The full transformation from stationary $qd0$ to rotating $qd0$, with the zero-sequence component included for completeness, is

$$\begin{bmatrix} i_q \\ i_d \\ i_0 \end{bmatrix} = \begin{bmatrix} \cos\theta & -\sin\theta & 0 \\ \sin\theta & \cos\theta & 0 \\ 0 & 0 & 1 \end{bmatrix} \begin{bmatrix} i_q^s \\ i_d^s \\ i_0 \end{bmatrix} \tag{5.145}$$

where $\theta = \omega t + \theta(0)$. In matrix notation, the above transformation may be expressed as

$$[i_{qd0}] = [\mathbf{T}_\theta][i_{qd0}^s] \tag{5.146}$$

In terms of the original abc currents,

$$[i_{qd0}] = [\mathbf{T}_\theta][\mathbf{T}_{qd0}^s][i_{abc}] \tag{5.147}$$

Denoting $[\mathbf{T}_\theta][\mathbf{T}_{qd0}^s]$ by $[\mathbf{T}_{qd0}]$, we obtain

$$[i_{qd0}] = [\mathbf{T}_{qd0}][i_{abc}] \tag{5.148}$$

Carrying out the matrix multiplication and simplifying, we can show that

$$[\mathbf{T}_{qd0}] = \frac{2}{3}\begin{bmatrix} \cos\theta & \cos\left(\theta - \dfrac{2\pi}{3}\right) & \cos\left(\theta - \dfrac{4\pi}{3}\right) \\ \sin\theta & \sin\left(\theta - \dfrac{2\pi}{3}\right) & \sin\left(\theta - \dfrac{4\pi}{3}\right) \\ \dfrac{1}{2} & \dfrac{1}{2} & \dfrac{1}{2} \end{bmatrix} \tag{5.149}$$

The inverse transformation can be shown to be

$$[\mathbf{T}_{qd0}]^{-1} = \begin{bmatrix} \cos\theta & \sin\theta & 1 \\ \cos\left(\theta - \dfrac{2\pi}{3}\right) & \sin\left(\theta - \dfrac{2\pi}{3}\right) & 1 \\ \cos\left(\theta - \dfrac{4\pi}{3}\right) & \sin\left(\theta - \dfrac{4\pi}{3}\right) & 1 \end{bmatrix} \tag{5.150}$$

As with $[\mathbf{T}_{qd0}^s]$, the transformation, $[\mathbf{T}_{qd0}]$, is not unitary in that $[\mathbf{T}_{qd0}]^t \neq [\mathbf{T}_{qd0}]^{-1}$; as such, the transformation is not power-invariant. This can be shown by starting with the total instantaneous power into the three-phase circuit in abc quantities and transforming the abc quantities to $qd0$ as follows:

$$P_{abc} = v_a i_a + v_b i_b + v_c i_c = \begin{bmatrix} v_a \\ v_b \\ v_c \end{bmatrix}^t \begin{bmatrix} i_a \\ i_b \\ i_c \end{bmatrix} \tag{5.151}$$

$$= \left[[\mathbf{T}_{qd0}]^{-1} \begin{bmatrix} v_q \\ v_d \\ v_0 \end{bmatrix} \right]^t \left[[\mathbf{T}_{qd0}]^{-1} \begin{bmatrix} i_q \\ i_d \\ i_0 \end{bmatrix} \right] \tag{5.152}$$

$$= [v_q \ \ v_d \ \ v_0] \left[[\mathbf{T}_{qd0}]^{-1} \right]^t [\mathbf{T}_{qd0}]^{-1} \begin{bmatrix} i_q \\ i_d \\ i_0 \end{bmatrix} \tag{5.153}$$

It can be shown that

$$\left[[\mathbf{T}_{qd0}]^{-1} \right]^t [\mathbf{T}_{qd0}]^{-1} = \begin{bmatrix} \frac{3}{2} & 0 & 0 \\ 0 & \frac{3}{2} & 0 \\ 0 & 0 & \frac{1}{3} \end{bmatrix} \tag{5.154}$$

Thus,

$$P_{abc} = \frac{3}{2}(v_q i_q + v_d i_d) + \frac{1}{3} v_0 i_0 \tag{5.155}$$

At this point, it is worthwhile to take stock of certain assumptions made which could restrict the use of the above transformation. Clearly, there is no restriction on the *abc* currents; they can be unbalanced or non-sinusoidal. The expressions concerning the rotating *qd* axes have been obtained without any restriction on the value of ω, the angular speed of the rotating *qd* axes. Implied in $[\mathbf{T}^s_{qd0}]$, however, is that the axes of the *abc* windings are spaced $(2\pi/3)$ electrical radians apart. Even though we have provided a mathematical basis for the *abc* to *qd0* transformation using the currents of a set of three-phase windings, we can show that the same transformation applies to other variables, such as the mmf, flux linkages, induced voltages, and terminal voltages.

5.11 PROJECTS

5.11.1 Project 1: *qd0* Transformation of Network Components

Figure 5.24 is a one-line diagram of a simple three-wire system consisting of a three-phase generator and a length of transposed three-phase line connecting the generator to a three-phase RL shunt load. On the load bus is a set of three-phase shunt capacitors that is added to compensate for the lagging power factor of the RL load. For simplicity, model the generator by an equivalent three-phase voltage source behind a source inductance, and the three-phase transmission line by a series RL line model. Since it is a three-wire system, there will not be any zero-sequence current components; as such, the zero component circuit connections may be omitted. Sketch the input-output relations between the *qd* components in a simulation of the given three-wire system.

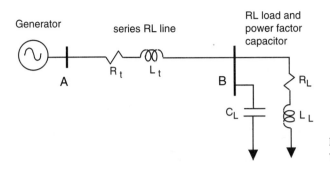

Figure 5.24   One-line diagram of a three-wire system.

5.11.2 Project 2: Space Vectors

In this project, we shall implement the transformation of instantaneous three-phase ac current to space vectors as given in Eqs. 5.116 and 5.118, and use it to examine the nature of the positive- and negative-sequence space vectors in the spatial domain.

Figure 5.25 shows the contents of SIMULINK file *s2* for transforming a set of *abc* currents to positive- and negative-sequence space vectors. The *abc* currents in *s2* are of the form:

$$i_a = 10e^{-\alpha t}\cos m(2\pi t) + \frac{10}{n}\cos n(2\pi t) \qquad A$$

$$i_b = 10e^{-\alpha t}\cos m\left(2\pi t - \frac{2\pi}{3}\right) + \frac{10}{n}\cos n\left(2\pi t - \frac{2\pi}{3}\right) \qquad A \qquad (5.156)$$

$$i_c = 10e^{-\alpha t}\cos m\left(2\pi t + \frac{2\pi}{3}\right) + \frac{10}{n}\cos n\left(2\pi t + \frac{2\pi}{3}\right) \qquad A$$

As programmed, the second component in Eq. 5.156 of each phase current will be switched off by switches controlled by *nSw* in *s2*, whenever the entered value of *n* is negative. The SIMULINK file, *s2*, has a masked block of the MATLAB script file *m2* for initializing the simulation and plotting the results of the simulation.

For example, entering the values of 1 for *m*, 0 for α, and -1 for *n*, the resultant *abc* currents will be a balanced set of three-phase currents of constant amplitude 10 A and frequency 1 Hz. As shown connected in Fig. 5.25, the two XY graphs will display the real part of the space vectors along the *x*-axis and the imaginary part of the space vectors along the *y*-axis. Based on the analytical results given in Eqs. 5.119 and 5.120 for the above set of currents, we expect to obtain an $\vec{i}_1$ of $15e^{j2\pi t}$ and an $\vec{i}_2$ of $15e^{-j2\pi t}$.

The final display of the two rotating space vectors is shown in Fig. 5.26. It is obtained using the *ode45* or *Linsim* numerical integration algorithm, a stop time of 0.95 second (0.95 of the basic period), and minimum and maximum step sizes of 5 msec. The XY graphs display the loci of the tips of both space vectors starting from +15 on the positive x-axis and tracing out circles of radius 15; the locus of the tip of $\vec{i}_1$ traces out the circle in the counter-clockwise direction, while that of $\vec{i}_2$ traces out the circle in the clockwise direction. The space vector represents a sinusoidal spatially distributed current sheet. As plotted, the length of its tip from the origin of the XY plot corresponds to the positive peak of the

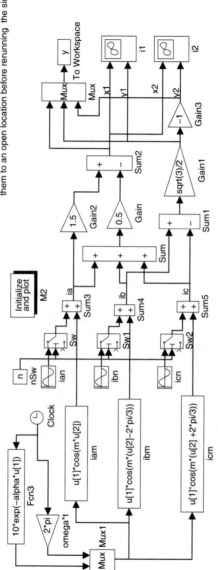

Figure 5.25 Simulation *s2* of *abc* to space vector transformation.

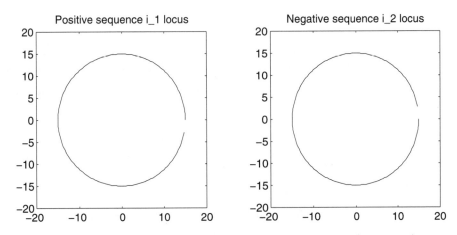

Figure 5.26   XY plots of positive- and negative-sequence space vectors: $\vec{i}_1$ (left plot), $\vec{i}_2$ (right plot).

TABLE 5.1   PROJECT 2 CASE STUDIES

Cases	Value of m	Value of n
1	$m = 3k+1, \ k = 0,1,2$	$n < 0$
2	$m = 3k, \ k = 1,2,3$	$n < 0$
3	$m = 3k-1, \ k = 1,2$	$n < 0$
4	$m = 1$	$n = 3k+1, \ k = 0,1,2$
5	$m = 1$	$n = 3k, \ k = 1,2,3$
6	$m = 1$	$n = 3k-1, \ k = 1,2$

spatially distributed current sheet. The trajectory of the tip displays fluctuation in amplitude and radial movement of the sinusoidal spatially distributed current sheet. With balanced fundamental phase currents, the trajectory of the tip is a circular locus at uniform speed, indicating that both magnitude and excitation frequency remain constant, as predicted by Eqs. 5.119 and 5.120.

For $m > 1$, the tips of the space vectors will trace over the previous circle when the peak values of the phase currents remain constant. For purposes of being able to tell them apart, an exponential decay factor to the peak value may be introduced to cause the locus to spiral inwards, so that the number of rounds that the tips of the vectors traveled in the same time interval may become evident.

1. Move the two XY graphs to an open position where you will be able to see both displays during a simulation run. Conduct simulation runs of $s2$ for values of n and m given in cases 1 through 6 in Table 5.1. For each run, jot down the direction of rotation of the loci of the positive- and negative-sequence space vectors. Save a sample of the plots from each case. Comment on the direction and speed of rotation of the two space

vectors in cases 1 and 3. Explain what you see for the odd and even triplens in cases 2 and 5.

Using values of $m = 1$, $\alpha = 0$, and $n < 0$, make modifications to the appropriate function blocks to simulate the following unbalanced conditions in the phase currents:

a. Create a magnitude imbalance by changing the magnitude of the i_{am} component of i_a only to 15 A.

b. Create a phase imbalance by changing i_{bm} to $10\cos(2\pi t - \pi/2)$ and i_{cm} to $10\cos(2\pi t + \pi/2)$ (i_{am} remains as $10\cos(2\pi t)$).

Sample results for a k of 2 in cases 4 and 6 with $\alpha = 0$ are given in Fig. 5.27.

5.11.3 Project 3: Sinusoidal and Complex Quantities in *qd*0

Figure 5.28a shows the overall diagram of *s3*. It is a SIMULINK implementation of the phase to qd transformations as given in Eqs. 5.130 and 5.145. In this project, we will use *s3* and its accompanying MATLAB script file, *m3*, to examine the waveforms of the $qd0$ components corresponding to sinusoidal and complex phase currents. As given in *s3*, the function blocks are programmed to generate sinusoidal *abc* currents of the form:

$$i_a = 10e^{-\alpha t}\cos m(2\pi t) \qquad A$$

$$i_b = 10e^{-\alpha t}\cos m\left(2\pi t - \frac{2\pi}{3}\right) \qquad A \qquad (5.157)$$

$$i_c = 10e^{-\alpha t}\cos m\left(2\pi t + \frac{2\pi}{3}\right) \qquad A$$

In *s3*, time is taken from the *clock* and multiplied by the value of angular frequency to obtain the value of ωt, which is then fed to the three *fcn* function modules that generate the sinusoidal three-phase currents. The template for the *fcn* function module can be found in the **Nonlinear** library block of SIMULINK. The value of m in the three function blocks is to be entered into the MATLAB workspace from the MATLAB command window. Figure 5.28b shows the inside of the *abc2qd0s* block. This block implements Eq. 5.130 to transform the input *abc* phase quantities to stationary $qd0$ quantities. Figure 5.28c shows the inside of the *qds2qd* block. This block implements Eq. 5.136 to transform the stationary qd quantities to another qd reference frame that is at an angle of θ to the stationary qds reference frame. The θ input to the *qds2qd* block can be made a multiple, *nframe*, of the ωt input of the phase quantities with or without some initial offset angle corresponding to the value of $\phi - \theta(0)$ in Eq. 5.142, ϕ being the phase angle of the phase quantities as defined in Eq. 5.133, and $\theta(0)$ being the initial angle of θ in Eq. 5.157. As programmed in *s3*, with the currents in Eq. 5.158, the value of ϕ is zero.

Shown in Fig. 5.29 are the $qd0$ currents for the case of $m = 1$ and $\phi = 0$ in Eq. 5.157, with an *nframe* of 1 that is rotating at synchronous speed in the forward direction, and an initial angle, $\theta(0)$, of zero. The first three traces are in the stationary reference frame. In

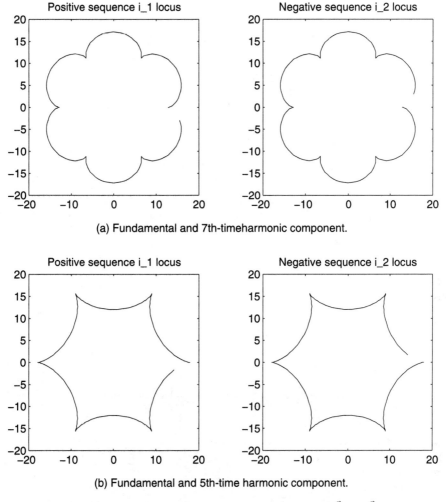

(a) Fundamental and 7th-timeharmonic component.

(b) Fundamental and 5th-time harmonic component.

Figure 5.27   XY plots of positive- and negative-sequence space vectors: $\vec{i}_1$ (left), $\vec{i}_2$ (right).

the synchronously rotating qd synchronous frame, the $qd0$ components of fundamental frequency abc currents are dc components. The form and magnitude of these waveforms agree with those predicted by the analytical expressions of the stationary $qd0$ currents given in Eq. 5.134 for $\phi = 0$, and of the qd components in the synchronously rotating frame given in Eq. 5.142.

1. Run the cases with $m = 1$ and $\phi = 0$, but with *nframe* first set to -1 and then to 2. Based on the results obtained, deduce what will be the frequency of the qd components for a balanced set of phase currents also having a frequency of 5ω in a qd reference frame that is also rotating in the forward direction at 5ω. Use the simulation to verify your answer by setting m and *nframe* to 5, and α to zero.

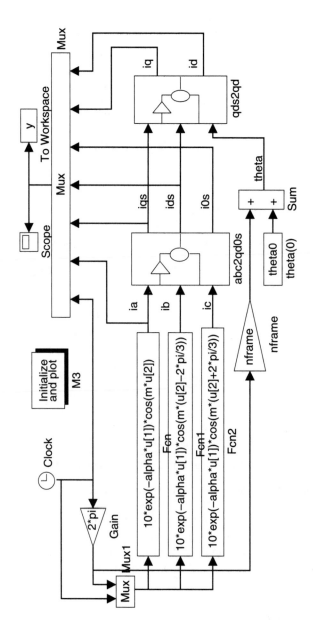

Figure 5.28 Simulation *s3* for phase to *qd0* transformation.

(a) Overall diagram of *s3*

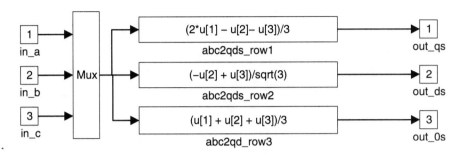

(b) Inside the *abc2qd0s* block

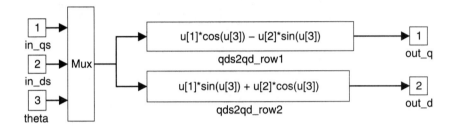

(c) Inside the *qds2qd* block

Figure 5.28   *(cont.)*

2. Repeat the above cases, but with a a α of 1, to see the form of the transformed attenuated sinusoids in the two rotating $qd0$ reference frames.

3. Run the case of $m = 1$ and *nframe* $= 1$ for $\theta(0) = \pm\pi/4$. Verify your results using the expression given in Eq. 5.142.

4. Run the case of $m = 2$ and *nframe* $= 2$ and examine the results obtained to see if you can explain them. (Hint: Determine the sequence of the *abc* currents and the direction of rotation of their resultant mmf or their space vector, $\vec{i}_1$).

5. Modify the *abc* current function blocks in a copy of *s3* to produce a set of complex *abc* phase currents with fundamental and *n*th time harmonic components of the form:

$$i_a = 10\cos(2\pi t) + \frac{10}{n}\cos n(2\pi t) \qquad A$$

$$i_b = 10\cos\left(2\pi t - \frac{2\pi}{3}\right) + \frac{10}{n}\cos n\left(2\pi t - \frac{2\pi}{3}\right) \qquad A \qquad (5.158)$$

$$i_c = 10\cos\left(2\pi t + \frac{2\pi}{3}\right) + \frac{10}{n}\cos n\left(2\pi t + \frac{2\pi}{3}\right) \qquad A$$

Obtain the $qd0$ components in the synchronously rotating reference frame, that is with *nframe* $= 1$ and $\theta(0) = 0$, for $n = 5$ and $n = 7$.

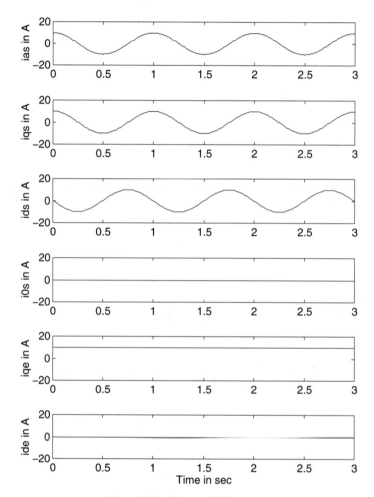

Figure 5.29   Phase and $qd0$ currents.

6

Three-phase
Induction Machines

6.1 INTRODUCTION

Three-phase induction machines are asynchronous speed machines, operating below synchronous speed when motoring and above synchronous speed when generating. They are comparatively less expensive to equivalent size synchronous or dc machines and range in size from a few watts to 10,000 hp. They, indeed, are the workhorses of today's industry. As motors, they are rugged and require very little maintenance. However, their speeds are not as easily controlled as with dc motors. They draw large starting currents, typically about six to eight times their full load values, and operate with a poor lagging power factor when lightly loaded.

6.1.1 Construction of Three-phase Induction Machines

Most induction motors are of the rotary type with basically a stationary stator and a rotating rotor. The stator has a cylindrical annulus magnetic core that is housed inside a metal frame onto which are mounted the bed-plates, end shields, and a terminal box. At the center of the end shields are bearings to support the rotor's shaft. The stator magnetic core is formed by stacking thin electrical steel laminations with uniformly spaced slots stamped in the inner circumference to accommodate the three distributed stator windings. For 60-Hz machines, the lamination is about 0.5 mm thick. The stator windings are formed by connecting coils

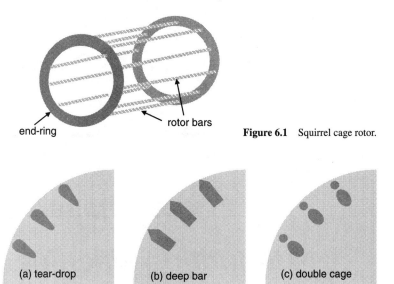

end-ring   rotor bars   **Figure 6.1** Squirrel cage rotor.

(a) tear-drop   (b) deep bar   (c) double cage

Figure 6.2   Slot shapes.

of copper or aluminum conductors. The coils are insulated from the slot walls. The axes of the stator windings of a P-pole machine are spaced $(2/P)(2\pi/3)$ mechanical radians apart, with each phase belt occupying the same number of slots. The terminals of the three stator phase windings can be star- or delta-connected.

The rotor consists of a cylindrical laminated iron core with uniformly spaced peripheral slots to accommodate the rotor windings. There are two main types of rotor windings: wound and squirrel-cage. In a wound rotor machine, rotor windings similar to those of the distributed windings on the stator are wound on a cylindrical laminated core with uniformly spaced slots on the outer periphery. The terminals of the rotor windings can be brought out via slip rings and brushes. These slip rings are insulated from one another and from the shaft. A variable amount of external resistance can be connected to the rotor windings via the slip rings to obtain higher starting torque or limited control of speed below synchronous speed. In very large and expensive drives, the terminals of the rotor windings may be connected to a slip frequency source to recover part of the power flowing into the rotor circuit.

Figure 6.1 shows the rotor winding of a squirrel-cage rotor. It has uniformly spaced axial bars that are soldered onto end rings at both ends. Single squirrel-cage slots are mostly oval and closed by a threshold 0.4 to 0.5 mm thick. Figure 6.2 shows some of the slot shapes used. Small rotors under 20 inches in diameter are usually die-cast. After the rotor core laminations are stacked in a mold, the mold is filled with molten aluminum. Rotor bars, rings, and cooling fan blades are cast in one economical process. In large machines, the rotor bars are of copper alloy and are driven into uniformly spaced slots on the periphery of the rotor; the bars are braced onto end rings at both ends. There is no insulation between the bars and walls of the rotor slots.

A difficult combination of high starting torque and good efficiency can be obtained using a double-cage rotor. The outer cage has a lower reactance of the two cages by virtue of its position. It is usually made of higher electrical resistance material, like brass or bronze, for it to provide most of the starting torque. The inner cage has a relatively higher reactance and is usually made of lower resistance material, like copper, for high efficiency at low slip.

The smaller the airgap, the better the mutual electromagnetic induction between stator and rotor windings. However, the extent to which the airgap can be reduced is dictated by manufacturing tolerances and costs, and by allowable core losses from the higher local saturation around the slots. The airgap is typically between 0.35 to 0.50 mm for motors up to 10 kW, and between 0.5 to 0.8 mm for motors in the range of 10 to 100 kW. Wider gaps, however, are used in motors which operate with exceptionally large peak load torque.

6.2 ROTATING MAGNETIC FIELD AND SLIP

It has been shown in Sections 5.4 and 5.10 of the previous chapter that a balanced set of three-phase currents flowing in a set of symmetrically placed, three-phase stator windings produces a rotating mmf field given by

$$F(\theta_a^e, t) = \frac{3}{2} \frac{4}{\pi} \frac{N}{P} I_m \cos(\theta_a^e - \omega_e t) \qquad A.turn \tag{6.1}$$

where θ_a^e is the electrical angle measured from the a-phase axis, $\omega_e \ (= 2\pi f_e)$ is the angular speed of the stator mmf in electrical radians per sec, and f_e is the frequency of the excitation currents. In mechanical radians per second, the synchronous speed is

$$\omega_{sm} = \frac{2}{P} \omega_e \qquad rad/s \tag{6.2}$$

In revolutions per minute, the synchronous speed is

$$N_s = \frac{60\omega_{sm}}{2\pi} = \frac{120 f_e}{P} \qquad rev/min \tag{6.3}$$

When the rotor is rotating at a steady speed of ω_{rm} mechanical radians per sec, the relative or slip speed between the rotor and synchronous rotating stator field, F, is

$$\text{slip speed} = \omega_{sm} - \omega_{rm} \tag{6.4}$$

The per unit slip, also referred to simply as the *slip*, is defined as the normalized slip speed, that is

$$s \triangleq \frac{\omega_{sm} - \omega_{rm}}{\omega_{sm}} = \frac{\omega_e - \omega_r}{\omega_e} \tag{6.5}$$

Slip s is negative in generating operation when the rotor rotates above synchronous speed.

The slip speed can be expressed as $s\omega_e$ or $s\omega_{sm}$, and the slip frequency as $s f_e$. When ω_{rm} is less than the synchronous speed, ω_{sm}, the rotor conductors are slipping backwards at a speed of $s\omega_{sm}$ relative to the moving airgap flux in the forward direction. As a result, the induced voltages in the rotor windings due to the synchronously rotating airgap flux are of slip frequency $s f_e$. When the rotor circuit is closed, the induced voltages will circulate currents in the rotor circuit. The magnitude of the currents flowing in the rotor circuit is determined by the magnitude of the rotor induced voltages and the rotor circuit impedance at slip frequency.

At standstill, ω_{rm} is zero or $s = 1$, thus the slip frequency when starting is f_e. As the motor accelerates towards the synchronous speed, the slip frequency decreases. As with the flow of currents in the stator, the rotor currents will establish their own revolving mmf field that rotates at $s\omega_{sm}$ rad/sec relative to the rotor. Since the rotor itself is rotating at ω_{rm}, the absolute speed of the rotor mmf field can be shown to be equal to the synchronous speed, that is

$$\omega_{rm} + s\omega_{sm} = \omega_{sm} \qquad mech.\,rad/sec \tag{6.6}$$

With the stator and rotor mmfs rotating at the same speed, a steady torque would be produced in steady-state when the magnitude and phase differences of these mmfs are constant.

6.3 CIRCUIT MODEL OF A THREE-PHASE INDUCTION MACHINE

6.3.1 Voltage Equations

Using the coupled circuit approach and motor notation, the voltage equations of the magnetically coupled stator and rotor circuits shown in Fig. 6.3 can be written as follows:

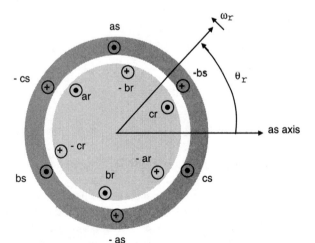

Figure 6.3 Idealized circuit model of a three-phase induction machine.

Stator voltage equations

$$v_{as} = i_{as}r_s + \frac{d\lambda_{as}}{dt} \quad V$$

$$v_{bs} = i_{bs}r_s + \frac{d\lambda_{bs}}{dt} \quad V \qquad (6.7)$$

$$v_{cs} = i_{cs}r_s + \frac{d\lambda_{cs}}{dt} \quad V$$

Rotor voltage equations

$$v_{ar} = i_{ar}r_r + \frac{d\lambda_{ar}}{dt} \quad V$$

$$v_{br} = i_{br}r_r + \frac{d\lambda_{br}}{dt} \quad V \qquad (6.8)$$

$$v_{cr} = i_{cr}r_r + \frac{d\lambda_{cr}}{dt} \quad V$$

Flux linkage equations.   In matrix notation, the flux linkages of the stator and rotor windings, in terms of the winding inductances and currents, may be written compactly as

$$\begin{bmatrix} \lambda_s^{abc} \\ \lambda_r^{abc} \end{bmatrix} = \begin{bmatrix} \mathbf{L}_{ss}^{abc} & \mathbf{L}_{sr}^{abc} \\ \mathbf{L}_{rs}^{abc} & \mathbf{L}_{rr}^{abc} \end{bmatrix} \begin{bmatrix} \mathbf{i}_s^{abc} \\ \mathbf{i}_r^{abc} \end{bmatrix} \quad Wb.\ turn \qquad (6.9)$$

where

$$\lambda_s^{abc} = (\lambda_{as}, \lambda_{bs}, \lambda_{cs})^t$$

$$\lambda_r^{abc} = (\lambda_{ar}, \lambda_{br}, \lambda_{cr})^t$$

$$\mathbf{i}_s^{abc} = (i_{as}, i_{bs}, i_{cs})^t \qquad (6.10)$$

$$\mathbf{i}_r^{abc} = (i_{ar}, i_{br}, i_{cr})^t$$

and the superscript t denotes the transpose of the array.

The submatrices of the stator-to-stator and rotor-to-rotor winding inductances are of the form:

$$\mathbf{L}_{ss}^{abc} = \begin{bmatrix} L_{ls}+L_{ss} & L_{sm} & L_{sm} \\ L_{sm} & L_{ls}+L_{ss} & L_{sm} \\ L_{sm} & L_{sm} & L_{ls}+L_{ss} \end{bmatrix} \quad H \qquad (6.11)$$

$$\mathbf{L}_{rr}^{abc} = \begin{bmatrix} L_{lr}+L_{rr} & L_{rm} & L_{rm} \\ L_{rm} & L_{lr}+L_{rr} & L_{rm} \\ L_{rm} & L_{rm} & L_{lr}+L_{rr} \end{bmatrix} \quad H \qquad (6.12)$$

Those of the stator-to-rotor mutual inductances are dependent on the rotor angle, that is

$$\mathbf{L}_{sr}^{abc} = \left[\mathbf{L}_{rs}^{abc}\right]^{t} = L_{sr} \begin{bmatrix} \cos\theta_r & \cos\left(\theta_r + \dfrac{2\pi}{3}\right) & \cos\left(\theta_r - \dfrac{2\pi}{3}\right) \\[2.5ex] \cos\left(\theta_r - \dfrac{2\pi}{3}\right) & \cos\theta_r & \cos\left(\theta_r + \dfrac{2\pi}{3}\right) \\[2.5ex] \cos\left(\theta_r + \dfrac{2\pi}{3}\right) & \cos\left(\theta_r - \dfrac{2\pi}{3}\right) & \cos\theta_r \end{bmatrix} H \quad (6.13)$$

where L_{ls} is the per phase stator winding leakage inductance, L_{lr} is the per phase rotor winding leakage inductance, L_{ss} is the self-inductance of the stator winding, L_{rr} is the self-inductance of the rotor winding, L_{sm} is the mutual inductance between stator windings, L_{rm} is the mutual inductance between rotor windings, and L_{sr} is the peak value of the stator-to-rotor mutual inductance.

When the reluctive drops in iron are neglected, we can express some of these inductances in terms of the winding turns, N_s and N_r, and the airgap permeance, P_g, using the expressions derived in Section 5.5:

$$L_{ss} = N_s^2 P_g \qquad\qquad\qquad L_{rr} = N_r^2 P_g$$

$$L_{sm} = N_s^2 P_g \cos\left(\frac{2\pi}{3}\right) \qquad\qquad L_{rm} = N_r^2 P_g \cos\left(\frac{2\pi}{3}\right) \qquad (6.14)$$

$$L_{sr} = N_s N_r P_g$$

Note that the idealized machine is described by six first-order differential equations, one for each winding. These differential equations are coupled to one another through the mutual inductances between the windings. In particular, the stator-to-rotor coupling terms are a function of rotor position; thus, when the rotor rotates, these coupling terms vary with time.

Mathematical transformations like the dq or $\alpha\beta$ can facilitate the computation of the transient solution of the above induction machine model by transforming the differential equations with time-varying inductances to differential equations with constant inductances.

6.4 MACHINE MODEL IN ARBITRARY $qd0$ REFERENCE FRAME

The idealized three-phase induction machine is assumed to have symmetrical airgap. The $qd0$ reference frames are usually selected on the basis of convenience or compatibility with the representations of other network components. The two common reference frames used in the analysis of induction machine are the stationary and synchronously rotating reference frames. Each has an advantage for some purpose. In the stationary rotating reference, the dq variables of the machine are in the same frame as those normally used for the supply network. It is a convenient choice of frame when the supply network is large or complex. In the synchronously rotating reference frame, the dq variables are steady in steady-state, a prerequisite when deriving the small-signal model about a chosen operating point.

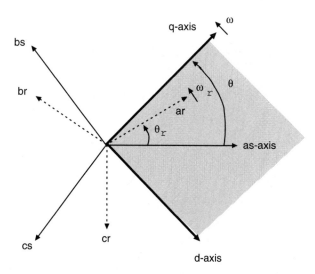

Figure 6.4 Relationship between *abc* and arbitrary $qd0$.

We shall first derive the equations of the induction machine in the arbitrary reference frame which is rotating at a speed of ω in the direction of the rotor rotation. Those of the induction machine in the stationary reference frame can then be obtained by setting $\omega = 0$, and those for the synchronously rotating reference frame are obtained by setting $\omega = \omega_e$. As before, we will begin with the voltage and torque equations of the machine in phase quantities. Applying the arbitrary $qd0$ reference transformation to these equations yields the corresponding $qd0$ equations. The relationship between the *abc* quantities and $qd0$ quantities of a reference frame rotating at an angular speed, ω, is shown in Fig. 6.4. The transformation equation from *abc* to this $qd0$ reference frame is given by

$$
\begin{bmatrix} f_q \\ f_d \\ f_0 \end{bmatrix} = \begin{bmatrix} \mathbf{T}_{qd0}(\theta) \end{bmatrix} \begin{bmatrix} f_a \\ f_b \\ f_c \end{bmatrix}
\tag{6.15}
$$

where the variable f can be the phase voltages, currents, or flux linkages of the machine.

For clarity, the first quadrant of the reference frame rotating at an arbitrary speed, $\omega(t)$, in Fig. 6.4 is shown as a shaded area. The transformation angle, $\theta(t)$, between the q-axis of the reference frame rotating at a speed of ω and the a-axis of the stationary stator winding may be expressed as

$$
\theta(t) = \int_0^t \omega(t)dt + \theta(0) \qquad elect.\,rad.
\tag{6.16}
$$

Likewise, the rotor angle, $\theta_r(t)$, between the axes of the stator and the rotor a-phases for a rotor rotating with speed $\omega_r(t)$ may be expressed as

$$
\theta_r(t) = \int_0^t \omega_r dt + \theta_r(0) \qquad elect.\,rad.
\tag{6.17}
$$

The angles, $\theta(0)$ and $\theta_r(0)$, are the initial values of these angles at the beginning of time t.

From Eq. 5.72, the $qd0$ transformation matrix, $[\mathbf{T}_{qd0}(\theta)]$, is

$$[\mathbf{T}_{qd0}(\theta)] = \frac{2}{3}\begin{bmatrix} \cos\theta & \cos\left(\theta - \frac{2\pi}{3}\right) & \cos\left(\theta + \frac{2\pi}{3}\right) \\ \sin\theta & \sin\left(\theta - \frac{2\pi}{3}\right) & \sin\left(\theta + \frac{2\pi}{3}\right) \\ \frac{1}{2} & \frac{1}{2} & \frac{1}{2} \end{bmatrix} \quad (6.18)$$

and its inverse is

$$[\mathbf{T}_{qd0}(\theta)]^{-1} = \begin{bmatrix} \cos\theta & \sin\theta & 1 \\ \cos\left(\theta - \frac{2\pi}{3}\right) & \sin\left(\theta - \frac{2\pi}{3}\right) & 1 \\ \cos\left(\theta + \frac{2\pi}{3}\right) & \sin\left(\theta + \frac{2\pi}{3}\right) & 1 \end{bmatrix} \quad (6.19)$$

6.4.1 $qd0$ Voltage Equations

In matrix notation, the stator winding abc voltage equations can be expressed as

$$\mathbf{v}_s^{abc} = p\lambda_s^{abc} + \mathbf{r}_s^{abc}\mathbf{i}_s^{abc} \quad (6.20)$$

Applying the transformation, $[\mathbf{T}_{qd0}(\theta)]$, to the voltage, flux linkage and current, Eq. 6.20 becomes

$$\mathbf{v}_s^{qd0} = [\mathbf{T}_{qd0}(\theta)]p[\mathbf{T}_{qd0}(\theta)]^{-1}[\lambda_s^{qd0}] + [\mathbf{T}_{qd0}(\theta)]\mathbf{r}_s^{abc}[\mathbf{T}_{qd0}(\theta)]^{-1}[\mathbf{i}_s^{qd0}] \quad (6.21)$$

The following time-derivative term may be expressed as

$$p[\mathbf{T}_{qd0}(\theta)]^{-1}[\lambda_s^{qd0}]$$

$$= \begin{bmatrix} -\sin\theta & \cos\theta & 0 \\ -\sin\left(\theta - \frac{2\pi}{3}\right) & \cos\left(\theta - \frac{2\pi}{3}\right) & 0 \\ -\sin\left(\theta + \frac{2\pi}{3}\right) & \cos\left(\theta + \frac{2\pi}{3}\right) & 0 \end{bmatrix}\frac{d\theta}{dt}[\lambda_s^{qd0}] + [\mathbf{T}_{qd0}(\theta)]^{-1}[p\lambda_s^{qd0}] \quad (6.22)$$

Substituting this back into Eq. 6.21 and rearranging, we will obtain

$$\mathbf{v}_s^{qd0} = \omega\begin{bmatrix} 0 & 1 & 0 \\ -1 & 0 & 0 \\ 0 & 0 & 0 \end{bmatrix}\lambda_s^{qd0} + p\lambda_s^{qd0} + \mathbf{r}_s^{qd0}\mathbf{i}_s^{qd0} \quad (6.23)$$

where

$$\omega = \frac{d\theta}{dt} \quad \text{and} \quad \mathbf{r}_s^{qd0} = r_s\begin{bmatrix} 1 & 0 & 0 \\ 0 & 1 & 0 \\ 0 & 0 & 1 \end{bmatrix} \quad (6.24)$$

Likewise, the rotor quantities must be transformed onto the same qd frame. From Fig. 6.4, we can see that the transformation angle for the rotor phase quantities is $(\theta - \theta_r)$. Using the transformation, $\mathbf{T}_{qd0}(\theta - \theta_r)$, on the rotor voltage equations, in the same manner as we have done with the stator voltage equations, we obtain the following $qd0$ voltage equations for the rotor windings:

$$\mathbf{v}_r^{qd0} = (\omega - \omega_r) \begin{bmatrix} 0 & 1 & 0 \\ -1 & 0 & 0 \\ 0 & 0 & 0 \end{bmatrix} \lambda_r^{qd0} + p\lambda_r^{qd0} + \mathbf{r}_r^{qd0}\mathbf{i}_r^{qd0} \tag{6.25}$$

6.4.2 $qd0$ Flux Linkage Relation

The stator $qd0$ flux linkages are obtained by applying $\mathbf{T}_{qd0}(\theta)$ to the stator abc flux linkages in Eq. 6.9, that is

$$\lambda_s^{qd0} = [\mathbf{T}_{qd0}(\theta)](\mathbf{L}_{ss}^{abc}\mathbf{i}_s^{abc} + \mathbf{L}_{sr}^{abc}\mathbf{i}_r^{abc}) \tag{6.26}$$

Using the appropriate inverse transformations to replace the abc stator and rotor currents by their corresponding $qd0$ currents, Eq. 6.26 becomes

$$\lambda_s^{qd0} = [\mathbf{T}_{qd0}(\theta)]\mathbf{L}_{ss}^{abc}[\mathbf{T}_{qd0}(\theta)]^{-1}\mathbf{i}_s^{qd0} + [\mathbf{T}_{qd0}(\theta)]\mathbf{L}_{sr}^{abc}[\mathbf{T}_{qd0}(\theta - \theta_r)]^{-1}\mathbf{i}_r^{qd0}$$

$$= \begin{bmatrix} L_{ls} + \frac{3}{2}L_{ss} & 0 & 0 \\ 0 & L_{ls} + \frac{3}{2}L_{ss} & 0 \\ 0 & 0 & L_{ls} \end{bmatrix} \mathbf{i}_s^{qd0} + \begin{bmatrix} \frac{3}{2}L_{sr} & 0 & 0 \\ 0 & \frac{3}{2}L_{sr} & 0 \\ 0 & 0 & 0 \end{bmatrix} \mathbf{i}_r^{qd0} \tag{6.27}$$

Similarly, the $qd0$ rotor flux linkages are given by

$$\lambda_r^{qd0} = [\mathbf{T}_{qd0}(\theta - \theta_r)]\mathbf{L}_{rs}^{abc}[\mathbf{T}_{qd0}(\theta)]^{-1}\mathbf{i}_s^{qd0} + [\mathbf{T}_{qd0}(\theta - \theta_r)]\mathbf{L}_{rr}^{abc}[\mathbf{T}_{qd0}(\theta - \theta_r)]^{-1}\mathbf{i}_r^{qd0}$$

$$= \begin{bmatrix} \frac{3}{2}L_{sr} & 0 & 0 \\ 0 & \frac{3}{2}L_{sr} & 0 \\ 0 & 0 & 0 \end{bmatrix} \mathbf{i}_s^{qd0} + \begin{bmatrix} L_{lr} + \frac{3}{2}L_{rr} & 0 & 0 \\ 0 & L_{lr} + \frac{3}{2}L_{rr} & 0 \\ 0 & 0 & L_{lr} \end{bmatrix} \mathbf{i}_r^{qd0} \tag{6.28}$$

The stator and rotor flux linkage relationships in Eqs. 6.27 and 6.28 can be expressed compactly as

$$\begin{bmatrix} \lambda_{qs} \\ \lambda_{ds} \\ \lambda_{0s} \\ \lambda'_{qr} \\ \lambda'_{dr} \\ \lambda'_{0r} \end{bmatrix} = \begin{bmatrix} L_{ls} + L_m & 0 & 0 & L_m & 0 & 0 \\ 0 & L_{ls} + L_m & 0 & 0 & L_m & 0 \\ 0 & 0 & L_{ls} & 0 & 0 & 0 \\ L_m & 0 & 0 & L'_{lr} + L_m & 0 & 0 \\ 0 & L_m & 0 & 0 & L'_{lr} + L_m & 0 \\ 0 & 0 & 0 & 0 & 0 & L'_{lr} \end{bmatrix} \begin{bmatrix} i_{qs} \\ i_{ds} \\ i_{0s} \\ i'_{qr} \\ i'_{dr} \\ i'_{0r} \end{bmatrix} \tag{6.29}$$

where the primed rotor quantities denote referred values to the stator side according to the following relationships:

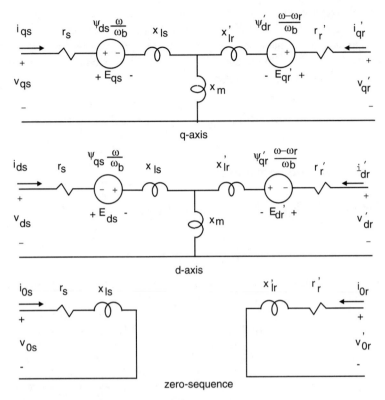

Figure 6.5 Equivalent circuit representation of an induction machine in the arbitrary reference frame.

$$\lambda'_{qr} = \frac{N_s}{N_r}\lambda_{qr} \qquad\qquad \lambda'_{dr} = \frac{N_s}{N_r}\lambda_{dr} \qquad\qquad (6.30)$$

$$i'_{qr} = \frac{N_r}{N_s}i_{qr} \qquad\qquad i'_{dr} = \frac{N_r}{N_s}i_{dr} \qquad\qquad (6.31)$$

$$L'_{lr} = \left(\frac{N_s}{N_r}\right)^2 L_{lr} \qquad\qquad\qquad (6.32)$$

and L_m, the magnetizing inductance on the stator side, is

$$L_m = \frac{3}{2}L_{ss} = \frac{3}{2}\frac{N_s}{N_r}L_{sr} = \frac{3}{2}\frac{N_s}{N_r}L_{rr} \qquad\qquad (6.33)$$

Substituting Eq. 6.29 into Eqs. 6.23 and 6.25, and then grouping the q, d, 0, and θ terms in the resulting voltage equations, we obtain the voltage equations that suggest the equivalent circuit shown in Fig. 6.5 for the induction machine in the arbitrary reference qd axes.

6.4.3 $qd0$ Torque Equation

The sum of the instantaneous input power to all six windings of the stator and rotor is given by

$$p_{in} = v_{as}i_{as} + v_{bs}i_{bs} + v_{cs}i_{cs} + v'_{ar}i'_{ar} + v'_{br}i'_{br} + v'_{cr}i'_{cr} \qquad W \qquad (6.34)$$

In terms of the $qd0$ quantities, the instantaneous input power is

$$p_{in} = \frac{3}{2}\left(v_{qs}i_{qs} + v_{ds}i_{ds} + 2v_{0s}i_{0s} + v'_{qr}i'_{qr} + v'_{dr}i'_{dr} + 2v'_{0r}i'_{0r}\right) \qquad W \qquad (6.35)$$

Using Eqs. 6.23 and 6.25 to substitute for the voltages on the right-hand side of Eq. 6.35, we obtain three kinds of terms: i^2r, $ip\lambda$, and $\omega\lambda i$ terms. The i^2r terms are the copper losses. The $ip\lambda$ terms represent the rate of exchange of magnetic field energy between windings. The $\omega\lambda i$ terms represent the rate of energy converted to mechanical work. The electromechanical torque developed by the machine is given by the sum of the $\omega\lambda i$ terms divided by the mechanical speed, that is

$$T_{em} = \frac{3}{2}\frac{P}{2\omega_r}\left[\omega(\lambda_{ds}i_{qs} - \lambda_{qs}i_{ds}) + (\omega - \omega_r)(\lambda'_{dr}i'_{qr} - \lambda'_{qr}i'_{dr})\right] \qquad N.m. \qquad (6.36)$$

Using the flux linkage relationship in Eq. 6.29, we can show that

$$\lambda_{ds}i_{qs} - \lambda_{qs}i_{ds} = -(\lambda'_{dr}i'_{qr} - \lambda'_{qr}i'_{dr}) = L_m(i_{dr}i_{qs} - i_{qr}i_{ds}) \qquad (6.37)$$

Equation 6.36 can thus also be expressed in the following forms:

$$
\begin{aligned}
T_{em} &= \frac{3}{2}\frac{P}{2}(\lambda'_{qr}i'_{dr} - \lambda'_{dr}i'_{qr}) \qquad N.m. \\
&= \frac{3}{2}\frac{P}{2}(\lambda_{ds}i_{qs} - \lambda_{qs}i_{ds}) \qquad N.m. \qquad (6.38)\\
&= \frac{3}{2}\frac{P}{2}L_m(i'_{dr}i_{qs} - i'_{qr}i_{ds}) \qquad N.m.
\end{aligned}
$$

For simulation purposes, the preference of one form over another is usually influenced by what variables are available in other parts of the simulation.

Another way to derive the torque expression is from the power absorbed by the speed voltages in the equivalent circuit representation given in Fig. 6.5. In the q, d, and 0 circuits, the resistive elements may be associated with copper losses, the reactances with the magnetic field energy, and the speed voltage terms associated with mechanical work. Defining the speed voltages in the same manner in which they have been defined in the literature on synchronous machines, that is

$$
\begin{array}{ll}
E_{qs} = \omega\lambda_{ds}, & E_{ds} = -\omega\lambda_{qs} \qquad V \\
E'_{qr} = (\omega - \omega_r)\lambda'_{dr}, & E'_{dr} = -(\omega - \omega_r)\lambda'_{qr} \qquad V
\end{array} \qquad (6.39)
$$

The real power absorbed by these four-speed voltage sources is the electromechanical power of the machine, that is

$$P_{em} = \frac{3}{2}\Re\left[(E_{qs} - jE_{ds})(i_{qs} - ji_{ds})^* + (E'_{qr} - jE'_{dr})(i'_{qr} - ji'_{dr})^*\right] \quad W \quad (6.40)$$

When the real part of Eq. 6.40 is divided by the rotor's mechanical speed, we will again obtain the expressions given in Eq. 6.38 for the electromagnetic torque.

A summary of the voltage, flux linkage, and torque equations of the induction machine in the arbitrary $qd0$ reference frame is given in Table 6.1.

TABLE 6.1 INDUCTION MACHINE EQUATIONS IN ARBITRARY REFERENCE FRAME

Stator $qd\,0$ voltage equations:

$$v_{qs} = p\lambda_{qs} + \omega\lambda_{ds} + r_s i_{qs}$$
$$v_{ds} = p\lambda_{ds} - \omega\lambda_{qs} + r_s i_{ds} \quad (6.41)$$
$$v_{0s} = p\lambda_{0s} + r_s i_{0s}$$

Rotor $qd\,0$ voltage equations:

$$v'_{qr} = p\lambda'_{qr} + (\omega - \omega_r)\lambda'_{dr} + r'_r i'_{qr}$$
$$v'_{dr} = p\lambda'_{dr} - (\omega - \omega_r)\lambda'_{qr} + r'_r i'_{dr} \quad (6.42)$$
$$v'_{0r} = p\lambda'_{0r} + r'_r i'_{0r}$$

where

$$\begin{bmatrix} \lambda_{qs} \\ \lambda_{ds} \\ \lambda_{0s} \\ \lambda'_{qr} \\ \lambda'_{dr} \\ \lambda'_{0r} \end{bmatrix} = \begin{bmatrix} L_{ls}+L_m & 0 & 0 & L_m & 0 & 0 \\ 0 & L_{ls}+L_m & 0 & 0 & L_m & 0 \\ 0 & 0 & L_{ls} & 0 & 0 & 0 \\ L_m & 0 & 0 & L'_{lr}+L_m & 0 & 0 \\ 0 & L_m & 0 & 0 & L'_{lr}+L_m & 0 \\ 0 & 0 & 0 & 0 & 0 & L'_{lr} \end{bmatrix} \begin{bmatrix} i_{qs} \\ i_{ds} \\ i_{0s} \\ i'_{qr} \\ i'_{dr} \\ i'_{0r} \end{bmatrix} \quad (6.43)$$

Torque equation:

$$T_{em} = \frac{3}{2}\frac{P}{2\omega_r}\left[\omega(\lambda_{ds}i_{qs} - \lambda_{qs}i_{ds}) + (\omega - \omega_r)(\lambda'_{dr}i'_{qr} - \lambda'_{qr}i'_{dr})\right] \quad N.m.$$
$$= \frac{3}{2}\frac{P}{2}(\lambda'_{qr}i'_{dr} - \lambda'_{dr}i'_{qr})$$
$$= \frac{3}{2}\frac{P}{2}(\lambda_{ds}i_{qs} - \lambda_{qs}i_{ds}) \quad (6.44)$$
$$= \frac{3}{2}\frac{P}{2}L_m(i'_{dr}i_{qs} - i'_{qr}i_{ds})$$

Base quantities. Oftentimes, machine equations are expressed in terms of the flux linkages per second, ψ's, and reactances, x's, instead of λ's and L's. These are related simply by the base or rated value of angular frequency, ω_b, that is

$$\psi = \omega_b \lambda \qquad V \text{ or per unit} \tag{6.45}$$

and

$$x = \omega_b L \qquad H \text{ or per unit} \tag{6.46}$$

where $\omega_b = 2\pi f_{rated}$ electrical radians per second, f_{rated} being the rated frequency in Hertz of the machine. With complex waveforms there may be justification to using the peak rather than the rms value of the rated phase voltage as the base value. The base quantities with peak rather than rms value of a P-pole, three-phase induction machine with rated line-to-line rms voltage, V_{rated}, and rated volt-ampere, S_{rated}, are as follows:

$$\text{base voltage } V_b = \sqrt{2/3} V_{rated}$$
$$\text{base volt} - \text{ampere } S_b = S_{rated}$$
$$\text{base peak current } I_b = 2S_b/3V_b$$
$$\text{base impedance } Z_b = V_b/I_b$$
$$\text{base torque } T_b = S_b/\omega_{bm}$$

where $\omega_{bm} = 2\omega_b/P$.

The equations of a symmetrical induction machine in the arbitrary reference frame in terms of the flux linkages per second and reactances at the base frequency are summarized in Table 6.2.

TABLE 6.2   ARBITRARY REFERENCE EQUATIONS IN TERMS OF ψ's AND X's

Stator and rotor $qd\,0$ voltage equations:

$$v_{qs} = \frac{p}{\omega_b}\psi_{qs} + \frac{\omega}{\omega_b}\psi_{ds} + r_s i_{qs}$$

$$v_{ds} = \frac{p}{\omega_b}\psi_{ds} - \frac{\omega}{\omega_b}\psi_{qs} + r_s i_{ds}$$

$$v_{0s} = \frac{p}{\omega_b}\psi_{0s} + r_s i_{0s}$$

$$v'_{qr} = \frac{p}{\omega_b}\psi'_{qr} + \left(\frac{\omega - \omega_r}{\omega_b}\right)\psi'_{dr} + r'_r i'_{qr} \tag{6.47}$$

$$v'_{dr} = \frac{p}{\omega_b}\psi'_{dr} - \left(\frac{\omega - \omega_r}{\omega_b}\right)\psi'_{qr} + r'_r i'_{dr}$$

$$v'_{0r} = \frac{p}{\omega_b}\psi'_{0r} + r'_r i'_{0r}$$

where

$$
\begin{bmatrix} \psi_{qs} \\ \psi_{ds} \\ \psi_{0s} \\ \psi'_{qr} \\ \psi'_{dr} \\ \psi'_{0r} \end{bmatrix} = \begin{bmatrix} x_{ls}+x_m & 0 & 0 & x_m & 0 & 0 \\ 0 & x_{ls}+x_m & 0 & 0 & x_m & 0 \\ 0 & 0 & x_{ls} & 0 & 0 & 0 \\ x_m & 0 & 0 & x'_{lr}+x_m & 0 & 0 \\ 0 & x_m & 0 & 0 & x'_{lr}+x_m & 0 \\ 0 & 0 & 0 & 0 & 0 & x'_{lr} \end{bmatrix} \begin{bmatrix} i_{qs} \\ i_{ds} \\ i_{0s} \\ i'_{qr} \\ i'_{dr} \\ i'_{0r} \end{bmatrix} \tag{6.48}
$$

Torque equation:

$$
T_{em} = \frac{3}{2}\frac{P}{2\omega_r}\left[\frac{\omega}{\omega_b}(\psi_{ds}i_{qs} - \psi_{qs}i_{ds}) + \frac{\omega - \omega_r}{\omega_b}(\psi'_{dr}i'_{qr} - \psi'_{qr}i'_{dr})\right] \qquad N.m.
$$

$$
= \frac{3}{2}\frac{P}{2\omega_b}(\psi'_{qr}i'_{dr} - \psi'_{dr}i'_{qr})
$$

$$
= \frac{3}{2}\frac{P}{2\omega_b}(\psi_{ds}i_{qs} - \psi_{qs}i_{ds}) \tag{6.49}
$$

$$
= \frac{3}{2}\frac{P}{2\omega_b}x_m(i'_{dr}i_{qs} - i'_{qr}i_{ds})
$$

6.5 $qd0$ STATIONARY AND SYNCHRONOUS REFERENCE FRAMES

As mentioned earlier, there is seldom a need to simulate an induction machine in the arbitrary rotating reference frame. For power system studies, induction machine loads, along with other types of power system components, are often simulated on a system's synchronously rotating reference frame. But for transient studies of adjustable-speed drives, it is usually more convenient to simulate an induction machine and its converter on a stationary reference frame. And for small-signal dynamic stability analysis about some operating condition, a synchronously rotating reference frame which yields steady values of steady-state voltages and currents under balanced conditions is used.

Since we have derived the equations of the induction machine for the general case, that is in the arbitrary rotating reference frame, the equations of the machine in the stationary and synchronously rotating reference frames can simply be obtained by setting the speed of the arbitrary reference frame, ω, to zero and ω_e, respectively. To distinguish among all these reference frames, the variables in the stationary and synchronously rotating reference frames will be identified by an additional superscript: s for variables in the stationary reference frame and e for variables in the synchronously rotating frame. The equations of a symmetrical induction machine in terms of ψ's and x's in the stationary and synchronous reference frames are summarized in Tables 6.3 and 6.4, respectively. The corresponding equivalent circuit representations are given in Figs. 6.6 and 6.7.

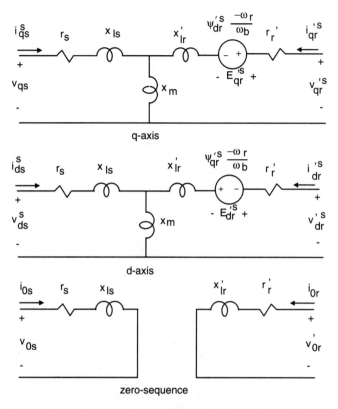

Figure 6.6   Equivalent circuit of an induction machine in the stationary reference frame.

TABLE 6.3   INDUCTION MACHINE EQUATIONS IN STATIONARY REFERENCE FRAME

Stator and rotor voltage equations:

$$v_{qs}^s = \frac{p}{\omega_b}\psi_{qs}^s + r_s i_{qs}^s$$

$$v_{ds}^s = \frac{p}{\omega_b}\psi_{ds}^s + r_s i_{ds}^s$$

$$v_{0s} = \frac{p}{\omega_b}\psi_{0s} + r_s i_{0s}$$

$$v_{qr}'^s = \frac{p}{\omega_b}\psi_{qr}'^s - \frac{\omega_r}{\omega_b}\psi_{dr}'^s + r_r' i_{qr}'^s$$

$$v_{dr}'^s = \frac{p}{\omega_b}\psi_{dr}'^s + \frac{\omega_r}{\omega_b}\psi_{qr}'^s + r_r' i_{dr}'^s$$

$$v_{0r}' = \frac{p}{\omega_b}\psi_{0r}' + r_r' i_{0r}'$$

(6.50)

Flux linkage equations:

$$
\begin{bmatrix}
\psi_{qs}^s \\
\psi_{ds}^s \\
\psi_{0s} \\
\psi_{qr}'^s \\
\psi_{dr}'^s \\
\psi_{0r}'
\end{bmatrix}
=
\begin{bmatrix}
x_{ls}+x_m & 0 & 0 & x_m & 0 & 0 \\
0 & x_{ls}+x_m & 0 & 0 & x_m & 0 \\
0 & 0 & x_{ls} & 0 & 0 & 0 \\
x_m & 0 & 0 & x_{lr}'+x_m & 0 & 0 \\
0 & x_m & 0 & 0 & x_{lr}'+x_m & 0 \\
0 & 0 & 0 & 0 & 0 & x_{lr}'
\end{bmatrix}
\begin{bmatrix}
i_{qs}^s \\
i_{ds}^s \\
i_{0s} \\
i_{qr}'^s \\
i_{dr}'^s \\
i_{0r}'
\end{bmatrix}
\tag{6.51}
$$

Torque equation:

$$
\begin{aligned}
T_{em} &= \frac{3}{2}\frac{P}{2\omega_b}(\psi_{qr}'^s i_{dr}'^s - \psi_{dr}'^s i_{qr}'^s) \qquad N.m.\\[6pt]
&= \frac{3}{2}\frac{P}{2\omega_b}(\psi_{ds}^s i_{qs}^s - \psi_{qs}^s i_{ds}^s) \\[6pt]
&= \frac{3}{2}\frac{P}{2\omega_b}x_m(i_{dr}'^s i_{qs}^s - i_{qr}'^s i_{ds}^s)
\end{aligned}
\tag{6.52}
$$

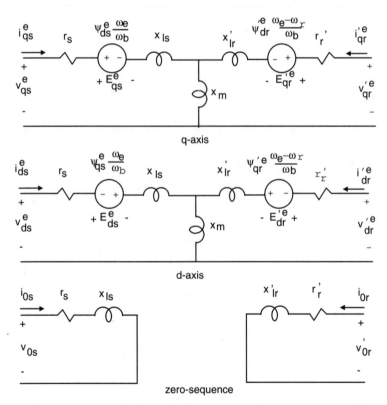

q-axis

d-axis

zero-sequence

Figure 6.7   Equivalent circuit representation of an induction machine in the synchronous rotating reference frame.

TABLE 6.4   INDUCTION MACHINE EQUATIONS IN SYNCHRONOUS REFERENCE FRAME

Stator and rotor $qd\,0$ voltage equations:

$$v_{qs}^e = \frac{p}{\omega_b}\psi_{qs}^e + \frac{\omega_e}{\omega_b}\psi_{ds}^e + r_s i_{qs}^e$$

$$v_{ds}^e = \frac{p}{\omega_b}\psi_{ds}^e - \frac{\omega_e}{\omega_b}\psi_{qs}^e + r_s i_{ds}^e$$

$$v_{0s} = \frac{p}{\omega_b}\psi_{0s} + r_s i_{0s}$$

$$v_{qr}^{\prime e} = \frac{p}{\omega_b}\psi_{qr}^{\prime e} + \left(\frac{\omega_e - \omega_r}{\omega_b}\right)\psi_{dr}^{\prime e} + r_r' i_{qr}^{\prime e} \qquad (6.53)$$

$$v_{dr}^{\prime e} = \frac{p}{\omega_b}\psi_{dr}^{\prime e} - \left(\frac{\omega_e - \omega_r}{\omega_b}\right)\psi_{qr}^{\prime e} + r_r' i_{dr}^{\prime e}$$

$$v_{0r}' = \frac{p}{\omega_b}\psi_{0r}' + r_r' i_{0r}'$$

Flux linkage equations:

$$\begin{bmatrix} \psi_{qs}^e \\ \psi_{ds}^e \\ \psi_{0s} \\ \psi_{qr}^{\prime e} \\ \psi_{dr}^{\prime e} \\ \psi_{0r}' \end{bmatrix} = \begin{bmatrix} x_{ls}+x_m & 0 & 0 & x_m & 0 & 0 \\ 0 & x_{ls}+x_m & 0 & 0 & x_m & 0 \\ 0 & 0 & x_{ls} & 0 & 0 & 0 \\ x_m & 0 & 0 & x_{lr}'+x_m & 0 & 0 \\ 0 & x_m & 0 & 0 & x_{lr}'+x_m & 0 \\ 0 & 0 & 0 & 0 & 0 & x_{lr}' \end{bmatrix} \begin{bmatrix} i_{qs}^e \\ i_{ds}^e \\ i_{0s} \\ i_{qr}^{\prime e} \\ i_{dr}^{\prime e} \\ i_{0r}' \end{bmatrix} \quad (6.54)$$

Torque equation:

$$T_{em} = \frac{3}{2}\frac{P}{2\omega_b}(\psi_{qr}^{\prime e} i_{dr}^{\prime e} - \psi_{dr}^{\prime e} i_{qr}^{\prime e}) \qquad N.m.$$

$$= \frac{3}{2}\frac{P}{2\omega_b}(\psi_{ds}^e i_{qs}^e - \psi_{qs}^e i_{ds}^e) \qquad (6.55)$$

$$= \frac{3}{2}\frac{P}{2\omega_b}x_m(i_{dr}^{\prime e} i_{qs}^e - i_{qr}^{\prime e} i_{ds}^e)$$

6.6 STEADY-STATE MODEL

For steady-state operation of the induction machine from a balanced three-phase sinusoidal supply, we can express the stator voltages and currents as

$$v_{as} = V_{ms}\cos(\omega_e t) \qquad\qquad i_{as} = I_{ms}\cos(\omega_e t - \phi_s)$$

$$v_{bs} = V_{ms}\cos\left(\omega_e t - \frac{2\pi}{3}\right) \qquad i_{bs} = I_{ms}\cos\left(\omega_e t - \frac{2\pi}{3} - \phi_s\right) \qquad (6.56)$$

$$v_{cs} = V_{ms}\cos\left(\omega_e t - \frac{4\pi}{3}\right) \qquad i_{cs} = I_{ms}\cos\left(\omega_e t - \frac{4\pi}{3} - \phi_s\right)$$

Similarly, the rotor voltages and currents with the rotor rotating at a slip of s may be expressed as

$$v_{ar} = V_{mr}\cos(s\omega_e t - \theta_r(0) - \delta) \qquad\qquad i_{ar} = I_{mr}\cos(s\omega_e t - \theta_r(0) - \delta - \phi_r)$$

$$v_{br} = V_{mr}\cos\left(s\omega_e t - \frac{2\pi}{3} - \theta_r(0) - \delta\right) \qquad i_{br} = I_{mr}\cos\left(s\omega_e t - \frac{2\pi}{3} - \theta_r(0) - \delta - \phi_r\right) \ (6.57)$$

$$v_{cr} = V_{mr}\cos\left(s\omega_e t - \frac{4\pi}{3} - \theta_r(0) - \delta\right) \qquad i_{cr} = I_{mr}\cos\left(s\omega_e t - \frac{4\pi}{3} - \theta_r(0) - \delta - \phi_r\right)$$

Transforming the above sets of stator and rotor abc variables to a stationary $qd0$ reference that has its q-axis aligned with the axis of the a-phase of the stator, we obtain

$$\vec{\mathbf{v}}_s = v_{qs}^s - j v_{ds}^s = V_{ms} e^{j\omega_e t}$$

$$\vec{\mathbf{i}}_s = i_{qs}^s - j i_{ds}^s = I_{ms} e^{-j\phi_s} e^{j\omega_e t}$$

$$\vec{\mathbf{v}}_r = (v_{qr}^r - j v_{dr}^r) e^{j\theta_r(t)} = (V_{mr} e^{j(s\omega_e t - \theta_r(0) - \delta)}) e^{j\theta_r(t)} \qquad (6.58)$$

$$\vec{\mathbf{i}}_r = (i_{qr}^r - j i_{dr}^r) e^{j\theta_r(t)} = (I_{mr} e^{j(s\omega_e t - \theta_r(0) - \delta - \phi_r)}) e^{j\theta_r(t)}$$

where the superscripts, s and r, are used to denote the $qd0$ components in the stationary and the rotor $qd0$ reference frame, respectively. In steady-state operation with the rotor rotating at a constant speed of $\omega_e(1-s)$,

$$\theta_r(t) = \omega_e(1-s)t + \theta_r(0) \qquad (6.59)$$

Substituting Eq. 6.59 into the expressions for the rotor's voltage and current space vectors of Eq. 6.58 and simplifying, we obtain

$$\vec{\mathbf{v}}_r = v_{qr}^s - j v_{dr}^s = V_{mr} e^{-j\delta} e^{j\omega_e t}$$

$$\vec{\mathbf{i}}_r = i_{qr}^s - j i_{dr}^s = I_{mr} e^{-j(\delta + \phi_r)} e^{j\omega_e t} \qquad (6.60)$$

Steady-state analysis with sinusoidal excitation is usually performed using rms time phasors. Denoting rms values of space vectors by bold upper-case letters with an arrow accent and rms time phasors by bold upper-case letters with a tilde accent, we have

$$\tilde{\mathbf{V}}_{as} = \frac{V_{ms}}{\sqrt{2}} e^{j0}$$

$$\tilde{\mathbf{I}}_{as} = \frac{I_{ms}}{\sqrt{2}} e^{-j\phi_s} \tag{6.61}$$

$$\tilde{\mathbf{V}}_{ar} = \frac{V_{mr}}{\sqrt{2}} e^{-j\delta}$$

$$\tilde{\mathbf{I}}_{ar} = \frac{I_{mr}}{\sqrt{2}} e^{-j(\delta+\phi_r)} \tag{6.62}$$

and

$$\vec{\mathbf{V}}_{qs}^s - j\vec{\mathbf{V}}_{ds}^s = \frac{v_{qs}^s - jv_{ds}^s}{\sqrt{2}} = \tilde{\mathbf{V}}_{as} e^{j\omega_e t}$$

$$\vec{\mathbf{I}}_{qs}^s - j\vec{\mathbf{I}}_{ds}^s = \frac{i_{qs}^s - ji_{ds}^s}{\sqrt{2}} = \tilde{\mathbf{I}}_{as} e^{j\omega_e t} \tag{6.63}$$

$$\vec{\mathbf{V}}_{qr}^s - j\vec{\mathbf{V}}_{dr}^s = \frac{v_{qr}^s - jv_{dr}^s}{\sqrt{2}} = \tilde{\mathbf{V}}_{ar} e^{j\omega_e t}$$

$$\vec{\mathbf{I}}_{qr}^s - j\vec{\mathbf{I}}_{dr}^s = \frac{i_{qr}^s - ji_{dr}^s}{\sqrt{2}} = \tilde{\mathbf{I}}_{ar} e^{j\omega_e t} \tag{6.64}$$

When the rotor voltages and currents are referred to the stator side, we have

$$\vec{\mathbf{V}}_{qr}^{\prime s} - j\vec{\mathbf{V}}_{dr}^{\prime s} = \left(\frac{N_s}{N_r}\right)\tilde{\mathbf{V}}_{ar} e^{j\omega_e t} = \tilde{\mathbf{V}}_{ar}' e^{j\omega_e t}$$

$$\vec{\mathbf{I}}_{qr}^{\prime s} - j\vec{\mathbf{I}}_{dr}^{\prime s} = \left(\frac{N_r}{N_s}\right)\tilde{\mathbf{I}}_{ar} e^{j\omega_e t} = \tilde{\mathbf{I}}_{ar}' e^{j\omega_e t} \tag{6.65}$$

The stationary qd voltage and flux linkage equations in Eqs. 6.50 and 6.51, expressed in terms of the above rms $qd0$ voltages and currents, can be assembled into the following complex rms space voltage vector equations:

$$\vec{\mathbf{V}}_{qs}^s - j\vec{\mathbf{V}}_{ds}^s = (r_s + j\omega_e(L_{ls} + L_m))(\vec{\mathbf{I}}_{qs}^s - j\vec{\mathbf{I}}_{ds}^s) + j\omega_e L_m(\vec{\mathbf{I}}_{qr}^{\prime s} - j\vec{\mathbf{I}}_{dr}^{\prime s})$$

$$\vec{\mathbf{V}}_{qr}^{\prime s} - j\vec{\mathbf{V}}_{dr}^{\prime s} = j(\omega_e - \omega_r)L_m(\vec{\mathbf{I}}_{qs}^s - j\vec{\mathbf{I}}_{ds}^s) \tag{6.66}$$

$$+ (r_r' + j(\omega_e - \omega_r)(L_{lr}' + L_m))(\vec{\mathbf{I}}_{qr}^{\prime s} - j\vec{\mathbf{I}}_{dr}^{\prime s})$$

Using the relationships between rms space vectors and rms time phasors given in Eqs. 6.63 through 6.65, writing $\omega_e - \omega_r$ as $s\omega_e$, and dropping the common $e^{j\omega_e t}$ term, Eq. 6.66 becomests

$$\tilde{\mathbf{V}}_{as} = (r_s + j\omega_e L_{ls})\tilde{\mathbf{I}}_{as} + j\omega_e L_m(\tilde{\mathbf{I}}_{as} + \tilde{\mathbf{I}}_{ar}')$$

$$\tilde{\mathbf{V}}_{ar}' = (r_r' + js\omega_e L_{lr}')\tilde{\mathbf{I}}_{ar}' + js\omega_e L_m(\tilde{\mathbf{I}}_{as} + \tilde{\mathbf{I}}_{ar}') \tag{6.67}$$

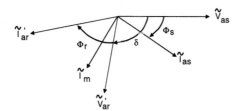

Figure 6.8 Phasor diagram of steady-state motoring condition.

Dividing through by the slip, the second equation in Eq. 6.67 becomes

$$\frac{\tilde{V}'_{ar}}{s} = (\frac{r'_r}{s} + j\omega_e L'_{lr})\tilde{I}'_{ar} + j\omega_e L_m(\tilde{I}_{as} + \tilde{I}'_{ar}) \qquad (6.68)$$

Equation 6.67 can thus be rewritten as

$$\tilde{V}_{as} = r_s\tilde{I}_{as} + j\frac{\omega_e}{\omega_b}x_{ls}\tilde{I}_{as} + j\frac{\omega_e}{\omega_b}x_m(\tilde{I}_{as} + \tilde{I}'_{ar})$$

$$\frac{\tilde{V}'_{ar}}{s} = \frac{r'_r}{s}\tilde{I}'_{ar} + j\frac{\omega_e}{\omega_b}x'_{lr}\tilde{I}'_{ar} + j\frac{\omega_e}{\omega_b}x_m(\tilde{I}_{as} + \tilde{I}'_{ar}) \qquad (6.69)$$

Figure 6.8 shows a phasor diagram of the stator and rotor variables, with $\tilde{I}_m = \tilde{I}_{as} + \tilde{I}'_{ar}$. An equivalent circuit which portrays the phasor equations of Eq. 6.69 is shown in Fig. 6.9a. The same equivalent circuit can be obtained through physical reasoning of the steady-state operation of the machine. Core losses, which may be considered to vary roughly with the square of the airgap flux, or $\hat{E}_m$, are often incorporated into the circuit by adding a core loss resistor r_c across $\hat{E}_m$, in parallel with the magnetizing reactance, x_m.

By adding and subtracting r'_r of the rotor branch in Fig. 6.9a and regrouping these terms differently, we can obtain the alternate equivalent circuit shown of Fig. 6.9b. When the rotor winding resistance, r'_r, is separated out, the remaining resistance of $r'_r(1-s)/s$ may be associated with the developed mechanical power, whereas the resistor, r'_r/s, is associated with the power through the airgap. If the primary interest is on the torque developed, the stator side can be replaced by a Thevenin's equivalent, as shown in Fig. 6.9c.

In steady-state, the average value of the developed power is

$$P_{em} = 3I'^2_{ar}\frac{1-s}{s}r'_r \qquad (6.70)$$

From which we can derive an expression for the average value of the electromagnetic torque developed by the P-pole machine:

$$T_{em} = \frac{P_{mech}}{\omega_{rm}} = 3I'^2_{ar}r'_r\frac{(1-s)}{s\omega_{sm}(1-s)}$$

$$= \frac{3I'^2_{ar}r'_r}{s\omega_{sm}} = \frac{3I'^2_{ar}r'_r}{\omega_{sm} - \omega_{rm}} \qquad (6.71)$$

With variable-frequency operations, the last form of the torque expression becomes useful when ω_e is zero and the value of slip s given by $(\omega_e - \omega_r)/\omega_e$ becomes undefined.

The operating characteristics of the induction machine operated from a current source supply is quite different from that when it is operated from a voltage supply. With a voltage

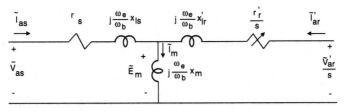

(a) Based on phasor equation Eq. 6.69

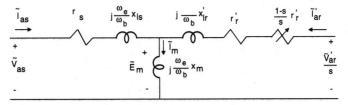

(b) Alternate representation showing rotor resistance

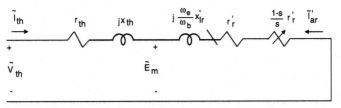

(c) Stator Thevenin's equivalent, with $\omega_e = \omega_b$ and $\tilde{V}'_{ar} = 0$

Figure 6.9   Equivalent circuit representations.

source, when the stator series impedance drop is small, the airgap voltage usually remains close to the supply voltage over a wide range of loading. With a current source, the terminal and airgap voltage could vary somewhat with loading when the stator current is held constant.

Consider first the case of the machine operated with a constant voltage supply to the stator windings, the terminals of the rotor windings shorted, that is $\tilde{V}'_{ar} = 0$. Figure 6.9c shows the resulting circuit where the stator-side portion of the circuit, including the magnetizing branch, has been replaced by a Thevenin's equivalent, as seen by the referred rotor current $\tilde{I}'_{ar}$. In terms of the original stator parameters, the Thevenin's circuit parameters are

$$\tilde{V}_{th} = \frac{jx_m}{r_s + j(x_{ls} + x_m)} \tilde{V}_{as}$$

$$\mathbf{Z}_{th} = r_{th} + jx_{th} = \frac{jx_m(r_s + jx_{ls})}{r_s + j(x_{ls} + x_m)}$$

(6.72)

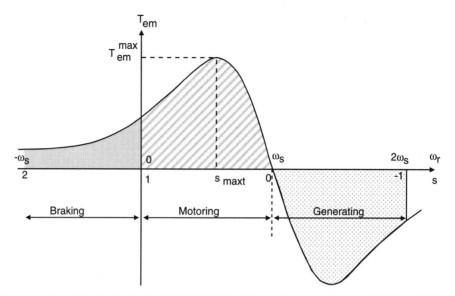

Figure 6.10   Average torque vs. slip with constant voltage supply.

In terms of the above Thevenin's circuit parameters, the expression for the average value of the electromagnetic torque developed by the P-pole machine with constant voltage supply is

$$T_{em} = \frac{3P}{2\omega_e} \frac{V_{th}^2(r_r'/s)}{(r_{th} + r_r'/s)^2 + (x_{th} + x_{lr}')^2} \tag{6.73}$$

Given the supply voltage, frequency, and the machine's parameters, we can use Eq. 6.73 to compute the average value of the torque developed at various slip values. Figure 6.10 is a plot of the average value of the torque developed vs. slip, showing what is normally termed the braking, motoring, and generating regions of the machine's operation. In the motoring region, the developed torque is in the direction of rotation. The generating region is above synchronous speed, and the developed torque is opposite to the direction of rotation. In the braking region, the developed torque is in the direction of rotation of the excitation mmf, but opposite to the direction of the rotor rotation.

Maximum torque is developed when the variable resistor, r_r'/s, draws maximum power from the source in Fig. 6.9a, that is when

$$\frac{r_r'}{s_{maxt}} = \sqrt{r_{th}^2 + (x_{th} + x_{lr}')^2} \tag{6.74}$$

Substituting this value of s_{maxt} in Eq. 6.73, the maximum torque developed with constant voltage supply is

$$T_{em}^{max} = \frac{3P}{4\omega_e} \frac{V_{th}^2}{r_{th} + \sqrt{r_{th}^2 + (x_{th} + x_{lr}')^2}} \qquad Nm \tag{6.75}$$

The above expression indicates that the value of the maximum torque developed is not dependent on the value of the rotor circuit resistance, r_r, even though the slip at which maximum torque is developed depends on the value of r_r.

Since the stator input impedance varies with slip, with constant voltage supply, the input current will vary accordingly. The stator input impedance is given by

$$\mathbf{Z}_{in} = r_s + jx_{ls} + \frac{jx_m(r_r'/s + jx_{lr}')}{r_r'/s + j(x_{lr}' + x_m)} \tag{6.76}$$

The stator input current and complex power can be computed from

$$\tilde{\mathbf{I}}_{as} = \frac{\tilde{\mathbf{V}}_{as}}{\mathbf{Z}_{in}} \tag{6.77}$$

$$\mathbf{S}_{in} = P_{in} + jQ_{in} = 3\tilde{\mathbf{V}}_{as}\tilde{\mathbf{I}}_{as}^* \tag{6.78}$$

With a constant current supply, the stator current is held fixed but the stator voltage will vary as the stator input impedance, given by Eq. 6.76, varies with slip. For a singly excited machine, $\tilde{\mathbf{V}}_{ar}'$ is zero, the average torque developed can be determined using Eq. 6.71 with the rotor current determined from

$$\tilde{\mathbf{E}}_m = j(\tilde{\mathbf{I}}_{as} + \tilde{\mathbf{I}}_{ar}')x_m = (r_r'/s + jx_{lr}')\tilde{\mathbf{I}}_{ar}' \tag{6.79}$$

that is

$$I_{ar}'^2 = \frac{x_m^2 I_{as}^2}{(r_r'/s)^2 + (x_{lr}' + x_m)^2} \tag{6.80}$$

Figures 6.11 and 6.12 illustrate the difference in the operating behavior of a 20-hp, 60-Hz, 220-V three-phase induction machine operated with constant voltage and with constant current supply, neglecting iron saturation. The equivalent circuit parameters of the 20-hp machine are as follows:

$r_s = 0.1062\ \Omega$	$x_{ls} = 0.2145\ \Omega$
$r_r' = 0.0764\ \Omega$	$x_{lr}' = 0.2145\ \Omega$
$x_m = 5.834\ \Omega$	$J_{rotor} = 2.8\ kgm^2$

With saturation, the effective value of x_m will decrease as saturation sets in; as such, the increases in stator voltage and torque in the saturation region will be lower than those shown in Figs. 6.11 and 6.12. The difference in the shape of the torque vs. speed curve is evident. With constant voltage supply, the airgap voltage E_m at large slip is higher than that with constant current supply. The considerably lower starting torque for the constant current supply case is because the values of airgap voltage (or flux) and rotor current are relatively lower than in the constant voltage supply at starting.

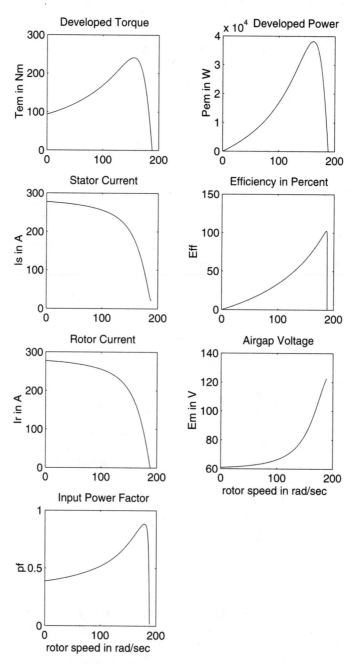

Figure 6.11 Operating characteristics with constant voltage supply.

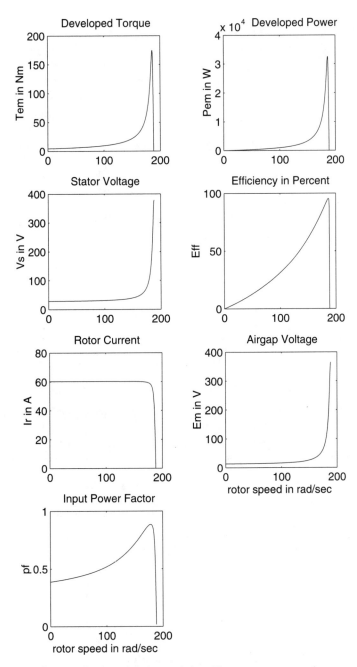

Figure 6.12   Operating characteristics with constant current supply.

6.7 TRANSIENT MODEL

The time constant of the rotor flux linkage, λ_r', can be quite long, especially in large or high-efficiency induction machines. For relatively short-term transients, it may be acceptable to use a simpler transient model that is obtained by assuming that the rotor winding flux linkage remains essentially constant over the brief period of the transient. Such a model is used in transient stability programs to represent induction motor loads.

We begin the derivation of the simple transient model by considering a perturbation to the variables in the following expressions of the flux linkages of the stator and rotor windings in the synchronously rotating qd axes:

$$\lambda_{qs}^e = L_s i_{qs}^e + L_m i_{qr}'^e$$

$$\lambda_{ds}^e = L_s i_{ds}^e + L_m i_{dr}'^e$$

$$\lambda_{qr}'^e = L_r' i_{qr}'^e + L_m i_{qs}^e \tag{6.81}$$

$$\lambda_{dr}'^e = L_r' i_{dr}'^e + L_m i_{ds}^e$$

With perturbations in the stator and rotor currents, the expressions of the perturbed stator and rotor flux linkages are

$$\lambda_{qso}^e + \Delta\lambda_{qs}^e = L_s(i_{qso}^e + \Delta i_{qs}^e) + L_m(i_{qro}'^e + \Delta i_{qr}'^e)$$

$$\lambda_{dso}^e + \Delta\lambda_{ds}^e = L_s(i_{dso}^e + \Delta i_{ds}^e) + L_m(i_{dro}'^e + \Delta i_{dr}'^e)$$

$$\lambda_{qro}'^e + \Delta\lambda_{qr}'^e = L_r'(i_{qro}'^e + \Delta i_{qr}'^e) + L_m(i_{qso}^e + \Delta i_{qs}^e) \tag{6.82}$$

$$\lambda_{dro}'^e + \Delta\lambda_{dr}'^e = L_r'(i_{dro}'^e + \Delta i_{dr}'^e) + L_m(i_{dso}^e + \Delta i_{ds}^e)$$

where the changes are denoted by Δ and the nominal or predisturbance values are denoted by an additional subscript, o.

If the rotor flux linkages, $\lambda_{qr}'^e$ and $\lambda_{dr}'^e$, are assumed to remain unchanged by the perturbation, that is the components, $\Delta\lambda_{qr}'^e$ and $\Delta\lambda_{dr}'^e$, are zero, or

$$\Delta\lambda_{qr}'^e = L_r' \Delta i_{qr}'^e + L_m \Delta i_{qs}^e = 0$$

$$\Delta\lambda_{dr}'^e = L_r' \Delta i_{dr}'^e + L_m \Delta i_{ds}^e = 0 \tag{6.83}$$

The above relationship between the changes in the stator and rotor qd currents can then be back substituted into the stator flux linkage expressions of Eq. 6.82 to obtain

$$\lambda_{qso}^e + \Delta\lambda_{qs}^e = L_s i_{qso}^e + L_m i_{qro}'^e + \left(L_s - \frac{L_m^2}{L_r'}\right)\Delta i_{qs}^e$$

$$\lambda_{dso}^e + \Delta\lambda_{ds}^e = L_s i_{dso}^e + L_m i_{dro}'^e + \left(L_s - \frac{L_m^2}{L_r'}\right)\Delta i_{ds}^e \tag{6.84}$$

Since the predisturbance values of flux linkages and currents also satisfy Eq. 6.81, Eq. 6.84 can be simplified to give the following relationships between the changes in stator flux linkages and changes in stator currents when the rotor flux linkages remain constant:

$$\Delta\lambda^e_{qs} = \left(L_s - \frac{L^2_m}{L'_r}\right)\Delta i^e_{qs} = L'_s\Delta i^e_{qs}$$

$$\Delta\lambda^e_{ds} = \left(L_s - \frac{L^2_m}{L'_r}\right)\Delta i^e_{ds} = L'_s\Delta i^e_{ds}$$

(6.85)

where L'_s is known as the stator transient inductance.

Let's define the qd component voltages behind the stator transient inductance as

$$E'_{qs} = \omega_e\left(\lambda^e_{ds} - L'_s i^e_{ds}\right)$$

$$E'_{ds} = -\omega_e\left(\lambda^e_{qs} - L'_s i^e_{qs}\right)$$

(6.86)

Using Eqs. 6.81 and 6.85, we can show that these voltage components behind the stator transient inductance are proportional to the rotor flux linkages, that is

$$
\begin{aligned}
E'_{qs} &= \omega_e\left(\lambda^e_{ds} - L'_s i^e_{ds}\right) \\
&= \omega_e\left(L_s i^e_{ds} + L_m i'^e_{dr} - L'_s i^e_{ds}\right) \\
&= \omega_e\frac{L_m}{L'_r}\lambda'^e_{dr} \\
E'_{ds} &= -\omega_e\left(\lambda^e_{qs} - L'_s i^e_{qs}\right) \\
&= -\omega_e\left(L_s i^e_{qs} + L_m i'^e_{qr} - L'_s i^e_{qs}\right) \\
&= -\omega_e\frac{L_m}{L'_r}\lambda'^e_{qr}
\end{aligned}
$$

(6.87)

Thus, when the rotor flux linkages are constant, the newly defined voltages behind the transient inductance, E'_{qs} and E'_{ds}, will be constant.

Consider now the following qd stator voltage equations in the synchronously rotating frame:

$$\frac{d\lambda^e_{qs}}{dt} = v^e_{qs} - r_s i^e_{qs} - \omega_e\lambda^e_{ds}$$

$$\frac{d\lambda^e_{ds}}{dt} = v^e_{ds} - r_s i^e_{ds} + \omega_e\lambda^e_{qs}$$

(6.88)

Using Eq. 6.86 to replace the speed voltage terms on the right-hand side of Eq. 6.88, we obtain the desired set of stator voltage equations of the simple transient model for operating conditions in which the rotor flux linkages can be assumed to remain constant:

$$\frac{d\lambda^e_{qs}}{dt} = v^e_{qs} - r_s i^e_{qs} - E'_{qs} + \omega_e L'_s i^e_{ds}$$

$$\frac{d\lambda^e_{ds}}{dt} = v^e_{ds} - r_s i^e_{ds} - E'_{ds} - \omega_e L'_s i^e_{qs}$$

(6.89)

For steady-state condition, we can put the above equation in the corresponding time phasor form. Setting the time derivative terms on the left to zero for steady-state,

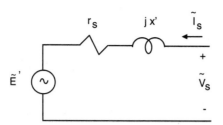

Figure 6.13 Simplified transient circuit representation.

multiplying the d equation of Eq. 6.89 by $-j$ and adding that to the q equation, and regrouping terms, we obtain

$$(v^e_{qs} - jv^e_{ds}) = (r_s + j\omega_e L'_s)(i^e_{qs} - ji^e_{ds}) + (E'_{qs} - jE'_{ds}) \tag{6.90}$$

In phasor notation, Eq. 6.90 becomes

$$\tilde{\mathbf{V}}_{as} = (r_s + j\omega_e L'_s)\tilde{\mathbf{I}}_{as} + \tilde{\mathbf{E}}' \tag{6.91}$$

An equivalent circuit representation of this equation is shown in Fig. 6.13. It consists of a single-phase voltage, $\tilde{\mathbf{E}}'$, behind the transient impedance, $\mathbf{Z}' = r_s + jx'$. In a simplified transient analysis of the induction machine, the value of E' before the disturbance can be determined from the operating condition just before the disturbance, that is, $\tilde{\mathbf{E}}'$ can be calculated from

$$\tilde{\mathbf{E}}' = \tilde{\mathbf{V}}_{aso} - (r_s + jx')\tilde{\mathbf{I}}_{aso} \tag{6.92}$$

where $\tilde{\mathbf{V}}_{aso}$ and $\tilde{\mathbf{I}}_{aso}$ are the phasor values just before the disturbance.

This simplified transient model can also be used to study the behavior of an induction machine over a brief period following a change in stator operating condition, such as a disconnection from the supply or a terminal short-circuit. In each of these cases, it may be assumed that the rotor flux linkage immediately after the disturbance is the same as that just before the disturbance. Subsequent decay of the trapped rotor flux linkage is at a rate that is dependent on the L/R ratio of the effective impedance as seen by the rotor current. In the case of a motor disconnected from its supply, the rotor flux linkage will decay at a rate corresponding to that determined by the open-circuit time constant:

$$T'_0 = \frac{x'_{lr} + x_m}{\omega_e r'_r} \tag{6.93}$$

When the stator terminals are shorted together, the corresponding time constant of the rotor flux linkage is known as the short-circuit transient time constant. It is given by

$$T' = \left(x'_{lr} + \frac{x_{ls} x_m}{x_{ls} + x_m}\right) \frac{1}{\omega_e r'_r} \tag{6.94}$$

For extended time studies, the allowance for changes in value of the rotor flux linkage, and $\tilde{\mathbf{E}}'$ for an open-circuit connection or a short-circuit fault at the stator terminals, can be approximated by an exponential decay of time constants, T'_0 or T', respectively. For example, the form of $\tilde{\mathbf{E}}'$ following an open-circuit may be approximated by

$$\tilde{\mathbf{E}}'(t) = \tilde{\mathbf{E}}'(0)e^{-\frac{t}{T_0'}} \tag{6.95}$$

where $t = 0$ is the time at which the disturbance occurs.

The phasor form of the simplified transient model shown in Fig. 6.13 can also be derived directly by perturbing the stator phasor voltage equation given in Eq. 6.69. With the perturbations in the electrical variables, the stator phasor voltage equation may be expressed as

$$\tilde{\mathbf{V}}_{aso} + \Delta\tilde{\mathbf{V}}_{as} = \left(r_s + j\frac{\omega_e}{\omega_b}(x_{ls} + x_m) \right)(\tilde{\mathbf{I}}_{aso} + \Delta\tilde{\mathbf{I}}_{as}) + j\frac{\omega_e}{\omega_b}x_m(\tilde{\mathbf{I}}_{aro}' + \Delta\tilde{\mathbf{I}}_{ar}) \tag{6.96}$$

From $\tilde{\Psi}_{aro}' = (x_{lr}' + x_m)\tilde{\mathbf{I}}_{aro}' + x_m\tilde{\mathbf{I}}_{aso}$ before the disturbance, $\Delta\tilde{\Psi}_{ar}' = 0$, and

$$\tilde{\Psi}_{aro}' + \Delta\tilde{\Psi}_{ar}' = (x_{lr}' + x_m)(\tilde{\mathbf{I}}_{aro}' + \Delta\tilde{\mathbf{I}}_{ar}') + x_m(\tilde{\mathbf{I}}_{aso} + \Delta\tilde{\mathbf{I}}_{as}) \tag{6.97}$$

we can determine that the change in rotor current is

$$\Delta\tilde{\mathbf{I}}_{ar}' = \frac{-x_m}{x_{lr}' + x_m}\Delta\tilde{\mathbf{I}}_{as} \tag{6.98}$$

Using the above relations to replace $(\tilde{\mathbf{I}}_{aro}' + \Delta\tilde{\mathbf{I}}_{ar})$ in the perturbed stator voltage equation, Eq. 6.96 can be simplified to

$$\begin{aligned}
\tilde{\mathbf{V}}_{aso} + \Delta\tilde{\mathbf{V}}_{as} &= \left(r_s + j\frac{\omega_e}{\omega_b}\left(\frac{x_{ls} + x_m - x_m^2}{x_{lr}' + x_m} \right) \right)(\tilde{\mathbf{I}}_{aso} + \Delta\tilde{\mathbf{I}}_{as}) \\
&\quad + j\frac{\omega_e}{\omega_b}\left(\frac{x_m}{x_{lr}' + x_m} \right)\tilde{\Psi}_{aro}'
\end{aligned} \tag{6.99}$$

Denoting the stator transient reactance by x' and the voltage behind the stator transient impedance x' by $\tilde{\mathbf{E}}'$, that is

$$x' = x_{ls} + x_m - \frac{x_m^2}{x_{lr}' + x_m} \tag{6.100}$$

$$\tilde{\mathbf{E}}' = j\frac{\omega_e}{\omega_b}\left(\frac{x_m}{x_{lr}' + x_m} \right)\tilde{\Psi}_{aro}' \tag{6.101}$$

the stator voltage equation when the rotor flux linkage may be assumed constant may be expressed as

$$\tilde{\mathbf{V}}_{as} = (r_s + jx')\tilde{\mathbf{I}}_{as} + \tilde{\mathbf{E}}' \tag{6.102}$$

It is the same as that given in Eq. 6.91.

6.8 SIMULATION OF AN INDUCTION MACHINE
IN THE STATIONARY REFERENCE FRAME

We shall next show how the developed model can be used to simulate a three-phase, P-pole, symmetrical induction machine in the stationary reference frame with winding connections as shown in Fig. 6.14. We start by considering the equations of its input voltages for the given neutral connections of the stator and rotor windings. The three applied voltages to the stator terminals, v_{ag}, v_{bg}, and v_{cg}, need not be balanced nor sinusoidal. In general, for the connection shown, the three stator phase voltages are

$$v_{as} = v_{ag} - v_{sg}$$

$$v_{bs} = v_{bg} - v_{sg} \tag{6.103}$$

$$v_{cs} = v_{cg} - v_{sg}$$

or

$$3v_{sg} = (v_{as} + v_{bs} + v_{cs}) - (v_{ag} + v_{bg} + v_{cg}) \tag{6.104}$$

In simulation, the voltage, v_{sg}, can be determined from the flow of phase currents in the neutral connection, that is

$$v_{sg} = R_{sg}(i_{as} + i_{bs} + i_{cs}) + L_{sg}\frac{d}{dt}(i_{as} + i_{bs} + i_{cs}) = 3\left(R_{sg} + L_{sg}\frac{d}{dt}\right)i_{0s} \tag{6.105}$$

where R_{sg} and L_{sg} are the resistance and inductance of the connection between the two neutral points, s and g. Clearly, when point s is solidly connected to point g, that is the connection impedance is zero, v_{sg} will be zero.

Equation 6.104 indicates that the voltage, v_{sg}, is also zero when the zero-sequence components of the source and stator phase voltages are both zero, as in a balanced operation of a symmetrical induction machine with a balanced set of applied voltages that are sinusoidal and balanced. That $v_{as} + v_{bs} + v_{cs}$ will be zero for balanced operation can be reached using the fact that i_{0s} is zero in Eq. 6.50.

In practice, a three-wire connection between supply and stator windings is quite common. With a three-wire connection, the stator zero-sequence current, i_{0s}, equal to $(i_{as} + i_{bs} + i_{cs})/3$, is zero by physical constraint, irrespective of whether the three-phase currents are balanced or not. Even though the sum of the phase currents is constrained by the physical connection to be zero, the phase currents may be unbalanced, as in a single-phasing situation where the a-phase is open and $i_{bs} + i_{cs} = 0$. But, the free-floating stator neutral voltage, v_{sg}, may not be zero, depending on whether the applied voltages are sinusoidal and balanced.

Where the applied voltages are non-sinusoidal, as from the output voltages of a six-step inverter, the zero-sequence component of the applied voltages may not be zero. Where the stator winding's neutral is free-floating, the voltage, v_{sg}, can be determined in simulation using Eq. 6.105 with $L_{sg} = 0$ and R_{sg} set as high as possible to approximate the actual open-circuit condition.

Consider next the transformation of stator phase voltages to $qd0$ stationary voltages. With the q-axis of the stationary qd reference always aligned with the stator a-phase axis

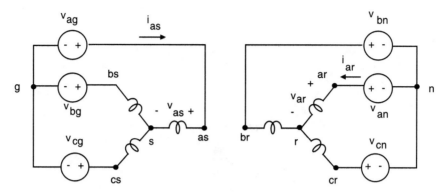

Figure 6.14   Stator and rotor connections.

and a ω of zero, we can obtain the following relationships by setting θ to zero in the transformation matrix of Eq. 6.18:

$$v_{qs}^s = \frac{2}{3}v_{as} - \frac{1}{3}v_{bs} - \frac{1}{3}v_{cs} = \frac{2}{3}v_{ag} - \frac{1}{3}v_{bg} - \frac{1}{3}v_{cg} - v_{sg}$$

$$v_{ds}^s = \frac{1}{\sqrt{3}}(v_{cs} - v_{bs}) = \frac{1}{\sqrt{3}}(v_{cg} - v_{bg}) \tag{6.106}$$

$$v_{0s} = \frac{1}{3}(v_{as} + v_{bs} + v_{cs}) = \frac{1}{3}(v_{ag} + v_{bg} + v_{cg}) - v_{sg}$$

The transformation of the abc rotor winding voltages to the same stationary $qd0$ reference frame can be taken in a single step using Eq. 6.18 with θ equal to $\theta_r(t)$, or in two separate steps as follows: First, transform the "referred" rotor phase voltages to a $qd0$ reference frame that is attached to the rotor with the q-axis of that frame aligned to the axis of the rotor's a-phase winding. The resulting equations of the rotor voltages after the first transformation are similar to those of the stator voltages, that is

$$v_{qr}^{\prime r} = \frac{2}{3}v_{ar}^\prime - \frac{1}{3}v_{br}^\prime - \frac{1}{3}v_{cr}^\prime = \frac{2}{3}v_{an}^\prime - \frac{1}{3}v_{bn}^\prime - \frac{1}{3}v_{cn}^\prime - v_{rn}^\prime$$

$$v_{dr}^{\prime r} = \frac{1}{\sqrt{3}}(v_{cr}^\prime - v_{br}^\prime) = \frac{1}{\sqrt{3}}(v_{cn}^\prime - v_{bn}^\prime) \tag{6.107}$$

$$v_{0r}^\prime = \frac{1}{3}(v_{ar}^\prime + v_{br}^\prime + v_{cr}^\prime) = \frac{1}{3}(v_{an}^\prime + v_{bn}^\prime + v_{cn}^\prime) - v_{rn}^\prime$$

where $v_{rn}^\prime$ is the voltage between the points, r and n. The prime here denotes values referred to the stator side. The voltages, $v_{qr}^{\prime r}$ and $v_{dr}^{\prime r}$, are still slip frequency voltages.

Next, perform a rotational transformation of the above qd rotor quantities onto the same stationary qd frame for the stator quantities, using either Eq. 5.139 or Eq. 5.145:

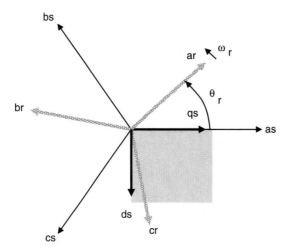

Figure 6.15 Axes of *abc* windings and *qd* stationary.

$$v'^s_{qr} = v''^r_{qr} \cos\theta_r(t) + v''^r_{dr} \sin\theta_r(t)$$

$$v'^s_{dr} = -v''^r_{qr} \sin\theta_r(t) + v''^r_{dr} \cos\theta_r(t) \tag{6.108}$$

$$\theta_r(t) = \int_0^t \omega_r(t)dt + \theta_r(0)$$

where, as shown in Fig. 6.15, $\theta_r(t)$ is the rotor angle angle, $\theta_r(0)$ is its initial value, and $\omega_r(t)$ is the instantaneous angular speed of the rotor.

The $qd0$ voltages at both the stator and rotor terminals, referred to the same stationary $qd0$ reference frame, can now be used as inputs, along with the load torque, to the model equations of the induction machine in the stationary $qd0$ frame to obtain the corresponding $qd0$ currents in the stationary reference frame. Usually, the stationary $qd0$ stator and rotor currents are then transformed back to phase currents for use in other parts of the simulation. The stator *abc* phase currents can be determined from the stator qd currents using the inverse transformation given by Eq. 6.19 with a θ of zero.

$$i_{as} = i^s_{qs} + i_{0s}$$

$$i_{bs} = -\frac{1}{2}i^s_{qs} - \frac{\sqrt{3}}{2}i^s_{ds} + i_{0s} \tag{6.109}$$

$$i_{cs} = -\frac{1}{2}i^s_{qs} + \frac{\sqrt{3}}{2}i^s_{ds} + i_{0s}$$

Again, the *abc* rotor currents can be obtained from the $qd0$ rotor currents using Eq. 6.19 with θ equal to $\theta_r(t)$ in one step, or in two separate steps as follows: First, transform the stationary qd rotor currents back to the qd frame attached to the rotor, using a rotational transformation:

$$i''^r_{qr} = i'^s_{qr} \cos\theta_r(t) - i'^s_{dr} \sin\theta_r(t)$$

$$i''^r_{dr} = i'^s_{qr} \sin\theta_r(t) + i'^s_{dr} \cos\theta_r(t) \tag{6.110}$$

Then, resolve the qd rotor currents in the rotor reference frame to the desired rotor abc phase currents, using the inverse $qd0$ to abc transformation of Eq. 6.19 with θ equal to zero:

$$i'_{ar} = i''_{qr} + i'_{0r}$$

$$i'_{br} = -\frac{1}{2}i''_{qr} - \frac{\sqrt{3}}{2}i''_{dr} + i'_{0r} \tag{6.111}$$

$$i'_{cr} = -\frac{1}{2}i''_{qr} + \frac{\sqrt{3}}{2}i''_{dr} + i'_{0r}$$

Observe that the machine's qd circuits have the same T-connection of inductors as in the equivalent circuit of the single-phase transformer. As such, the same technique to handle the cut-set inductors in the single-phase transformer simulation can be used here.

The model equations of the induction machine in the stationary $qd0$ reference frame may be rearranged into the following form for simulation:

$$\psi^s_{qs} = \omega_b \int \left\{ v^s_{qs} + \frac{r_s}{x_{ls}}(\psi^s_{mq} - \psi^s_{qs}) \right\} dt$$

$$\psi^s_{ds} = \omega_b \int \left\{ v^s_{ds} + \frac{r_s}{x_{ls}}(\psi^s_{md} - \psi^s_{ds}) \right\} dt \tag{6.112}$$

$$i_{0s} = \frac{\omega_b}{x_{ls}} \int \{v_{0s} - i_{0s}r_s\} dt$$

$$\psi'^s_{qr} = \omega_b \int \left\{ v'^s_{qr} + \frac{\omega_r}{\omega_b}\psi'^s_{dr} + \frac{r'_r}{x'_{lr}}(\psi^s_{mq} - \psi'^s_{qr}) \right\} dt$$

$$\psi'^s_{dr} = \omega_b \int \left\{ v'^s_{dr} - \frac{\omega_r}{\omega_b}\psi'^s_{qr} + \frac{r'_r}{x'_{lr}}(\psi^s_{md} - \psi'^s_{dr}) \right\} dt \tag{6.113}$$

$$i'_{0r} = \frac{\omega_b}{x'_{lr}} \int \{v'_{0r} - i'_{0r}r'_r\} dt$$

$$\psi^s_{mq} = x_m(i^s_{qs} + i'^s_{qr})$$

$$\psi^s_{md} = x_m(i^s_{ds} + i'^s_{dr}) \tag{6.114}$$

$$\psi^s_{qs} = x_{ls}i^s_{qs} + \psi^s_{mq} \qquad\qquad i^s_{qs} = \frac{\psi^s_{qs} - \psi^s_{mq}}{x_{ls}}$$

$$\psi^s_{ds} = x_{ls}i^s_{ds} + \psi^s_{md} \qquad\qquad i^s_{ds} = \frac{\psi^s_{ds} - \psi^s_{md}}{x_{ls}}$$

$$\tag{6.115}$$

$$\psi'^s_{qr} = x'_{lr}i'^s_{qr} + \psi^s_{mq} \qquad\qquad i'^s_{qr} = \frac{\psi'^s_{qr} - \psi^s_{mq}}{x'_{lr}}$$

$$\psi'^s_{dr} = x'_{lr}i'^s_{dr} + \psi^s_{md} \qquad\qquad i'^s_{dr} = \frac{\psi'^s_{dr} - \psi^s_{md}}{x'_{lr}}$$

where

$$\frac{1}{x_M} = \frac{1}{x_m} + \frac{1}{x_{ls}} + \frac{1}{x'_{lr}} \tag{6.116}$$

and

$$\psi^s_{mq} = x_M \left(\frac{\psi^s_{qs}}{x_{ls}} + \frac{\psi'^s_{qr}}{x'_{lr}} \right)$$

$$\psi^s_{md} = x_M \left(\frac{\psi^s_{ds}}{x_{ls}} + \frac{\psi'^s_{dr}}{x'_{lr}} \right) \tag{6.117}$$

The torque equation is

$$T_{em} = \frac{3}{2}\frac{P}{2\omega_b}(\psi^s_{ds}i^s_{qs} - \psi^s_{qs}i^s_{ds}) \qquad N.m. \tag{6.118}$$

The equation of motion of the rotor is obtained by equating the inertia torque to the accelerating torque, that is

$$J\frac{d\omega_{rm}}{dt} = T_{em} + T_{mech} - T_{damp} \qquad N.m. \tag{6.119}$$

In Eq. 6.119, T_{mech} is the externally-applied mechanical torque in the direction of the rotor speed and T_{damp} is the damping torque in the direction opposite to rotation. The value of T_{mech} will be negative for the motoring condition, as in the case of a load torque, and will be positive for the generating condition, as in the case of an applied shaft torque from a prime mover.

When used in conjunction with Eqs. 6.112 and 6.113, the per unit speed, ω_r/ω_b, needed for building the speed voltage terms in the rotor voltage equations, can be obtained by integrating

$$\frac{2J\omega_b}{P}\frac{d(\omega_r/\omega_b)}{dt} = T_{em} + T_{mech} - T_{damp} \qquad N.m. \tag{6.120}$$

Often, the above equation of motion is written in terms of the inertia constant, H, defined as the ratio of the kinetic energy of the rotating mass at base speed to the rated power, that is

$$H = \frac{J\omega^2_{bm}}{2S_b} \tag{6.121}$$

Expressed in per unit values of the machine's own base power and voltage, Eq. 6.120 can be rewritten as

$$2H\frac{d(\omega_r/\omega_b)}{dt} = T_{em} + T_{mech} - T_{damp} \qquad \text{in per unit} \tag{6.122}$$

Figure 6.16 shows the flow of variables in both the q- and d-axes circuit simulation of a three-phase induction machine. Note the cross-coupling speed voltage terms between the parts of the q-axis and d-axis circuit simulation. The zero-sequence components, however, are not coupled to qd circuits.

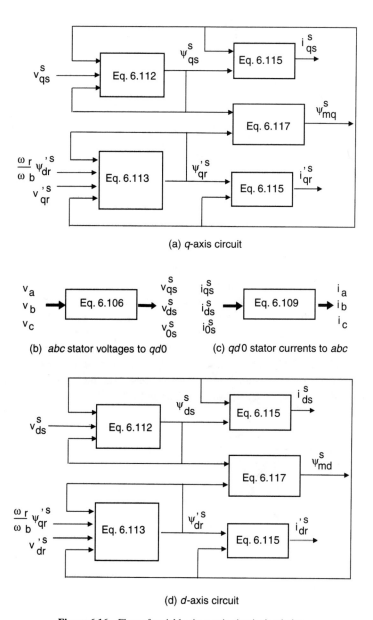

Figure 6.16   Flow of variables in q-axis circuit simulation.

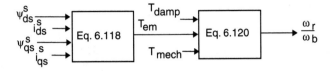

(e) Developed torque, speed and angle

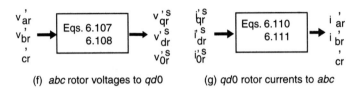

(f) *abc* rotor voltages to *qd*0   (g) *qd*0 rotor currents to *abc*

Figure 6.16 *(cont.)* Flow of variables in *d*-axis circuit simulation.

6.8.1 Saturation of Mutual Flux

In reality, iron saturation affects both the leakage and mutual flux components, the latter more so than the former. The leakage flux paths of the machine windings are very complicated, compared to those of the mutual flux path. Thus, accounting for the effects of iron saturation on the leakage is not an easy task. In this text, we will confine our attention to account for the saturation effects on the magnetizing flux path. For machines with a uniform airgap, we can reasonably assume that iron saturation affects the q- and d-axis components in the same manner, so that we can then use a common saturation characteristic like the one shown in Fig. 6.17 for both these components, even though localized saturation can actually affect the qd components non-uniformly. Denoting the saturated value of the mutual flux linkages by the superscript, *sat*, the saturated value of the mutual flux linkage per second in

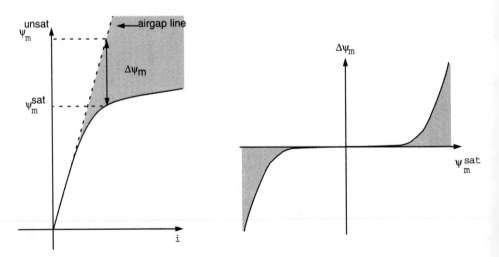

Figure 6.17 Saturation characteristic.

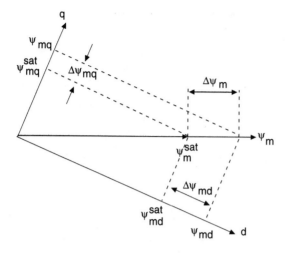

Figure 6.18   Approximation of saturation in qd components.

the q-axis is given by

$$\psi_{mq}^{sat} = \psi_{mq}^{s} - \Delta\psi_{mq}^{s}$$

$$= x_m \left(\frac{\psi_{qs}^{s} - \psi_{mq}^{sat}}{x_{ls}} + \frac{\psi_{qr}^{'s} - \psi_{mq}^{sat}}{x_{lr}^{'}} \right) - \Delta\psi_{mq}^{s} \qquad (6.123)$$

Grouping like terms and using Eq. 6.116, the above expression can be written as

$$\psi_{mq}^{sat} = \frac{x_M}{x_{ls}}\psi_{qs}^{s} + \frac{x_M}{x_{lr}^{'}}\psi_{qr}^{'s} - \frac{x_M}{x_m}\Delta\psi_{mq}^{s} \qquad (6.124)$$

Similarly, the saturated value of the d-axis mutual flux linkage is given by

$$\psi_{md}^{sat} = \frac{x_M}{x_{ls}}\psi_{ds}^{s} + \frac{x_M}{x_{lr}^{'}}\psi_{dr}^{'s} - \frac{x_M}{x_m}\Delta\psi_{md}^{s} \qquad (6.125)$$

Assuming a proportional reduction in the flux linkages of the q- and d-axes, we obtain from Fig. 6.18 the following relationships:

$$\Delta\psi_{mq}^{s} = \frac{\psi_{mq}^{sat}}{\psi_{m}^{sat}}\Delta\psi_m$$

$$\qquad (6.126)$$

$$\Delta\psi_{md}^{s} = \frac{\psi_{md}^{sat}}{\psi_{m}^{sat}}\Delta\psi_m$$

where

$$\psi_{m}^{sat} = \sqrt{(\psi_{mq}^{sat})^2 + (\psi_{md}^{sat})^2} \qquad (6.127)$$

The relationship between $\Delta\psi_m$ and ψ_{m}^{sat} can be determined from the no-load test curve of the machine, as shown in Fig. 6.17. Figure 6.19 highlights the part of the induction machine simulation that is affected by the inclusion of mutual flux saturation. The value of $\Delta\psi_m$ can be determined from ψ_{m}^{sat} using piece-wise segments or a look-up table as in the representation of the transformer core saturation described in Chapter 4.

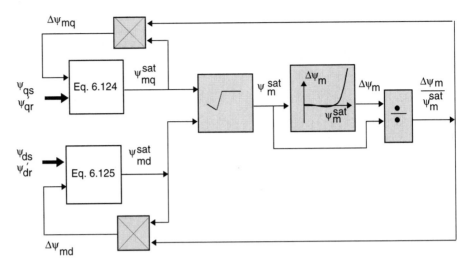

Figure 6.19   Simulation of mutual flux saturation.

6.9 LINEARIZED MODEL

Quite often, a linearized model of a nonlinear system may be needed in preliminary control design, which employs well-established linear control techniques. Linearized models of an induction machine can be obtained by performing a small perturbation on all the machine variables, of either the full nonlinear model in the synchronously rotating reference frame, or of the phasor equations where these variables in steady-state are not time-varying [44, 53].

The combined differential-algebraic equations of Eqs. 6.53, 6.54, and 6.55 in Table 6.4, and that of the equation of rotor motion, Eq. 6.120, is of the general form:

$$f(\dot{x}, x, u, y) = 0 \qquad (6.128)$$

where x is a vector of state variables, $(\psi_{qs}; \psi'_{qr}; \psi_{ds}; \psi'_{dr}; \omega_r/\omega_b)$; u is the vector of inputs, $(v^e_{qs}; v^e_{ds}; T_{mech})$; and y is the vector of desired outputs, such as i^e_{qs}, i^e_{ds}, and T_{em}. The dot accent on the x vector denotes the operator d/dt.

When a small displacement, denoted by Δ, is applied to each of the components of the x, u, and y vectors, the perturbed variables will still satisfy the governing differential-algebraic equations, that is

$$f(\dot{x}_{x=x_o} + \Delta\dot{x}, x_o + \Delta x, u_o + \Delta u, y_o + \Delta y) = 0 \qquad (6.129)$$

where the subscript, o, denotes the steady-state value before the disturbance. In steady-state, $\dot{x}_{x=xo} = 0$ and

$$f(x_o, u_o, y_o) = 0 \qquad (6.130)$$

The above steady-state relations can be back-substituted into Eq. 6.129 to simply it. And when higher order Δ terms are ignored, the remaining terms can be regrouped and written in the following standard state-variable form:

$$\Delta\dot{x} = A\,\Delta x + B\,\Delta u$$
$$\Delta y = C\,\Delta x + D\,\Delta u \tag{6.131}$$

With MATLAB/SIMULINK, the whole perturbation and regrouping process of determining the values of the $[A, B, C, D]$ matrices of a nonlinear system that you have simulated in SIMULINK can be done numerically. The procedure to obtain the numerical values of the $[A, B, C, D]$ matrices about some desired operating point can be accomplished in three steps: First, the complete steady-state of the SIMULINK system at some desired operating point must be determined using the SIMULINK **trim** function:

$$[x_o, u_o, y_o] = trim['s4eig', x_o^g, u_o^g, y_o^g] \tag{6.132}$$

where *s4eig* is the filename of the SIMULINK simulation of the system with all its inputs u defined by input ports numbered in the same sequence as in u, all its outputs y defined by output ports numbered in the same sequence as in y, and x the vector of the state variables. The actual ordering of the states in x by SIMULINK can be determined using the MATLAB function:

$$[sizes, x0, xstr] = s4eig([], [], [], 0) \tag{6.133}$$

The **trim** function uses a quadratic nonlinear programming algorithm. Good initial guesses of x_o, u_o, and y_o are to be provided to the function through x_o^g, u_o^g, and y_o^g. Index variables can be used to specify which elements of x_o^g, u_o^g, and y_o^g are to be held fixed and which are allowed to change during the iterations for the steady-state.

When the initial values are obtained, we can then proceed to use the MATLAB **linmod** function to determine the $[A, B, C, D]$ matrices of the small-signal model of the nonlinear system about the chosen steady-state operating point.

Figure 6.20a shows an overall diagram of the simulation *s4eig* of a three-phase induction machine in the synchronously rotating reference frame. The details inside the *Qaxis*, *Daxis*, and *Rotor* blocks are shown in Figs. 6.20b through 6.20d, respectively. Note that the inputs and outputs are connected to sequentially numbered input and output ports as required by **trim** and **linmod** to obtain the steady-state and small-signal models.

The synchronously rotating reference frame is chosen because the steady-state voltages, fluxes, currents, and torque of the machine in this reference frame are all non-time-varying. Thus, the $[A, B, C, D]$ matrices will also be time-invariant.

The MATLAB Control Toolbox [45] has several useful functions that can directly make use of the $[A, B, C, D]$ of the small-signal model. Among them are block-building functions for interconnecting two or more subsystems, functions to determine transfer functions, functions to determine the Nyquist and root locus plots, and functions to help design compensators and estimators.

For example, given below is the MATLAB *m4.m* script file which performs the computation of the $(\Delta\omega_r/\omega_b)/\Delta T_{mech}$ transfer functions at several T_{mech} levels. The main tasks performed by the script file are:

1. Load the file containing the machine parameters and rating.
2. Obtain the actual ordering of x.

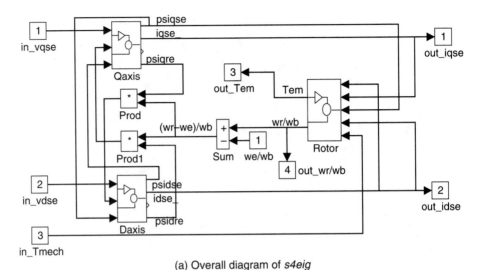

(a) Overall diagram of *s4eig*

Figure 6.20   Simulation *s4eig* of a three-phase induction machine in the synchronously rotating reference frame.

3.   **a.** Setting up the Tmech loading in the T vector.
     b. Determining the complete steady-state at each Tmech loading.
     c. Determining the [A, B, C, D] matrices of the small-signal model. Determine the transfer functions by:

4. Plot the poles of the transfer functions at various loading levels.

```
% M-file for Project 4 on linearized analysis in Chapter 6
% To be used in the same directory containing the SIMULINK
% file, s4eig, of an induction machine in the synchronous
% reference frame

% It does the following:

%    (a) loads parameters and rating of machine;
%
%    (b) sets up Tmech loading levels in T vector for tasks (i) thru (iv)

%
%    (i) uses Simulink trim function to determine
% steady-state of a desired operating point of the
% Simulink system s4eig;

%   (ii) uses Matlab linmod to determine A,B,C, and D;

% (iii) uses Matlab ss2tf to determine
% the speed-torque transfer function

%   (iv) uses Matlab tf2zp to determine
```

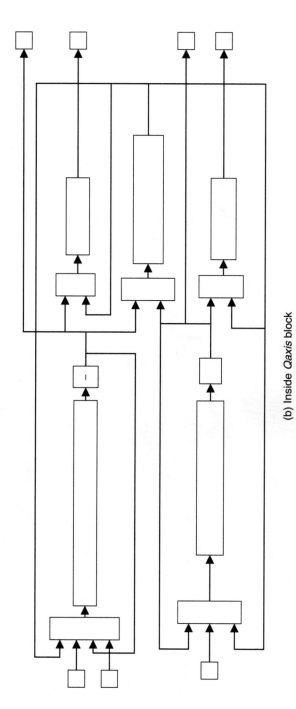

(b) Inside *Qaxis* block

Figure 6.20 *(cont.)*: Simulation *s4eig* of a three-phase induction machine in the synchronously rotating reference frame.

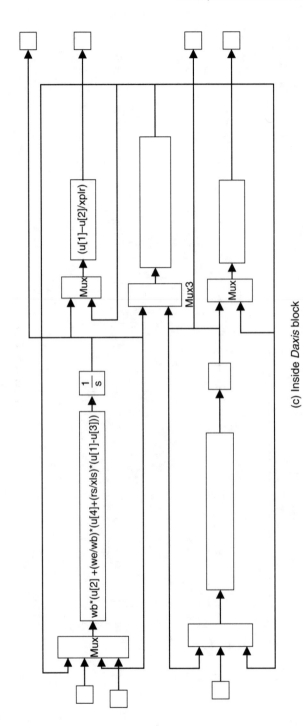

(c) Inside *Daxis* block

Figure 6.20 *(cont.):* Simulation *s4eig* of a three-phase induction machine in the synchronously rotating reference frame.

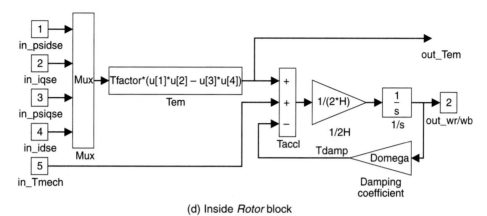

(d) Inside *Rotor* block

Figure 6.20 *(cont.):* Simulation *s4eig* of a three-phase induction machine in the synchronously rotating reference frame.

```
% the poles and zeros of the speed-torque transfer function

%   (c) generates a plot of poles for changing
% Tmech

% Enter script file to load parameters and rating of motor

p20hp % parameters of 20 hp three-phase induction motor

% After initializing the parameters of the imwe
% in the Matlab workspace, make initial guess of all
% values in the state (x), input (u), and output (y)
% vectors. From the diagram of imwe, we see that
% u = [vqe; vde; Tmech]
% y = [iqse; idse; Tem; wr/wb]

% Since the ordering is dependent on how Simulink assembles
% the system equation, the current ordering of the state
% variables can be determined using

% define all initial value variables
Psiqso = Vm;
Psipqro = Vm;
Psidso = 0;
Psipdro = 0;
wrbywbo = 1;

[sizes,x0,xstr] = s4eig([], [], [], 0)

% which yields xstr =

%/s4eig/Daxis/psids_
%/s4eig/Daxis/psidr'_
%/s4eig/Qaxis/psiqs_
```

```
%/s4eig/Qaxis/psiqr'_
%/s4eig/Rotor/1/s

% or  x = [ psids; psipdr; psiqs; psipqr; wr/wb ]

% Input the following guesses:

Vm = Vrated*sqrt(2/3) % peak voltage per phase
T = [0:-Tb:-Tb]% specify range and increment of
% external mech torque, negative for
% motoring
xg = [Psidso; Psipdro; Psiqso; Psipqro; wrbywbo];
% IMPORTANT!! must update xg in accordance with
% actual ordering of state variables
yg=[0; 0; -0; 1];

index = 0;

% For loop to compute transfer function at various levels
% of Tmech loading specified in T vector

for Tmech = T
index = index + 1;     % set index

u = [Vm; 0; Tmech];
x=xg;
y=yg;

% use index variables to specify which of the above
% inputs in the initial guess should be held fixed

iu=[1; 2; 3]; % all input variables held fixed
ix = []; % all state variables can vary
iy = []; % all output free

% Use Simulink trim function to determine the desired
% steady-state operating point. For more details
% type help trim after the matlab prompt.
% Results from trim must be verified.

[x,u,y,dx] = trim('s4eig',x,u,y,ix,iu,iy);

xg = x; % store current steady-state to use as guesses for next
yg = y; % increment in loading

% Use Matlab linmod function to determine the state-space
% representation at the chosen operating point
%
% dx/dt = [A] x + [B] u
%    y   = [C] x + [D] u

[A, B, C, D] = linmod('s4eig', x, u);
```

```matlab
% For transfer function (Dwr/wb)/DTmech

bt=B(:,3); % select third column input
ct=C(4,:); % select fourth row output
dt=D(4,3); % select fourth row and third column

% Use Matlab ss2tf to determine transfer function
% of the system at the chosen operating point.
% If desired, transfer function can be printed using

[numt(index,:),dent(index,:)] = ss2tf(A,bt,ct,dt,1);

 printsys(numt(index,:),dent(index,:),'s')

% Use Matlab tf2zp to determine the poles and zeros
% of system transfer function

[zt(:,index),pt(:,index),kt(index)] = tf2zp(numt(index,:),dent(index,:));

% z,p column vectors
% For transfer function (Dwr/wb)/Dvqse

bv=B(:,1); % select third column input
cv=C(4,:); % select fourth row output
dv=D(4,1); % select fourth row and third column

% Use Matlab ss2tf to determine transfer function
% of the system at the chosen operating point.


[numv(index,:),denv(index,:)] = ss2tf(A,bv,cv,dv,1);

printsys(numv(index,:),denv(index,:),'s')

% Use Matlab tf2zp to determine the poles and zeros
% of system transfer function

[zv(:,index),pv(:,index),kv(index)] = tf2zp(numv(index,:),denv(index,:));
% z,p column vectors

end % end of for Tmech loop

% Print loading level and corresponding gain,zeros, and poles
% of (Dwr/wb)/DTmech

index = 0;
for Tmech = T
index = index + 1;

fprintf('\n (Dwr/wb)/DTmech \n')
fprintf('Tmech loading is %10.2e\n',Tmech )
fprintf('Gain is %10.2e\n',kt(index))
[mzero,nzero] = size(zt);
```

```
fprintf('\nZeros are: \n')
for m = 1:mzero
fprintf('%12.3e %12.3ei\n',real(zt(m,index)), imag(zt(m,index)))
end
[mpole,npole] = size(pt);
fprintf('\nPoles are: \n')
for m = 1:mpole
fprintf('%12.3e %12.3ei\n',real(pt(m,index)), imag(pt(m,index)))
end
end

% Print loading level and corresponding gain,zeros, and poles
% of (Dwr/wb)/Dvqse

index = 0;
for Tmech = T
index = index + 1;

fprintf('\n (Dwr/wb)/Dvqse \n')
fprintf('Tmech loading is %10.2e\n',Tmech )
fprintf('Gain is %10.2e\n',kv(index))
[mzero,nzero] = size(zv);
fprintf('\nZeros are: \n')
for m = 1:mzero
fprintf('%12.3e %12.3ei\n',real(zv(m,index)), imag(zv(m,index)))
end
[mpole,npole] = size(pv);
fprintf('\nPoles are: \n')
for m = 1:mpole
fprintf('%12.3e %12.3ei\n',real(pv(m,index)), imag(pv(m,index)))
end
end

% Root Locus Plot of (Dwr/wb)/Dvqse with a sensor TF of
% 1/(0.05s+1)
clf;
index =2 % pick the rated torque condition
numG = numv(index,:); % numerator of (Dwr/wb)/Dvqse
denG = denv(index,:);    % denominator of (Dwr/wb)/Dvqse
numH = 1; % numerator of sensor's TF
denH = [0.05 1]; % denominator of sensor's TF
[numGH,denGH] = series(numG,denG,numH,denH);
k=logspace(0,5,600); % define gain array
[r,k]= rlocus(numGH,denGH,k); % store loci

% customize plot over the region of interest
ymax = 400;
xmax = 400;
plot(r,'-') % plot loci using a black continuous line
title('Root Locus of (Dwr/wb)/Dvqse')
xlabel('real axis')
ylabel('imag axis')
axis square; % equal length axis
```

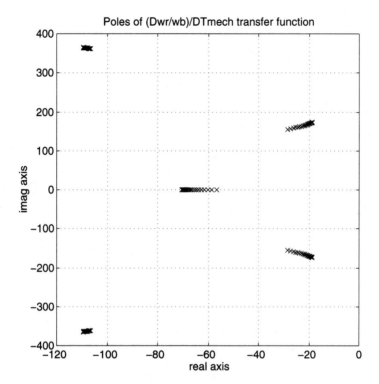

Figure 6.21   Migration of poles of $(\Delta\omega_r/\omega_b)/\Delta T_{mech}$ from no-load to twice base torque.

```
hold on;
grid on;
axis('equal') %equal scaling for both axis
axis([-xmax,xmax/10,-ymax/10,ymax]) % resize
plot(real(r(1,:)), imag(r(1,:)),'x') % mark poles with x's
hold off
% Transfer to keyboard for printing of plot
% disp('When ready to return to program type ''return''');
% keyboard
% In keyboard mode, that is K >>, you can use the Matlab command
% kk = rlocfind(numGH,denGH) and the mouse to determine
% the gain of any point on the root locus.
```

Figure 6.21 shows the poles of the $(\Delta\omega_r/\omega_b)/\Delta T_{mech}$ transfer functions of the 20-hp motor of Projects 2 and 4 about steady-state operating points ranging from zero to twice base torque. Since base torque is defined as base volt ampere divided by base mechanical speed, where base mechanical speed is the mechanical synchronous speed, the base torque for an induction motor will be less than its rated torque. The gain, zeros, and poles of the transfer function vary from the values at no-load of

```
Tmech loading is   0.00e+00
Gain is   1.50e-01
```

```
Zeros are:
  -9.630e+01      3.603e+02i
  -9.630e+01     -3.603e+02i
  -6.706e+01      1.667e+01i
  -6.706e+01     -1.667e+01i

Poles are:
  -1.095e+02      3.648e+02i
  -1.095e+02     -3.648e+02i
  -1.864e+01      1.735e+02i
  -1.864e+01     -1.735e+02i
  -7.050e+01      0.000e+00i
```

to values at twice base load torque of

```
Tmech loading is  -1.58e+02
Gain is    1.50e-01

Zeros are:
  -9.653e+01      3.590e+02i
  -9.653e+01     -3.590e+02i
  -6.683e+01      4.309e+01i
  -6.683e+01     -4.309e+01i

Poles are:
  -1.066e+02      3.617e+02i
  -1.066e+02     -3.617e+02i
  -2.840e+01      1.558e+02i
  -2.840e+01     -1.558e+02i
  -5.680e+01      0.000e+00i
```

6.10 SINGLE-PHASE INDUCTION MOTOR

A three-phase symmetrical induction motor upon losing one of its stator phase supplies while running may continue to operate as essentially a single-phase motor with the remaining line-to-line voltage across the other two connected phases. The same machine might not, however, restart with a single-phase supply unless a rotating magnetic field is somehow produced at standstill. In very small single-phase induction motors meant for operating on single-phase voltage supply, a rotating field component is produced using a shaded pole. The short-circuited turn on the shaded pole shown in Fig. 6.22a introduces a delay to the time-varying flux component through the shaded portion of the stator pole; as a result, the flux under the unshaded portion peaks before that under the shaded portion of the pole. An auxiliary winding, however, is required to produce the stronger revolving field needed in larger motors. The axis of the auxiliary winding is placed about 90° ahead of that of the main winding. Of a smaller conductor size and requiring less turns than those of the main stator winding, the auxiliary winding has a higher resistance to inductance ratio than the main winding. When connected in parallel with the main winding across the same

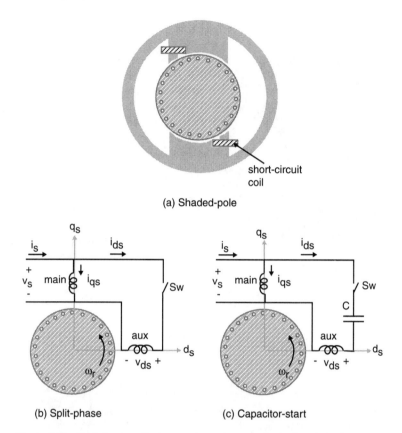

(a) Shaded-pole

(b) Split-phase   (c) Capacitor-start

Figure 6.22   Shaded-pole, split-phase, and capacitor-start single-phase induction motors.

ac voltage, as in the split-phase single-phase induction motor of Fig. 6.22b, the current of the auxiliary winding, i_{ds}, leads i_{qs} of the main winding. For even larger single-phase induction motors, that lead can be further increased by connecting a capacitor in series with the auxiliary winding, as in the capacitor-start motor shown in Fig. 6.22c.

Since the axes of the main and auxiliary stator windings are already orthogonal, for simplicity in analysis, the stationary qd axes may be aligned with the orthogonal axes of the physical windings, as shown in Fig. 6.22. If the coil sides of the main and auxiliary windings do not occupy the same slots, there will be negligible leakage coupling between them. The cage rotor may be represented by a pair of equivalent qd symmetrical rotor windings. Using the coupled circuit approach and motor notation, the voltage equations of the magnetically coupled stator and rotor circuits shown in Fig. 6.23 can be written as follows:

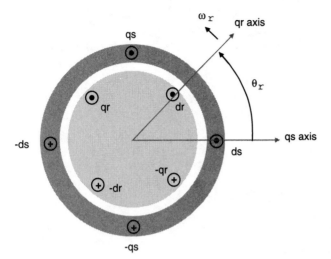

Figure 6.23 Idealized circuit model of a single-phase induction machine.

$$v_{qs} = i_{qs}r_{qs} + \frac{d\lambda_{qs}}{dt} \qquad V$$

$$v_{ds} = i_{ds}r_{ds} + \frac{d\lambda_{ds}}{dt} \qquad V$$

$$v_{qr} = i_{qr}r_{qr} + \frac{d\lambda_{qr}}{dt} \qquad V$$

$$v_{dr} = i_{dr}r_{dr} + \frac{d\lambda_{dr}}{dt} \qquad V$$

(6.134)

Flux linkages. In matrix notation, the flux linkages of the stator and rotor windings can be written compactly in terms of the winding inductances and currents as

$$\begin{bmatrix} \lambda_{qs} \\ \lambda_{ds} \\ \lambda_{qr} \\ \lambda_{dr} \end{bmatrix} = \begin{bmatrix} L_{qsqs} & & L_{qsqr}\cos\theta_r & L_{qsdr}\sin\theta_r \\ & L_{dsds} & -L_{dsqr}\sin\theta_r & L_{dsdr}\cos\theta_r \\ L_{qrqs}\cos\theta_r & -L_{qrds}\sin\theta_r & L_{qrqr} & \\ L_{drqs}\sin\theta_r & L_{drds}\cos\theta_r & & L_{drdr} \end{bmatrix} \begin{bmatrix} i_{qs} \\ i_{ds} \\ i_{qr} \\ i_{dr} \end{bmatrix}$$

(6.135)

If the reluctive drops in iron are neglected and the permeance of the uniform airgap is P_g, the winding inductances may be expressed in terms of their effective turns, N_{qs}, N_{ds}, N_{qr}, and N_{dr}, and the permeance of the airgap:

$$L_{qsqs} = L_{lqs} + N_{qs}^2 P_g \qquad\qquad L_{dsds} = L_{lds} + N_{ds}^2 P_g$$

$$L_{qsqr} = N_{qs}N_{qr}P_g \qquad\qquad L_{qsdr} = N_{qs}N_{dr}P_g$$

$$L_{dsqr} = N_{ds}N_{qr}P_g \qquad\qquad L_{dsdr} = N_{ds}N_{dr}P_g$$

$$L_{qrqr} = L_{lqr} + N_{qr}^2 P_g \qquad\qquad L_{drdr} = L_{ldr} + N_{dr}^2 P_g$$

(6.136)

where $L_{lqs}, L_{lds}, L_{lqr}$, and L_{ldr} are the leakage inductances of the stator and rotor windings. For a symmetrical cage rotor, $N_{qr} = N_{dr}$ and $L_{lqr} = L_{ldr}$; henceforth, they will be denoted by N_r and L_{lr}, respectively.

Choosing the qd reference frame on the asymmetrical stator, such that its q-axis is aligned with the axis of the qs winding and its d-axis with that of the ds winding, the rotor qd windings may be transformed to the chosen qd stationary reference using

$$\begin{bmatrix} f_{qr}^s \\ f_{dr}^s \end{bmatrix} = \underbrace{\begin{bmatrix} \cos\theta_r & \sin\theta_r \\ -\sin\theta_r & \cos\theta_r \end{bmatrix}}_{\mathbf{T}_{qd}(\theta_r)} \begin{bmatrix} f_{qr}^r \\ f_{dr}^r \end{bmatrix} \tag{6.137}$$

where the variable f can be the phase voltages, currents, or flux linkages of the qr and dr windings, and θ_r is the angle between the axes of the qr and qs windings, measured from the stationary qs axis as given by Eq. 6.17. The inverse of $[\mathbf{T}_{qd}(\theta_r)]$ is

$$[\mathbf{T}_{qd}(\theta_r)]^{-1} = \begin{bmatrix} \cos\theta_r & -\sin\theta_r \\ \sin\theta_r & \cos\theta_r \end{bmatrix} \tag{6.138}$$

Besides the above rotational transformation to be applied to the variables of the rotating qr and dr windings, we also would like to refer, or scale, the transformed variables and rotor winding parameters to the stator or input side. Since the qs and ds windings are unequal, one of them will also have to be referred to the other. We will refer the actual variables of the auxiliary ds winding to the main qs winding. To simplify the notation, let's use the superscript s and prime to denote the transformed rotor variables in the stationary qd reference frame and the referred quantities to the stator qs winding, respectively. In terms of the transformed rotor variables and of the ds variables referred to the qs winding, Eq. 6.135 for the flux linkages becomes

$$\begin{aligned}
\lambda_{qs} &= L_{lqs}i_{qs} + L_{mq}(i_{qs} + i_{qr}'^s) \\
\lambda_{ds}' &= L_{lds}'i_{ds}' + L_{mq}(i_{ds}' + i_{dr}'^s) \\
\lambda_{qr}'^s &= L_{lr}'i_{qr}'^s + L_{mq}(i_{qs} + i_{qr}'^s) \\
\lambda_{dr}'^s &= L_{lr}'i_{dr}'^s + L_{mq}(i_{ds}' + i_{dr}'^s)
\end{aligned} \tag{6.139}$$

The voltage equations of the qs and ds stator windings may be expressed in terms of the transformed and referred quantities as

$$\begin{aligned}
v_{qs} &= r_{qs}i_{qs} + \frac{d\lambda_{qs}}{dt} \\
v_{ds}' &= r_{ds}'i_{ds}' + \frac{d\lambda_{ds}'}{dt}
\end{aligned} \tag{6.140}$$

Applying the transformation, $[\mathbf{T}_{qd}(\theta_r)]$, to the voltages, flux linkages, and currents of the qr and dr windings in the qr and dr voltage equations, we will obtain

$$\begin{bmatrix} v_{qr}^s \\ v_{dr}^s \end{bmatrix} = [\mathbf{T}_{qd}(\theta_r)] \begin{bmatrix} r_r & 0 \\ & r_r \end{bmatrix} [\mathbf{T}_{qd}(\theta_r)]^{-1} \begin{bmatrix} i_{qr}^s \\ i_{dr}^s \end{bmatrix}$$

$$+ [\mathbf{T}_{qd}(\theta_r)] p [\mathbf{T}_{qd}(\theta_r)]^{-1} \begin{bmatrix} \lambda_{qr}^s \\ \lambda_{dr}^s \end{bmatrix} \tag{6.141}$$

It can be shown that

$$p[\mathbf{T}_{qd}(\theta_r)]^{-1}\begin{bmatrix}\lambda^s_{qr}\\\lambda^s_{dr}\end{bmatrix}=\begin{bmatrix}-\sin\theta_r & \cos\theta_r\\-\cos\theta_r & -\sin\theta_r\end{bmatrix}\begin{bmatrix}\lambda^s_{qr}\dfrac{d\theta_r}{dt}\\[2ex]\lambda^s_{dr}\dfrac{d\theta_r}{dt}\end{bmatrix}$$

$$+[\mathbf{T}_{qd}(\theta_r)]^{-1}\begin{bmatrix}\dfrac{d\lambda^s_{qr}}{dt}\\[2ex]\dfrac{d\lambda^s_{dr}}{dt}\end{bmatrix} \tag{6.142}$$

Substituting this back into the previous voltage equations and simplifying, we obtain

$$v'^s_{qr}=r'_r i'^s_{qr}-\lambda'^s_{dr}\frac{d\theta_r}{dt}+\frac{d\lambda'^s_{qr}}{dt}$$

$$v'^s_{dr}=r'_r i'^s_{dr}+\lambda'^s_{qr}\frac{d\theta_r}{dt}+\frac{d\lambda'^s_{dr}}{dt} \tag{6.143}$$

Using ψ for $\omega_b\lambda$ in Eqs. 6.140 and 6.143, we can deduce from the resulting equations the equivalent circuit shown in Fig. 6.24. Variables and parameters in the shaded portion of the equivalent circuit are referred to the main qs winding.

$$v'_{ds}=\frac{N_{qs}}{N_{ds}}v_{ds} \qquad\qquad\qquad\qquad i'_{ds}=\frac{N_{ds}}{N_{qs}}i_{ds}$$

$$L_{mq}=N^2_{qs}P_g \qquad\qquad\qquad L'_{md}=\left(\frac{N_{qs}}{N_{ds}}\right)^2 N^2_{ds}P_g=L_{mq}$$

$$L'_{lds}=\left(\frac{N_{qs}}{N_{ds}}\right)^2 L_{lds} \qquad\qquad r'_{ds}=\left(\frac{N_{qs}}{N_{ds}}\right)^2 r_{ds}$$

$$L'_{lr}=\left(\frac{N_{qs}}{N_r}\right)^2 L_{lr} \qquad\qquad\quad r'_r=\left(\frac{N_{qs}}{N_r}\right)^2 r_r$$

$$v'^s_{qr}=\frac{N_{qs}}{N_r}v^s_{qr} \qquad\qquad\qquad v'^s_{dr}=\frac{N_{qs}}{N_r}v^s_{dr}$$

$$i'^s_{qr}=\frac{N_r}{N_{qs}}i^s_{qr} \qquad\qquad\qquad i'^s_{dr}=\frac{N_r}{N_{qs}}i^s_{dr}$$

$$\lambda'^s_{qr}=\frac{N_{qs}}{N_r}\lambda^s_{qr} \qquad\qquad\qquad \lambda'^s_{dr}=\frac{N_{qs}}{N_r}\lambda^s_{dr}$$

Torque equation.   The sum of the instantaneous input power to all four windings of the stator and rotor may be expressed as

$$p_{in}=v_{qs}i_{qs}+v'_{ds}i'_{ds}+v'^s_{qr}i'^s_{qr}+v'^s_{dr}i'^s_{dr} \qquad W \tag{6.144}$$

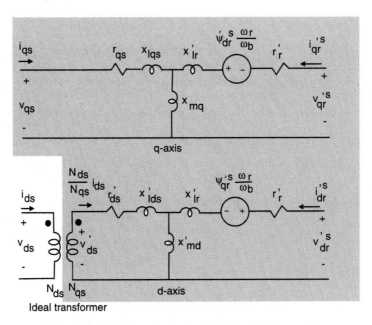

Figure 6.24   Equivalent circuit of a single-phase induction machine.

Using Eqs. 6.140 and 6.143 to substitute for the voltages and eliminating terms associated with copper losses and the rate of change of magnetic field energy, we can equate the remaining terms to the rate of mechanical work, that is

$$T_{em}\frac{d\theta_{rm}}{dt} = (\lambda_{qr}'^s i_{dr}'^s - \lambda_{dr}'^s i_{qr}'^s)\frac{d\theta_r}{dt} \qquad N.m. \tag{6.145}$$

Replacing the mechanical speed, $d\theta_{rm}/dt$, by $(2/P)\,d\theta_r/dt$, the developed torque may be expressed as

$$T_{em} = \frac{P}{2}(\lambda_{qr}'^s i_{dr}'^s - \lambda_{dr}'^s i_{qr}'^s) \qquad N.m.$$
$$= \frac{P}{2}L_{mq}(i_{dr}'^s i_{qs} - i_{qr}'^s i_{ds}') \tag{6.146}$$

Simulation of a single-phase induction motor.   For simulation purposes, the equations of a single-phase induction motor may be rearranged as shown in Table 6.5.

TABLE 6.5   SINGLE-PHASE INDUCTION MACHINE EQUATIONS

Voltage and flux equations:

$$\psi_{qs} = \omega_b \int \left\{ v_{qs} + \frac{r_{qs}}{x_{lqs}} (\psi_{mq} - \psi_{qs}) \right\} d(t)$$

$$\psi'_{ds} = \omega_b \int \left\{ v'_{ds} + \frac{r'_{ds}}{x'_{lds}} (\psi'_{md} - \psi'_{ds}) \right\} d(t)$$

$$(6.147)$$

$$\psi'^{s}_{qr} = \omega_b \int \left\{ v'^{s}_{qr} + \frac{\omega_r}{\omega_b} \psi'^{s}_{dr} + \frac{r'_r}{x'_{lr}} (\psi_{mq} - \psi'^{s}_{qr}) \right\} d(t)$$

$$\psi'^{s}_{dr} = \omega_b \int \left\{ v'^{s}_{dr} - \frac{\omega_r}{\omega_b} \psi'^{s}_{qr} + \frac{r'_r}{x'_{lr}} (\psi'_{md} - \psi'^{s}_{dr}) \right\} d(t)$$

$$(6.148)$$

$$\psi_{mq} = x_{mq} (i_{qs} + i'^{s}_{qr})$$

$$\psi'_{md} = x_{mq} (i'_{ds} + i'^{s}_{dr})$$

$$(6.149)$$

$$\psi_{qs} = x_{lqs} i_{qs} + \psi_{mq} \qquad\qquad i_{qs} = \frac{\psi_{qs} - \psi_{mq}}{x_{lqs}}$$

$$\psi'_{ds} = x'_{lds} i'_{ds} + \psi'_{md} \qquad\qquad i'_{ds} = \frac{\psi'_{ds} - \psi'_{md}}{x'_{lds}}$$

$$(6.150)$$

$$\psi'^{s}_{qr} = x'_{lr} i'^{s}_{qr} + \psi_{mq} \qquad\qquad i'^{s}_{qr} = \frac{\psi'^{s}_{qr} - \psi_{mq}}{x'_{lr}}$$

$$\psi'^{s}_{dr} = x'_{lr} i'^{s}_{dr} + \psi'_{md} \qquad\qquad i'^{s}_{dr} = \frac{\psi'^{s}_{dr} - \psi'_{md}}{x'_{lr}}$$

where

$$\frac{1}{x_{Mq}} = \frac{1}{x_{mq}} + \frac{1}{x_{lqs}} + \frac{1}{x'_{lr}}$$

$$\frac{1}{x_{Md}} = \frac{1}{x_{mq}} + \frac{1}{x'_{lds}} + \frac{1}{x'_{lr}}$$

$$(6.151)$$

and

$$\psi_{mq} = x_{Mq} \left(\frac{\psi_{qs}}{x_{lqs}} + \frac{\psi'^{s}_{qr}}{x'_{lr}} \right)$$

$$\psi'_{md} = x_{Md} \left(\frac{\psi'_{ds}}{x'_{lds}} + \frac{\psi'^{s}_{dr}}{x'_{lr}} \right)$$

$$(6.152)$$

Torque and rotor motion equations:

$$T_{em} = \frac{P}{2\omega_b}(\psi'_{ds}i_{qs} - \psi_{qs}i'_{ds}) \qquad N.m. \tag{6.153}$$

$$J\frac{d\omega_{rm}}{dt} = T_{em} + T_{mech} - T_{damp} \qquad N.m.$$

$$2H\frac{d(\omega_r/\omega_b)}{dt} = T_{em} + T_{mech} - T_{damp} \qquad \text{in per unit} \tag{6.154}$$

Steady-state analysis.   Unlike the steady-state condition of a balanced three-phase induction machine where the airgap flux is a uniform rotating field and the torque is constant, the airgap flux and torque of a single-phase induction machine with non-symmetric stator windings, in general, are pulsating even in steady-state. As shown in Section 5.10, the pulsating mmf from a single-phase excitation may be viewed as the resultant of forward and backward revolving components. In terms of space vectors, the fundamental component of the airgap mmf that is established by the stator qs and ds currents at a point θ_q counter-clockwise from the qs axis may be expressed as

$$
\begin{aligned}
\vec{F}_s(\theta_q) &= \vec{F}_{qs} + \vec{F}_{ds} \\
&= \frac{N_{qsine}}{2}\vec{i}_{qs} + \frac{N_{dsine}}{2}\vec{i}_{ds} \\
&= \frac{N_{qsine}}{2}i_{qs}\cos\theta_q + \frac{N_{dsine}}{2}i_{ds}\cos(\underbrace{\theta_q + \frac{\pi}{2}}_{\theta_d}) \\
&= \frac{N_{qsine}}{2}i_{qs}\left(\frac{e^{j\theta_q} - e^{-j\theta_q}}{2}\right) + \frac{N_{dsine}}{2}i_{ds}\left(\frac{e^{j\theta_q} - e^{-j\theta_q}}{2j}\right) \\
&= \frac{N_{qsine}}{2}\left\{\underbrace{\left(\frac{i_{qs}}{2} + j\frac{N_{dsine}}{N_{qsine}}\frac{i_{ds}}{2}\right)e^{j\theta_q}}_{\vec{i}_{2s}} + \underbrace{\left(\frac{i_{qs}}{2} - j\frac{N_{dsine}}{N_{qsine}}\frac{i_{ds}}{2}\right)e^{-j\theta_q}}_{\vec{i}_{1s}}\right\}
\end{aligned}
\tag{6.155}
$$

where $\vec{i}_{1s}$ and $\vec{i}_{2s}$ are the positive- and negative-sequence space vector components of the stator currents, i_{qs} and i_{ds}, and N_{qsine} and N_{dsine} are the turns of equivalent sinusoidally distributed qs and ds windings. As shown in Section 5.4, N_{qsine} and N_{dsine} are related to the effective number of turns, N_{qs} and N_{ds}, by the same constants, thus the ratio, N_{qsine}/N_{dsine}, is equal to N_{qs}/N_{ds}.

For steady-state analysis of operating conditions where the voltages and currents are sinusoidal, for instance $i_{qs} = \Re\{\sqrt{2}I_{qrms}e^{j(\omega_e t + \phi_q)}\}$ and $i_{ds} = \Re\{\sqrt{2}I_{drms}e^{j(\omega_e t + \phi_d)}\}$, the relationship between the rms time phasors of the positive and negative sequence current components and the rms time phasors of the phase currents are

$$\begin{bmatrix} \tilde{\mathbf{I}}_{1s} \\ \tilde{\mathbf{I}}_{2s} \end{bmatrix} = \frac{1}{\sqrt{2}} \begin{bmatrix} 1 & -j \\ 1 & j \end{bmatrix} \begin{bmatrix} \tilde{\mathbf{I}}_{qs} \\ \frac{N_{dsine}}{N_{qsine}} \tilde{\mathbf{I}}_{ds} \end{bmatrix} = \frac{1}{\sqrt{2}} \begin{bmatrix} 1 & -j \\ 1 & j \end{bmatrix} \begin{bmatrix} \tilde{\mathbf{I}}_{qs} \\ \tilde{\mathbf{I}}'_{ds} \end{bmatrix} \qquad (6.156)$$

When the two winding currents form an orthogonal balanced set, that is $I_{qrms} = I_{drms}$ and $\phi_d = \phi_q - \pi/2$, the positive-sequence rms phasor, $\tilde{\mathbf{I}}_{1s}$, is $I_{qrms}\underline{/0}$ and the negative-sequence rms phasor, $\tilde{\mathbf{I}}_{2s}$, is zero. The inverse transformation can be readily shown to be

$$\begin{bmatrix} \tilde{\mathbf{I}}_{qs} \\ \tilde{\mathbf{I}}'_{ds} \end{bmatrix} = \frac{1}{\sqrt{2}} \begin{bmatrix} 1 & 1 \\ j & -j \end{bmatrix} \begin{bmatrix} \tilde{\mathbf{I}}_{1s} \\ \tilde{\mathbf{I}}_{2s} \end{bmatrix} \qquad (6.157)$$

Since the inverse transformation matrix is equal to the conjugate transpose of the transformation matrix, the transformation is power-invariant. For example, the total complex power into the two stator windings may be expressed as

$$\mathbf{S}_{in} = \begin{bmatrix} \tilde{\mathbf{V}}_{qs} & \tilde{\mathbf{V}}'_{ds} \end{bmatrix} \begin{bmatrix} \tilde{\mathbf{I}}_{qs} \\ \tilde{\mathbf{I}}'_{ds} \end{bmatrix}^{*} = \begin{bmatrix} \tilde{\mathbf{V}}_{1s} & \tilde{\mathbf{V}}_{2s} \end{bmatrix} \begin{bmatrix} \tilde{\mathbf{I}}_{1s} \\ \tilde{\mathbf{I}}_{2s} \end{bmatrix}^{*} \qquad (6.158)$$

The rms time phasors of the other variables of the rotor windings in the qd stationary reference frame, such as $\tilde{\mathbf{V}}'^{s}_{qr}$, $\tilde{\mathbf{V}}'^{s}_{dr}$, $\tilde{\mathbf{I}}'^{s}_{qr}$, $\tilde{\mathbf{I}}'^{s}_{dr}$, $\tilde{\lambda}'^{s}_{qr}$, and $\tilde{\lambda}'^{s}_{dr}$, are related to their corresponding positive- and negative-sequence phasors by the same set of transformations.

The flux linkage equations of the stator and rotor windings, when expressed in terms of positive- and negative-sequence components, are as follows:

$$\begin{bmatrix} \tilde{\lambda}_{1s} \\ \tilde{\lambda}_{2s} \end{bmatrix} = \frac{1}{2} \begin{bmatrix} (L_{lqs} + L'_{lds}) & (L_{lqs} - L'_{lds}) \\ (L_{lqs} - L'_{lds}) & (L_{lqs} + L'_{lds}) \end{bmatrix} \begin{bmatrix} \tilde{\mathbf{I}}_{1s} \\ \tilde{\mathbf{I}}_{2s} \end{bmatrix} + L_{mq} \begin{bmatrix} \tilde{\mathbf{I}}_{1s} + \tilde{\mathbf{I}}'_{1r} \\ \tilde{\mathbf{I}}_{2s} + \tilde{\mathbf{I}}'_{2r} \end{bmatrix} \qquad (6.159)$$

In general, when the stator windings are dissimilar, the off-diagonal leakage term will not be zero. In our equivalent circuit, the mutual inductances, L_{mq} and L_{md}, are equal in value, as shown in the table above, because the airgap is uniform and all other windings have been referred to the qs winding. The flux linkage equations of the symmetrical rotor winding can thus be simplified to

$$\begin{bmatrix} \tilde{\lambda}'_{1r} \\ \tilde{\lambda}'_{2r} \end{bmatrix} = L'_{lr} \begin{bmatrix} \tilde{\mathbf{I}}'_{1r} \\ \tilde{\mathbf{I}}'_{2r} \end{bmatrix} + L_{mq} \begin{bmatrix} \tilde{\mathbf{I}}_{1s} + \tilde{\mathbf{I}}'_{1r} \\ \tilde{\mathbf{I}}_{2s} + \tilde{\mathbf{I}}'_{2r} \end{bmatrix} \qquad (6.160)$$

For steady-state, the time derivative operator, d/dt, on variables that are varying sinusoidally at an angular frequency of ω_e can be replaced by $j\omega_e$, and the rotor speed, $d\theta_r/dt$, by ω_r. With these replacements, the steady-state voltage equations of the stator and rotor windings become

$$\begin{bmatrix} \tilde{\mathbf{V}}_{1s} \\ \tilde{\mathbf{V}}_{2s} \end{bmatrix} = \frac{1}{2} \begin{bmatrix} (r_{qs} + r'_{ds}) & (r_{qs} - r'_{ds}) \\ (r_{qs} - r'_{ds}) & (r_{qs} + r'_{ds}) \end{bmatrix} \begin{bmatrix} \tilde{\mathbf{I}}_{1s} \\ \tilde{\mathbf{I}}_{2s} \end{bmatrix} + \begin{bmatrix} j\omega_e \tilde{\lambda}_{1s} \\ j\omega_e \tilde{\lambda}_{2s} \end{bmatrix} \qquad (6.161)$$

$$\begin{bmatrix} \tilde{\mathbf{V}}'_{1r} \\ \tilde{\mathbf{V}}'_{2r} \end{bmatrix} = r'_r \begin{bmatrix} \tilde{\mathbf{I}}'_{1r} \\ \tilde{\mathbf{I}}'_{2r} \end{bmatrix} + \begin{bmatrix} j(\omega_e - \omega_r)\tilde{\lambda}'_{1r} \\ j(\omega_e + \omega_r)\tilde{\lambda}'_{2r} \end{bmatrix} \qquad (6.162)$$

With a cage rotor, $\tilde{\mathbf{V}}'_{1r}$ and $\tilde{\mathbf{V}}'_{2r}$ are zero. Dividing the steady-state equations of $\tilde{\mathbf{V}}'_{1r}$ and $\tilde{\mathbf{V}}'_{2r}$ by $(\omega_e - \omega_r)/\omega_e$ and $(\omega_e + \omega_r)/\omega_e$, respectively, and substituting s for $(\omega_e - \omega_r)/\omega_e$ and $2 - s$ for $(\omega_e + \omega_r)/\omega_e$, the resulting equations for the rotor windings are

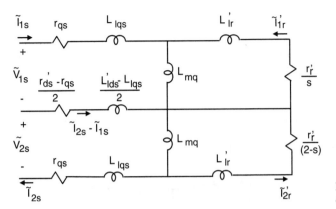

Figure 6.25 Steady-state equivalent circuit of a single-phase induction machine.

$$0 = \frac{r_r'}{s}\tilde{\mathbf{I}}_{1r} + j\omega_e\left\{ L_{lr}'\tilde{\mathbf{I}}_{1r} + L_{mq}(\tilde{\mathbf{I}}_{1s} + \tilde{\mathbf{I}}_{1r}') \right\}$$

$$0 = \frac{r_r'}{(2-s)}\tilde{\mathbf{I}}_{2r} + j\omega_e\left\{ L_{lr}'\tilde{\mathbf{I}}_{2r} + L_{mq}(\tilde{\mathbf{I}}_{2s} + \tilde{\mathbf{I}}_{2r}') \right\} \tag{6.163}$$

The sequence network representation in Fig. 6.25 for a single-phase induction motor with dissimilar stator windings may be deduced from the form of the above stator and rotor winding voltage equations. The coupling between the positive- and negative-sequence components from the off-diagonal resistive and leakage inductive terms in the stator voltage equations may be expressed as follows:

$$\tilde{\mathbf{V}}_{1s} = r_{qs}\tilde{\mathbf{I}}_{1s} + \frac{(r_{ds}' - r_{qs})}{2}(\tilde{\mathbf{I}}_{1s} - \tilde{\mathbf{I}}_{2s})$$

$$+ j\omega_e\left\{ L_{lqs}\tilde{\mathbf{I}}_{1s} + \frac{(L_{lds}' - L_{lqs})}{2}(\tilde{\mathbf{I}}_{1s} - \tilde{\mathbf{I}}_{2s}) + L_{mq}(\tilde{\mathbf{I}}_{1s} + \tilde{\mathbf{I}}_{1r}') \right\}$$

$$\tilde{\mathbf{V}}_{2s} = r_{qs}\tilde{\mathbf{I}}_{2s} + \frac{(r_{ds}' - r_{qs})}{2}(\tilde{\mathbf{I}}_{2s} - \tilde{\mathbf{I}}_{1s}) \tag{6.164}$$

$$+ j\omega_e\left\{ L_{lqs}\tilde{\mathbf{I}}_{2s} + \frac{(L_{lds}' - L_{lqs})}{2}(\tilde{\mathbf{I}}_{2s} - \tilde{\mathbf{I}}_{1s}) + L_{mq}(\tilde{\mathbf{I}}_{2s} + \tilde{\mathbf{I}}_{2r}') \right\}$$

Average torque in steady-state.   Denoting $j\omega_e\tilde{\lambda}_{1r}'$ by $\tilde{\mathbf{E}}_{1r}$ and $j\omega_e\tilde{\lambda}_{2r}'$ by $\tilde{\mathbf{E}}_{2r}$, the total complex power flowing across the airgap from the stator to the rotor is given by

$$S_{ag} = -\tilde{\mathbf{E}}_{1r}\tilde{\mathbf{I}}_{1r}^{\star} - \tilde{\mathbf{E}}_{2r}\tilde{\mathbf{I}}_{2r}^{\star} \tag{6.165}$$

The real part of S_{ag} corresponds to the synchronous watt flow. The average torque developed by the motor can be obtained by dividing the positive- and negative-sequence synchronous watts by their respective synchronous speed in mechanical radians per second, that is

$$T_{avg} = \Re\left(\frac{-\tilde{\mathbf{E}}_{1r}\tilde{\mathbf{I}}_{1r}^{*}}{\omega_{bm}} - \frac{\tilde{\mathbf{E}}_{2r}\tilde{\mathbf{I}}_{2r}^{*}}{-\omega_{bm}}\right)$$

$$= \frac{P}{2\omega_e}\left\{\tilde{\mathbf{I}}_{1r}^{2}\left(\frac{r_r'}{s}\right) - \tilde{\mathbf{I}}_{2r}^{2}\left(\frac{r_r'}{2-s}\right)\right\} \tag{6.166}$$

6.11 PROJECTS

6.11.1 Project 1: Operating Characteristics

In this project, we will implement a simulation of a stationary reference $qd0$ model of a singly excited, three-phase induction motor. Following which, we will use the simulation to examine the motoring, generating, and braking characteristics of a test motor supplied with sinusoidal voltages.

(a) Figure 6.26a shows an overall diagram of the simulation $s1$ of an induction machine in the stationary reference frame. The details inside the main blocks of $s1$ are given in Fig. 6.26b through 6.26e, with the exception of the *Daxis* block, which by symmetry, is similar to in the *Qaxis* block. In this simulation, the stator neutral is floating, its voltage with respect to the system reference point, g in Fig. 6.14, is obtained by connecting a small fictitious capacitor, C_{sg}, between points s and g. Also included is the zero-sequence circuit simulation, which can be disabled when the operating condition is balanced.

The given $m1$ M-file uses the machine parameters in $p1hp$ to set up the following machine parameters of a 1-hp, three-phase, 60-Hz, four-pole, 200-V induction motor:

$$r_s = 3.35\ \Omega \qquad\qquad\qquad L_{ls} = L_{lr}' = 6.94\ mH$$

$$L_m = 163.73\ mH \qquad\qquad\qquad r_r' = 1.99\ \Omega$$

$$J_{rotor} = 0.1\ kgm^2$$

When the value of C_{sg} is kept small, to say $1/(50Z_b\omega_b)$, the simulation will run satisfactorily with a minimum step size of $2e^{-4}$, a maximum step size of $1e^{-2}$, and a tolerance of $1e^{-7}$ using the *ode15s* or Adams/Gear method. When the stator neutral is grounded, C_{sg} is infinite and the gain in the *1/Csg* block should be set to zero.

(b) Using a stop time of 0.8 seconds, simulate the motor starting from rest with rated voltage applied and no mechanical load. Plot the torque vs. speed curve of the run-up and note the values of the starting ($s = 1$) and breakdown ($s = s_{maxt}$) torques. Compare the torque-speed curve obtained under dynamic conditions to the corresponding steady-state curve shown in Fig. 6.27 that is obtained from Eq. 6.73 for steady-state operation. Compare the values of s_{maxt} and T_{em}^{max} computed from the expressions given in Eqs. 6.74 and 6.75 with those observed on the dynamic torque vs. speed curve.

(c) As implemented in the rotor simulation shown in Fig. 6.26(c), the machine will be motoring when T_{mech} is negative, and generating when T_{mech} is positive. Determine the base torque, T_b, of the machine that you are simulating, then apply step changes of load

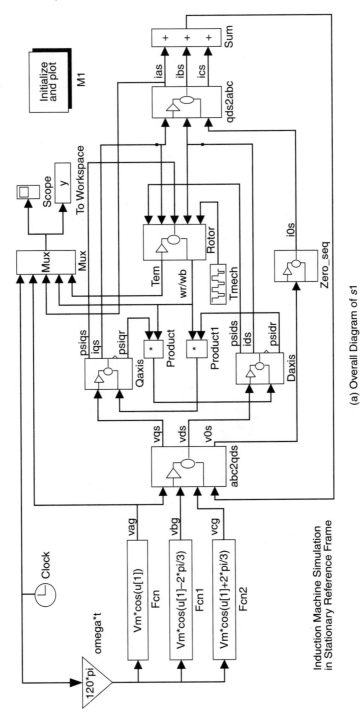

(a) Overall Diagram of *s1*

Figure 6.26 Simulation *s1* of an induction machine in the stationary reference frame.

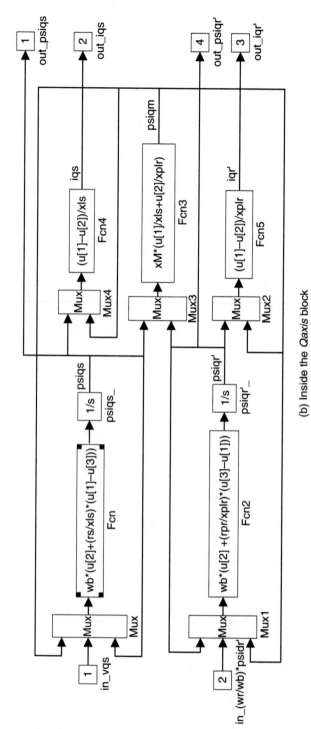

(b) Inside the *Qaxis* block

Figure 6.26 *(cont.):* Simulation *s1* of an induction machine in the stationary reference frame.

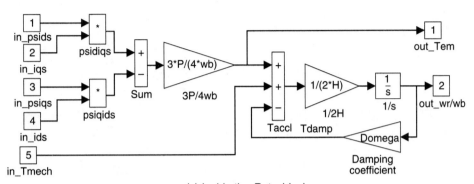

(c) Inside the *Rotor* block

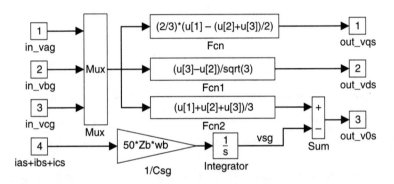

(d) Inside the *abc2qds* block

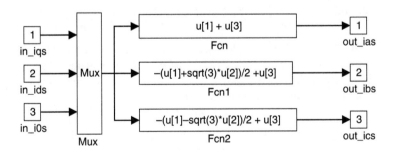

(e) Inside the *qds2abc* block

Figure 6.26 *(cont.):* Simulation *s1* of an induction machine in the stationary reference frame.

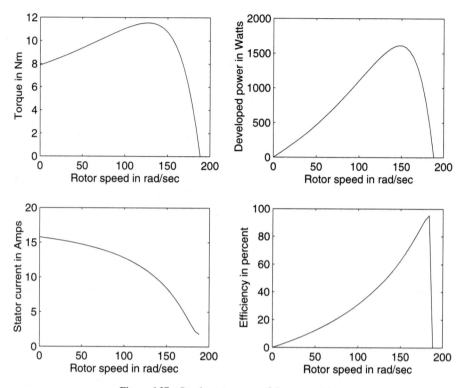

Figure 6.27   Steady-state curves of the test machine.

torque from zero to the base value of T_b in two increments. Plot the transient response of i_{as}, T_{em}, and ω_r to the applied step change in load torque. Cross-check your results with the sample results shown in Fig. 6.28 of the machine starting under no-load and with step changes in motoring a T_{mech} of zero to 50 percent, 50 to 100 percent, and from 100 back down to 50 percent of rated torque applied at $t = 0.8$, 1.2, and 1.6 seconds, respectively.

(d) Repeat Part (c) with negative values of load torque. Check the rotor speed obtained when steady-state is reached after each increment against that predicted by Eq. 6.73. Check also the phasing between v_{as} and i_{as} to ensure that it is consistent with the condition that the machine is generating.

(e) Plugging is a form of braking where the field rotation of the stator's mmf is opposite to that of the rotor's. Modify a copy of $s1$ so that you can reverse the connection of the b and c-phase supply voltages to the stator on the fly. Start the motor up as in Part (a). When the motor has run up to its no-load speed and the transients are over, reverse the sequence of the supply voltages to the stator windings, causing the airgap field to reverse in rotation, opposite to that of the rotor's rotation. Figure 6.29 shows the motoring characteristic for a forward rotating field in a solid curve and the braking characteristic after the field reversal in a dotted curve. The marked trajectory shows the machine motoring initially at no-load speed. The rotor speed remains the same at the switch-

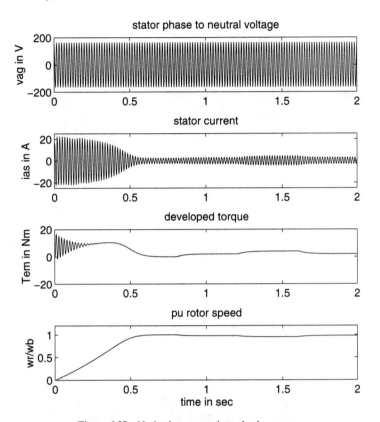

Figure 6.28   No-load startup and step load response.

over from motoring to braking mode because the speed is continuous for finite rotor inertia. Immediately following the reversal, the slip is almost 2 and the stator current will be relatively large compared to its rated value. Plot the values i_{as}, T_{em}, and ω_r as the rotor slows down to zero and accelerates in the other direction.

6.11.2 Project 2: Starting Methods

It can be seen from the equivalent circuit of Fig. 6.9 that when the slip is large, the effective rotor resistance, r_r'/s, is small. Thus, starting a motor from standstill with the rated stator voltage applied can result in large starting current, typically as much as six to eight times the rated current of the motor. The starting current of a large induction motor could result in excessive voltage drop along the feeder that is disruptive or objectionable to other loads on that same feeder. Various methods can be used to reduce the current drawn by the motor during starting. One popular method is to reduce the voltage applied to the stator windings during starting, using an autotransformer or star-delta contactors. Since the torque developed is proportional to the square of the applied stator voltage, the starting torque will be correspondingly reduced during the period when the stator voltage is reduced; as

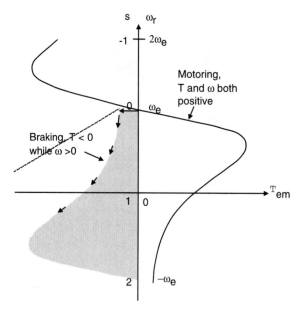

Figure 6.29 Trajectory of a plugging operation.

such, the acceleration during that period will be slower. In some applications, such as in compressor drives, the requirement of large starting torque may necessitate the use of another method of reducing the starting current, that of using external rotor resistance during starting. Figure 6.30 shows the circuit and torque vs. speed curve for external rotor resistance starting of a wound-rotor machine. In this project, we will examine the starting performance of some of these methods on a 20-hp, four-pole, 220-V, three-phase, 60-Hz induction motor machine with the following parameters:

$$r_s = 0.1062 \ \Omega \qquad\qquad\qquad x_{ls} = 0.2145 \ \Omega$$

$$r'_r = 0.0764 \ \Omega \qquad\qquad\qquad x'_{lr} = 0.2145 \ \Omega$$

$$x_m = 5.834 \ \Omega \qquad\qquad\qquad J_{rotor} = 2.8 \ kgm^2$$

The rated speed of the motor is 1748.3 rev/min at a slip of 0.0287, and the rated current is 49.68 A.

Direct-on-line starting. Make a copy of the MATLAB and SIMULINK files, *m1* and *s1*, given in Project 1. Modify *m1* to use the parameters of the 20-hp induction motor instead of those of the 1-hp motor. For this part of the project on direct-on-line starting, you can use *s1* as given.

Start the motor from standstill with full-rated supply voltage and a load torque of T_b (T_b being the base torque of the machine). Stop the simulation when steady-state is reached. Plot the starting current drawn i_{as} and ω_r. Note the magnitude of the starting current at the beginning of starting, and the time and rotor speed at which the starting current drops significantly from its initial starting value.

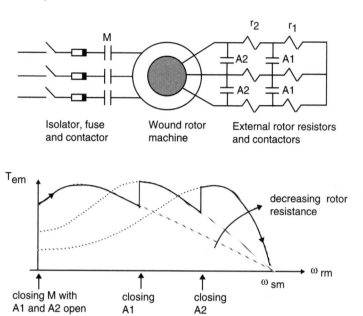

Figure 6.30   Wound rotor machine with external rotor resistors.

Reduced voltage starting.   With a reduced voltage method, the supply voltage is temporarily reduced during the initial phase of starting to keep the initial starting current drawn low. When the rotor accelerates up, the slip decreases and the stator current drops. When the starting current has fallen significantly, the supply voltage can then be raised to the rated value to complete the run-up. Using an autotransformer, the initial starting voltage can be adjusted by selecting the appropriate turns ratio of the autotransformer. With a machine that is designed to run normally with its stator windings delta-connected, the terminals of the three stator windings, if available externally, can be temporarily reconnected as wye to reduce the current drawn during the initial phase of starting. With the same rated supply voltages, the phase voltage across each phase of the wye-connected stator windings will be only $1/\sqrt{3}$ of that when the windings are delta-connected.

Modify your copy of $s1$ so that the magnitude of the input voltages to each stator phase can be changed from $1/\sqrt{3}$ rated to rated in response to a timer during a run. Repeat the startup run as in Part (a) with a load torque of T_b, but with the magnitude of the supply voltage kept at $1/\sqrt{3}$ of rated, and note how long it takes the starting current to start decreasing significantly. Set the timer to the time noted above and repeat the run with the reduced voltage supply, and with the input voltage to the stator phases reverting back to full value after the scheduled delay. Plot v_{as}, i_{as}, T_{em}, and ω_r and compare this set of results with that obtained in Part (a) with the direct-on-line starting method. With wye-delta or star-delta starting, besides changing the magnitude of the input voltages to the stator windings, there may also be a change in phase of the input voltages when the stator windings are being reconnected in delta.

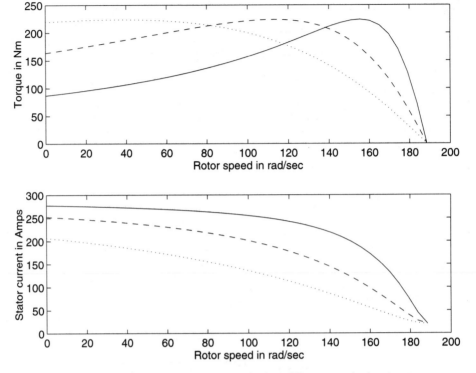

Figure 6.31   Torque and stator current curves for three different rotor circuit resistances.

Rotor resistance starting.   Induction machines with a wound rotor construction, where the rotor windings terminals are accessible via the slip rings, can be started with external resistances introduced to limit the starting current drawn. Note from the discussion of Eq. 6.75 that the maximum torque developed is not dependent on the rotor-circuit resistance, but the slip at which the maximum torque is developed can be varied by changing the rotor circuit resistance, r_r'.

Use Eq. 6.74 to determine the values of rotor circuit resistance that would cause maximum torque to be developed by the test motor at slips of 0.8 and 0.4, respectively. Figure 6.31 shows the torque and stator current vs. rotor speed curves computed from steady-state expressions using the nominal value of rotor circuit resistance and the two values computed above.

Modify a copy of *s1* so that the rotor circuit resistance can be step changed during starting to three different values: that which will give maximum torque at $s = 0.8$ in steady-state; that which give maximum torque at $s = 0.4$; and, the nominal rotor winding resistance given. Using the information given in Fig. 6.31, determine by trial and error a suitable timing to step change the rotor circuit resistance using rotor speed as the feedback variable to initiate these step changes for the case where peak starting current is kept below $170\sqrt{2}$A when starting with a load torque of T_b. Run the simulation of the motor starting from standstill with rated supply voltages and a load torque of T_b using the above scheme of rotor resistance adjustment. Plot i_{as}, T_{em}, and ω_r and compare the results obtained with those obtained using the direct-on-line starting method in Part (a) and those obtained using the reduced voltage method in Part (b).

Comment on the differences in starting current drawn and the starting time that the machine takes to reach full-load speed among the three methods studied in Parts (a) through (c).

6.11.3 Project 3: Open-circuit Conditions

Since the developed induction machine simulation has voltage input and current output, the simulation of a short-circuit condition at the machine terminals is straightforward, but the simulation of an open-circuit condition is not. The objective of this project is to examine two specific cases: that where only one out of three stator phases is open and that where all three phases are open for a while and then reclosed onto another supply bus.

Single-phasing.   The condition in which the motor operates with one of its stator phases open-circuited is referred to as single-phasing. In practice, single-phasing of three-phase motors can often be the result of one of the supply fuses being blown. Momentary single-phasing is also encountered in simulating some adjustable-speed drive in which the stator terminals of the motor are cyclically switched by an inverter, connecting only two phases of the motor to the dc supply, one phase to the positive rail and the other phase to the negative rail, and leaving the third phase open-circuited briefly every sixth of a switching cycle.

Method a.   During single-phasing, the terminal voltage of the open stator phase is the same as that of its internal phase voltage. Consider the stator supply connection shown in Fig. 6.14 where the neutral point of the stator winding is floating and let the a-phase of the stator be the open phase, that is $i_{as} = 0$. With the neutral point of the stator windings floating and the a-phase open, $i_{bs} + i_{cs} = 0$ and $i_{0s} = 0$. Since $i_{0s} = 0$, we can readily show that v_{0s} will also be zero. Show that for the above single-phasing operation:

$$i_{qs} = 0$$

$$\psi_{qs}^s = \psi_{mq}^s$$

$$v_{qs}^s = \frac{x_{mq}}{\omega_b} \frac{di_{qr}^{\prime s}}{dt} \tag{6.167}$$

$$i_{qr}^{\prime s} = \frac{\psi_{qr}^{\prime s}}{x_{lr}^{\prime} + x_{mq}}$$

$$v_{as} = v_{qs}^s = \frac{1}{\omega_b} \frac{x_{mq}}{x_{lr}^{\prime} + x_{mq}} \frac{d\psi_{qr}^{\prime s}}{dt}$$

If the value of x_m may be considered constant, the last expression in Eq. 6.167 may then be used to generate the input voltage, v_{as}, for the open a-phase when simulating a single-phasing operation of the motor. In the simulation, the time derivative term, $d\psi_{qr}^{\prime s}/dt$, can be taken from the input to the integrator that produces $\psi_{qr}^{\prime s}$ in Fig. 6.26b, to avoid taking the time derivative.

Method b.   On the other hand, if the value of x_m varies with operating condition, as with iron saturation, the above open-circuit voltage expression will not be very useful. For such a case, the open-circuit voltage at the terminal of a-phase can be obtained from the voltage across a very large resistance, R_H, between the external terminal of the a-phase stator winding and the system ground, g, and the voltage, v_{sg}, between the floating neutral of the stator windings and the system ground.

$$v_{as} = v_{ag} - v_{sg}$$
$$v_{ag} = -i_{as} R_H$$
(6.168)

To avoid creating a high-gain algebraic loop, the value of v_{sg} can be obtained from a small capacitor, C_{sg}, instead of another large resistor connected across points, s and g, that is

$$v_{sg} = \frac{1}{C_{sg}} \int (i_{as} + i_{bs} + i_{cs}) dt$$
(6.169)

In the first half of this project, we will experiment with the two methods of implementing an open-circuit condition on the a-phase of a three-phase induction machine and examine the results. Start by modifying a copy of $s1$ used in Project 1 of this chapter using Method A to simulate a single-phasing operation of a three-phase motor having the same parameters as those given in Project 1. Modify a copy of the MATLAB M-file, $m1$, to set up the simulation and to plot the desired results. Start the simulation with a three-phase voltage supply and a load torque of less than one-half the base torque, T_b, of the motor. You can get the motor simulation to settle quicker to the steady-state by initializing the rotor speed with the synchronous value. When steady-state is reached, activate the changeover from three-phase to single-phasing operation to open the a-phase connection to the three-phase supply at the next current zero of i_{as}. Explain why the precaution is taken to open the phase connection at the moment of its current zero? Plot i_{as}, v_{as}, v_{sg}, T_{em}, and ω_r vs. time. Compare these plots with those obtained for the same load torque in Part (b) of Project 1.

Next, modify another copy of $s1$, this time using Method B to simulate the same single-phasing operation of the three-phase motor as with Method A. Determine whether it is necessary for you to include the equation of the stator's zero-sequence component in your simulation of this condition. Repeat the simulation as in the case with Method A.

Bus transfer.   An important operation where all three phases of the motor are open for a short period arises when a critical motor must be transferred from one bus to another to maintain continuity in its operation. Figure 6.32 shows the condition in a nuclear power plant where the induction motor driving the coolant pump may on occasion have to be transferred from the auxiliary bus to the system supply bus. The same situation occurs again when transferring back from the system supply bus to the auxiliary bus to restore normal operation. During the short time that the stator is disconnected from its supply, its rotor will decelerate and the internal voltages of the stator due to the trapped rotor flux will decrease in magnitude and retard in phase. Frequent questions related to such a bus transfer are

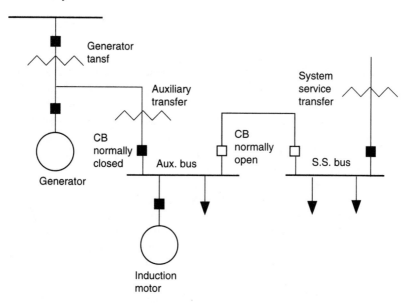

Figure 6.32 Bus transfer of induction motor in a power plant.

- How long can the motor be left open-circuited and yet be able to complete the transfer in a reasonably smooth manner?
- What is the best point of the wave to reclose onto the new supply bus voltage for minimal switching transients?

1. Modify your simulation to conduct a bus transfer run where the disconnection of each stator phase from its supply bus is open at the zero of the corresponding stator current. The stator circuit can be left open for some arbitrary length of time and the reconnection of all three stator stator phases to their new supply bus voltages can be closed immediately by manual control or can be automatically synchronized to some recognizable event.

2. With the machine motoring a load torque of T_b and using supply voltages of rated magnitude, vary the duration of the state where the stator phases are completely open to see how quickly the terminal voltage of the stator decays in magnitude and retards in phase relative to the steady supply voltage.

3. Repeat the run with rotor inertia changed by a factor of 0.5 and 2.

4. Conduct runs with the reclosing at the instant when the supply voltage and internal stator voltage of the corresponding phases are in phase, 90°, and 180° out of phase.

6.11.4 Project 4: Linearized Analysis

A linearized model of an induction machine can be useful in small-signal stability prediction and in control design that uses linear control techniques. In this project, we will learn how to obtain a linear model of an induction machine from its SIMULINK simulation and use it to determine the root-locus of some transfer functions.

First, set up the SIMULINK simulation, *s4eig*, shown in Fig. 6.20 of an induction machine in the synchronously rotating reference frame. Compared to the simulation of the same machine in the stationary reference frame, the main changes are those associated with the speed voltage terms and the requirement of the SIMULINK **trim** and MATLAB **linmod** functions that the inputs and outputs be defined by sequentially numbered input and output ports.

Examine the given *m4* script file that has been programmed to determine the transfer functions of $(\Delta\omega_r/\omega_b)/\Delta T_{mech}$ and $(\Delta\omega_r/\omega_b)/\Delta v_{qs}^e$ at just the no-load and full-load operating points of the 20-hp, 4-pole, 220-V motor of Project 2. Run *m4* once to check the ordering of the state variables by SIMULINK on your computer, as the ordering may be different from what is shown, depending on how SIMULINK assembles the model on your computer. If the ordering is different from what is given in *m4*, the ordering of the state variables in *m4* will have to be changed to conform with the new ordering on your computer. A check of the outputs from **trim** is also recommended as the outputs can be off the mark when the initial guess is not close.

The MATLAB script file, *m4*, is to be used in conjunction with the SIMULINK *s4eig* file to determine the open-loop transfer function of a system where the forward transfer function is $(\Delta\omega_r/\omega_b)/\Delta v_{qs}^e$ and the feedback function of the speed sensor is $1/(0.05s + 1)$. It also performs a root-locus plot of the open-loop system. Figure 6.33 shows the root-locus plot from *m4* of the open-loop transfer function, $\Delta(\omega_r/\omega_b)/\Delta v_{qs}^e$, of the 20-hp motor at base torque condition. Each branch of the locus starts from an open-loop transfer pole with the gain equal to zero and moves toward an open-loop zero as the gain approaches infinity. Note that parts of the two branches of the locus reentering the real axis have been omitted. That is because points corresponding to the discrete values of gain in the *k* vector are joined by lines in the above plot. If need be, these parts can be filled in by inserting a small range of closely spaced gain values spanning the omitted region. Check your SIMULINK simulation, *s4eig*, and your MATLAB M-file, *m4*, using the sample root-locus plot.

Step response.   Use *m4* to insert the machine parameters and compute the transfer function, $numG/denG = \Delta(\omega_r/\omega_b)/\Delta v_{qs}^e$. With the above information in the MATLAB workspace, you can use the following MATLAB script given in *m4ustp* to determine the step response of the transfer function, $numG/denG$, of the induction machine:

```
% M-file for the second part of Project 4 on linearized analysis
%   in Chapter 6 to obtain the unit step response of the motor
%   transfer function, numG/denG.
% It can only be used after the transfer function has been
% determined by m4, that is numG/denG must be already
% defined in the MATLAB workspace.

% m4ustp computes the step response using the step function
% and plots the unit step response of the transfer function
% numG/denG.
```

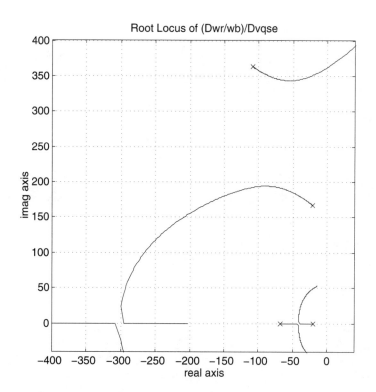

Figure 6.33   Root-locus for $\dfrac{K}{0.05s+1}\,\dfrac{\Delta(\omega_r/\omega_b)}{\Delta v_{qs}^e}$.

```
% Requires MATLAB Control System Toolbox function step

%
t =[0:0.0005:0.25]; % set up the regularly spaced time points
% Obtain the step response of the transfer function
[y,x,ti]=step(numG,denG,t); % step is a Control System Toolbox
% function
plot(ti,y);
title('step response of numG/denG')
```

Figure 6.34 shows a sample plot of the unit step response of the transfer function, $\Delta(\omega_r/\omega_b)/\Delta v_{qs}^e$.

Cross-check the above step response output of the transfer function against that obtained from the SIMULINK simulation of the induction machine in the synchronously rotating reference by applying a unit step voltage disturbance to v_{qs}^e of the induction machine operating at rated voltage and motoring a load torque of T_b. Figure 6.35 shows the simulation inside the SIMULINK file, *s4stp*, for conducting such a step input. The inputs to the *s4eig* block are the fixed value of T_{mech} at the value of base torque, T_b, and the *repeating sequence* signal of v_{qs}^e with time values = [0 1. 1. 1.2] and output values of

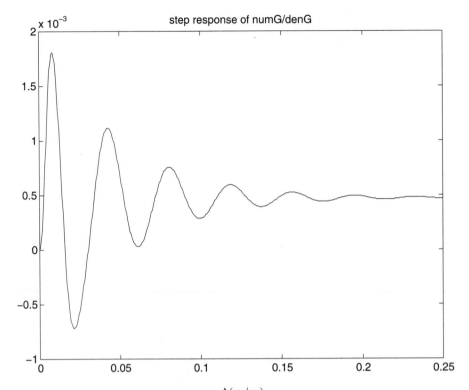

Figure 6.34 Step response of $\dfrac{\Delta(\omega_r/\omega_b)}{\Delta v_{qs}^e}$ at rated motoring condition.

$[V_m\ V_m\ V_m+1\ V_m+1]$. The initial delay of a second in the *repeating sequence* signal for v_{qs}^e is to allow time for the simulation starting from some approximate initial condition to settle to steady-state before applying the step input voltage. Note that the post-disturbance level of v_{qs}^e is the same as that of the calculated unit step response. The length of the **To Workspace** is adjusted to capture the tail end of the simulation run where the disturbance is in effect. Shorter run-times can be used if you are using the computed steady-state condition from **trim** to initialize your simulation. Figure 6.36 shows the step response from the above simulation for a nominal condition in which machine is motoring with rated stator voltage and a load torque of T_b. Except for the difference in steady-state level, this response curve is identical to that shown in Fig. 6.34.

Obtain the root-locus and check out the step response plot of the following transfer functions for the motor operated with rated voltage, that is rated v_{qs}^e:

- $\Delta(\omega_r/\omega_b)/\Delta v_{qs}^e$ when motoring zero and base load torque.
- $\Delta(\omega_r/\omega_b)/\Delta T_{mech}$ when motoring zero and base load torque.

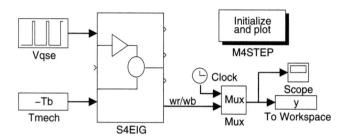

Figure 6.35   Setup for checking $\dfrac{\Delta(\omega_r/\omega_b)}{\Delta v_{qs}^e}$ in *s4stp*.

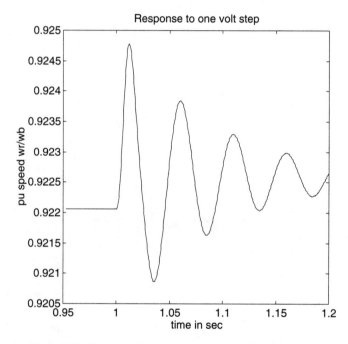

Figure 6.36   Simulated (ω_r/ω_b) response to a unit disturbance in v_{qs}^e.

6.11.5 Project 5: Some Non-zero v_{sg} Conditions

As indicated earlier, when the stator windings are connected by a three-wire connection to the source, the voltage, v_{sg}, can be non-zero even though the sums, $i_{as} + i_{bs} + i_{cs}$ and $v_{as} + v_{bs} + v_{cs}$, are both zero. In this project, we will examine a few kinds of source voltages which can result in a non-zero v_{sg} using the MATLAB and SIMULINK files, *m5A* and *s5a*. Aside from the changes in the *abc2qds* block to handle the case of a non-zero v_{sg} shown in Fig. 6.37b, the rest of *s5a* is the same as that in *s1* described earlier.

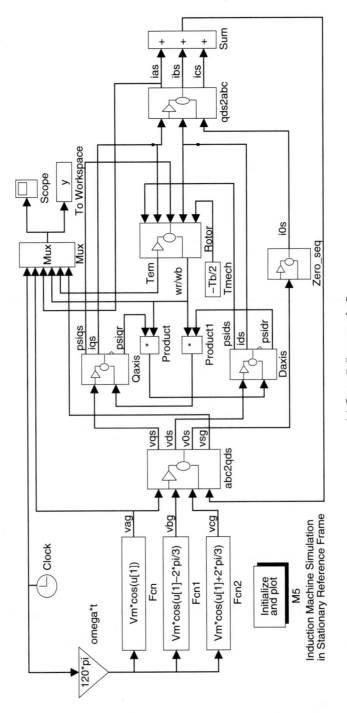

Figure 6.37   Simulation *s5a* of an induction machine.

(a) Overall diagram of *s5a*

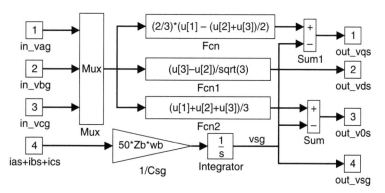

(b) Inside the *abc2qds* block of *s5a*

Figure 6.37   *(cont.)*

We will use the same 1-hp motor as in Project 1 so that the results obtained here can be compared against those obtained with balanced three-phase, sinusoidal voltages. It is apparent from Eq. 6.104 that when the source voltages are unbalanced in a manner that $v_{ag} + v_{bg} + v_{cg}$ is not equal to zero, the value of v_{sg} will not be zero. If you are not familiar with such unbalanced conditions for sinusoidal source voltages, experiment with various combinations of magnitude and phase unbalance by inserting the appropriate expressions in the corresponding function generator blocks. With extreme unbalanced conditions, the motor may not develop sufficient torque to run up with the value of mechanical torque, T_{mech}, that is set at $-T_b/2$ in *s5a*.

Aside from unbalanced sinusoidal waveforms, we could also use *s5a* to examine the following cases of complex voltage waveforms. For instance, to each phase of the function generators for v_{ag}, v_{bg}, and v_{cg} we can add harmonic components, such as

$$v_{ag} = V_m \cos \omega_e t + \frac{V_m}{h} \cos(h\omega_e t + \alpha_n)$$

$$v_{bg} = V_m \cos\left(\omega_e t - \frac{2\pi}{3}\right) + \frac{V_m}{h} \cos\left(h\omega_e t - \frac{2h\pi}{3} + \alpha_n\right) \qquad (6.170)$$

$$v_{cg} = V_m \cos\left(\omega_e t + \frac{2\pi}{3}\right) + \frac{V_m}{h} \cos\left(h\omega_e t + \frac{2h\pi}{3} + \alpha_n\right)$$

Experiment with odd values of $h = 3n$ and $3n \pm 1$ where n is an integer. Determine the relationship between the value of h and the condition for zero v_{sg}.

Next, we will apply the above observations to the case where the source voltages to the induction motor are square waves, each phase being displaced 120° apart from the others and of magnitude such that their rms value is the same as that of the three-phase sine voltages used in Project 1. Figure 6.38 shows the inside of the SIMULINK file, *s5b*. The set of quasi-square, three-phase voltages is generated by connecting a **sign** function block from the **Nonlinear** library block after the sinusoidal output of each function generator block, and by amplifying the outputs of the **sign** blocks back to the desired level. Use *m5* to set up the parameters of the 1-hp motor in the MATLAB workspace and conduct a simulation run

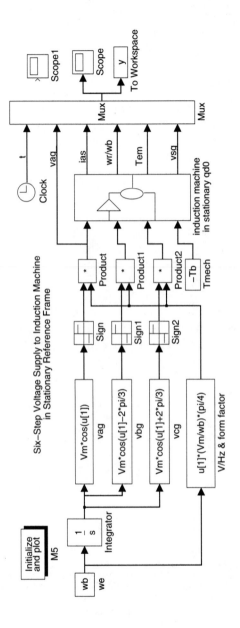

Figure 6.38 Overall diagram of *s5b*.

with $s5b$ of the startup of the motor with a constant load torque of $T_{mech} = -T_b$. Compare the results of this run with those obtained with sinusoidal voltages of the same rms value in Project 1.

6.11.6 Project 6: Single-phase Induction Motor

In this project, we will implement the SIMULINK simulation, $s6$, of a single-phase induction motor and use it to study the characteristics of a single-phase induction motor operating with no external capacitor, as in a split-phase motor, with a starting capacitor, and with both starting and run capacitors.

When operated as a split-phase motor, the slight lead of the ds stator winding current to that of the qs winding is from the difference in L/R ratio of the two windings. Ideally, when the magnitude of ds and qs currents is the same and the ds winding current leads the qs winding current by 90°, there will not be any negative-sequence component. At any operating point, the negative-sequence component can be minimized, even though the magnitude of the winding currents are not equal if the phase difference between the ds and qs winding currents is made to approach 90° by connecting an appropriately-sized capacitor in series with the ds winding to the supply voltage. The output torque will have a higher average value and less pulsations when the negative-sequence component of the stator currents is reduced. For a capacitor-start motor, the starting capacitor remains connected until the rotor speed reaches a predetermined cut-off speed, after which the capacitor and ds winding are both disconnected from the supply voltage. The starting capacitor is sized to give the optimum lead between the ds and qs winding at the beginning of starting. A capacitor-start and capacitor-run motor uses two capacitors, the smaller run capacitor is permanently connected in series with the ds winding and is sized to give optimal output torque at some nominal running condition. The larger starting capacitor, sized to improve the output torque at starting, is connected in parallel with the run capacitor during starting, but is disconnected after the rotor speed reaches the predetermined cut-off speed.

The details of the SIMULINK simulation, $s6$, of a single-phase induction motor using the model equations given in Table 6.5 are shown in Figure 6.39. The overall diagram of $s6$ is shown in Fig. 6.39a. The single-phase voltage supply to the motor is represented as an ideal ac voltage using a *sine wave* source block. Figure 6.39b shows the inside of the *ExtConn* block that produces the input voltages to the qs and ds circuits. It contains the simulation for inserting the appropriate capacitor voltage or open-circuit voltage to the ds winding. For example, in the case of a capacitor-start motor, the switching logic will simultaneously disconnect the start capacitor at current zero and change the applied voltage to the ds winding terminal from that of the supply voltage minus the capacitor voltage to the open-circuit voltage to simulate the open ds winding immediately after the rotor has attained the cut-off speed. The input voltage to the ds winding to simulate the open condition is obtained using Method A described in Project 3. It is based on the following equations for the open-circuit condition:

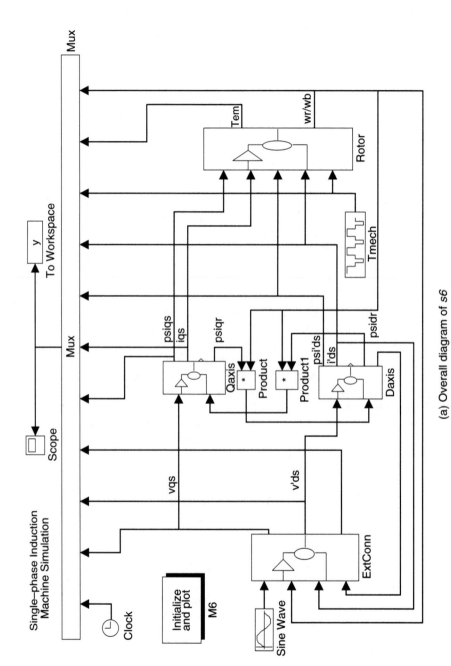

(a) Overall diagram of s6

Figure 6.39 Simulation s6 of a single-phase induction motor.

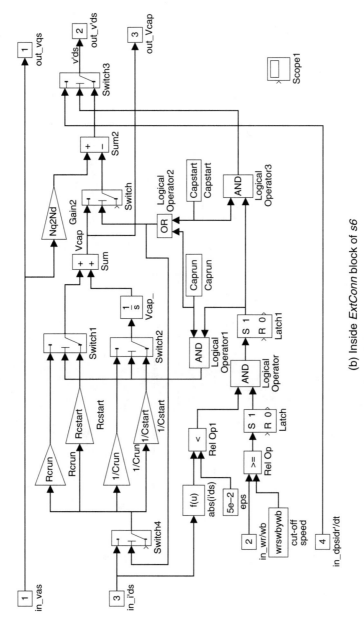

Figure 6.39 (cont.): Simulation *s6* of a single-phase induction motor.

(b) Inside *ExtConn* block of *s6*

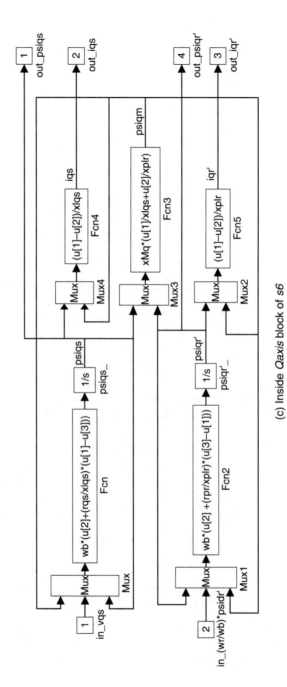

(c) Inside *Qaxis* block of *s6*

Figure 6.39 (cont.): Simulation *s6* of a single-phase induction motor.

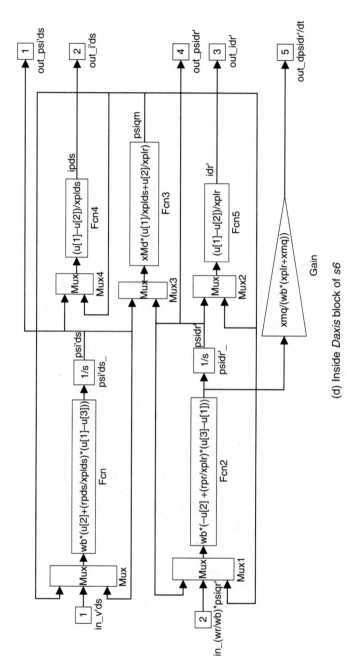

(d) Inside *Daxis* block of *s6*

Figure 6.39 (cont.): Simulation *s6* of a single-phase induction motor.

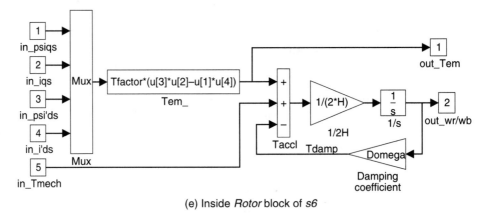

(e) Inside *Rotor* block of *s6*

Figure 6.39   *(cont.):* Simulation *s6* of a single-phase induction motor.

$$i'_{ds} = 0$$

$$\psi'_{ds} = \psi'_{md} = x_{mq}i'_{dr}$$

$$v'_{ds} = \frac{x_{mq}}{\omega_b}\frac{di'_{dr}}{dt} \tag{6.171}$$

$$i'_{dr} = \frac{\psi'_{dr}}{x'_{lr} + x_{mq}}$$

$$v'_{ds} = \frac{1}{\omega_b}\frac{x_{mq}}{x'_{lr} + x_{mq}}\frac{d\psi'_{dr}}{dt}$$

As can be seen from Figs. 6.39c and 6.39d, the simulations of the qs and ds circuits in the *Qaxis* and *Daxis* blocks are similar to those of a three-phase induction machine, except for the unequal parameters of the qs and ds stator windings. The input voltage to the ds winding to simulate an open condition is generated inside the *Daxis* block. The *Rotor* block shown in Fig. 6.39e is where the torque equation and rotor's equation of motion are simulated.

Implement the SIMULINK file, *s6*, on your computer and use the accompanying MATLAB M-file, *m6*, to study the starting and load response behavior of a 1/4-hp, 110-V, four-pole, 60-Hz single-phase induction motor. The parameters of the 1/4-hp motor and external capacitor values given in the *psph.m* file are from [43]. If you examine the listing of the M-file, *m6*, you will see that it has been programmed to determine and plot the steady-state characteristics of the motor, set up the motor parameters, initial values, and the sequence of step loading for the motor starting from standstill condition, and plot certain simulated results. The steady-state characteristics can be helpful when you need to cross-check your simulation.

Running up with no mechanical loading.   First, we will use the simulation to study the behavior of the motor starting from standstill with no mechanical loading. Use *m6* and the motor parameter given in *psph* to determine the steady-state characteristics and to set up the condition for simulating the startup of the motor operating as a split-phase. Repeat the study with the same motor parameters, but with it operating as a capacitor-start and then as a capacitor-run motor.

Examine the three sets of steady-state characteristics and compare their average output torque, efficiency, winding current magnitude, and phase difference between the *qs* and *ds* winding currents. Note the speed at which the phase difference between the *qs* and *ds* winding currents in the capacitor-start and capacitor-run is close to 90°.

Conduct a simulation of the split-phase motor running up from standstill with no mechanical loading. Repeat the study with the motor operating as a capacitor-start motor and then with the motor as a capacitor-start and capacitor-run motor. Explain observed differences in acceleration, magnitude of torque pulsations, and final rotor speed with the help of the steady-state characteristics.

Response to step changes in loading.   Note the programmed sequence of step changes in T_{mech} given in *m6*. Conduct a simulation of the motor operating as a split-phase motor starting from standstill, this time followed by step changes in external mechanical torque after the initial run-up from standstill with no loading. Repeat the study with the motor operating as a capacitor-start motor and then with the motor as a capacitor-start and capacitor-run motor. Note the variations in rotor speed, winding currents, and torque pulsations with the change in loading.

Sample results for the case when the motor is operated as a split-phase motor are given in Figs. 6.40, 6.41, and 6.42; those for the same motor operated as a capacitor-start motor are given in Figs. 6.43, 6.44, and 6.45; and, those when it is operated as a capacitor-start, capacitor-run motor are given in Figs. 6.46, 6.47, and 6.48.

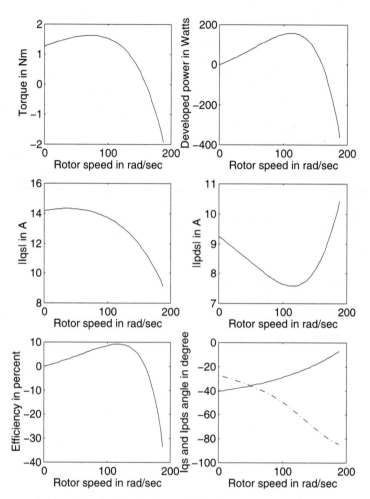

Figure 6.40   Steady-state characteristics of 1/4-hp split-phase motor.

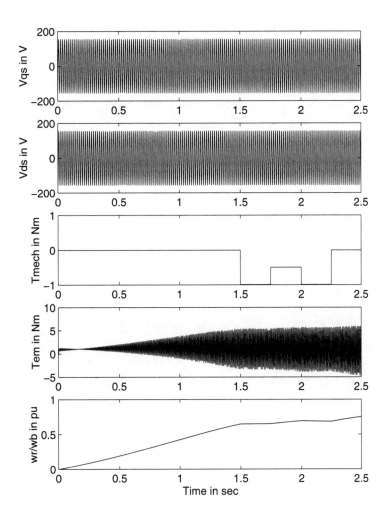

Figure 6.41   Startup and load response of 1/4-hp split-phase motor.

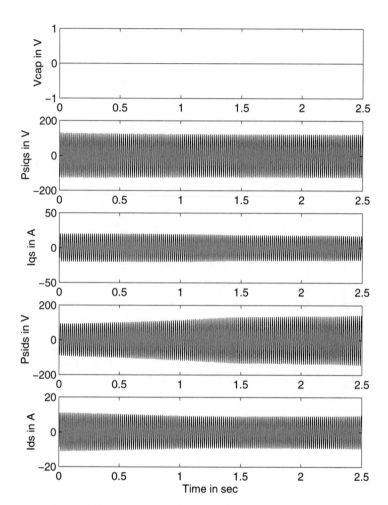

Figure 6.42 Startup and load response of 1/4-hp split-phase motor.

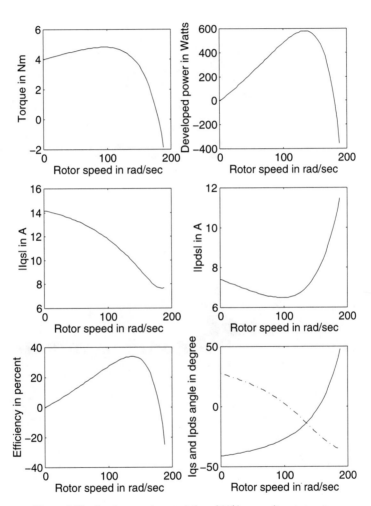

Figure 6.43   Steady-state characteristics of 1/4-hp capacitor-start motor.

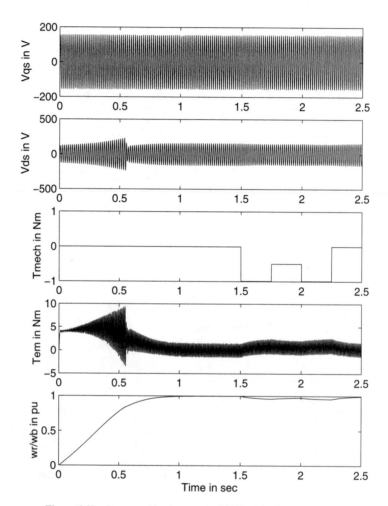

Figure 6.44   Startup and load response of 1/4-hp capacitor-start motor.

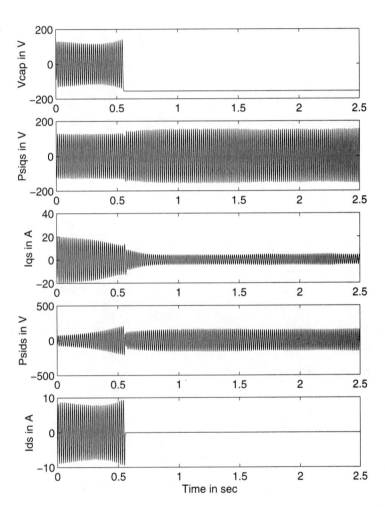

Figure 6.45   Startup and load response of 1/4-hp capacitor-start motor.

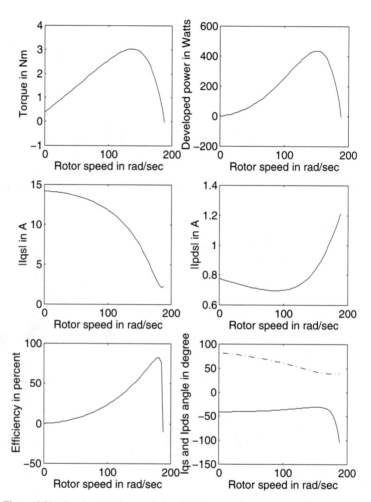

Figure 6.46   Steady-state characteristics of 1/4-hp capacitor-start, capacitor-run motor.

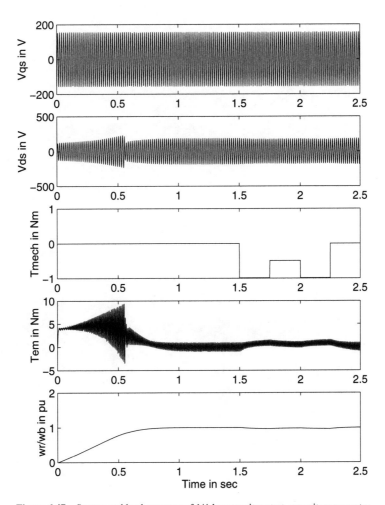

Figure 6.47   Startup and load response of 1/4-hp capacitor-start, capacitor-run motor.

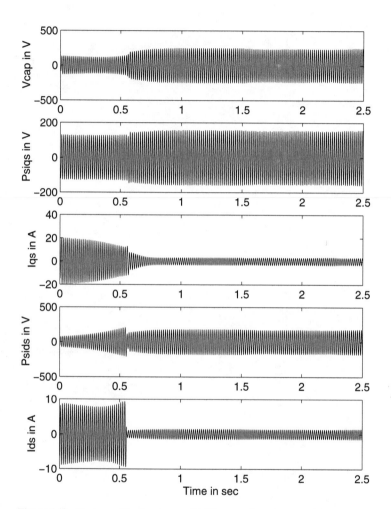

Figure 6.48   Startup and load response of 1/4-hp capacitor-start, capacitor-run motor.

7

Synchronous
Machines

7.1 INTRODUCTION

Of the electric machines that run at synchronous speed, the largest and also perhaps the most common are the three-phase synchronous machines. Although the construction of three-phase synchronous machines is relatively more expensive than that of induction machines, their higher efficiency is an advantage at a higher power rating. Three-phase synchronous machines are widely used for power generation and large motor drives.

The stator of a synchronous machine consists of a stack of laminated ferromagnetic core with internal slots, a set of three-phase distributed stator windings placed in the core slots, and an outer framework with end shields and bearings for the rotor shaft. The turns of the stator windings are equally distributed over pole-pairs, and the phase axes are spaced $2\pi/3$ electrical radians apart.

The cross-sectional shape of the rotor can be salient or cylindrical. Salient pole construction is mostly used in low-speed applications where the diameter to length ratio of the rotor can be made larger to accommodate the high pole number. Salient pole synchronous machines are often used in hydro generators to match the low operating speed of the hydraulic turbines. The short, pancake-like rotor has separate pole pieces bolted onto the periphery of a spider-web-like hub. Salient here, refers to the protruding poles; the alternating arrangement of pole iron and interpolar gap results in preferred directions of

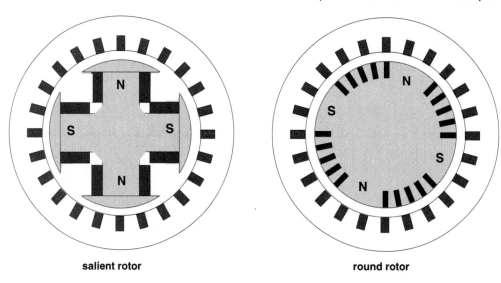

salient rotor   **round rotor**

Figure 7.1   Cross-sections of salient and cylindrical four-pole machines.

magnetic flux paths or magnetic saliency. Figure 7.1 shows the cross-sections of salient and cylindrical rotor synchronous machines with four poles.

The cylindrical or round rotor construction is favored in high-speed applications where the diameter to length ratio of the rotor has to be kept small to keep the mechanical stresses from centrifugal forces within acceptable limits. Two- and four-pole cylindrical rotor synchronous machines are used in steam generators to efficiently match the high operating speed of steam turbines. The long cylindrical rotors are usually machined from solid castings of chromium-nickel-molybdenum steel, with axial slots for the field winding on both sides of a main pole. Direct current excitation to the field winding can be supplied through a pair of insulated slip rings mounted on the rotor shaft. Alternatively, the dc excitation can be obtained from the rectified output of a small alternator mounted on the same rotor shaft of the synchronous machine. The second excitation method dispenses with the slip rings and is called brushless excitation.

In the basic two-pole representation of a synchronous machine, the axis of the north pole is called the direct or d-axis. The quadrature, or q-axis, is defined in the direction 90 electrical degrees ahead of the direct axis. Under no-load operation with only field excitation, the field mmf will be along the d-axis, and the stator internal voltage, $d\lambda_{af}/dt$, will be along the q-axis.

The mathematical description or model developed in this section is based on the concept of an ideal synchronous machine with two basic poles. The fields produced by the winding currents are assumed to be sinusoidally distributed around the airgap. This assumption of sinusoidal field distribution ignores the space harmonics, which may have secondary effects on the machine's behavior. It is also assumed that the stator slots cause no appreciable variation of any of the rotor winding inductances with rotor angle. Although saturation is not explicitly taken into account in this model, it can be accounted for by adjusting the reactances along the two axes with saturation factors or by introducing a compensating component to the main field excitation.

Since three-phase synchronous generators have been widely used for power generation, there is a wealth of literature on their modeling and the determination of their parameters. Although very elaborate models of these machines can be established by considering the magnetic circuits of the machines, in general, the coupled-circuit approach is simpler to understand and is more widely adopted. To properly simulate a machine we need to have not only a representative model but also accurate parameters for the model. Whether a model is representative or not would depend on the purpose and how adequately it serves that purpose. Elaborate models will generally require more data than what is usually available, more effort to program, and more computer time to run. Complicated models requiring parameters that cannot be readily obtainable from tests are to be avoided.

Although the rotor may have only one physically identifiable field winding, additional windings are often used to represent the damper windings and the effects of current flow in the rotor iron. For the salient-pole rotor machine, usually two such additional windings are used, one on the d-axis and the other on the q-axis. Over the years, experience on power system simulation has shown that most synchronous generators can be adequately represented by a model that is based on an equivalent idealized machine with one or two sets of damper windings, besides the field winding. Damper windings in the equivalent machine model can be used to represent physical armortisseur windings, or the damping effects of eddy currents in the solid iron portion of the rotor poles. Figure 7.2 shows a circuit representation of an idealized machine model of the synchronous machine commonly used in analysis.

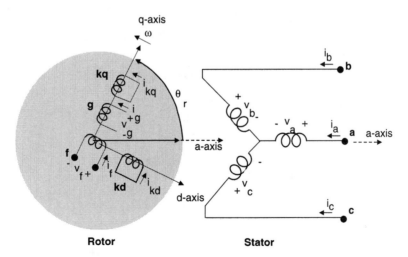

Figure 7.2   Circuit representation of an idealized machine.

7.2 MATHEMATICAL MODEL

Before we derive the mathematical equations of the circuit model shown in Fig. 7.2, let's take a brief look at the variation of inductances with rotor positions. In general, the permeances along the d- and q-axes are not the same. Whereas the mmfs of the rotor windings are always directed along the d- or q-axis, the direction of the resultant mmf of the stator windings relative to these two axes will vary with the power factor. A common approach to handling the magnetic effect of the stator's resultant mmf is to resolve it along the d- and q-axes, where it could be dealt with systematically. For example, let's consider the magnetic effect of just the a-phase current flowing in the stator. As shown in Fig. 7.3, the resolved components of the a-phase mmf, F_a, produce the flux components, $\phi_d = P_d F_a \sin\theta_r$ and $\phi_q = P_q F_a \cos\theta_r$ along the d- and q-axes, respectively.

The linkage of these resolved flux components with the a-phase winding is

$$
\begin{aligned}
\lambda_{aa} &= N_s(\phi_d \sin\theta_r + \phi_q \cos\theta_r) \qquad Wb.turn \\
&= N_s F_a (P_d \sin^2\theta_r + P_q \cos^2\theta_r) \\
&= N_s F_a \left(\frac{P_d + P_q}{2} - \frac{P_d - P_q}{2} \cos 2\theta_r \right)
\end{aligned}
\tag{7.1}
$$

The above expression of λ_{aa} is of the form $A - B \cos 2\theta_r$.

Similarly, the linkage of the flux components, ϕ_d and ϕ_q, by the b-phase winding that is $2\pi/3$ ahead may be written as

$$
\begin{aligned}
\lambda_{ba} &= N_s F_a \left\{ P_d \sin\theta_r \sin\left(\theta_r - \frac{2\pi}{3}\right) + P_q \cos\theta_r \cos\left(\theta_r - \frac{2\pi}{3}\right) \right\} \qquad Wb.turn \\
&= N_s F_a \left\{ -\frac{P_d + P_q}{4} - \frac{P_d - P_q}{2} \cos 2\left(\theta_r - \frac{\pi}{3}\right) \right\}
\end{aligned}
\tag{7.2}
$$

The expression of the mutual flux linkage, λ_{ba}, is of the form $-(A/2) - B \cos 2(\theta_r - \frac{\pi}{3})$. The magnitude of its second harmonic component in θ_r is the same as that of λ_{aa}, but the constant part is half that of λ_{aa}.

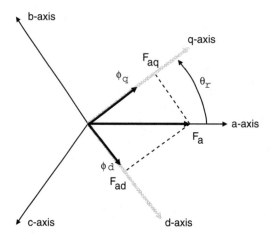

Figure 7.3 mmf components along dq axes.

Based on the functional relationship of λ_{aa} with the rotor angle, θ_r, we can deduce that the self-inductance of the stator a-phase winding, excluding the leakage, has the form:

$$L_{aa} = L_0 - L_{ms} \cos 2\theta_r \qquad H \qquad (7.3)$$

Those of the b- and c-phases, L_{bb} and L_{cc}, are similar to that of L_{aa} but with θ_r replaced by $(\theta_r - \frac{2\pi}{3})$ and $(\theta_r - \frac{4\pi}{3})$, respectively.

Similarly, we can deduce from Eq. 7.2 that the mutual inductance between the a- and b-phases of the stator is of the form

$$L_{ab} = L_{ba} = -\frac{L_0}{2} - L_{ms} \cos 2\left(\theta_r - \frac{\pi}{3}\right) \qquad H \qquad (7.4)$$

Similar expressions for L_{bc} and L_{ca} can be obtained by replacing θ_r in Eq. 7.4 with $(\theta_r - \frac{2\pi}{3})$ and $(\theta_r - \frac{4\pi}{3})$, respectively.

Using the motoring convention, the applied voltage to each of the seven windings shown in Fig. 7.2 is balanced by a resistive drop and a $d\lambda/dt$ term. The voltage equations for the stator and rotor windings can be arranged into the form:

$$\begin{bmatrix} \mathbf{v}_s \\ \mathbf{v}_r \end{bmatrix} = \begin{bmatrix} \mathbf{r}_s & 0 \\ 0 & \mathbf{r}_r \end{bmatrix} \begin{bmatrix} \mathbf{i}_s \\ \mathbf{i}_r \end{bmatrix} + \frac{d}{dt} \begin{bmatrix} \Lambda_s \\ \Lambda_r \end{bmatrix} \qquad V \qquad (7.5)$$

where

$$\mathbf{v}_s = [v_a, \ v_b, \ v_c]^t$$

$$\mathbf{v}_r = [v_f, \ v_{kd}, \ v_g, \ v_{kq}]^t$$

$$\mathbf{i}_s = [i_a \ i_b \ i_c]^t$$

$$\mathbf{i}_r = [i_f \ i_{kd} \ i_g \ i_{kq}]^t$$

$$\mathbf{r}_s = diag[r_a \ r_b \ r_c]$$

$$\mathbf{r}_r = diag[r_f \ r_{kd} \ r_g \ r_{kq}]$$

$$\Lambda_s = [\lambda_a, \ \lambda_b, \ \lambda_c]^t$$

$$\Lambda_r = [\lambda_f, \ \lambda_{kd}, \ \lambda_g, \ \lambda_{kq}]^t$$

The symbols of the per phase parameters are as follows:

r_s	armature or stator winding resistance
r_f	d-axis field winding resistance
r_g	q-axis field winding resistance
r_{kd}	d-axis damper winding resistance
r_{kq}	q-axis damper winding resistance
L_{ls}	armature or stator winding leakage inductance
L_{lf}	d-axis field winding leakage inductance
L_{lg}	q-axis field winding leakage inductance
L_{lkd}	d-axis damper winding leakage inductance

L_{lkq} $\quad$ q-axis damper winding leakage inductance
L_{md} $\quad$ d-axis stator magnetizing inductance
L_{mq} $\quad$ q-axis stator magnetizing inductance
L_{mf} $\quad$ d-axis field winding magnetizing inductance
L_{mg} $\quad$ q-axis field winding magnetizing inductance
L_{mkd} $\quad$ d-axis damper winding magnetizing inductance
L_{mkq} $\quad$ q-axis damper winding magnetizing inductance

The equations for the flux linkages of the stator and rotor windings can be expressed as

$$\Lambda_s = \mathbf{L}_{ss}\mathbf{i}_s + \mathbf{L}_{sr}\mathbf{i}_r \qquad Wb.\ turn$$
$$\Lambda_r = [\mathbf{L}_{sr}]^t\mathbf{i}_s + \mathbf{L}_r\mathbf{i}_r \tag{7.6}$$

where

$$\mathbf{L}_{ss} = \begin{bmatrix} L_{ls} + L_0 - L_{ms}\cos 2\theta_r & -\frac{1}{2}L_0 - L_{ms}\cos 2\left(\theta_r - \frac{\pi}{3}\right) & -\frac{1}{2}L_0 - L_{ms}\cos 2\left(\theta_r + \frac{\pi}{3}\right) \\ -\frac{1}{2}L_0 - L_{ms}\cos 2\left(\theta_r - \frac{\pi}{3}\right) & L_{ls} + L_0 - L_{ms}\cos 2\left(\theta_r - \frac{2\pi}{3}\right) & -\frac{1}{2}L_0 - L_{ms}\cos 2(\theta_r - \pi) \\ -\frac{1}{2}L_0 - L_{ms}\cos 2\left(\theta_r + \frac{\pi}{3}\right) & -\frac{1}{2}L_0 - L_{ms}\cos 2(\theta_r + \pi) & L_{ls} + L_0 - L_{ms}\cos 2\left(\theta_r + \frac{2\pi}{3}\right) \end{bmatrix} \tag{7.7}$$

$$\mathbf{L}_{rr} = \begin{bmatrix} L_{lf} + L_{mf} & L_{fkd} & 0 & 0 \\ L_{kdf} & L_{lkd} + L_{mkd} & 0 & 0 \\ 0 & 0 & L_{lg} + L_{mg} & L_{gkq} \\ 0 & 0 & L_{kqg} & L_{lkq} + L_{mkq} \end{bmatrix} \tag{7.8}$$

$$\mathbf{L}_{sr} = \begin{bmatrix} L_{sf}\sin\theta_r & L_{skd}\sin\theta_r & L_{sg}\cos\theta_r & L_{skq}\cos\theta_r \\ L_{sf}\sin\left(\theta_r - \frac{2\pi}{3}\right) & L_{skd}\sin\left(\theta_r - \frac{2\pi}{3}\right) & L_{sg}\cos\left(\theta_r - \frac{2\pi}{3}\right) & L_{skq}\cos\left(\theta_r - \frac{2\pi}{3}\right) \\ L_{sf}\sin\left(\theta_r + \frac{2\pi}{3}\right) & L_{skd}\sin\left(\theta_r + \frac{2\pi}{3}\right) & L_{sg}\cos\left(\theta_r + \frac{2\pi}{3}\right) & L_{skq}\cos\left(\theta_r + \frac{2\pi}{3}\right) \end{bmatrix} \tag{7.9}$$

It is evident from Eqs. 7.7 and 7.9 that the elements of $\mathbf{L}_{ss}$ and $\mathbf{L}_{sr}$ are a function of the rotor angle which varies with time at the rate of the speed of rotation of the rotor. These time-dependent coefficients present computational difficulty when Eq. 7.5 is being used to solve for the phase quantities directly. To obtain the phase currents from the flux linkages, the inverse of the time-varying inductance matrix will have to be computed at every time step. The computation of the inverse at every time step is time-consuming and could produce numerical stability problems.

7.2.1 Transformation to the Rotor's $qd0$ Reference Frame

We shall show in this subsection that when the stator quantities are transformed to a $qd0$ reference frame that is attached to the machine's rotor the resulting voltage equation has time-invariant coefficients. In the idealized machine, the axes of the rotor windings are already along the q- and d-axes, and the $qd0$ transformation need only be applied to the stator winding quantities. In vector notation, we define the augmented transformation matrix:

$$\mathbf{C} = \begin{bmatrix} \mathbf{T}_{qd0}(\theta_r) & 0 \\ 0 & \mathbf{U} \end{bmatrix} \tag{7.10}$$

where $\mathbf{U}$ is a unit matrix and

$$\mathbf{T}_{qd0}(\theta_r) = \frac{2}{3} \begin{bmatrix} \cos\theta_r & \cos\left(\theta_r - \frac{2\pi}{3}\right) & \cos\left(\theta_r + \frac{2\pi}{3}\right) \\ \sin\theta_r & \sin\left(\theta_r - \frac{2\pi}{3}\right) & \sin\left(\theta_r + \frac{2\pi}{3}\right) \\ \frac{1}{2} & \frac{1}{2} & \frac{1}{2} \end{bmatrix} \tag{7.11}$$

For convenience, we will denote the transformed $qd0$ voltages, currents, and flux linkages of the stator, that are

$$\begin{aligned} \mathbf{v}_{qd0} &= \mathbf{T}_{qd0}(\theta_r)\mathbf{v}_s \\ \mathbf{i}_{qd0} &= \mathbf{T}_{qd0}(\theta_r)\mathbf{i}_s \\ \mathbf{\Lambda}_{qd0} &= \mathbf{T}_{qd0}(\theta_r)\mathbf{\Lambda}_s \end{aligned} \tag{7.12}$$

by

$$\begin{aligned} \mathbf{v}_{qd0} &= [v_q,\ v_d,\ v_0]^t \\ \mathbf{i}_{qd0} &= [i_q,\ i_d,\ i_0]^t \\ \mathbf{\Lambda}_{qd0} &= [\lambda_q,\ \lambda_d,\ \lambda_0]^t \end{aligned} \tag{7.13}$$

Applying the transformation $\mathbf{T}_{qd0}(\theta_r)$ to only the stator quantities in Eq. 7.5, the stator voltage equations become

$$\mathbf{v}_{qd0} = \mathbf{T}_{qd0}\mathbf{r}_s\,\mathbf{T}_{qd0}^{-1}\mathbf{i}_{qd0} + \mathbf{T}_{qd0}\frac{d}{dt}\mathbf{T}_{qd0}^{-1}\mathbf{\Lambda}_{qd0} \tag{7.14}$$

If $r_a = r_b = r_c = r_s$, the resistive drop term in the above equation reduces to

$$\mathbf{T}_{qd0}\mathbf{r}_s\,\mathbf{T}_{qd0}^{-1}\mathbf{i}_{qd0} = r_s\mathbf{i}_{qd0} \tag{7.15}$$

The second term on the right side of Eq. 7.14 can be expanded as follows:

$$\mathbf{T}_{qd0}\frac{d}{dt}\,\mathbf{T}_{qd0}^{-1}\mathbf{\Lambda}_{qd0} = \mathbf{T}_{qd0}\left[\left(\frac{d}{dt}\,\mathbf{T}_{qd0}^{-1}\right)\mathbf{\Lambda}_{qd0} + \mathbf{T}_{qd0}^{-1}\frac{d}{dt}\mathbf{\Lambda}_{qd0}\right] \tag{7.16}$$

Substituting in the transformation matrix from Eq. 7.12 and simplifying, we can show that

$$\frac{d}{dt}\mathbf{T}_{qd0}^{-1}\Lambda_{qd0} = \omega_r \begin{bmatrix} -\sin\theta_r & \cos\theta_r & 0 \\ -\sin\left(\theta_r - \dfrac{2\pi}{3}\right) & \cos\left(\theta_r - \dfrac{2\pi}{3}\right) & 0 \\ -\sin\left(\theta_r + \dfrac{2\pi}{3}\right) & \cos\left(\theta_r + \dfrac{2\pi}{3}\right) & 0 \end{bmatrix}\Lambda_{qd0}$$

and that

$$\mathbf{T}_{qd0}\left[\frac{d}{dt}\mathbf{T}_{qd0}^{-1}\right]\Lambda_{qd0} = \omega_r \begin{bmatrix} 0 & 1 & 0 \\ -1 & 0 & 0 \\ 0 & 0 & 0 \end{bmatrix}\Lambda_{qd0}$$

where ω_r denotes $d\theta_r/dt$ in electrical radians/sec.

The last term in Eq. 7.16 simplifies to

$$\mathbf{T}_{qd0}\mathbf{T}_{qd0}^{-1}\frac{d}{dt}\Lambda_{qd0} = \frac{d}{dt}\Lambda_{qd0}$$

Back-substituting these results into Eq. 7.14, the stator voltage equations of the idealized synchronous machine in its own rotor qd reference frame simplifies to

$$\mathbf{v}_{qd0} = r_s\mathbf{i}_{qd0} + \omega_r \begin{bmatrix} 0 & 1 & 0 \\ -1 & 0 & 0 \\ 0 & 0 & 0 \end{bmatrix}\Lambda_{qd0} + \frac{d}{dt}\Lambda_{qd0}$$

7.2.2 Flux Linkages in Terms of Winding Currents

The corresponding relationship between flux linkage Λ_{qd0} and the $qd0$ currents can be obtained by transforming only the stator quantities, that is

$$\Lambda_{qd0} = \mathbf{T}_{qd0}\mathbf{L}_{ss}\mathbf{T}_{qd0}^{-1} + \mathbf{T}_{qd0}\mathbf{L}_{sr} \qquad Wb.\ turn \qquad (7.17)$$

The algebra in simplifying the right-hand side of the above equation is too lengthy to be included here, but it can be shown that Eq. 7.17 yields the following expressions for the stator $qd0$ flux linkages in which all the inductances shown are independent of the rotor angle, θ_r:

$$\lambda_q = \left\{L_{ls} + \frac{3}{2}(L_0 - L_{ms})\right\}i_q + L_{sg}i_g + L_{skq}i_{kq}$$

$$\lambda_d = \left\{L_{ls} + \frac{3}{2}(L_0 + L_{ms})\right\}i_d + L_{sfd}i_f + L_{skd}i_{kd} \qquad (7.18)$$

$$\lambda_0 = L_{ls}i_0$$

With the chosen rotor dq reference frame, the rotor winding variables need no rotational transformation. The expressions for the flux linkages of the rotor windings are

$$\lambda_f = \frac{3}{2}L_{sf}i_d + L_{ff}i_f + L_{fkd}i_{kd}$$

$$\lambda_{kd} = \frac{3}{2}L_{skd}i_d + L_{fkd}i_f + L_{kdkd}i_{kd}$$

$$\lambda_g = \frac{3}{2}L_{sg}i_q + L_{gg}i_g + L_{gkq}i_{kq}$$

$$\lambda_{kq} = \frac{3}{2}L_{skq}i_q + L_{gkq}i_g + L_{kqkq}i_{kq}$$

(7.19)

7.2.3 Referring Rotor Quantities to the Stator

Observe that the terms in Eq. 7.19 associated with the stator current components, i_d and i_q, have a 3/2 factor which will render the inductance coefficient matrices for the d- and q-axes windings non-symmetric when equations of Eq. 7.19 are combined with those of Eq. 7.19. Replacing the actual currents of the rotor windings by the following equivalent rotor currents will result in flux linkage equations with symmetric inductance coefficient matrices:

$$\underline{i}_f = \frac{2}{3}\,i_f$$

$$\underline{i}_{kd} = \frac{2}{3}\,i_{kd}$$

$$\underline{i}_g = \frac{2}{3}\,i_g$$

$$\underline{i}_{kq} = \frac{2}{3}\,i_{kq}$$

(7.20)

Let's also denote the equivalent magnetizing inductances of the d- and q-axes stator windings in Eq. 7.19 by L_{md} and L_{mq}, that is

$$L_{md} = \frac{3}{2}(L_0 + L_{ms})$$

$$= \frac{3}{2}\left\{N_s^2\frac{P_d + P_q}{2} - N_s^2\frac{P_d - P_q}{2}\right\} = \frac{3}{2}\,N_s^2\,P_d$$

(7.21)

and

$$L_{mq} = \frac{3}{2}(L_0 - L_{ms}) = \frac{3}{2}N_s^2 P_q$$

(7.22)

Expressing the stator and rotor flux linkages in terms of the equivalent rotor currents and magnetizing inductances given by Eqs. 7.20 through 7.22, we have

$$\lambda_q = (L_{ls} + L_{mq})i_q + \frac{3}{2}L_{sg}\underline{i}_g + \frac{3}{2}L_{skq}\underline{i}_{kq}$$

$$\lambda_d = (L_{ls} + L_{md})i_d + \frac{3}{2}L_{sf}\underline{i}_f + \frac{3}{2}L_{skd}\underline{i}_{kd}$$

$$\lambda_0 = L_{ls}i_0$$

$$\lambda_f = \frac{3}{2}L_{sf}i_d + \frac{3}{2}(L_{lf} + L_{mf})\underline{i}_f + \frac{3}{2}L_{fkd}\underline{i}_{kd} \tag{7.23}$$

$$\lambda_{kd} = \frac{3}{2}L_{skd}i_d + \frac{3}{2}L_{fkd}\underline{i}_f + \frac{3}{2}(L_{lkd} + L_{mkd})\underline{i}_{kd}$$

$$\lambda_g = \frac{3}{2}L_{sg}i_q + \frac{3}{2}(L_{lg} + L_{mg})\underline{i}_g + \frac{3}{2}L_{gkq}\underline{i}_{kq}$$

$$\lambda_{kq} = \frac{3}{2}L_{skq}i_q + \frac{3}{2}L_{gkq}\underline{i}_g + \frac{3}{2}(L_{lkq} + L_{mkq})\underline{i}_{kq}$$

Next, we shall refer the rotor quantities to the stator using the appropriate turns ratios, denoting the equivalent rotor currents referred to the stator by a prime superscript:

$$i'_f = \frac{N_f}{N_s}\underline{i}_f = \frac{2}{3}\frac{N_f}{N_s}\,i_f$$

$$i'_{kd} = \frac{N_{kd}}{N_s}\underline{i}_{kd} = \frac{2}{3}\frac{N_{kd}}{N_s}\,i_{kd} \tag{7.24}$$

$$i'_{kq} = \frac{N_{kq}}{N_s}\underline{i}_{kq} = \frac{2}{3}\frac{N_{kq}}{N_s}\,i_{kq}$$

$$v'_f = \frac{N_s}{N_f}\,v_f \qquad\qquad v'_{kd} = \frac{N_s}{N_{kd}}\,v_{kd}$$

$$\tag{7.25}$$

$$v'_g = \frac{N_s}{N_g}\,v_g \qquad\qquad v'_{kq} = \frac{N_s}{N_{kq}}\,v_{kq}$$

$$\lambda'_f = \frac{N_s}{N_f}\,\lambda_f \qquad\qquad \lambda'_{kd} = \frac{N_s}{N_{kd}}\,\lambda_{kd}$$

$$\tag{7.26}$$

$$\lambda'_g = \frac{N_s}{N_g}\,\lambda_g \qquad\qquad \lambda'_{kq} = \frac{N_s}{N_{kq}}\,\lambda_{kq}$$

$$r'_f = \frac{3}{2}\left(\frac{N_s}{N_f}\right)^2 r_f \qquad\qquad r'_{kd} = \frac{3}{2}\left(\frac{N_s}{N_{kd}}\right)^2 r_{kd}$$

$$\tag{7.27}$$

$$r'_g = \frac{3}{2}\left(\frac{N_s}{N_g}\right)^2 r_g \qquad\qquad r'_{kq} = \frac{3}{2}\left(\frac{N_s}{N_{kq}}\right)^2 r_{kq}$$

Using Eqs. 7.21 and 7.22, we can express the winding inductances as

$$L_{sf} = N_s N_f P_d = \frac{2}{3}\frac{N_f}{N_s} L_{md} \qquad\qquad L_{skd} = N_s N_{kd} P_d = \frac{2}{3}\frac{N_{kd}}{N_s} L_{md}$$

$$L_{sg} = N_s N_g P_q = \frac{2}{3}\frac{N_g}{N_s} L_{mq} \qquad\qquad L_{skq} = N_s N_{kq} P_q = \frac{2}{3}\frac{N_{kq}}{N_s} L_{mq}$$

$$L'_{ff} = \frac{3}{2}\left(\frac{N_s}{N_f}\right)^2 L_{lf} + L_{md} \qquad\qquad L_{mf} = N_f^2 P_d = \frac{2}{3}\left(\frac{N_f}{N_s}\right)^2 L_{md}$$

$$L'_{kdkd}\frac{3}{2}\left(\frac{N_s}{N_{kd}}\right)^2 L_{lkd} + L_{md} \qquad\qquad L_{mkd} = N_{kd}^2 P_d = \frac{2}{3}\left(\frac{N_{kd}}{N_s}\right)^2 L_{md} \qquad (7.28)$$

$$L_{fkd} = N_f N_{kd} P_d = \frac{2}{3}\left(\frac{N_f N_{kd}}{N_s^2}\right) L_{md} \qquad\qquad L_{gkq} = N_g N_{kq} P_q = \frac{2}{3}\left(\frac{N_g N_{kq}}{N_s^2}\right) L_{mq}$$

$$L'_{gg} = \frac{3}{2}\left(\frac{N_s}{N_g}\right)^2 L_{lg} + L_{mq} \qquad\qquad L_{mg} = N_g^2 P_q = \frac{2}{3}\left(\frac{N_g}{N_s}\right)^2 L_{mq}$$

$$L'_{kqkq} = \frac{3}{2}\left(\frac{N_s}{N_{kq}}\right)^2 L_{lkq} + L_{mq} \qquad\qquad L_{mkq} = N_{kq}^2 P_q = \frac{2}{3}\left(\frac{N_{kq}}{N_s}\right)^2 L_{mq}$$

In using the values of L_{md} and L_{mq} as the common mutual inductances on the d-axis and q-axis circuits, we have essentially defined their corresponding fluxes as the mutual fluxes in these axes; any additional flux linked by a current is considered a leakage component in the corresponding current path. Traditionally, the sums, $(L_{md} + L_{ls})$ and $(L_{mq} + L_{ls})$, are referred to as the d-axis and q-axis synchronous inductance, respectively, that is

$$L_d = L_{md} + L_{ls}$$

$$L_q = L_{mq} + L_{ls}$$

$$(7.29)$$

7.2.4 Voltage Equations in the Rotor's $qd0$ Reference Frame

A summary of the winding equations for the synchronous machine in the rotor's qd reference frame with all rotor quantities referred to the stator is given below:

$qd\,0$ Equations of a Synchronous Machine:

$$v_q = r_s i_q + \frac{d\lambda_q}{dt} + \lambda_d \frac{d\theta_r}{dt} \qquad V$$

$$v_d = r_s i_d + \frac{d\lambda_d}{dt} - \lambda_q \frac{d\theta_r}{dt}$$

$$v_0 = r_s i_0 + \frac{d\lambda_0}{dt}$$

$$v_f' = r_f' i_f' + \frac{d\lambda_f'}{dt} \qquad\qquad (7.30)$$

$$v_{kd}' = r_{kd}' i_{kd}' + \frac{d\lambda_{kd}'}{dt}$$

$$v_g' = r_g' i_g' + \frac{d\lambda_g'}{dt}$$

$$v_{kq}' = r_{kq}' i_{kq}' + \frac{d\lambda_{kq}'}{dt}$$

where the flux linkages are given by

$$\lambda_q = L_q i_q + L_{mq} i_g' + L_{mq} i_{kq}' \qquad Wb.\ turn$$

$$\lambda_d = L_d i_d + L_{md} i_f' + L_{md} i_{kd}'$$

$$\lambda_0 = L_{ls} i_0$$

$$\lambda_f' = L_{md} i_d + L_{md} i_{kd}' + L_{ff}' i_f' \qquad\qquad (7.31)$$

$$\lambda_{kd}' = L_{md} i_d + L_{md} i_f' + L_{kdkd}' i_{kd}'$$

$$\lambda_g' = L_{mq} i_q + L_{gg}' i_g' + L_{mq} i_{kq}'$$

$$\lambda_{kq}' = L_{mq} i_q + L_{mq} i_g' + L_{kqkq}' i_{kq}'$$

Figure 7.4 shows an equivalent circuit representation of the synchronous machine that is based on the above set of voltage and flux relations. It should be remembered that the winding equations and equivalent circuit diagram have been derived for an idealized machine. One of the limitations of this description is the assumption of a common mutual inductance for the stator and rotor windings. With proper parameters, such an assumption has been shown to have very little consequence insofar as the stator quantities are concerned, but the discrepancies between measured and simulated rotor variables, especially the field current, can be significant [58]. More refined models [58, 62, 63] with allowances for unequal stator and rotor windings' mutual inductances and the damping effects in the pole iron will be discussed later in Section 7.8.

7.2.5 Electromagnetic Torque

The expression for the electromagnetic torque developed by the machine can be obtained from the component of the input power that is transferred across the airgap. The total input

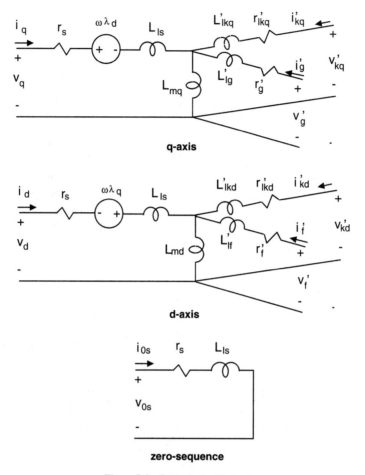

Figure 7.4   Equivalent $qd0$ circuits.

power into the machine is given by

$$P_{in} = v_a i_a + v_b i_b + v_c i_c + v_f i_f + v_g i_g \qquad W \qquad (7.32)$$

When the stator phase quantities are transformed to the rotor $qd0$ reference frame that rotates at a speed of $\omega_r = d\theta_r/dt$, Eq. 7.32 becomes

$$
\begin{aligned}
P_{in} &= \frac{3}{2}(v_q i_q + v_d i_d) + 3 v_0 i_0 + v_f i_f + v_g i_g \qquad W \\
&= \frac{3}{2}\left(r_s(i_q^2 + i_d^2) + i_q \frac{d\lambda_q}{dt} + i_d \frac{d\lambda_d}{dt} + \omega_r(\lambda_d i_q - \lambda_q i_d) \right) \qquad (7.33) \\
&\quad + 3 i_0^2 r_0 + 3 i_0 \frac{d\lambda_0}{dt} + i_f^2 r_f + i_f \frac{d\lambda_f}{dt} + i_g^2 r_g + i_g \frac{d\lambda_g}{dt}
\end{aligned}
$$

Eliminating terms in Eq. 7.34 identifiable with the ohmic losses and the rate of change in magnetic energy, the above expression of the electromechanical power developed reduces to

$$P_{em} = \frac{3}{2}\omega_r(\lambda_d i_q - \lambda_q i_d) \qquad W \tag{7.34}$$

For a P-pole machine, $\omega_r = (P/2)\omega_{rm}$, with ω_{rm} being the rotor speed in mechanical radians per second. Thus, Eq. 7.34 for a P-pole machine can also be written as

$$P_{em} = \frac{3}{2}\frac{P}{2}\omega_{rm}(\lambda_d i_q - \lambda_q i_d) \qquad W \tag{7.35}$$

Dividing the electromechanical power by the mechanical speed of the rotor, we obtain the following expression for the electromechanical torque developed by a P-pole machine:

$$T_{em} = \frac{3}{2}\frac{P}{2}(\lambda_d i_q - \lambda_q i_d) \qquad N.m \tag{7.36}$$

7.3 CURRENTS IN TERMS OF FLUX LINKAGES

Often a synchronous machine is simulated using the flux linkages of the windings as the state variables. To handle the cut set of inductors formed by the T-circuit of leakage and mutual inductances, we introduce the following mutual flux linkages in the q- and d-axes:

$$\lambda_{mq} = L_{mq}(i_q + i'_g + i'_{kq}) \qquad Wb.\ turn$$
$$\lambda_{md} = L_{md}(i_d + i'_f + i'_{kd}) \tag{7.37}$$

The currents can then be expressed simply as

$$i_q = \frac{1}{L_{ls}}(\lambda_q - \lambda_{mq}) \qquad\qquad i_d = \frac{1}{L_{ls}}(\lambda_d - \lambda_{md}) \qquad A$$

$$i'_g = \frac{1}{L'_{lg}}(\lambda'_g - \lambda_{mq}) \qquad\qquad i'_f = \frac{1}{L'_{lf}}(\lambda'_f - \lambda_{md}) \tag{7.38}$$

$$i'_{kq} = \frac{1}{L'_{lkq}}(\lambda'_{kq} - \lambda_{mq}) \qquad\qquad i'_{kd} = \frac{1}{L'_{lkd}}(\lambda'_{kd} - \lambda_{md})$$

Using the above expressions of the d-axis winding currents in Eq. 7.37 and collecting all λ_{md} terms to the left, we will obtain

$$\lambda_{md} = \frac{L_{MD}}{L_{ls}}\lambda_d + \frac{L_{MD}}{L'_{lf}}\lambda'_f + \frac{L_{MD}}{L'_{lkd}}\lambda'_{kd} \qquad Wb.\ turn \tag{7.39}$$

where

$$\frac{1}{L_{MD}} \triangleq \frac{1}{L_{ls}} + \frac{1}{L'_{lf}} + \frac{1}{L'_{lkd}} + \frac{1}{L_{md}} \tag{7.40}$$

Back-substituting the above expression for λ_{md} into the expressions of the d-axis currents given in Eq. 7.38 and regrouping, we obtain the following matrix equation of the currents in terms of the total flux linkages for the d-axis windings:

$$
\begin{bmatrix} i_d \\ i'_f \\ i'_{kd} \end{bmatrix} = \begin{bmatrix} \left(1 - \dfrac{L_{MD}}{L_{ls}}\right)\dfrac{1}{L_{ls}} & -\dfrac{L_{MD}}{L_{ls}L'_{lf}} & -\dfrac{L_{MD}}{L_{ls}L'_{lkd}} \\[3mm] -\dfrac{L_{MD}}{L'_{lf}L_{ls}} & \left(1 - \dfrac{L_{MD}}{L'_{lf}}\right)\dfrac{1}{L'_{lf}} & -\dfrac{L_{MD}}{L'_{lf}L'_{lkd}} \\[3mm] -\dfrac{L_{MD}}{L'_{lkd}L_{ls}} & -\dfrac{L_{MD}}{L'_{lkd}L'_{lf}} & \left(1 - \dfrac{L_{MD}}{L'_{lkd}}\right)\dfrac{1}{L'_{lkd}} \end{bmatrix} \begin{bmatrix} \lambda_d \\ \lambda'_f \\ \lambda'_{kd} \end{bmatrix} \quad (7.41)
$$

A set of expressions similar to those given in Eqs. 7.38 through 7.41 for the windings on the q-axis can be derived in a similar manner. The above expressions of the currents in terms of the flux linkages can be substituted into the voltage and torque equations to obtain a mathematical model of the synchronous machine in which the state variables are the flux linkages of the rotor windings.

7.4 STEADY-STATE OPERATION

In this section, we shall examine the equations and phasor diagrams for steady-state motoring and generating operations to help us better understand the operational behavior of the machine. When performing simulations, the steady-state equations can be used to determine the proper steady-state values to initialize the simulation so that it starts with the desired steady-state condition. More importantly, a knowledge of the steady-state behavior is indispensable for checking the correctness of the simulation results.

We shall assume balanced steady-state condition with the rotor rotating at the synchronous speed, ω_e, and the field excitation held constant. For ease of referral, we will refer to the q-axis on the rotor as the q_r axis, the q-axis of the synchronously rotating frame as the q_e-axis, and the q-axis of the stationary reference frame which is along the axis of the a-phase winding as q_s. When dealing with just one machine, it is convenient to use the a-phase terminal voltage of its stator as the synchronous reference phasor, from which all phase angles are measured; in other words, the stator phase voltages can be expressed as

$$
v_a = V_m \cos(\omega_e t) \qquad V
$$

$$
v_b = V_m \cos\left(\omega_e t - \frac{2\pi}{3}\right) \qquad (7.42)
$$

$$
v_c = V_m \cos\left(\omega_e t - \frac{4\pi}{3}\right)
$$

The space vectors and phasors of the above phase voltages are referred to the q_e-axis of a synchronously rotating frame, which we will assume to start with an initial angle of $\theta_e(0) = 0$ from the q_s-axis.

For balanced operation, the steady-state stator currents flowing into the machine may be expressed as

$$i_a = I_m \cos(\omega_e t + \phi) \qquad A$$

$$i_b = I_m \cos\left(\omega_e t + \phi - \frac{2\pi}{3}\right) \qquad\qquad (7.43)$$

$$i_c = I_m \cos\left(\omega_e t + \phi - \frac{4\pi}{3}\right)$$

It is evident from the above voltage and current expressions that the terminal power factor angle is ϕ; the value of ϕ is positive for a leading terminal power factor condition and negative for a lagging terminal power factor.

At this stage, we do not know the orientation of the rotor's q_r-axis with respect to the synchronously rotating q_e-axis. Since the rotor is also rotating at synchronous speed in steady-state, we do know that the angle between the q_r- and q_e-axes will have a steady value that is not varying with time. To locate the q_r-axis, we first transform the phase voltages and currents to the synchronously rotating frame. From the material presented earlier in Section 5.10.2, in particular Eq. 5.142, the qd components in the synchronously rotating reference frame are

$$v_q^e - j v_d^e = V_m + j0 = V_m e^{j0}$$

$$i_q^e - j i_d^e = I_m \cos\phi + j I_m \sin\phi = I_m e^{j\phi} \qquad\qquad (7.44)$$

Note that, in steady-state, the stator qd voltage and current components in the synchronously rotating frame are constant. The zero-sequence components of the balanced sets of phase voltages and currents are zero.

7.4.1 Steady-state Equations of the Stator

Usually only the f field winding is externally excited, that is $v_f' \neq 0$, and the other rotor windings have no external input, that is $v_{kd}' = v_g' = v_{kq}' = 0$. In steady-state, the rotor is rotating at synchronous speed, that is $\omega_r(t) = d\theta_r(t)/dt = \omega_e$. The relative speed of the rotor to the synchronously rotating resultant field in the airgap is zero; as such, there will be no speed voltages in the rotor windings. Therefore, $i_f' = (v_f'/r_f')$, and the other rotor currents, i_g', i_{kd}', and i_{kq}', are zero. Since both stator and rotor currents are constant, the flux linkages, λ_d and λ_q, will also be constant and the $d\lambda_d/dt$ and $d\lambda_q/dt$ terms in Eq. 7.30 will be zero. Thus, in steady-state, the qd voltage equations of the stator windings in the rotor qd reference frame will reduce to

$$v_q = r_s i_q + \omega_e L_d i_d + E_f \qquad V$$

$$v_d = r_s i_d - \omega_e L_q i_q \qquad\qquad (7.45)$$

where E_f, referred to as the steady-state field excitation voltage on the stator side, is

$$E_f \triangleq \omega_e L_{md} \left(\frac{v_f'}{r_f'}\right) \qquad V \qquad\qquad (7.46)$$

Note that E_f is along the q_r-axis of the rotor.

7.4.2 Locating the Rotor's q_r-axis

Let's now define the angle, $\delta(t)$, between the q_r- and q_e-axes as

$$\delta(t) = \theta_r(t) - \theta_e(t) \qquad elect.\,rad.$$

$$= \int_0^t \{\omega_r(t) - \omega_e\}dt + \theta_r(0) - \theta_e(0) \tag{7.47}$$

where $\theta_r(t)$ is the angle between the rotor's q_r-axis and the axis of the stator a-phase winding, and $\theta_e(t)$ is the angle between the q_e-axis of the synchronously rotating reference frame and the same a-phase axis. As defined, δ is the angle between the q_r-axis of the rotor and the q_e-axis of the synchronously rotating reference frame, measured with respect to the q_e-axis. In steady-state, with the rotor rotating at the same synchronous speed, that is $\omega_r(t) = \omega_e$, the angle, δ, will be constant. The steady-state value of δ will have accumulated contributions from all three terms on the right side of the second equation in Eq. 7.47.

The value of δ can be determined as follows: When the qd component equations in Eq. 7.45 are put in the complex form, that is

$$(v_q - jv_d) = (r_s + j\omega_e L_q)(i_q - ji_d) + \omega_e(L_d - L_q)i_d + E_f \qquad V \tag{7.48}$$

it is apparent that the last two terms on the right-hand side are real, that is along the q_r-axis of the rotor. Thus, the resultant of the remaining terms must also be real, or along the q_r-axis of the rotor. Traditionally, the resultant voltage from the remaining terms is denoted by $\mathbf{E}_q$, that is

$$\mathbf{E}_q = (v_q - jv_d) - (r_s + j\omega_e L_q)(i_q - ji_d) = |E_q|e^{j0} \qquad V \tag{7.49}$$

Both E_q and E_f are along the rotor's q-axis, that is the q_r-axis and, for motoring operation, differ only in magnitude, that is

$$E_f = E_q - i_d(x_d - x_q) \qquad V \tag{7.50}$$

The same relationship for generating operation, with current direction reversed, say $i_d^g = -i_d$, is $E_f = E_q + i_d^g(x_d - x_q)$.

The stator qd components of Eq. 7.44 in the synchronously rotating reference can be transformed to the rotor's qd axes by a forward rotational transformation of angle δ. Using the relations given in Eqs. 5.140 and 5.142, we can obtain

$$v_q - jv_d = (v_q^e - jv_d^e)e^{-j\delta} = V_m e^{-j\delta} \qquad V$$

$$i_q - ji_d = (i_q^e - ji_d^e)e^{-j\delta} = I_m e^{j(\phi-\delta)} \qquad A \tag{7.51}$$

Using the relationships of Eqs. 7.51 and 7.44 in Eq. 7.49, we will obtain

$$|E_q|e^{j0} = (v_q^e - jv_d^e)e^{-j\delta} - (r_s + j\omega_e L_q)(i_q^e - ji_d^e)e^{-j\delta} \tag{7.52}$$

Multiplying through by $e^{-j\delta}$, the above equation becomes

$$|E_q|e^{j\delta} = V_m - (r_s + j\omega_e L_q)(I_m \cos\phi + jI_m \sin\phi) \tag{7.53}$$

Taking the ratio of the imaginary part to real part on both sides, we obtain

$$\tan\delta = -\frac{r_s I_m \sin\phi + \omega_e L_q I_m \cos\phi}{V_m - r_s I_m \cos\phi + \omega_e L_q I_m \sin\phi} \tag{7.54}$$

When using the above equation to determine δ, we should take note that it has been derived assuming that stator currents are flowing into the machine and that ϕ is positive for leading terminal power factor condition. It can easily be adapted to generating convention by flipping the sign of I_m.

7.4.3 Time Phasors and Space Vectors

It is instructive to see how time phasors, especially those of the stator windings, are related to the corresponding space vectors. The stator current space vector in the stationary qd axes can be expressed as

$$\vec{i}_s^s = i_q^s - ji_d^s = \frac{2}{3}(i_a + \mathbf{a}i_b + \mathbf{a}^2 i_c) \quad A \tag{7.55}$$

where the superscript, s, is to denote stationary qd variables. Using the identity of $\cos(\omega_e t + \phi) = (e^{j(\omega_e t + \phi)} + e^{-j(\omega_e t + \phi)})/2$ and $\mathbf{a} = e^{j\frac{2\pi}{3}}$, the above expression for $\vec{i}_s^s$ reduces to

$$i_q^s - ji_d^s = I_m e^{j\phi} e^{j\omega_e t} \quad A \tag{7.56}$$

Denoting the rms time phasor of the a-phase stator current by $\tilde{I}_a$, we have

$$\tilde{I}_a = \frac{I_m}{\sqrt{2}} e^{j\phi} \quad \text{or} \quad \frac{I_m}{\sqrt{2}}\underline{/\phi} \quad A \tag{7.57}$$

In terms of the current phasor, the space vector, $\vec{i}_s$, can be expressed as

$$\vec{i}_s^s = i_q^s - ji_d^s = \sqrt{2}\tilde{I}_a e^{j\omega_e t} \quad A \tag{7.58}$$

The relationship between the instantaneous a-phase current and its phasor can, therefore, be expressed as

$$i_a = \Re\left[\sqrt{2}\tilde{I}_a e^{j\omega_e t}\right] \quad A \tag{7.59}$$

Likewise, for the set of balanced stator voltages given in Eq. 7.42, we have

$$\vec{v}_s^s = v_q^s - jv_d^s = \sqrt{2}\tilde{V}_a e^{j\omega_e t} \quad V \tag{7.60}$$

where

$$\tilde{V}_a = \frac{V_m}{\sqrt{2}} e^{j0} \quad \text{or} \quad \frac{V_m}{\sqrt{2}}\underline{/0} \quad V \tag{7.61}$$

Using the results in Eqs. 7.51 and 7.44, we can deduce the following relationships:

$$(v_q^e - jv_d^e) = \sqrt{2}\tilde{V}_a \qquad (v_q - jv_d) = \sqrt{2}\tilde{V}_a e^{-j\delta}$$

$$(i_q^e - ji_d^e) = \sqrt{2}\tilde{I}_a \qquad (i_q - ji_d) = \sqrt{2}\tilde{I}_a e^{-j\delta} \tag{7.62}$$

or

$$\tilde{\mathbf{V}}_a e^{-j\delta} = \left(\frac{v_q}{\sqrt{2}} - j\frac{v_d}{\sqrt{2}}\right) = \vec{\mathbf{V}}_q - j\vec{\mathbf{V}}_d$$

$$\tilde{\mathbf{I}}_a e^{-j\delta} = \left(\frac{i_q}{\sqrt{2}} - j\frac{i_d}{\sqrt{2}}\right) = \vec{\mathbf{I}}_q - j\vec{\mathbf{I}}_d \tag{7.63}$$

The rms space vectors are denoted in bold upper-case letters to distinguish them from the instantaneous space vectors.

Equation 7.63 indicates that the conventional rms time phasor is, in this case, equal to $(1/\sqrt{2})$ times the corresponding space vector. The above relations suggest that dividing Eq. 7.45 by $(1/\sqrt{2})$ will yield the following steady-state qd voltage equations of the synchronous machine in rms quantities, that is

$$\vec{\mathbf{V}}_q = r_s\vec{\mathbf{I}}_q + \omega_e L_d\vec{\mathbf{I}}_d + \vec{\mathbf{E}}_f$$

$$\vec{\mathbf{V}}_d = r_s\vec{\mathbf{I}}_d - \omega_e L_q\vec{\mathbf{I}}_q \qquad V \tag{7.64}$$

where $\vec{\mathbf{E}}_f$, the rms space vector of the stator field excitation voltage, E_f, has a value of $(\omega_e L_{md}/\sqrt{2})(v_f'/r_f')$ along the q_r-axis.

7.4.4 Steady-state Torque Expression

The total complex power into all three phases of the stator windings is given by

$$\mathbf{S} = 3(\vec{\mathbf{V}}_q - j\vec{\mathbf{V}}_d)(\vec{\mathbf{I}}_q - j\vec{\mathbf{I}}_d)^* \qquad VA \tag{7.65}$$

The electromagnetic power developed by the machine is obtained by subtracting from the input real power the losses in the stator, which in this model is just the copper losses in the stator windings. Thus, subtracting $3(I_q^2 + I_d^2)r_s$ from the real part of the input power, the expression for electromagnetic power is

$$P_{em} = \Re\left[3(\omega_e L_d\vec{\mathbf{I}}_d + \vec{\mathbf{E}}_f + j\omega_e L_q\vec{\mathbf{I}}_q)(\vec{\mathbf{I}}_d + j\vec{\mathbf{I}}_q)\right] \qquad W$$

$$= 3\{\mathbf{E}_f\mathbf{I}_q + \omega_e(L_d - L_q)\mathbf{I}_d\mathbf{I}_q\} \tag{7.66}$$

The expression for the electromagnetic torque developed by the machine is obtained by dividing the expression for the electromagnetic power by the actual rotor speed, that is

$$T_{em} = \frac{P_{em}}{\omega_{sm}} = \left(\frac{2}{P\omega_e}\right)P_{em} \qquad N.m$$

$$= 3\left(\frac{2}{P\omega_e}\right)\{\mathbf{E}_f\mathbf{I}_q + \omega_e(L_d - L_q)\mathbf{I}_d\mathbf{I}_q\} \tag{7.67}$$

The first torque component is the main torque component in a synchronous machine with field excitation. The second component is referred to as the reluctance torque component. It is present only when there is rotor saliency, that is $L_d \neq L_q$. Small three-phase reluctance motors are designed to operate on reluctance torque alone. They have simple and robust salient rotors that require no field excitation.

Alternate expressions for P_{em} and T_{em} in terms of the terminal voltage can be obtained by replacing the stator current components in Eqs. 7.66 and 7.67 using Eq. 7.64 with

$\vec{\mathbf{V}}_q = \mathbf{V}_a \cos\delta$ and $\vec{\mathbf{V}}_d = \mathbf{V}_a \sin\delta$. For large machines, where the small resistive drop may be neglected, such expressions for the electromagnetic power and torque can be shown to reduce to

$$P_{em} = -3\left\{\frac{E_f V_a}{X_d}\sin\delta + \frac{V_a^2}{2}\left(\frac{1}{X_q}-\frac{1}{X_d}\right)\sin2\delta\right\} \qquad W$$

$$T_{em} = -3\left(\frac{2}{P\omega_e}\right)\left\{\frac{E_f V_a}{X_d}\sin\delta + \frac{V_a^2}{2}\left(\frac{1}{X_q}-\frac{1}{X_d}\right)\sin2\delta\right\} \qquad N.m \quad (7.68)$$

where $\mathbf{V}_a$ is the rms value of the stator phase voltage, that is $V_m/\sqrt{2}$, $X_d = \omega_e L_d$ is the d-axis synchronous reactance and $X_q = \omega_e L_q$ is the q-axis synchronous reactance. As given in motoring convention, both P_{em} and T_{em} are positive for motoring and negative for generating since the value of δ, as defined, is positive for generating and negative for motoring.

7.4.5 Phasor and Space Vector Diagrams

Based on the equations derived for steady-state operation, we can construct the space vector or phasor diagrams shown in Figs. 7.5 through 7.9. Figure 7.5 is a phasor diagram for motoring with a leading terminal power factor. As shown, the armature mmf, $\tilde{\mathbf{F}}_a$, has a d-axis component that bucks the field mmf, $\tilde{\mathbf{F}}_f$. Figure 7.6 is also for motoring, but with a lagging terminal power factor. Here, $\tilde{\mathbf{F}}_a$ reinforces $\tilde{\mathbf{F}}_f$, essentially requiring a lower $\tilde{\mathbf{F}}_f$ for the same resultant field in the airgap.

In motoring operations, the stator currents are impressed by external voltages applied to the stator terminals as illustrated in Fig. 7.7a. But in generating operations, the stator currents are driven by the internally induced stator voltages to flow against the external voltages at the terminals as shown in Fig. 7.7b. The equations that we have derived using motoring convention can easily be adapted to generating operations by reversing the sign of the stator currents, $\tilde{\mathbf{I}}_a$, $\tilde{\mathbf{I}}_q$, and $\tilde{\mathbf{I}}_d$. Thus, in generating convention, with stator currents flowing out of the machine terminals and the machine variables in rms values, Eq. 7.64 becomes

$$\vec{\mathbf{V}}_q = -r_s\vec{\mathbf{I}}_q - \omega_e L_d\vec{\mathbf{I}}_d + \vec{\mathbf{E}}_f \qquad V$$
$$\vec{\mathbf{V}}_d = -r_s\vec{\mathbf{I}}_d + \omega_e L_q\vec{\mathbf{I}}_q \qquad\qquad (7.69)$$

Figures 7.8 and 7.9 show the phasor diagrams for leading and lagging power factor conditions, respectively, for generating operation. The angle δ is measured with respect to the q_e-axis of the synchronously rotating reference vector, which in the cases shown, is also aligned with the phasor of the a-phase stator voltage. Note that value of δ is positive for generating and negative for motoring.

7.5 SIMULATION OF THREE-PHASE SYNCHRONOUS MACHINES

The winding equations of the synchronous machine model derived in Section 7.2 can be implemented in a simulation that uses voltages as input and currents as output. The main

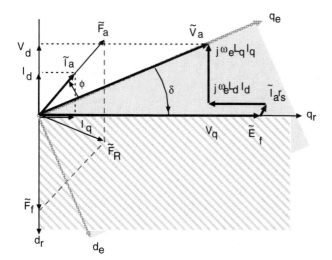

Figure 7.5 Motoring with a leading power factor.

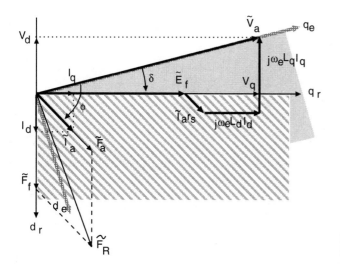

Figure 7.6 Motoring with a lagging power factor.

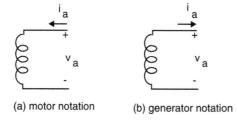

(a) motor notation   (b) generator notation

Figure 7.7 Assumed direction of winding current.

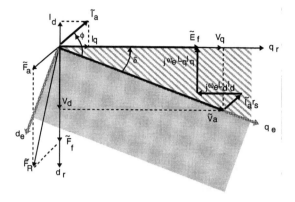

Figure 7.8 Generating with a leading power factor.

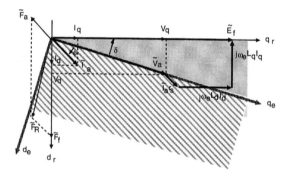

Figure 7.9 Generating with a lagging power factor.

inputs to the machine simulation are the stator abc phase voltages, the excitation voltage to the field windings, and the applied mechanical torque to the rotor.

The abc phase voltages of the stator windings must be transformed to the qd reference frame attached to the rotor. Although the angle, $\theta_r(t)$, increases with time, $\cos \theta_r(t)$ and $\sin \theta_r(t)$ will remain bounded. In simulation, the values of $\cos \theta_r(t)$ and $\sin \theta_r(t)$ can be obtained from a variable-frequency oscillator circuit which has a provision for setting the proper initial value of θ_r. The transformation from abc to qd rotor may be performed in a single step or two separate steps. In two steps, the intermediate outputs from the first step are the stator voltages in the stationary qd reference frame, that is

$$v_q^s = \frac{2}{3}v_a - \frac{1}{3}v_b - \frac{1}{3}v_c \qquad V$$

$$v_d^s = \frac{1}{\sqrt{3}}(v_c - v_b) \qquad\qquad (7.70)$$

$$v_0 = \frac{1}{3}(v_a + v_b + v_c)$$

The second step yields

$$v_q = v_q^s \cos\theta_r(t) - v_d^s \sin\theta_r(t)$$

$$v_d = v_q^s \sin\theta_r(t) + v_d^s \cos\theta_r(t) \tag{7.71}$$

where

$$\theta_r(t) = \int_0^t \omega_r(t)dt + \theta_r(0) \qquad elect.\,rad. \tag{7.72}$$

Alternatively, the transformation can be performed in a single step:

$$v_q = \frac{2}{3}\left\{ v_a \cos\theta_r(t) + v_b \cos\left(\theta_r(t) - \frac{2\pi}{3}\right) + v_c \cos\left(\theta_r(t) - \frac{4\pi}{3}\right)\right\}$$

$$v_q = \frac{2}{3}\left\{ v_a \sin\theta_r(t) + v_b \sin\left(\theta_r(t) - \frac{2\pi}{3}\right) + v_c \sin\left(\theta_r(t) - \frac{4\pi}{3}\right)\right\} \tag{7.73}$$

$$v_0 = \frac{1}{3}(v_a + v_b + v_c)$$

Expressing the $qd0$ voltage equations as integral equations of the flux linkages of the windings, the above stator $qd0$ voltages along with other inputs can then be used in the integral equations to solve for the flux linkages of the windings. For the case of a machine with only one field winding in the d-axis and a pair of damper windings in the d- and q-axes, the integral equations of the winding flux linkages are as follows:

$$\psi_q = \omega_b \int \left\{ v_q - \frac{\omega_r}{\omega_b}\psi_d + \frac{r_s}{x_{ls}}(\psi_{mq} - \psi_q)\right\} dt \qquad Wb.\,turn/s\,V$$

$$\psi_d = \omega_b \int \left\{ v_d + \frac{\omega_r}{\omega_b}\psi_q + \frac{r_s}{x_{ls}}(\psi_{md} - \psi_d)\right\} dt$$

$$\psi_0 = \omega_b \int \left(v_0 - \frac{r_s}{x_{ls}}\psi_0 \right) dt$$

$$\psi_{kq}' = \frac{\omega_b r_{kq}'}{x_{lkq}'} \int (\psi_{mq} - \psi_{kq}')dt \tag{7.74}$$

$$\psi_{kd}' = \frac{\omega_b r_{kd}'}{x_{lkd}'} \int (\psi_{md} - \psi_{kd}')dt$$

$$\psi_f' = \frac{\omega_b r_f'}{x_{md}} \int \left\{ E_f + \frac{x_{md}}{x_{lf}'}(\psi_{md} - \psi_f')\right\} dt$$

where

$$\psi_{mq} = \omega_b L_{mq}(i_q + i_{kq}')$$

$$\psi_{md} = \omega_b L_{md}(i_d + i_{kd}' + i_f') \tag{7.75}$$

$$E_f = x_{md}\frac{v_f'}{r_f'}$$

$$\psi_q = x_{ls}i_q + \psi_{mq}$$

$$\psi_d = x_{ls}i_d + \psi_{md}$$

$$\psi_0 = x_{ls}i_0$$

$$\psi_f' = x_{lf}'i_f' + \psi_{md}$$
(7.76)

$$\psi_{kd}' = x_{lkd}'i_{kd}' + \psi_{md}$$

$$\psi_{kq}' = x_{lkq}'i_{kq}' + \psi_{mq}$$

Note that the above equations are in motoring convention, that is with the currents, i_q and i_d, into the positive polarity of the stator windings' terminal voltages. As before, to handle the cut set of inductors in the q- and d-axis circuits, we will express the mutual flux linkages in terms of the total flux linkages of the windings as

$$\psi_{mq} = x_{MQ}\left(\frac{\psi_q}{x_{ls}} + \frac{\psi_{kq}'}{x_{lkq}'}\right)$$
(7.77)

$$\psi_{md} = x_{MD}\left(\frac{\psi_d}{x_{ls}} + \frac{\psi_{kd}'}{x_{lkd}'} + \frac{\psi_f'}{x_{lf}'}\right)$$

where

$$\frac{1}{x_{MQ}} = \frac{1}{x_{mq}} + \frac{1}{x_{lkq}'} + \frac{1}{x_{ls}}$$
(7.78)

$$\frac{1}{x_{MD}} = \frac{1}{x_{md}} + \frac{1}{x_{lkd}'} + \frac{1}{x_{lf}'} + \frac{1}{x_{ls}}$$

Having the values of the flux linkages of the windings and those of the mutual flux linkages along the d- and q-axes, we can determine the winding currents using

$$i_q = \frac{\psi_q - \psi_{mq}}{x_{ls}} \quad A$$

$$i_d = \frac{\psi_d - \psi_{md}}{x_{ls}}$$

$$i_{kd}' = \frac{\psi_{kd}' - \psi_{md}}{x_{lkd}'}$$
(7.79)

$$i_{kq}' = \frac{\psi_{kq}' - \psi_{mq}}{x_{lkq}'}$$

$$i_f' = \frac{\psi_f' - \psi_{md}}{x_{lf}'}$$

The stator winding qd currents can be transformed back to abc winding currents using the following rotor to stationary qd and stationary $qd0$ to abc transformations:

$$i_q^s = i_q \cos\theta_r(t) + i_d \sin\theta_r(t)$$
$$i_d^s = -i_q \sin\theta_r(t) + i_d \cos\theta_r(t)$$

(7.80)

$$i_a = i_q^s + i_0$$
$$i_b = -\frac{1}{2}i_q^s - \frac{1}{\sqrt{3}}i_d^s + i_0$$
$$i_c = -\frac{1}{2}i_q^s + \frac{1}{\sqrt{3}}i_d^s + i_0$$

(7.81)

7.5.1 Torque Expression

As given earlier in Eq. 7.36, the electromechanical torque developed by a machine with P-poles in motoring convention is

$$T_{em} = \frac{P_{em}}{\omega_{rm}} = \frac{3}{2}\frac{P}{2}(\lambda_d i_q - \lambda_q i_d) \qquad N.m$$
$$= \frac{3}{2}\frac{P}{2\omega_b}(\psi_d i_q - \psi_q i_d) \qquad N.m.$$

(7.82)

The value of T_{em} from the above expression is positive for motoring operation and negative for generating operation.

7.5.2 Equation of Motion of the Rotor Assembly

In motoring convention, the net acceleration torque, $T_{em} + T_{mech} - T_{damp}$, is in the direction of the rotor's rotation. Here, T_{em}, the torque developed by the machine, is positive when the machine is motoring and negative when the machine is generating; T_{mech}, the externally-applied mechanical torque in the direction of rotation, will be negative when the machine is motoring a load and will be positive when the rotor is being driven by a prime mover as in generating; and, T_{damp}, the frictional torque, acts in a direction opposite to the rotor's rotation. Equating the net acceleration torque to the inertia torque, we have

$$T_{em} + T_{mech} - T_{damp} = J\frac{d\omega_{rm}(t)}{dt} = \frac{2J}{P}\frac{d\omega_r(t)}{dt} \qquad N.m$$

(7.83)

The rotor angle, δ, is defined as the angle of the q_r-axis of the rotor with respect to the q_e-axis of the synchronously rotating reference frame, that is

$$\delta(t) = \theta_r(t) - \theta_e(t) \qquad elect.\,rad.$$
$$= \int_0^t \{\omega_r(t) - \omega_e\}dt + \theta_r(0) - \theta_e(0)$$

(7.84)

Since ω_e is constant,

$$\frac{d\{\omega_r(t) - \omega_e\}}{dt} = \frac{d\omega_r(t)}{dt}$$

(7.85)

Using Eq. 7.85 to replace $d\omega_r(t)/dt$ in Eq. 7.83, the slip speed can be determined from an integration of

$$\omega_r(t) - \omega_e = \frac{P}{2J} \int_0^t (T_{em} + T_{mech} - T_{damp})dt \qquad elect.\,rad/s \qquad (7.86)$$

Note that $\theta_r(t)$ and $\theta_e(t)$ are the angles of the q_r- and q_e-axes of the rotor and synchronously rotating reference frame, respectively, measured with respect to the stationary axis of the a-phase stator winding. The angle, δ, will be equal to the conventional power angle defined as that between the q_r-axis of the rotor and the terminal voltage phasor if the phasor of v_a is aligned with the q_e-axis of the reference synchronously rotating frame, that is $v_a = V_m \cos(\omega_e t + \theta_e(0))$ and $\theta_e(0)$ is zero. If $\theta_e(0)$ is not zero, as with a sine wave excitation of $v_a = V_m \sin(\omega_e t) = V_m \cos(\omega_e t - \pi/2)$ where $\theta_e(0) = -\pi/2$, the no-load steady-state value of δ will be $-\pi/2$ instead of zero. In this case, the a-phase phasor voltage, $\tilde{V}_{as}$, will still be aligned with the q_r-axis of the rotor at no-load, but both of them will be lagging $\pi/2$ behind the q_e-axis of the reference synchronously rotating frame. In a multi-machine system, the reference axis can be the q_r-axis of one of the generators in the system or the q_e-axis of an infinite bus voltage.

The initial values of $\theta_e(0)$ for the bus voltage, $\theta_r(0)$ for the variable frequency oscillator, and $\delta(0)$ for the rotor angle must be consistent if the machine's simulation is to begin with the desired operating condition.

7.5.3 Per-unit Expressions for Torque and the Equation of Motion

For studying power systems where there are many transformers and a wide range of equipment ratings, there are advantages to choosing an appropriate per unit system which will eliminate the bother of having to keep track of the primary and secondary side quantities of the transformers and also provide a rough but quick check on the values of the parameters. In the case of a study involving just one synchronous machine, the use of a per unit system offers no such advantage, other than perhaps the convenience of having the per unit parameters of the machine already available in terms of a set of base values that correspond to those of the rating of the machine. In such a situation, the base power, S_b, is the rated kVA of the machine.

For transient studies, the peak value rather than the rms value of the rated phase voltage is to be chosen as the base voltage, that is the base voltage, V_b, is $\left(\sqrt{2}V_{line-to-line}/\sqrt{3}\right)$. Similarly, choosing the peak value of the rated current as the base current, I_b, that is $I_b = 2S_b/3V_b$, the base values for the stator impedance and torque are given by

$$\text{base impedance:} \quad Z_b = \frac{V_b}{I_b} \quad \Omega$$

$$\text{base torque:} \quad T_b = \frac{S_b}{\omega_{bm}} \quad N.m \qquad (7.87)$$

The base mechanical angular frequency, ω_{bm}, is $2\omega_b/P$, where ω_b is the base electrical angular frequency and P is the number of poles. Using the second expression given in Eq. 7.82, the per unit electromagnetic torque developed is

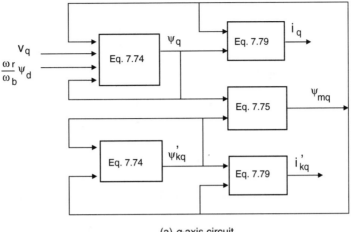

(a) *q*-axis circuit

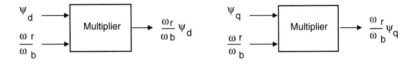

(b) Speed voltages

Figure 7.10   Flow of variables in the *q*-axis circuit simulation.

$$T_{em(pu)} = \frac{T_{em}}{T_b} = \frac{\frac{3}{2}\frac{P}{2\omega_b}(\psi_d i_q - \psi_q i_d)}{\frac{3}{2}\left(\frac{V_b I_b}{\frac{2}{P}\omega_b}\right)} \qquad pu \qquad (7.88)$$

Since the base for the flux linkages, ψ_q and ψ_d, is the same as V_b for the stator voltage, the above expression for the torque in per unit reduces to

$$T_{em(pu)} = \psi_{d(pu)} i_{q(pu)} - \psi_{q(pu)} i_{d(pu)} \qquad (7.89)$$

Equation 7.83 for the motion of the rotor assembly, expressed in per unit, is

$$T_{em(pu)} + T_{mech(pu)} - T_{damp(pu)} = \left(\frac{1}{T_b}\right)\left(\frac{2J}{P}\right)\frac{d\omega_r}{dt} \qquad pu \qquad (7.90)$$

In terms of the inertia constant, H, that is defined as $H = \frac{1}{2}J\omega_{bm}^2/S_b$ *sec.*, we have

$$T_{em(pu)} + T_{mech(pu)} - T_{damp(pu)} = 2H\frac{d(\omega_r/\omega_b)}{dt} = 2H\frac{d\{(\omega_r - \omega_e)/\omega_b\}}{dt} \qquad pu \qquad (7.91)$$

Figure 7.10 shows the flow of variables in a simulation of a three-phase synchronous machine in the rotor reference frame. Figure 7.11 shows an overall diagram of the SIMULINK simulation, *s1*.

The inputs to the simulation are the stator *abc* voltages, the excitation voltage, E_{ex}, to the field winding, and the externally-applied mechanical torque, T_{mech}, on the rotor. The

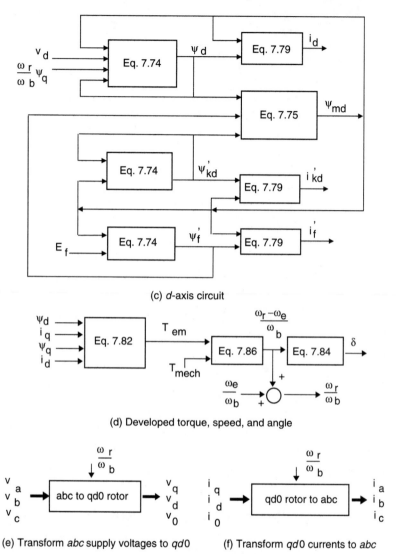

(c) *d*-axis circuit

(d) Developed torque, speed, and angle

(e) Transform *abc* supply voltages to *qd*0   (f) Transform *qd*0 currents to *abc*

Figure 7.10   *(cont.)*   Flow of variables in the *d*-axis circuit simulation.

equations given earlier are in the motoring convention, in which the assumed direction of the stator currents are into positive polarity of the stator windings' terminal voltages. Since we will be mainly using the simulation for generating operation, Fig. 7.11 shows the stator $qd0$ currents opposite in sign to those used for motoring notation. This step is taken merely for convenience since we will be using the simulation mainly for generating operations.

In Fig. 7.11, the transformation of the input stator *abc* voltages to the rotor qd reference frame are performed inside the *abc2qd0* block. Details of the *abc2qd0* block are

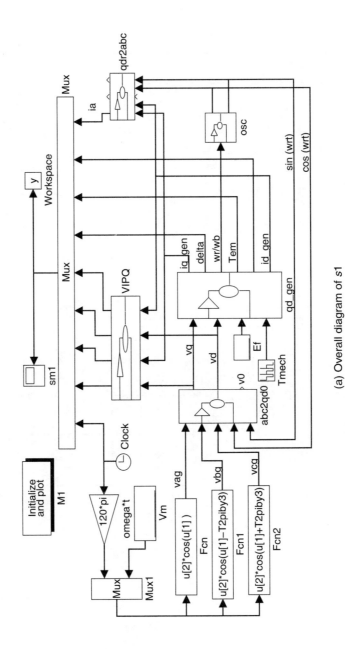

(a) Overall diagram of *s1*

Figure 7.11   Simulation *s1* of a synchronous generator.

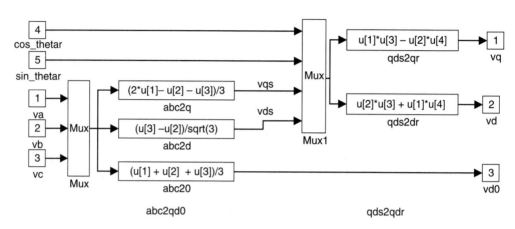

(b) Inside the *abc2qd0* block

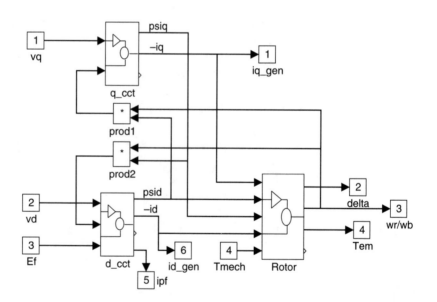

(c) Inside the *qd_gen* block

Figure 7.11   *(cont.):* Simulation *s1* of a synchronous generator.

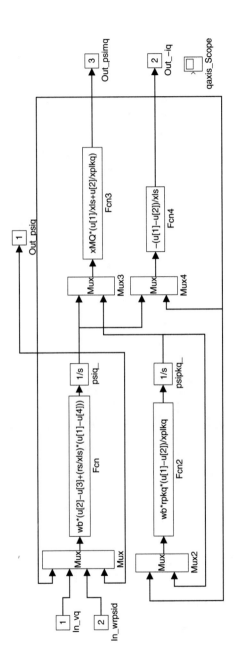

(d) Inside the *q–cct* block

Figure 7.11 (cont.): Simulation *s1* of a synchronous generator.

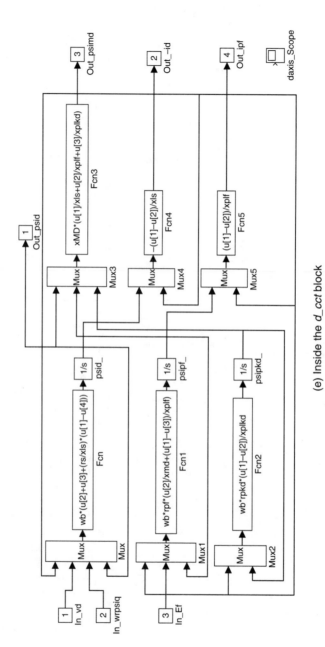

(e) Inside the *d_cct* block

Figure 7.11 (cont.): Simulation *s1* of a synchronous generator.

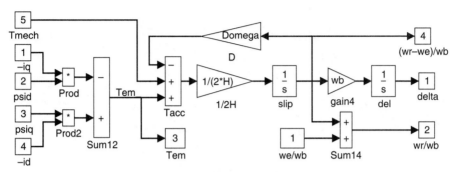

(f) Inside the *Rotor* block

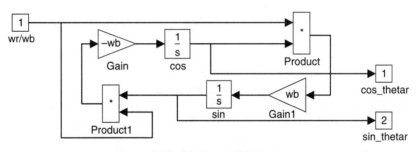

(g) Inside the *osc* block

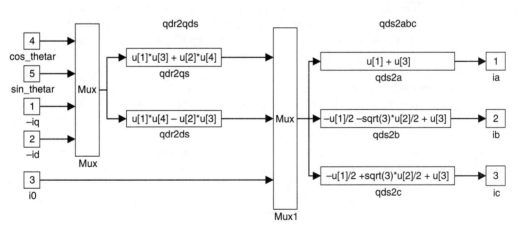

(h) Inside the *qdr2abc* block

Figure 7.11   *(cont.):* Simulation *s1* of a synchronous generator.

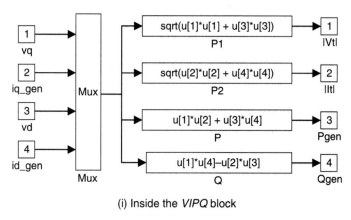

(i) Inside the *VIPQ* block

Figure 7.11   *(cont.)*: Simulation *s1* of a synchronous generator.

shown in Fig. 7.11b. The transformation uses the $\cos\theta_r(t)$ and $\sin\theta_r(t)$ generated by the *oscillator* block.

The block *qd_gen* contains the simulation of the generator proper in its rotor reference frame. The simulation of the q-axis circuit equations with one damper winding on the rotor is performed inside the *q_cct* block, that of the d-axis circuit with two rotor windings inside *d_cct* block. The details of the *q_cct* and *d_cct* blocks are shown in Figs. 7.11d and 7.11e, respectively. A close examination of the manner in which the rotor winding equations are implemented in the simulation will reveal that the addition of another rotor winding having a common magnetizing reactance with other windings of the same axis can be done quite easily.

It is evident from simulation exercises of the previous chapters that startup transients occur whenever the variables of the simulation are not properly initialized. Waiting for the startup transients of the simulation to settle can be rather time-consuming when the simulated system has some rather long time constants. Since the field and mechanical time constants of a synchronous machine are much larger than those of the stator transients, to minimize the startup transients it is important that we at least try to initialize the corresponding states with good initial estimates. For the given simulation, the initial values of the integrators, *intd, intf, intkd*, should be set with values as close to those of the desired operating condition as possible to minimize the initial simulation transients.

The equations associated with rotor motion and rotor angle are implemented inside the *Rotor* block. The details of the *Rotor* block are shown in Fig. 7.11f. In simulation, startup transients can also be minimized by temporarily setting the value of the damping coefficient, D_ω, in Fig. 7.11f high enough to provide a large amount of damping; however, when the desired steady-state condition is established, it should be reset to correspond to the actual value before conducting the study.

When using a two-stage transformation between *abc* and *qd0* variables, the $\cos\theta_r(t)$ and $\sin\theta_r(t)$ terms are generated by a variable-frequency oscillator. Details of such an oscillator circuit are shown in Fig. 7.11g. The initial values of the *cos* and *sin* integrators in this block can be set to give any desired initial value of $\theta_r(0)$. For example, using initial values of $\cos\theta_r(0) = 1$ and $\sin\theta_r(0) = 0$ corresponds to starting the rotor qd axes with an initial value of $\theta_r(0) = 0$ to the axis of the stator winding and $\theta_r(t)$ will be equal to $\omega_r t$.

The transformation of the $qd0$ rotor reference currents back to abc stator currents are performed inside the $qdr2abc$ block. The details of the $qdr2abc$ block are shown in Fig. 7.11h.

Figure 7.11i shows the inside of the *VIPQ* block, in which the instantaneous magnitude of the stator voltage, stator current, and stator real and reactive power at the generator's terminal are computed. The instantaneous values of the real and reactive power in per unit flowing out of the stator terminals of the generator are computed from the following:

$$P = \Re(v_q - jv_d)(i_q - ji_d)* = v_q i_q + v_d i_d$$
$$Q = \Im(v_q - jv_d)(i_q - ji_d)* = v_q i_d - v_d i_q \tag{7.92}$$

7.6 MACHINE PARAMETERS

The machine data from manufacturers are usually in the form of reactances, time constants, and resistances; most are derived from measurements taken from the stator windings. The usual methods of extracting the required parameters, especially those of the rotor windings from stator measurements, hinges on the observation that the effective time constants of the various rotor currents are significantly different. A commonly used demonstration of this phenomena is the short-circuit oscillogram of the stator currents when a three-phase short-circuit is applied to the machine whose stator is initially open-circuited and its field excitation held constant. Besides the dc offset, the symmetrical portion of the short-circuit current typically exhibits two distinctly different decay periods: generally referred to as the sub-transient and transient periods. The sub-transient period refers to the first few cycles of the short-circuit when the current decay is very rapid, attributable mainly to the changes in the currents of the damper windings. The rate of current decay in the transient period is slower and is attributed mainly to changes in the currents of the rotor field windings.

The theorem of constant flux linkage is useful in determining the initial values of transient fluxes in inductively coupled circuits. Briefly stated, the flux linkage of any inductive circuit with finite resistance and emf cannot change instantly; in fact, if there is no resistance or emf in the circuit, its flux linkage would remain constant. The constant flux linkage theorem can thus be used to determine the currents immediately after a change in terms of the currents before the change. The theorem can be used to explain the typical flux distributions shown in Fig. 7.12 of a synchronous machine during the so-called sub-transient, transient, and steady-state periods after a stator-side disturbance. In the sub-transient period that immediately follows the disturbance, changes in currents in the outer damping windings limit the stator-induced flux from penetrating the rotor. As these damper winding currents decay, we enter the transient period where current changes in the field windings react in the same manner, albeit slower. Finally, in steady-state, the stator-induced flux penetrates both field and damper windings of the rotor. These pictures of the flux conditions in the machine during the sub-transient, transient, and steady-state periods are helpful to understanding the constraint conditions that we will be using next in the derivation of the effective inductances of the machine for these same three periods.

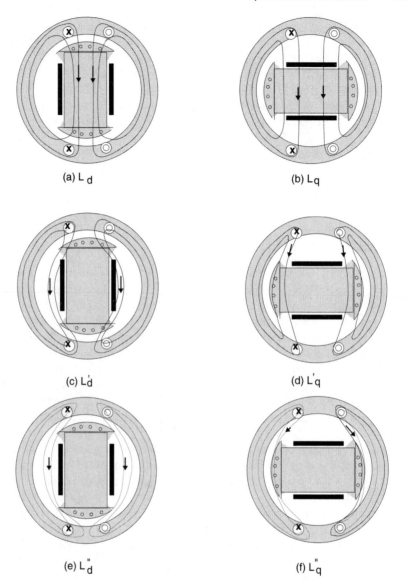

(a) L_d

(b) L_q

(c) L_d'

(d) L_q'

(e) L_d''

(f) L_q''

Figure 7.12   Flux paths for the steady-state, transient, and sub-transient inductances.

7.6.1 Synchronous Inductances

In general, inductance is defined as the ratio of flux linked to current. When the peak of the rotating mmf is aligned with the d-axis, the ratio of the stator flux linkage to stator current is referred to as the d-axis synchronous inductance, L_d. Likewise, when the peak of the rotating mmf is aligned with the q-axis, the ratio of the stator flux linkage to stator current

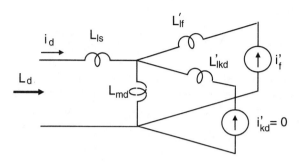

Figure 7.13 Equivalent d-axis circuit for steady-state.

is the q-axis synchronous inductance, L_q. Even in so-called round rotor machines, the value of L_d may be slightly greater than that of L_q because of the rotor slots.

A change in λ_d or λ_q in synchronous operation where all rotor winding currents are unchanged must be accompanied by a corresponding change in i_d or i_q, that is

$$\Delta\lambda_d = L_d\Delta i_d \qquad Wb.\ turn$$
$$\Delta\lambda_q = L_q\Delta i_q \tag{7.93}$$

In other words, the effective inductances to i_d and i_q are L_d and L_q, respectively. Fig. 7.13 shows the equivalent circuit for the steady-state condition.

For such a change in stator condition, the flux linkages on both the d- and q-axes may be viewed as consisting of two components: that which remains unchanged when stator currents vary and that which changes along with the stator current. That is, $\lambda_d = \lambda_d^{st} + \Delta\lambda_d$ and $\lambda_q = \lambda_q^{st} + \Delta\lambda_q$; λ_d^{st} and λ_q^{st} being the components which remain unchanged or stationary through the change in stator currents, and $\Delta\lambda_d$ and $\Delta\lambda_q$ being the corresponding components which change with stator currents. Since the damper winding currents, i_{kd}' and i_{kq}', are zero in steady-state, the unchanged components may be expressed as

$$\lambda_d^{st} = \lambda_d - \Delta\lambda_d = \lambda_d - L_d i_d = L_{md} i_f' \qquad Wb.\ turn$$
$$\lambda_q^{st} = \lambda_q - \Delta\lambda_q = \lambda_q - L_q i_q = L_{mq} i_g'. \tag{7.94}$$

We can also identify the speed voltages that correspond to the two stationary flux linkage components of Eq. 7.94 using the relationship

$$E_f - jE_g = j\omega_r(\lambda_q^{st} - j\lambda_d^{st}) \qquad V \tag{7.95}$$

Using Eq. 7.94 to replace the stationary components of the flux linkages, we obtain

$$E_f = \omega_r L_{md} i_f' = \omega_r L_{md}\left(\frac{v_f'}{r_f'}\right)$$

$$E_g = -\omega_r L_{mq} i_g' = -\omega_r L_{mq}\left(\frac{v_g'}{r_g'}\right) \tag{7.96}$$

In steady-state, the rotor speed, ω_r, is equal to ω_e, thus the above expressions for E_f and E_g are the same as those for steady-state operation. E_f and E_g are the field excitation voltages on the stator side in the q- and d-axis, respectively.

7.6.2 Transient Inductances

Since the resistances of the damper windings are usually greater than those of the field windings, the induced currents in the damper decay much more rapidly than those of the field windings. For the transient period, we can assume that the transients in the damper circuits, being highly damped, are over while the induced currents in the field windings are still changing to oppose the change in flux linkage caused by the stator currents.

Let's first consider the changes in flux linkages of the windings on the d-axis. With $\Delta\lambda_f = 0$ and $\Delta i'_{kd} = \Delta i'_{kq} = 0$,

$$\Delta\lambda_f = L_{md}\Delta i_d + L'_{ff}\Delta i_f = 0 \qquad Wb.\ turn$$
$$\Delta\lambda_d = L_d\Delta i_d + L_{md}\Delta i'_f \tag{7.97}$$

Eliminating the change in field current to express $\Delta\lambda_d$ in terms of Δi_d only, we have

$$\Delta\lambda_d = \left(L_d - \frac{L^2_{md}}{L'_{ff}}\right)\Delta i_d \tag{7.98}$$

The ratio of $\Delta\lambda_d$ to Δi_d in this situation is referred to as the *d-axis transient inductance*, that is

$$L'_d \triangleq \frac{\Delta\lambda_d}{\Delta i_d} = L_d - \frac{L^2_{md}}{L'_{ff}} \tag{7.99}$$

Likewise, by considering the changes in flux linkages of the windings on the q-axis, we can show that the *q-axis transient inductance* is given by

$$L'_q \triangleq \frac{\Delta\lambda_q}{\Delta i_q} = L_q - \frac{L^2_{mq}}{L'_{gg}} \tag{7.100}$$

Figure 7.14 shows the equivalent d-axis circuit applicable during the transient period.

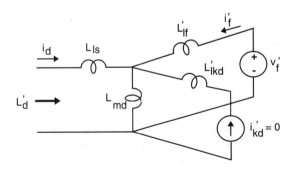

Figure 7.14   Equivalent d-axis circuit condition during the transient period.

7.6.3 Transient Flux Linkages and Voltages behind the Transient Inductances

Equation 7.98 gives the change in λ_d to a change in i_d when both i_d and i_f are free to change. Likewise, we can derive a similar relationship between $\Delta\lambda_q$ and Δi_q when both i_q and i_g are free to change. Again, we can identify the components in λ_d and λ_q which will remain unchanged by transient stator currents under the transient condition where both λ_f and λ_g are being held constant by changes in i'_f and i'_g. They are given by

$$\lambda'_d = \lambda_d - \Delta\lambda_d = \lambda_d - L'_d i_d \qquad Wb.\ turn$$

$$\lambda'_q = \lambda_q - \Delta\lambda_q = \lambda_q - L'_q i_q \tag{7.101}$$

The definitions and alternate forms of the speed voltage terms associated with the transient flux linkages, λ'_q and λ'_d, are

$$E'_q \triangleq \omega_r \lambda'_d = \omega_r \left\{ L_d i_d + L_{md} i'_f - \left(L_d - \frac{L_{md}^2}{L'_{ff}} \right) i_d \right\} \qquad V$$

$$= \omega_r L_{md} \left(\frac{L'_{ff} i'_f + L_{md} i_d}{L'_{ff}} \right) \tag{7.102}$$

$$= \frac{\omega_r L_{md} \lambda'_f}{L'_{ff}}$$

$$E'_d \triangleq -\omega_r \lambda'_q = -\omega_r \left\{ L_q i_q + L_{md} i'_g - \left(L_q - \frac{L_{mq}^2}{L'_{gg}} \right) i_q \right\} \qquad V$$

$$= -\omega_r L_{mq} \left(\frac{L'_{gg} i'_g + L_{mq} i_q}{L'_{gg}} \right) \tag{7.103}$$

$$= -\frac{\omega_r L_{mq} \lambda'_g}{L'_{gg}}$$

E'_q and E'_d are the *voltages behind the transient inductances* in the q- and d-axis, respectively.

7.6.4 Sub-transient Inductances

For the sub-transient period, transient currents induced in the rotor windings will keep the flux linkage of every rotor circuit initially constant. With the d-axis rotor flux linkages held constant, that is $\Delta\lambda'_f = \Delta\lambda'_{kd} = 0$, we obtain

$$\Delta\lambda'_f = L_{md}\Delta i_d + L'_{ff}\Delta i'_f + L_{md}\Delta i'_{kd} = 0$$

$$\Delta\lambda'_{kd} = L_{md}\Delta i_d + L_{md}\Delta i'_f + L'_{kdkd}\Delta i'_{kd} = 0 \tag{7.104}$$

Equation 7.104 can be used to express the changes in the rotor currents in terms of Δi_d, that is

$$\begin{bmatrix} \Delta i'_f \\ \Delta i'_{kd} \end{bmatrix} = \frac{-L_{md}\Delta i_d}{L'_{ff}L'_{kdkd} - L^2_{md}} \begin{bmatrix} L'_{kdkd} - L_{md} \\ L'_{ff} - L_{md} \end{bmatrix} = \frac{-L_{md}\Delta i_d}{L'_{ff}L'_{kdkd} - L^2_{md}} \begin{bmatrix} L'_{lkd} \\ L'_{lf} \end{bmatrix} \qquad (7.105)$$

The corresponding change in the d-axis stator flux linkage is given by

$$\Delta \lambda_d = L_d \Delta i_d + L_{md} \Delta i'_f + L_{md} \Delta i'_{kd} \qquad (7.106)$$

Substituting Eq. 7.105 into Eq. 7.106 and defining the ratio of $\Delta \lambda_d$ to Δi_d in this situation as the *d-axis sub-transient inductance* L''_d, we obtain

$$L''_d \triangleq \frac{\Delta \lambda_d}{\Delta i_d} = L_d - \frac{L^2_{md}(L'_{lkd} + L'_{lf})}{L'_{ff}L'_{kdkd} - L^2_{md}} \qquad H$$

$$= L_{ls} + \frac{L_{md}L'_{lkd}L'_{lf}}{L'_{lkd}L'_{lf} + L_{md}(L'_{lkd} + L'_{lf})} \qquad (7.107)$$

$$= L_{ls} + \frac{L_{md}\frac{L'_{lkd}L'_{lf}}{L'_{lkd}+L'_{lf}}}{L_{md} + \frac{L'_{lkd}L'_{lf}}{L'_{lkd}+L'_{lf}}}$$

The last form of Eq. 7.107 indicates that the value of the d-axis sub-transient inductance is the Thevenin's equivalent inductance of the d-axis stator circuit as viewed from the stator d-axis terminals shown in Fig 7.15. It is the combined inductance of L_{ls} in series with the parallel combination of L_{md}, L'_{lf}, and L'_{lkd}.

Similarly, the expression for the q-axis sub-transient inductance, defined as $\Delta \lambda_q / \Delta i_q = L''_q$, can be shown to be

$$L''_q = L_{ls} + \frac{L_{mq}\frac{L'_{lkq}L'_{lg}}{L'_{lkq}+L'_{lg}}}{L_{mq} + \frac{L'_{lkq}L'_{lg}}{L'_{lkq}+L'_{lg}}} \qquad H \qquad (7.108)$$

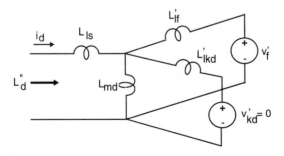

Figure 7.15 Equivalent d-axis circuit condition during sub-transient period.

7.6.5 Sub-transient Flux Linkages and Voltages Behind the Sub-transient Inductances

As in steady-state and transient conditions, we can identify the components in λ_d and λ_q which will remain unchanged by transient stator currents under the sub-transient condition when the flux linkages of the rotor field and damper windings are constant.

$$
\begin{aligned}
\lambda_d'' &= \lambda_d - \Delta\lambda_d = \lambda_d - L_d'' i_d \\
\lambda_q'' &= \lambda_q - \Delta\lambda_q = \lambda_q - L_q'' i_q
\end{aligned}
\tag{7.109}
$$

Likewise, we can also define the speed voltage components associated with λ_d'' and λ_q'' as

$$
\begin{aligned}
E_q'' \triangleq \omega_r \lambda_d'' &= \omega_r \left(L_{md} i_f' + L_{md} i_{kd}' + \frac{L_{md}^2 (L_{lkd}' + L_{lf}') i_d}{L_{ff}' L_{kdkd}' - L_{md}^2} \right) \qquad V \\
&= E_q + \omega_r L_{md} i_{kd}' + \omega_r (L_d - L_d'') i_d
\end{aligned}
\tag{7.110}
$$

and

$$
\begin{aligned}
E_d'' \triangleq -\omega_r \lambda_q'' &= -\omega_r \left(L_{mq} i_g' + L_{mq} i_{kq}' + \frac{L_{mq}^2 (L_{lkq}' + L_{lg}') i_q}{L_{gg}' L_{kqkq}' - L_{mq}^2} \right) \\
&= E_d - \omega_r L_{mq} i_{kq}' - \omega_r (L_q - L_q'') i_q
\end{aligned}
\tag{7.111}
$$

where

$$
\begin{aligned}
E_q &= \omega_r L_{md} i_f' \\
E_d &= -\omega_r L_{mq} i_g'
\end{aligned}
\tag{7.112}
$$

E_q'' and E_d'' are the *voltages behind the sub-transient inductances* in the q- and d-axis, respectively. Note that, unlike in Eq. 7.96 for steady-state condition where i_f' and i_g' are constant, the values of i_f' and i_g' in Eq. 7.112 can vary with time.

7.6.6 Transient Time Constants

Associated with the two sets of rotor windings in the machine are two different sets of time constants. The set with the larger values is the transient time constants, and the set with the smaller values is the sub-transient time constants. In general, the damper windings, which have a higher resistance than the field windings, are associated with the sub-transient time constants.

When the stator is open-circuited and the effects of the higher winding resistance damper windings are disregarded, the change in field currents in response to a change in excitation voltages is governed by the open-circuit field time constants, defined as

$$
T_{do}' \triangleq \frac{L_{ff}'}{r_f'} \qquad s
$$

$$
T_{qo}' \triangleq \frac{L_{gg}'}{r_g'}
\tag{7.113}
$$

T'_{do} is typically of the order of 2 to 11 seconds and is greater than the other time constants discussed below. However, when both the field and stator windings are short-circuited and the effects of the damper windings ignored, the apparent inductance of the field winding changes with the external connection. It can be shown that the ratio of the time constants of the field winding with short-circuited stator windings to that with open-circuited stator windings is equal to the ratio of the apparent inductance seen by the stator current with field short-circuited to that with field open-circuited. For the d-axis, that ratio is (L'_d/L_d). Therefore, the time constant of the field windings under these conditions is given by the relation:

$$\frac{T'_d}{T'_{do}} = \frac{L'_d}{L_d} \qquad (7.114)$$

Using

$$\frac{L_{md}}{L'_{lf}} = \frac{L_d - L'_d}{L'_d - L_{ls}} \quad \text{and} \quad L_{md} = L_d - L_{ls}$$

$$L'_{lf} = \frac{(L_d - L_{ls})(L'_d - L_{ls})}{(L_d - L'_d)} \qquad (7.115)$$

we can also express the d-axis transient time constant in Eq. 7.113 as

$$T'_{do} = \frac{1}{r'_f}\left(\frac{(L_d - L_{ls})^2}{L_d - L'_d}\right) \qquad (7.116)$$

7.6.7 Sub-transient Time Constants

The open-circuit sub-transient time constant in the d-axis, T''_{do}, is the time constant of the kd damper winding current when the terminals of the field winding, f, are shorted and the stator windings are open-circuited. It is also defined as the time in seconds required for the rapidly decreasing initial d-axis component of the symmetrical voltage to decrease to $(1/e)$ of its initial value when a short-circuit on the armature terminals of a machine running at rated speed is suddenly removed. During this period of initial decay of the open-circuit stator voltage, the resistance of the field winding is neglected. The effective inductance to the kd damper winding current under this condition is given by

$$L''_{kdo} = L'_{lkd} + \frac{L_{md}L'_{lf}}{L_{md} + L'_{lf}} \qquad H \qquad (7.117)$$

Therefore,

$$T''_{do} = \frac{L''_{kdo}}{r'_{kd}} \qquad s \qquad (7.118)$$

It would be interesting to verify that the time constant of the kd damper winding current with the stator open-circuited is indeed T''_{do}. To do this, we begin with the following d-axis flux linkage relations for an open-circuited stator condition, that is $i_d = 0$:

$$\lambda'_{kd} = L'_{kdkd}i'_{kd} + L_{md}i'_f$$
$$\lambda'_f = L_{md}i'_{kd} + L'_{ff}i'_f \tag{7.119}$$

Next we consider a step change in the excitation voltage. Currents induced in the damper windings will oppose a change in field current. Immediately following the step change, the flux linkage of the kd damper winding is assumed to remain unchanged, that is $\lambda'_{kd} = 0$. From the first equation of Eq. 7.119, we obtain

$$i'_f = -\frac{L'_{kdkd}}{L_{md}}i'_{kd} \quad \text{or} \quad \frac{di'_f}{dt} = -\frac{L'_{kdkd}}{L_{md}}\frac{di'_{kd}}{dt} \tag{7.120}$$

From the voltage equations of the f and kd winding, we have

$$\frac{v'_f}{L'_{ff}} = \frac{r'_f i'_f}{L'_{ff}} + \frac{d i'_f}{dt} + \frac{L_{md}}{L'_{ff}}\frac{di'_{kd}}{dt}$$
$$0 = \frac{r'_{kd}i'_{kd}}{L_{md}} + \frac{di'_f}{dt} + \frac{L'_{kdkd}}{L_{md}}\frac{di'_{kd}}{dt} \tag{7.121}$$

Using Eq. 7.120 to replace i'_f, the damper winding equation in Eq. 7.121 can be written as

$$\frac{di'_{kd}}{dt} + \frac{r'_{kd}}{L'_{kdkd} - \frac{L^2_{md}}{L'_{ff}}}i'_{kd} = -\frac{\frac{v'_f L_{md}}{L'_{ff}}}{L'_{kdkd} - \frac{L^2_{md}}{L'_{ff}}} \tag{7.122}$$

From Eq. 7.122, we can show that the time constant of i'_{kd} for the open-circuited stator condition is the same as that given earlier in Eq. 7.118. The time constant can be expressed in the following forms:

$$T''_{d0} = \frac{L'_{kdkd} - \frac{L^2_{md}}{L'_{ff}}}{r'_{kd}}$$
$$= \frac{1}{r'_{kd}}\left(L'_{lkd} + \frac{L'_{lf}L_{md}}{L_{md} + L'_{lf}}\right) \tag{7.123}$$

By symmetry, we can write down the open-circuit sub-transient time constant of the q-axis rotor windings as

$$T''_{qo} = \frac{1}{r'_{kq}}\left(L'_{lkq} + \frac{L'_{lg}L_{mq}}{L_{mq} + L'_{lg}}\right) \tag{7.124}$$

The short-circuit sub-transient time constant of the d-axis, T''_d, is the time constant of the damper winding kd current when the field and stator circuits are all shorted. The effective inductance to the kd winding current under this condition is given by

$$L''_{kd} = L'_{lkd} + \frac{L'_{lf}\left(\frac{L_{ls}L_{md}}{L_{ls}+L_{md}}\right)}{L_{ls} + \left(\frac{L_{ls}L_{md}}{L_{ls}+L_{md}}\right)} \tag{7.125}$$

Therefore,

$$T_d'' = \frac{L_{kd}''}{r_{kd}'} \qquad (7.126)$$

7.7 CALCULATING MACHINE PARAMETERS

In this section, we shall describe a procedure to calculate the parameters for the developed model from data normally available from the manufacturers. The primes associated with the rotor quantities indicate parameters referred to the stator by the appropriate turns ratio. The values used should be of consistent units, either engineering or per unit.

Often, the armature leakage reactance is given directly; if not given, the value of the zero-sequence reactance may be used instead, that is $x_{ls} = x_0$. Subtracting the value of leakage reactance from the direct and quadrature synchronous reactances yields the direct and quadrature magnetizing reactances:

$$x_{mq} = x_q - x_{ls}$$
$$x_{md} = x_d - x_{ls} \qquad (7.127)$$

The field leakage reactances can be determined from the definition of L_d' given in Eq. 7.99, which can be written as

$$x_d' = x_{ls} + \frac{x_{md}x_{lf}'}{x_{md} + x_{lf}'} \qquad (7.128)$$

from which we can obtain

$$x_{lf}' = \frac{x_{md}(x_d' - x_{ls})}{x_{md} - (x_d' - x_{ls})} \qquad (7.129)$$

Or, if $x_{md} \gg x_{lf}'$,

$$x_{lf}' \approx x_d' - x_{ls} \qquad (7.130)$$

The leakage reactance of the d-axis damper winding can be determined from the expression for L_d'' given in Eq. 7.107, that is

$$x_d'' = x_{ls} + \frac{x_{md}x_{lf}'x_{lkd}'}{x_{md}x_{lf}' + x_{md}x_{lkd}' + x_{lf}'x_{lkd}'} \qquad (7.131)$$

from which we obtain

$$x_{lkd}' = \frac{(x_d'' - x_{ls})x_{md}x_{lf}'}{x_{lf}'x_{md} - (x_d'' - x_{ls})(x_{md} + x_{lf}')} \qquad (7.132)$$

Or, if $x_{md} \gg x_{lf}'$,

$$x_{lkd}' \approx \frac{(x_d'' - x_{ls})x_{lf}'}{x_{lf}' - (x_d'' - x_{ls})} \qquad (7.133)$$

With no g-winding on the q-axis, the leakage reactance of the q-axis damper winding can be determined from the expression for L_q'' in Eq. 7.108 by setting $L_{lg} \rightarrow \infty$, that is

$$x_q'' = x_{ls} + \frac{x_{mq}x_{lkq}'}{x_{mq} + x_{lkq}'} \tag{7.134}$$

from which we obtain

$$x_{lkq}' = \frac{x_{mq}(x_q'' - x_{ls})}{x_{mq} - (x_q'' - x_{ls})} \tag{7.135}$$

Alternatively, if $x_{mq} \gg x_{lkq}'$,

$$x_{lkq}' \approx x_q'' - x_{ls} \tag{7.136}$$

The stator phase resistance may be assumed to be the same as the positive-sequence resistance of the stator that is normally given. The winding resistances of the rotor windings are to be determined from the time constants given. The field resistance can be calculated from the d-axis transient open-circuit time constant. Eq. 7.113 can be written as

$$T_{do}' = \frac{1}{\omega_b r_f'}(x_{lf}' + x_{md}) \tag{7.137}$$

Rearranging, we obtain

$$r_f' = \frac{1}{\omega_b T_{do}'}(x_{lf}' + x_{md}) \tag{7.138}$$

Given the value of T_{do}'', we can determine r_{kd}'. Equation 7.118 can be written either as

$$T_{do}'' = \frac{1}{\omega_b r_{kd}'}\left(x_{lkd}' + \frac{x_{md}x_{lf}'}{x_{md} + x_{lf}'}\right) \tag{7.139}$$

or

$$T_{do}'' = \frac{1}{\omega_b r_{kd}'}(x_{lkd}' + x_d' - x_{ls}) \tag{7.140}$$

Rearranging, we obtain

$$r_{kd}' = \frac{1}{\omega_b T_{do}''}(x_{lkd}' + x_d' - x_{ls}) \tag{7.141}$$

Similarly, using Eq. 7.124 for T_{qo}'', we can determine r_{kq}' as follows:

$$T_{q0}'' = \frac{1}{\omega_b r_{kq}'}(x_{lkq}' + x_{mq}) \tag{7.142}$$

from which we obtain

$$r_{kq}' = \frac{1}{\omega_b T_{qo}''}(x_{lkq}' + x_{mq}) \tag{7.143}$$

Alternatively, if the short-circuit time constants instead of the open-circuit time constants are given, the resistances can be found from

$$r'_f = \frac{1}{\omega_b T'_d}\left(x'_{lf} + \frac{x_{md}x_{ls}}{x_{md}+x_{ls}}\right)$$

$$r'_{kd} = \frac{1}{\omega_b T''_d}\left(x'_{lkd} + \frac{x_{md}x_{ls}x'_{lf}}{x_{md}x_{ls}+x_{md}x'_{lf}+x_{ls}x'_{lf}}\right) \qquad (7.144)$$

$$r'_{kq} = \frac{1}{\omega_b T''_q}\left(x'_{lkq} + \frac{x_{mq}x_{ls}}{x_{mq}+x_{ls}}\right)$$

7.7.1 Sample Calculation

Given the following data in per unit of the machine base, determine the rest of the parameters required for the simulation model:

$x_d = 1.63\ pu$	$T'_{do} = 4.3\ s$
$x_q = 1.56$	$T''_{do} = .032$
$x'_d = .174$	$T'_d = ?$
$x''_d = .123$	$T''_d = .023$
$x''_q = .124$	$T''_q = .023$
$x_{ls} = .093$	$T''_{qo} = .066?$
$r_s = .0032$	

$x_{md} = 1.63 - .093 = 1.54$

$x_{mq} = 1.56 - .093 = 1.47$

$$x'_{lf} = \frac{(1.54)(.081)}{1.54-.081} = .0855 \qquad \text{or } x'_{lf} = .174 - .093 = .081\ pu$$

$$x'_{lkd} = \frac{(.030)(1.54)(.0855)}{(.0855)(1.54)-(.03)(1.6255)} = .0478 \quad \text{or } x'_{lkd} = \frac{(.123-.093)(.0855)}{(.0855)-(.123-.093)} = 0.046$$

$$x'_{lkq} = \frac{(1.47)(.031)}{1.44} = .0316 \qquad \text{or } x'_{lkq} = .124 - .093 = .031$$

$$r'_f = \frac{1}{377(4.3)}(.0855+1.54) = .001$$

$$r'_{kd} = \frac{1}{377(.032)}(.0855 + \frac{(1.54)((.093)}{1.63}) = .0107$$

$$r'_{kq} = \frac{1}{377(.023)}(.0316 + \frac{(1.47)((.093)}{1.56}) = .014$$

7.8 HIGHER-ORDER MODELS

As simulation capability improved and problem interest expanded beyond those of transients of the stator quantities and the first swing stability of the rotor electromechanical oscillations, it was recognized that simulated responses using the traditional circuit model of Fig. 7.4 with machine parameters as calculated previously often did not agree well with actual measured responses, especially in the rotor winding quantities [58] of machines with solid iron rotors. Better agreement can be obtained with refinements in the structure of the d-axis model and in the techniques used to determine the machine parameters. Depending on the rotor construction, the shielding effects of the damper winding currents and the eddy currents induced in a solid rotor can affect the transient characteristics of the rotor significantly.

One approach is to model such effects by a low-order ladder network of RL branches to accommodate the complex frequency-dependent nature of the rotor's surface impedance. Another approach to synthesizing the rotor circuit is to model it along the lines that we have traditionally viewed the partition of current paths from the physical construction. Besides the physical field and damper circuits, we could also add a third circuit for the eddy currents induced in the pole surface. These three current paths are in close proximity in the rotor slots. Figure 7.16 shows only the slot leakage flux components that couple two or more current paths in a rotor slot. The field current path is deep in the slot and it links all the slot leakage flux components shown [63], whereas the second current path associated with

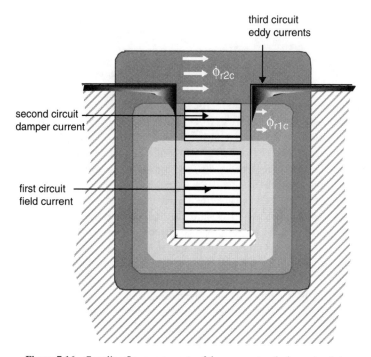

Figure 7.16   Coupling flux components of three current paths in a rotor slot.

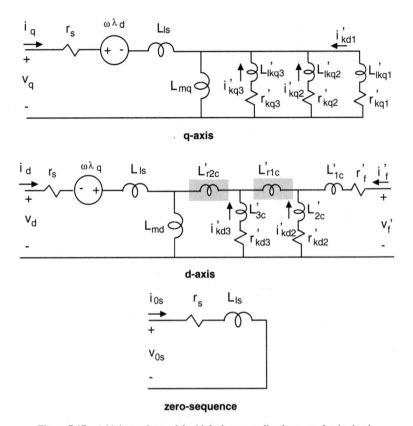

Figure 7.17   A higher-order model with leakage coupling between d-axis circuits.

the damper current is higher up the slot and it links ϕ_{r1c} partially and all of ϕ_{r2c} that crosses above it. Similarly, the third current path which is associated with the eddy currents flowing on the pole face links ϕ_{r2c} partially, but not ϕ_{r1c}. Using the relation between magnetic and electric circuits described in Section 3.2.2, each of the flux components coupling two or more of the current paths may be translated to a mutual inductance in an equivalent circuit representation of the rotor where the identity of these current paths are retained. For example, Fig. 7.17 shows a refined $qd0$ circuit model [63] with three q and three d rotor circuits. The inductances, L_{r2c} and L_{r1c}, are the coupling inductances associated with flux components, such as ϕ_{r2c} and ϕ_{r1c} in a rotor slot shown in Fig. 7.16. The inductances, L'_{1c}, L'_{2c}, and L'_{3c}, are from the self-leakages of the field, damper, and eddy currents. It is of interest to note that similar circuit representation has been used for the coupling between the rotor circuits of a double-cage induction motor [64].

Equally important to the improvement in accuracy are the development of methods to measure operational impedances and transfer functions and the determination of the machine parameters that fit over a significant frequency range using standstill and on-line frequency response tests [57]. In reference [62] Canay has described a method to determine the model parameters from measured data of a standstill frequency response test for circuit models such as that shown in Fig. 7.17, which accounts for the unequal mutual inductances between stator and rotor circuits.

7.8.1 Equations for simulation

As before, the equations for an equivalent circuit with unequal stator and rotor mutual inductances and coupling inductances between d-axis rotor circuits can be arranged with ψ_{mq} and ψ_{md} maintained, so that core saturation can be included in the manner described in Section 6.8.1. Those equations for the equivalent circuit with three rotor circuits in Fig. 7.17 can be shown to be as follows:

$$
\begin{aligned}
\psi_q - \psi_{mq} &= x_{ls}i_q \\
\psi_d - \psi_{md} &= x_{ls}i_d \\
\psi_0 &= x_{ls}i_0 \\
\psi'_{kd3} - \psi_{md} &= (x'_{3c} + x'_{r2c})i'_{kd3} + x'_{r2c}i'_{kd2} + x'_{r2c}i'_f \\
\psi'_{kd2} - \psi_{md} &= x'_{r2c}i'_{kd3} + (x'_{2c} + x'_{r1c} + x'_{r2c})i'_{kd2} + (x'_{r2c} + x'_{r1c})i'_f \\
\psi'_f - \psi_{md} &= x'_{r2c}i'_{kd3} + (x'_{r1c} + x'_{r2c})i'_{kd2} + (x'_{1c} + x'_{r2c} + x'_{r1c})i'_f \\
\psi'_{kq3} - \psi_{mq} &= x'_{lkq3}i'_{kq3} \\
\psi'_{kq2} - \psi_{mq} &= x'_{lkq2}i'_{kq2} \\
\psi'_{kq1} - \psi_{mq} &= x'_{lkq1}i'_{kq1}
\end{aligned}
\tag{7.145}
$$

The flux linkage equations of the three rotor circuits on the d-axis with mutual coupling may be written in matrix form as

$$
\begin{bmatrix} \psi'_{kd3} - \psi_{md} \\ \psi'_{kd2} - \psi_{md} \\ \psi'_f - \psi_{md} \end{bmatrix} = \underbrace{\begin{bmatrix} (x'_{3c} + x'_{r2c}) & x'_{r2c} & x'_{r2c} \\ x'_{r2c} & (x'_{2c} + x'_{r1c} + x'_{r2c}) & (x'_{r1c} + x'_{r2c}) \\ x'_{r2c} & (x'_{r1c} + x'_{r2c}) & (x'_{1c} + x'_{r1c} + x'_{r2c}) \end{bmatrix}}_{X_r} \begin{bmatrix} i'_{kd3} \\ i'_{kd2} \\ i'_f \end{bmatrix}
\tag{7.146}
$$

A suitable machine model is determined by first selecting the appropriate circuit representation before determining its parameters to fit the frequency response test data. Thus, ignoring the leakage couplings of the three d-axis rotor circuits is not equivalent to dropping the off-diagonal terms of the full X matrix in Eq. 7.146, as the diagonal values from a parameter fitting without off-diagonal elements could be different from those obtained with off-diagonal elements. When the off-diagonal, x_{r2c} and x_{r1c}, terms are ignored, the mutual inductances of the stator and rotor windings will again be equal and all three d-axis rotor circuits will be decoupled from one another.

The inverse relation of Eq. 7.181 with $B = X^{-1}$ allows us to determine i'_{kd3}, i'_{kd2}, and i'_f from the values of $\psi'_{kd3}, \psi'_{kd2}, \psi'_f$, and ψ_{md}.

$$\begin{bmatrix} i'_{kd3} \\ i'_{kd2} \\ i'_f \end{bmatrix} = \underbrace{\begin{bmatrix} b_{11} & b_{12} & b_{13} \\ b_{21} & b_{22} & b_{23} \\ b_{31} & b_{32} & b_{33} \end{bmatrix}}_{B} \begin{bmatrix} \psi'_{kd3} - \psi_{md} \\ \psi'_{kd2} - \psi_{md} \\ \psi'_f - \psi_{md} \end{bmatrix} \qquad (7.147)$$

The equations for the stator qd currents and q-axis rotor circuit currents can be determined in a similar way as before, that is

$$i_q = \frac{\psi_q - \psi_{mq}}{x_{ls}}$$

$$i_d = \frac{\psi_d - \psi_{md}}{x_{ls}}$$

$$i'_{kq3} = \frac{\psi'_{kq3} - \psi_{mq}}{x'_{lkq3}} \qquad (7.148)$$

$$i'_{kq2} = \frac{\psi'_{kq2} - \psi_{mq}}{x'_{lkq2}}$$

$$i'_{kq1} = \frac{\psi'_{kq1} - \psi_{mq}}{x'_{lkq1}}$$

The flux linkages, ψ_{md} and ψ_{mq}, may be expressed in terms of the total flux linkages of the windings, that is

$$\psi_{md} = x_{md}(i_d + i'_{kd3} + i'_{kd2} + i'_f)$$

$$= x_{MD}\left(\frac{\psi_d}{x_{ls}} + \sum_{j=1}^{3} b_{j1}\psi'_{kd3} + \sum_{j=1}^{3} b_{j2}\psi'_{kd2} + \sum_{j=1}^{3} b_{j3}\psi'_f\right)$$

$$\psi_{mq} = x_{mq}(i_q + i'_{kq3} + i'_{kq2} + i'_{kq1}) \qquad (7.149)$$

$$\psi_{mq} = x_{MQ}\left(\frac{\psi_q}{x_{ls}} + \frac{\psi'_{kq3}}{x'_{lkq3}} + \frac{\psi'_{kq2}}{x'_{lkq2}} + \frac{\psi'_{kq1}}{x'_{lkq1}}\right)$$

where

$$\frac{1}{x_{MD}} = \frac{1}{x_{md}} + \sum_{i}^{3}\sum_{j}^{3} b_{ij} + \frac{1}{x_{ls}}$$

$$\frac{1}{x_{MQ}} = \frac{1}{x_{mq}} + \frac{1}{x'_{lkq3}} + \frac{1}{x'_{lkq2}} + \frac{1}{x'_{lkq1}} + \frac{1}{x_{ls}} \qquad (7.150)$$

The stator and rotor circuit flux linkages are obtained by integrating their respective voltage equations, that is

$$\psi_q = \omega_b \int \left\{ v_q - \frac{\omega_r}{\omega_b}\psi_d + \frac{r_s}{x_{ls}}(\psi_{mq} - \psi_q) \right\} dt \qquad Wb. \ turn/s \ V$$

$$\psi_d = \omega_b \int \left\{ v_d + \frac{\omega_r}{\omega_b}\psi_q + \frac{r_s}{x_{ls}}(\psi_{md} - \psi_d) \right\} dt$$

$$\psi_0 = \omega_b \int \left(v_0 - \frac{r_s}{x_{ls}}\psi_0 \right) dt$$

$$\psi'_{kq3} = \frac{\omega_b r'_{kq3}}{x'_{lkq3}} \int (\psi_{mq} - \psi'_{kq3})dt$$

$$\psi'_{kq2} = \frac{\omega_b r'_{kq2}}{x'_{lkq2}} \int (\psi_{mq} - \psi'_{kq2})dt \qquad\qquad (7.151)$$

$$\psi'_{kq1} = \frac{\omega_b r'_{kq1}}{x'_{lkq1}} \int (\psi_{mq} - \psi'_{kq1})dt$$

$$\psi'_{kd3} = -\omega_b r'_{kd3} \int \left\{ b_{11}\psi'_{kd3} + b_{12}\psi'_{kd2} + b_{13}\psi'_f - (b_{11}+b_{12}+b_{13})\psi_{md} \right\} dt$$

$$\psi'_{kd2} = -\omega_b r'_{kd2} \int \left\{ b_{21}\psi'_{kd3} + b_{22}\psi'_{kd2} + b_{23}\psi'_f - (b_{21}+b_{22}+b_{23})\psi_{md} \right\} dt$$

$$\psi'_f = \frac{\omega_b r'_f}{x_{md}} \int \left\{ E_f - x_{md} \left(b_{31}\psi'_{kd3} + b_{32}\psi'_{kd2} + b_{33}\psi'_f - (b_{31}+b_{32}+b_{33})\psi_{md} \right) \right\} dt$$

where $E_f = x_{md}v'_f/r'_f$.

The rotor speed, $\omega_r(t)$, is determined from the following expression of the slip speed:

$$\omega_r(t) - \omega_e = \frac{P}{2J} \int_0^t (T_{em} + T_{mech} - T_{damp})dt \qquad elect. \ rad/s \qquad (7.152)$$

where T_{em}, positive for motoring operation and negative for generating operation, is computed from

$$T_{em} = \frac{3}{2}\frac{P}{2\omega_b}(\psi_d i_q - \psi_q i_d) \qquad N.m. \qquad (7.153)$$

7.9 PERMANENT MAGNET SYNCHRONOUS MOTORS

The dc excitation of the field winding in a synchronous machine can be provided by permanent magnets. One obvious change with replacing the electrical excitation with a permanent magnet is the elimination of copper losses. Machines so excited can offer simpler construction, lower weight and size for the same performance, with reduced losses, and thus higher efficiency. The disadvantages are that present prices of permanent magnet materials (except for ferrites) are relatively high, and that magnet characteristics change with time. The choice of permanent magnets for a motor is influenced by factors such

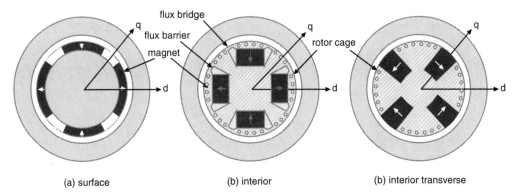

Figure 7.18   Arrangements of permanent magnets in synchronous motors.

as the motor's performance, weight, size, and efficiency, and economic factors regarding the material and production. Machinability, tolerance, and ease of handling of various permanent magnet materials can also affect production costs significantly. Figure 7.18 shows three common arrangements of permanent magnets in synchronous motors. The surface mount arrangement is popular with modern high H_c magnetic materials that are not easily susceptible to demagnetization. Since the leakage of the magnet with regard to airgap flux generation is small, this arrangement can result in a smaller volume of permanent magnet material being used. With material of lesser H_c and where the risk of demagnetization is high, the magnet is usually embedded in the rotor. The embedded arrangement will allow a longer length of magnet than the surface mount arrangement. Furthermore, with material of low remanence, there is room for the flux focusing that is needed.

In general, a *line-start* motor will have a rotor cage to help start the motor on a fixed frequency supply. The induction motor torque component of a *line-start* motor has to overcome the pulsating torque caused by the magnets during starting. An *inverter-fed* motor may or may not have a rotor cage as the inverter frequency can be synchronized to the rotor speed. *Inverter-fed* motors can be fed with pwm sinusoidal voltage waveforms or quasi-square voltage waveforms of 120° or 180° wide.

A circuit model of a permanent magnet motor, which is used for predicting its transient behavior, can be obtained using either of the two equivalent circuit representations for permanent magnets given in Fig. 3.8. Using the equivalent circuit of Fig. 3.8b for permanent magnets, the equivalent $qd0$ circuit representation shown in Fig. 7.19 of a permanent magnet synchronous motor with damper cage windings but no g winding can be obtained by modifying the $qd0$ circuit given in Fig. 7.4. For modeling purposes, the permanent magnet inductance, L_{rc}, that is associated with its recoil slope, can be lumped with the common d-axis mutual inductance of the stator and damper windings and the combined d-axis mutual inductance denoted still by L_{md}. The current, i'_m, is the equivalent magnetizing current of the permanent magnets, referred to the stator side. The corresponding $qd0$ equations for the above equivalent $qd0$ circuits of the permanent magnet motor are summarized in Table 7.1.

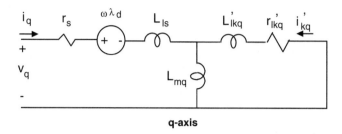

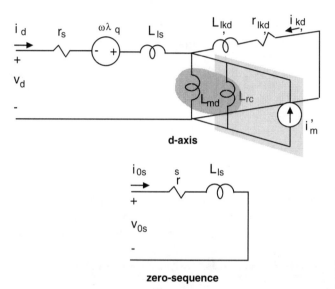

Figure 7.19 Equivalent $qd0$ circuits of a permanent magnet synchronous motor.

TABLE 7.1 $qd0$ EQUATIONS OF A PERMANENT MAGNET MOTOR

$qd\,0$ Voltage equations:

$$v_q = r_s i_q + \frac{d\lambda_q}{dt} + \lambda_d \frac{d\theta_r}{dt} \qquad V$$

$$v_d = r_s i_d + \frac{d\lambda_d}{dt} - \lambda_q \frac{d\theta_r}{dt}$$

$$v_0 = r_s i_0 + \frac{d\lambda_0}{dt} \qquad (7.154)$$

$$0 = r'_{kd} i'_{kd} + \frac{d\lambda'_{kd}}{dt}$$

$$0 = r'_{kq} i'_{kq} + \frac{d\lambda'_{kq}}{dt}$$

Flux linkages:

$$\lambda_q = L_q i_q + L_{mq} i'_{kq} \qquad Wb. \ turn$$

$$\lambda_d = L_d i_d + L_{md} i'_{kd} + \underbrace{L_{md} i'_m}_{\lambda'_m}$$

$$\lambda_0 = L_{ls} i_0 \tag{7.155}$$

$$\lambda'_{kq} = L_{mq} i_q + L'_{kqkq} i'_{kq}$$

$$\lambda'_{kd} = L_{md} i_d + L'_{kdkd} i'_{kd} + L_{md} i'_m$$

Electromagnetic torque:

$$T_{em} = \frac{3}{2}\frac{P}{2}(\lambda_d i_q - \lambda_q i_d) \qquad N.m \tag{7.156}$$

or,

$$T_{em} = \underbrace{\frac{3}{2}\frac{P}{2}(L_d - L_q)i_d i_q}_{reluctance \ torque} + \underbrace{\frac{3}{2}\frac{P}{2}(L_{md} i'_{kd} i_q - L_{mq} i'_{kq} i_d)}_{induction \ torque} + \underbrace{\frac{3}{2}\frac{P}{2}L_{md} i'_m i_q}_{excitation \ torque} \tag{7.157}$$

In Eq. 7.157, the developed electromagnetic torque is separated into three components: a reluctance component, which is negative when $L_d < L_q$; an induction component which is an asynchronous torque; and an excitation component from the field of the permanent magnet.

The mutual flux linkages in the q- and d-axes may be expressed by

$$\lambda_{mq} = L_{mq}(i_q + i'_{kq}) \qquad Wb. \ turn$$

$$\lambda_{md} = L_{md}(i_d + i'_m + i'_{kd}) \tag{7.158}$$

As before, the winding currents can be expressed as

$$i_q = \frac{1}{L_{ls}}(\lambda_q - \lambda_{mq}) \qquad\qquad i_d = \frac{1}{L_{ls}}(\lambda_d - \lambda_{md}) \qquad A$$

$$i'_{kq} = \frac{1}{L'_{lkq}}(\lambda'_{kq} - \lambda_{mq}) \qquad\qquad i'_{kd} = \frac{1}{L'_{lkd}}(\lambda'_{kd} - \lambda_{md}) \tag{7.159}$$

With the above relations for the d-axis winding currents substituted into Eq. 7.158 and simplifying, we will obtain

$$\lambda_{md} = L_{MD}\left(\frac{\lambda_d}{L_{ls}} + \frac{\lambda'_{kd}}{L'_{lkd}} + i'_m\right) \qquad Wb. \ turn \tag{7.160}$$

where

$$\frac{1}{L_{MD}} = \frac{1}{L_{ls}} + \frac{1}{L'_{lkd}} + \frac{1}{L_{md}} \tag{7.161}$$

Similar expressions for λ_{mq} and L_{MQ} can be written for the q-axis.

For steady-state conditions, where $\omega = \omega_e$, as in the case of E_f in a wound field machine, $\omega_e \lambda'_m$ or $x_{md} i'_m$ can be denoted by E_m, the magnet's excitation voltage on the stator side. And if stator resistance is disregarded, replacing E_f by E_m in Eq. 7.68, the developed torque of a permanent magnet motor in terms of the rms phase voltage V_a at its terminal is

$$T_{em} = -3 \left(\frac{2}{P\omega_e} \right) \left\{ \frac{V_a E_m}{X_d} \sin\delta + \frac{V_a^2}{2} \left(\frac{1}{X_q} - \frac{1}{X_d} \right) \sin 2\delta \right\} \qquad N.m \qquad (7.162)$$

7.10 SYNCHRONOUS MACHINE PROJECTS

7.10.1 Project 1: Operating Characteristics and Parameter Variations

The purpose of this project is to implement *s1*, the SIMULINK simulation of a three-phase synchronous machine shown in Fig. 7.11, and use the simulation to determine the operational characteristics of the generator with the parameters given below. In addition, we will examine the effects that a change in damper winding resistance has on the damping of electromechanical oscillations, and also how a change in rotor inertia affects the duration of the first swing.

Since the full $qd0$ model already has damper windings included, the damping coefficient, D_ω, may be used to represent the damping from windage and friction on the rotor. Since the field and rotor mechanical time constants of a synchronous machine are usually large, to minimize the initial startup transients, the d-axis winding fluxes, ψ'_f and ψ_d, and the rotor slip speed, $(\omega_r - \omega_e)/\omega_b$, should be initialized with values that are closed to the starting condition. The best initial values to use are the steady-state values corresponding to the starting condition. But a rough setting of ψ'_f and ψ_d to one per unit and the rotor speed, ω_r, to synchronous value may get the machine simulation to synchronize in reasonable time, if the operating condition is relatively stable.

For this project, the MATLAB file, *m1*, may be used to determine the resistances and reactances of the circuit model from the given data, and to set up the desired initial condition for the SIMULINK simulation, *s1*, of the machine connected to a fixed voltage supply. As implemented in *s1*, the bus voltage magnitude, V_m, the applied excitation voltage, E'_{ex}, and the externally applied mechanical torque, T_{mech}, are all represented by *repeating sequence source* signals, which can be programmed to produce the desired values over the run-time of the simulation.

(a) Check the given MATLAB file, *m1*, to make sure that it establishes the desired starting condition of the generator delivering one per unit power at unity power factor to the external bus of voltage magnitude at one per unit. Note that the applied excitation voltage, E'_{ex}, in per unit referred to the stator is equal to E_f. Set the *repeating sequence source* signals for V_m, E'_{ex}, and T_{mech} for a run where the magnitudes of the bus voltage and externally applied mechanical torque are both held constant at their starting values, but that for E'_{ex} to take a step increment of 10 percent at $t = 0.2$ sec., that is

Ex_time = [0 0.2 0.2 tstop]

Ex_value = [1 1 1.1 1.1]*Efo

Efo being the applied excitation for the starting condition of generating one per unit power at unity power factor. Record the response of $|V_t|$, $|I_t|$, P, Q, δ, T_{em}, i'_f, and i_a to the step change in excitation voltage. Note the long response time to the change in field excitation and the sign of Q with the increased field excitation.

(b) Repeat Part (a). This time, use a step decrement of 10 percent in E'_{ex} at $t = 0.2$ sec., that is

Ex_time = [0 0.2 0.2 tstop]

Ex_value = [1 1 0.9 0.9]*Efo

Comment on whether the machine is producing or absorbing reactive power after the change and cross-check by examining the relative phasing between the v_a and i_a.

(c) Obtain the machine response to the following step changes in the externally applied mechanical torque: from its starting value when the machine is generating rated power at unity power factor into the bus first to zero at $t = 0.5$ sec., and then to the negative of its starting value at $t = 3$ sec. The time and value arrays of the T_{mech} *repeating sequence source* signal are

tmech_time = [0 0.5 0.5 3 3 tstop]

tmech_value = [1 1 0 0 -1 -1]*Tmech

Tmech being the initial condition value of the externally applied torque when the machine is generating rated power at unity power factor.

Comment on whether the machine is generating or absorbing real power after the two step changes and cross-check your comments by examining the sign of the power angle, δ. Comment on the electromechanical oscillation frequency and damping.

(d) Repeat Part (c) with different values of the kd and kq damper winding resistances, r'_{kd} and r'_{kq}. First reduce r'_{kd} and r'_{kq} to 60 percent of their base values, respectively. Then, increase r'_{kd} and r'_{kq} to 140 percent of their base values, respectively. Comment on how the change in damper winding resistances affects the damping of the rotor oscillations. Also, determine which damper winding resistance has a greater effect on the damping of the rotor oscillations. Why?

(e) Change the value of H, the inertia constant of the machine, from the base value of 3.77 seconds to 5 seconds and repeat Part (c). Compare the oscillation frequency of δ obtained in Part (c) with that in Part (e) for the same disturbance and deduce from the comparison the effect that larger rotor inertia has on the time it takes the rotor to go through the first swing.

Figure 7.20 shows sample results for Part (c) for the machine with nominal circuit parameters and inertia. It can be seen that the initial condition established by *m1* starts the simulation off with the machine generating rated power at unity power factor. At $t = 0.5$ seconds, the externally applied mechanical torque is dropped from one per unit to zero for the next 2.5 seconds. At $t = 3$ seconds, the mechanical torque again is changed, this

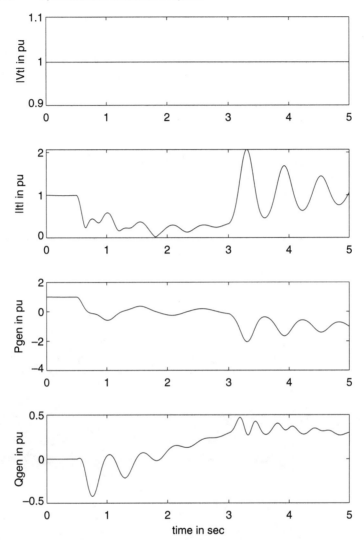

Figure 7.20   Response of *set*1 generator to step changes in T_{mech}.

time from zero to -1 per unit torque. The applied excitation voltage, E'_{ex}, is held constant throughout this run.

7.10.2 Project 2: Terminal Faults on a Synchronous Generator

The purpose of this project is to examine the response of a synchronous generator operating under fixed rotor speed and fixed excitation voltage to several kinds of electrical faults at its stator terminals.

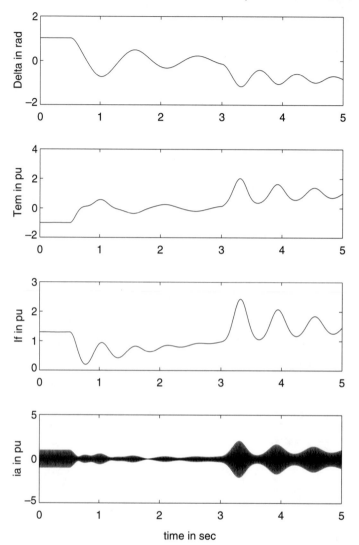

Figure 7.20 *(cont.):* Response of *set*1 generator to step changes in T_{mech}.

Make a copy of the Project 1 SIMULINK file *s1* and modify the copy to enable you to simulate the following electrical faults on its stator terminals:

1. Three-phase bolted fault.
2. Phase-to-phase bolted fault between phases *b* and *c*.
3. Single-phase-to-ground fault on phase *a*.

TABLE 7.2   SYNCHRONOUS GENERATOR
SET 1 RATINGS AND PARAMETERS

$S_{rated} = 920.35$ MVA	$V_{rated} = 18$ kV
$N_{rated} = 1800$ rpm	Rated power factor $= 0.9$
$x_{ls} = 0.215$ per unit	$r_s = 0.0048$ per unit
$x_d = 1.790$ per unit	$x_q = 1.660$ per unit
$x'_d = 0.355$ per unit	$x'_q = 0.570$ per unit
$x''_d = 0.275$ per unit	$x''_q = 0.275$ per unit
$T'_{do} = 7.9$ s	$T'_{qo} = 0.41$ s
$T''_{do} = 0.032$ s	$T''_{qo} = 0.055$ s
$H = 3.77$ s	$D_\omega = 0$ per unit

Test your simulation using the *set1* parameters of the synchronous generator given in Table 7.2 with the help of the MATLAB file, *m1*.

The modified simulation should allow you to simulate the condition of applying any of the above faults directly to the machine's stator terminals which are initially open-circuited. Set the inertia constant, H, to some large value, 999 say, to keep rotor speed constant over a brief period before and during the short-circuit. Edit the initial condition specified in *m1* for your simulation to start with the machine running at synchronous speed, no-load, and rated terminal voltage.

(a) When the initial electrical transients are over, apply the three-phase short circuit at the peak of v_a. Record the response of i_a, i_b, i_c, i'_f, and T_{em}. Compare the waveforms of the fault currents with those given in standard machine text, such as [26]. Repeat the run with the short-circuit applied at the voltage zero, v_a. Comment on the differences in the dc offset of the waveforms of short-circuit currents obtained at the two different points of the wave.

(b) When the initial electrical transients are over, apply a phase-to-phase short circuit across phases b and c at the peak of v_a. Record the response of i_a, i_b, i_c, i'_f, and T_{em}.

(c) When the initial electrical transients are over, apply a single-phase-to-ground short circuit to phase a at the peak of v_a. Record the response of i_a, i_b, i_c, i'_f, and T_{em}.

7.10.3 Project 3: Linearized Analysis of a Synchronous Generator

The objectives of this project are to learn how we can determine a linear model from a SIMULINK simulation of the machine and use it to determine the small-signal characteristics of the machine. Specifically, we shall determine the linear models of a synchronous machine that is generating directly into an infinite bus at two different operating conditions, use these linear models to determine the transfer functions, $\Delta P_{gen}/\Delta T_{mech}$ and $\Delta Q_{gen}/\Delta E_f$, and examine the effect of a change in rotor inertia and damper winding resistance on the system eigenvalues.

Figure 7.21 shows the SIMULINK simulation of a synchronous machine connected directly to an infinite bus given in the SIMULINK file, *s3eig*. In the given simulation, v_q^e and v_d^e are the infinite bus voltage components in a synchronously rotating reference. The transformation of the synchronously rotating reference input stator voltages to the reference frame rotating with the rotor can be accomplished using Eq. 7.51 in the form of

$$v_q = v_q^e \cos\delta - v_d^e \sin\delta$$
$$v_d = v_q^e \sin\delta + v_d^e \cos\delta$$
(7.163)

The rest of the simulation of the synchronous machine in Fig. 7.21 is in the rotor reference frame. Again, in using the SIMULINK **trim** and MATLAB **linmod** functions, the inputs and outputs must be defined by sequentially numbered input and output ports.

Use the MATLAB script file,*m3*, to perform the following tasks:

1. Load the parameters and rating of the synchronous machine given in Project 1 onto the MATLAB workspace. Make sure that the units used are consistent with those employed in your SIMULINK representation of the system.

2. Have in mind the operating conditions for which you would like to obtain linear models of the system. Modify this part of *m3* accordingly. For this project, we would like to obtain the transfer functions for the following two operating conditions:

a. $P_{gen} = 0$ and $Q_{gen} = 0$ when $v_{qs}^e = 1$ and $v_{ds}^e = 0$.
b. $P_{gen} = 0.8$ and $Q_{gen} = 0.6$ when $v_{qs}^e = 1$ and $v_{ds}^e = 0$.

3. The outer `for` loop in *m3* performs the following subtasks for each of the above operating conditions:

a. Uses the SIMULINK **trim** function to determine the desired steady-state operating point of the SIMULINK system, *s3eig*.
b. Uses the MATLAB **linmod** function to determine the state variable model, that is the A,B,C, and D matrices of the linear model. Obtains the system eigenvalues using the MATLAB **eig** function on the system matrix, A, that is from eig(A).
c. Uses the MATLAB **ss2tf** to determine the transfer functions of $\Delta Q_{gen}/\Delta E_f$ and $\Delta P_{gen}/\Delta T_{mech}$ at each of the above operating points.
d. Uses the MATLAB **tf2zp** to determine the poles and zeros of the two sets of transfer functions.
e. Plots the root-loci of the transfer functions, $\Delta Q_{gen}/\Delta E_f$ and $\Delta P_{gen}/\Delta T_{mech}$, at the operating point where $P_{gen} = 0.8$, $Q_{gen} = 0.6$, $v_{qs}^e = 1$, and $v_{ds}^e = 0$.

Sample results of the transfer functions are given in Tables 7.3 and 7.4. A sample root-locus plot of $\Delta Q_{gen}/\Delta E_f$ at the operating point where $P_{gen} + j Q_{gen} = 0.8 + j0.6$ and $v_{qs}^e - j v_{ds}^e = 1 + j0$ is given in Fig. 7.22.

Edit the parameter values in a copy of the machine parameter file, *set1.m*, that is used by *m3* to perform parameter sensitivity studies of the system eigenvalues at the same

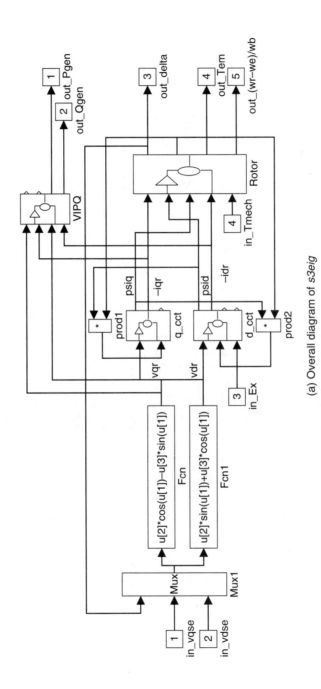

(a) Overall diagram of *s3eig*

Figure 7.21 Simulation *s3eig* of a synchronous generator.

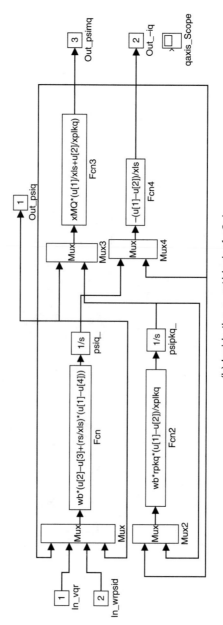

(b) Inside the *q_cct* block of *s3eig*

Figure 7.21 (cont.): Simulation *s3eig* of a synchronous generator.

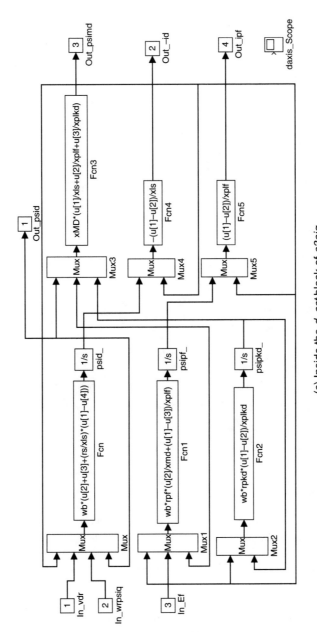

(c) Inside the *d_cct* block of *s3eig*

Figure 7.21 (cont.): Simulation *s3eig* of a synchronous generator.

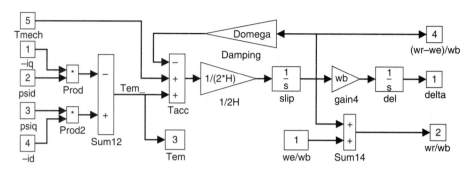

(d) Inside the Rotor block of *s3eig*

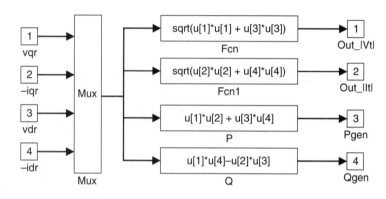

(e) Inside the *VIPQ* block of s3eig

Figure 7.21   *(cont.):* Simulation *s3eig* of a synchronous generator.

two operating conditions: first to a change in rotor inertia, H, and then to changes in rotor damper winding resistances, r'_{kd} and r'_{kq}. The system eigenvalues at each of the operating conditions can be obtained using the MATLAB function **eig()** on the system matrix, A. In addition to the results for the base case parameters, determine the system eigenvalues for the cases when H alone is changed to 60% and to 140% of the nominal value. Repeat the study for the following two changes in the damper winding resistances, r'_{kd} and r'_{kq}: first with both changed to 60% and then to 140% of their nominal values. Examine the system eigenvalues obtained with and without the changes and determine whether there is a change in frequency and damping of the rotor dynamics. Confirm these changes from simulated responses with the appropriate machine parameters to a small perturbation in mechanical input torque about the corresponding operating condition using the SIMULINK system, *s3*.

TABLE 7.3 $\Delta p_{gen}/\Delta t_{mech}$

Output P+jQ = 0+j0 Gain is $-5.68 - 14$	Output P+jQ = 0.8+j0.6 Gain is 8.53e$-$14
Zeros are:	Zeros are:
3.199e+15 0.000e+00i	-2.492e+15 0.000e+00i
-3.283e+00 3.769e+02i	-3.285e+00 3.761e+02i
-3.283e+00 -3.769e+02i	-3.285e+00 -3.761e+02i
-4.071e+01 0.000e+00i	-6.090e+01 0.000e+00i
-1.818e+01 0.000e+00i	-3.283e+01 0.000e+00i
-6.327e$-$01 0.000e+00i	-3.972e$-$01 0.000e+00i
Poles are:	Poles are:
-6.373e+00 3.762e+02i	-6.374e+00 3.762e+02i
-6.373e+00 -3.762e+02i	-6.374e+00 -3.762e+02i
-1.088e+02 0.000e+00i	-1.092e+02 0.000e+00i
-4.071e+01 0.000e+00i	-4.042e+01 0.000e+00i
-6.980e$-$01 5.479e+00i	-7.468e$-$01 9.817e+00i
-6.980e$-$01 -5.479e+00i	-7.468e$-$01 -9.817e+00i
-6.327e$-$01 0.000e+00i	-3.958e$-$01 0.000e+00i

TABLE 7.4 $\Delta q_{gen}/\Delta e_f$

Output P+jQ = 0+j0 Gain is 1.97e$-$01	Output P+jQ = 0.8+j0.6 Gain is 1.64e$-$01
Zeros are:	Zeros are:
-3.077e+00 3.762e+02i	-2.579e+00 3.742e+02i
-3.077e+00 -3.762e+02i	-2.579e+00 -3.742e+02i
-1.088e+02 0.000e+00i	-1.098e+02 0.000e+00i
-7.292e+01 0.000e+00i	-7.292e+01 0.000e+00i
-6.981e$-$01 5.479e+00i	-7.062e$-$01 9.389e+00i
-6.981e$-$01 -5.479e+00i	-7.062e$-$01 -9.389e+00i
Poles are:	Poles are:
-6.373e+00 3.762e+02i	-6.374e+00 3.762e+02i
-6.373e+00 -3.762e+02i	-6.374e+00 -3.762e+02i
-1.088e+02 0.000e+00i	-1.092e+02 0.000e+00i
-4.071e+01 0.000e+00i	-4.042e+01 0.000e+00i
-6.980e$-$01 5.479e+00i	-7.468e$-$01 9.817e+00i
-6.980e$-$01 -5.479e+00i	-7.468e$-$01 -9.817e+00i
-6.327e$-$01 0.000e+00i	-3.958e$-$01 0.000e+00i

7.10.4 Project 4: Permanent Magnet Synchronous Motor

The *line-start* permanent magnet motor is a high-efficiency synchronous motor with self-starting capability when operating from a fixed frequency voltage source. The permanent magnets embedded in its rotor provide the synchronous excitation and the rotor cage provides the induction motor torque for starting. The difference in permeability between the magnet and rotor core also results in significant magnetic saliency and reluctance torque at synchronous speed. At asynchronous speeds, the dc excitation and saliency of the permanent magnets will cause pulsating torque components. When the field strength of the

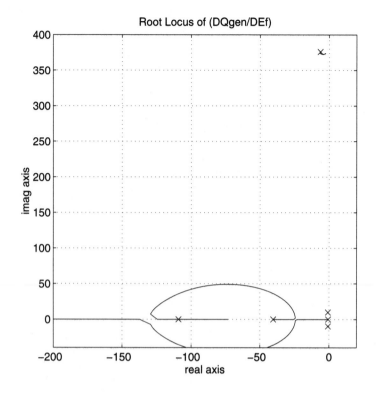

Figure 7.22   Root-locus plot of $\Delta Q_{gen}/\Delta E_f$.

magnet is too strong, a line-start permanent magnet motor may fail to synchronize because of the excessive pulsating torque component from the dc excitation of the magnet.

The objective of this project is to implement the SIMULINK simulation, *s4*, of a *line-start* permanent magnet motor and use the simulation to explore the behavior of the torque components during a starting run of the motor from standstill. In particular, we will examine the ability of the motor to synchronize with various values of magnet field strength, mechanical loading, and rotor inertia.

Figure 7.23 shows the SIMULINK simulation, *s4*, of a line-start permanent magnet motor connected to a fixed frequency supply source. The equations used for the permanent magnet motor are given in Table 7.5.

TABLE 7.5   SIMULATED $qd0$ EQUATIONS OF PERMANENT MAGNET MOTOR

Flux linkages:

$$\psi_q = \omega_b \int \left\{ v_q - \frac{\omega_r}{\omega_b}\psi_d + \frac{r_s}{x_{ls}}(\psi_{mq} - \psi_q) \right\} dt$$

$$\psi_d = \omega_b \int \left\{ v_d + \frac{\omega_r}{\omega_b}\psi_q + \frac{r_s}{x_{ls}}(\psi_{md} - \psi_d) \right\} dt$$

$$\psi_0 = \omega_b \int \left(v_0 + \frac{r_s}{x_{ls}}\psi_0 \right) dt \qquad (7.164)$$

$$\psi'_{kq} = \frac{\omega_b r'_{kq}}{x'_{lkq}} \int (\psi_{mq} - \psi'_{kq}) dt$$

$$\psi'_{kd} = \frac{\omega_b r'_{kd}}{x'_{lkd}} \int (\psi_{md} - \psi'_{kd}) dt$$

where

$$\psi_{mq} = x_{MQ} \left(\frac{\psi_q}{x_{ls}} + \frac{\psi'_{kq}}{x'_{lkq}} \right)$$

$$\psi_{md} = x_{MD} \left(\frac{\psi_d}{x_{ls}} + \frac{\psi'_{kd}}{x'_{lkd}} + i'_m \right) \qquad (7.165)$$

$$\frac{1}{x_{MQ}} = \frac{1}{x_{mq}} + \frac{1}{x'_{lkq}} + \frac{1}{x_{ls}}$$

$$\frac{1}{x_{MD}} = \frac{1}{x_{md}} + \frac{1}{x'_{lkd}} + \frac{1}{x_{ls}} \qquad (7.166)$$

Winding currents:

$$i_q = \frac{\psi_q - \psi_{mq}}{x_{ls}} \qquad\qquad i'_{kq} = \frac{\psi'_{kq} - \psi_{mq}}{x'_{lkq}}$$

$$i_d = \frac{\psi_d - \psi_{md}}{x_{ls}} \qquad\qquad i'_{kd} = \frac{\psi'_{kd} - \psi_{md}}{x'_{lkd}} \qquad (7.167)$$

Rotor motion:

$$T_{em} = (\psi_d i_q - \psi_q i_d) \qquad pu$$

$$= \underbrace{(x_d - x_q)i_d i_q}_{\text{reluctance torque}} + \underbrace{x_{md}i'_{kd}i_q - x_{mq}i'_{kq}i_d}_{\text{induction torque}} + \underbrace{x_{md}i'_m i_q}_{\text{excitation torque}} \qquad (7.168)$$

$$T_{em} + T_{mech} - T_{damp} = 2H\frac{d\{(\omega_r - \omega_e)/\omega_b\}}{dt} \qquad pu \qquad (7.169)$$

$$\delta(t) = \theta_r(t) - \theta_e(t) \qquad elect.\,rad.$$

$$= \omega_b \int_0^t \{(\omega_r(t) - \omega_e)/\omega_b\}dt + \theta_r(0) - \theta_e(0) \tag{7.170}$$

The SIMULINK file, *s4*, contains an implementation of a line-start, three-phase permanent magnet synchronous motor connected directly to a 60-Hz, three-phase supply

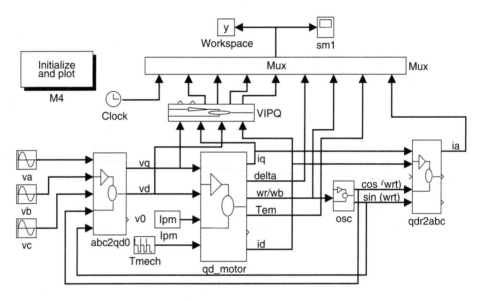

(a) Overall diagram of *s4*

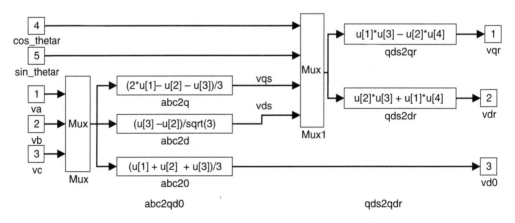

(b) Inside the *abc2qd0* block of *s4*

Figure 7.23   Simulation *s4* of a line-start permanent magnet motor.

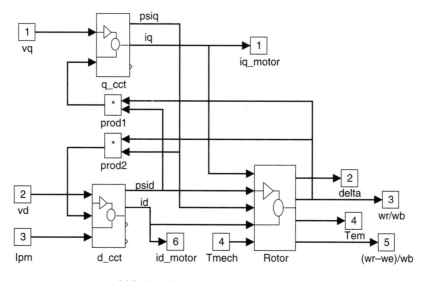

(c) Inside the *qd_motor* block of *s4*

Figure 7.23   *(cont.)*: Simulation *s4* of a line-start permanent magnet motor.

voltage of rated magnitude. The MATLAB file, *m4*, has been programmed to work with *s4*; it will set up in the MATLAB workspace the parameters of the 230-V, 4-hp, two-pole, 60-Hz, three-phase line-start permanent magnet given in Table 7.6 [67].

The MATLAB script file, *m4*, will prompt the user to input the desired equivalent magnet current, i'_m, or the terminal input power condition for it to establish the equivalent magnetizing current. It also allows the user two options to initialize the integrators in *s4*: that with the steady-state currents, fluxes, and rotor speed corresponding to the terminal condition specified by the user, and that with just the magnet's *d*-axis field flux, zero stator current, and rotor speed as in starting from standstill. As programmed, it will allow the user to repeat anew the calculations to determine the equivalent magnetizing current for a fresh terminal condition or to make repeated simulations with changes in parameters, external loading, or stop time. The file, *m4*, is also set up to provide a plot of the input real and reactive power, P_m and Q_m, the power angle, δ, the per unit speed of the rotor, ω_r/ω_b, the

TABLE 7.6   PARAMETERS OF LINE-START MOTOR IN PER UNIT

$x_{ls} = 0.065$ per unit	$r_s = 0.017$ per unit
$x_d = 0.543$ per unit	$x_q = 1.086$ per unit
$r'_{kd} = 0.054$ per unit	$r'_{kq} = 0.108$ per unit
$x'_{lkd} = 0.132$ per unit	$x'_{lkq} = 0.132$ per unit
$H = 0.3$ s	$D_\omega = 0$ per unit

Adapted from M. A. Rahman and T. A. Little, "Dynamic Performance of Permanent Magnet Synchronous Motors," IEEE Trans. of Apparatus and Systems, Vol. 103, June 1984, (© 1984 IEEE).

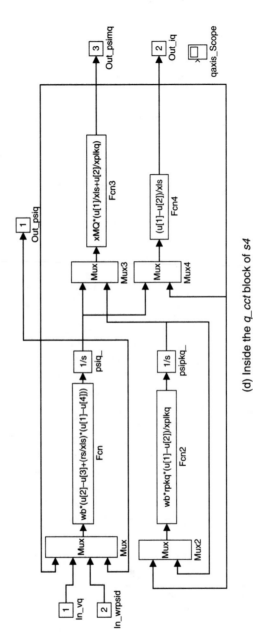

(d) Inside the *q_cct* block of *s4*

Figure 7.23 (cont.): Simulation *s4* of a line-start permanent magnet motor.

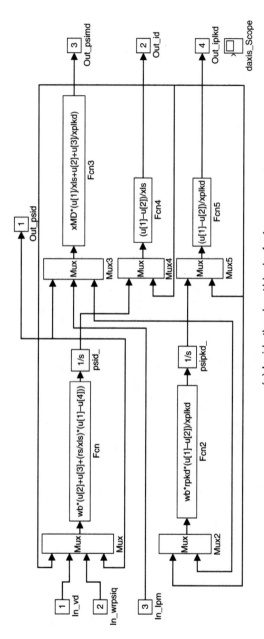

Figure 7.23 (cont.): Simulation *s4* of a line-start permanent magnet motor.

(e) Inside the *d_cct* block of *s4*

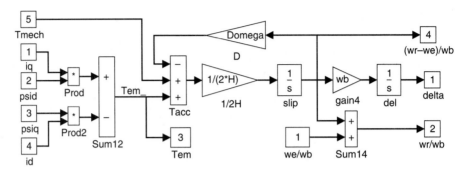

(f) Inside the *Rotor* block of *s4*

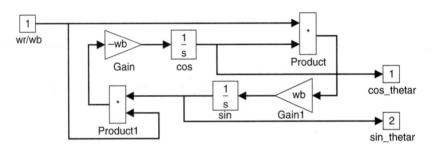

(g) Inside the *osc* block of *s4*

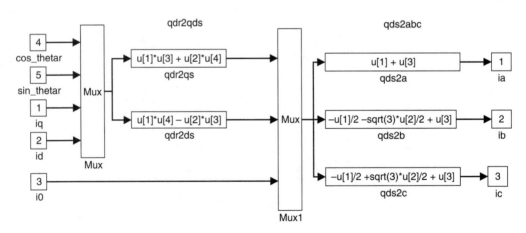

(h) Inside the *qdr2abc* block of *s4*

Figure 7.23 *(cont.):* Simulation *s4* of a line-start permanent magnet motor.

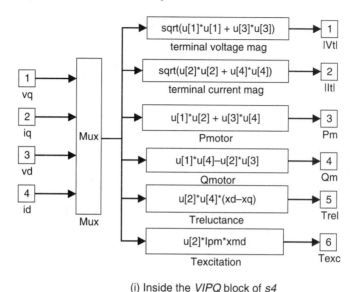

(i) Inside the *VIPQ* block of *s4*

Figure 7.23   *(cont.):* Simulation *s4* of a line-start permanent magnet motor.

TABLE 7.7   MAGNET STRENGTH FOR GIVEN TERMINAL CONDITIONS

Case	Rotor inertia, H	External loading, T_{mech}	i'_m for S_m to be	Remarks
1	Nominal	0 pu	Set $i'_m = 0$	Demagnetized
2	Nominal	0 pu	$S_m = 0.9 + j0.436$	$i'_m = 1.6203$
3	Nominal	0 pu	$S_m = 0.9 - j0.4366$	$i'_m = 2.6985$
4	Nominal	0 pu	$S_m = 1.0 + j0$	$i'_m = 2.2223$
5	Nominal	-0.5 pu	$S_m = 1 + j0$	$i'_m = 2.2223$
6	2× Nominal	-0.5 pu	$S_m = 1 + j0$	$i'_m = 2.2223$
7	2× Nominal	0 pu	$S_m = 1 + j0$	$i'_m = 2.2223$
8	2× Nominal	-0.5 pu	$S_m = 0.9 + j0.436$	$i'_m = 1.6203$

electromagnetic torque, T_{em}, and the stator *a*-phase current, i_a, in one figure. In a separate figure, it will plot T_{em} along with its reluctance, induction and excitation components.

First, we will examine the reluctance, induction, and excitation components of the instantaneous torque developed by the motor during a starting run from standstill. These runs are to be conducted with the motor starting up from rest, with flux components other than those of the magnet, rotor speed, and angle all set initially to zero. The strength of the permanent magnets is to be changed in these simulations by adjusting the value of the equivalent magnetizing current, i'_m, denoted by *Ipm* in *m4* according to the terminal conditions specified under Column 4 of Table 7.7.

Where the magnet strength is the same, such as in Cases 4 through 6, changes in H or mechanical loading from one case to another can be entered directly into the MATLAB workspace, instead of rerunning *m4*. For example, changing the mechanical loading from no-load in Case 4 to a constant motoring load of -0.5 pu in Case 5 can be done by entering

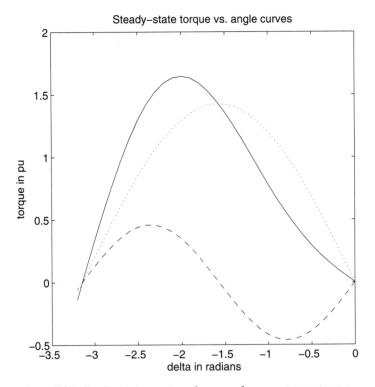

Figure 7.24   Steady-state torque vs. angle curves of permanent magnet motor.

in the new value of *tmech_value* of the *repeating sequence* in the MATLAB window, that is *tmech_value* = [−0.5 −0.5]. Similarly, changing the rotor inertia from nominal value in Case 5 to twice nominal value in Case 6 can be accomplished by typing in $H = 2 * H$ in the MATLAB workspace. Figure 7.24 shows the steady-state torque vs. angle curves of the permanent magnet motor for Case 4 when i'_m is 1.6203 A. The curves are obtained using a torque equation similar to that given in Eq. 7.68 where stator resistance has been neglected. It shows the reluctance and excitation torque components, and their resultant. The lower $\partial T/\partial \delta$ slope, or synchronizing torque coefficient on the stable side of the resultant torque curve, is due to the reluctance torque component. Sample results of the startup run for Case 8 in Table 7.7 are shown in Fig. 7.25. From the subplot of δ vs. time in Fig. 7.25, we can see that the accumulated value of δ settled to a steady value indicating that the machine finally achieved synchronous operation. An accumulated value of δ that is larger than 2π indicates that the rotor ran asynchronously during the initial phase of the startup run. Figure 7.26 shows a plot of the same δ vs. time over the principal value of $\pm\pi$, in which the pole slip is displayed more graphically by the transitions from $-\pi$ to π.

Examine the results you obtained, that is check for correctness in operating mode, such as whether the sign of δ is consistent with the motoring mode, the change in excitation with terminal power factor, the rotor speed, and the sign and value of the developed torque, real power, and reactive power when the motor is synchronized. Next, examine the behavior

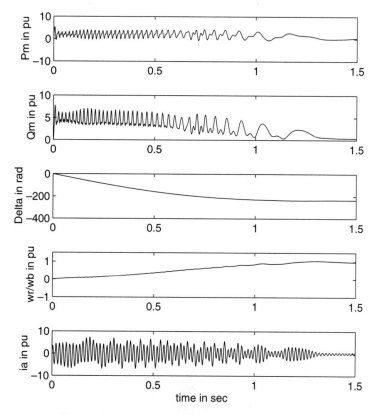

Figure 7.25 Starting transients of permanent magnet motor.

and values of the torque components during asynchronous and synchronous operations. Identify, if you can, which of these components contribute to the acceleration of the rotor during starting and which of them contribute to steady-state operation.

Next, we will use the simulation to obtain the dynamic response of the permanent magnet motor to a sequence of step-load changes. We can use *m4* to initialize the simulation in *s4* with the condition corresponding to Case 2 in Table 7.7. In this case, we will select to initialize the integrators in simulation with the computed steady-state currents, fluxes, and rotor speed to minimize the simulation startup transients. Some startup transients will be present because not all variables are properly initialized in this manner. Before performing the simulation, we will have to set up the *repeating sequence* signal source for the mechanical loading by typing in the following statements in the MATLAB window:

$$tmech\_value = [0\ 0\ -1.\ -1.\ -0.5\ -0.5\ -1.\ -1.\ ]$$

$$tmech\_time = [0.\ 0.4\ 0.4\ 0.8\ 0.8\ 1.2\ 1.2\ 1.5]$$

Figure 7.27 shows the simulated response of the permanent magnet motor with an i'_m 1.6203 pu to the above sequence of step changes in mechanical loading.

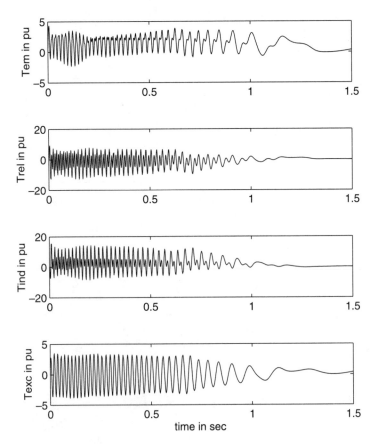

Figure 7.25   *(cont.):* Starting transients of permanent magnet motor.

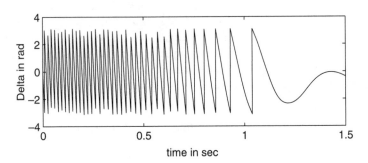

Figure 7.26   Pole slipping during the startup run.

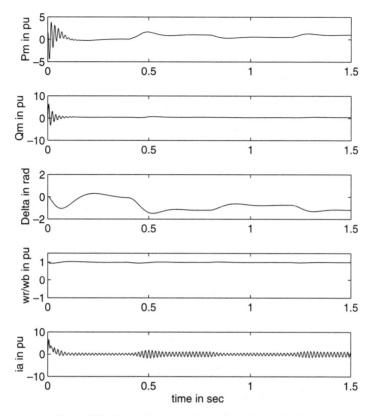

Figure 7.27   Response to step changes in mechanical loading.

7.10.5 Project 5: Simulation of Synchronous Machine Model with Unequal Stator and Rotor Mutual and Coupling of Rotor Circuits

The objectives of this project are to implement a simulation of the synchronous model described in Section 7.8 and to compare its simulated output to a three-phase short-circuit, especially the rotor quantities, with the output obtained using two other simpler machine models.

Figure 7.28a shows the overall diagram of *s5*, a SIMULINK simulation of the 2x3 synchronous machine model described in Section 7.8. Figures 7.28b and 7.28c show the *d*- and *q*-axis circuit blocks inside the *qd_gen* block of *s5* for the case with three rotor circuits in each axis. Not shown are the other parts of the simulation which are the same as those given earlier in Fig. 7.11. The *b* coefficient terms in the *d*-axis block are computed in the MATLAB m-file, *m5*, as part of the initialization procedure that also establishes the desired steady-state starting condition for a simulation run.

Along with *s5* and *m5* are two M-files, *set3a* and *set3b*, which are programmed to set up the machine parameters of the 722.222 MVA, 60-Hz, two-pole test machine: *set3a* uses

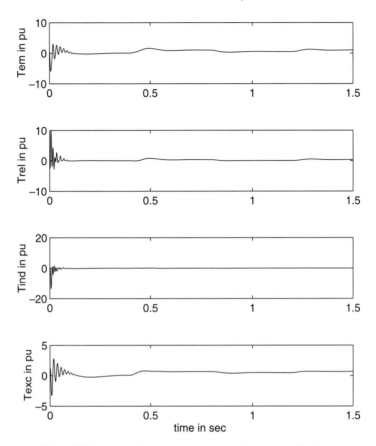

Figure 7.27   *(cont.):* Response to step changes in mechanical loading.

the parameters for the 2x3 equivalent circuit model given in [62] whereas *set3b* uses the parameters given in the discussion of [63] for a 2x3 equivalent circuit model where x_{r1c} is zero. To complete the comparison, we have also included in *set3c* the machine parameters of the same 722.222 MVA generator for the simpler model described in Section 7.5. The M-file, *set3c*, is to be used by *m1* to set up the simulation, *s1*, for the same short-circuit test. The machine model used in *s1* has one damper winding on the *q*-axis and one field and one damper winding on the *d*-axis, but there is no leakage coupling between the field and damper windings in the *d*-axis.

Both *m5* and *m1* are programmed to start the simulation with the machine generating rated power at unity power factor into the external bus of one per unit voltage magnitude. For the short-circuit case, the run-time is 1.5 seconds and the external voltage bus voltage magnitude, V_m, is stepped down from one per unit to zero at $t = 0.1$ seconds for a number of cycles specified by the user.

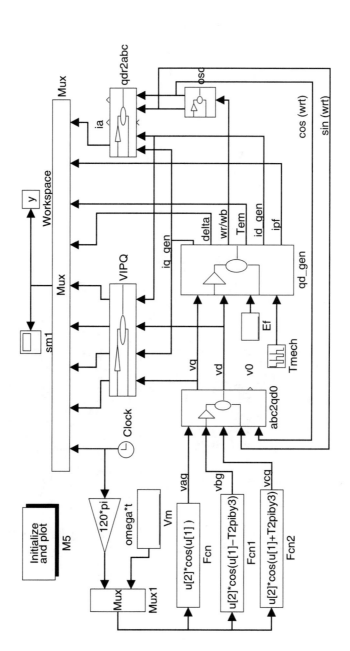

(a) Overall diagram of *s5*

Figure 7.28 Simulation *s5* of a synchronous generator model with unequal mutuals.

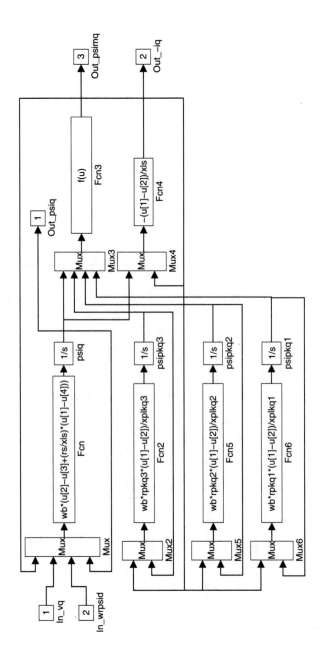

Figure 7.28 (cont.): Simulation *s5* of a synchronous generator model with unequal mutuals.

(b) Inside the *d-axis* block of *s5*

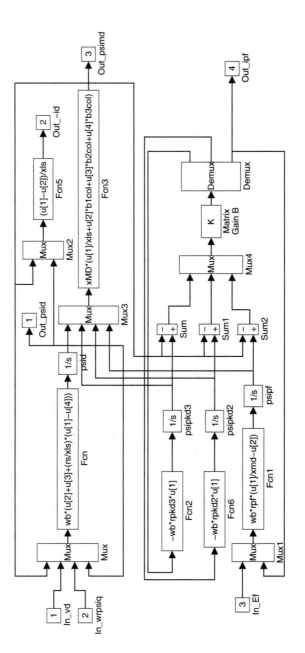

Figure 7.28 (cont.): Simulation *s5* of a synchronous generator model with unequal mutuals.

(c) Inside the *q-axis* block of *s5*

(a) Set up simulation *s5* to conduct the short-circuit simulation using the machine parameters in *set3a*. Select a step disturbance in V_m and enter 10 for the fault duration in number of cycles. Save the two sets of plots. Save the *Workspace y* arrays in a data file using the MATLAB **save** command so that you can later plot i_f' and δ of this run with those obtained in Parts (b) and (c).

(b) Repeat Part (a) using the parameter file, *set3b*. Save the two sets of plots. Reassign the *Workspace y* arrays in your MATLAB workspace with a unique variable name and store them in a separate data file for plotting later.

(c) Set up the simulation *s1* using *m1* to conduct the same short-circuit run with the parameter file, *set3c*, and save the two sets of plots. Again, reassign the *Workspace y* arrays in your MATLAB workspace with a unique variable name and store them in a separate data file for plotting later.

(d) Reload the stored data files from Parts (a), (b), and (c) into the MATLAB workspace and use *plot5c* to plot the three δ curves on one plot and the three i_f' curves on separate plots to avoid overcrowding. Comment on any observable differences in δ and i_f' from these three different machine models.

Sample results from the run in Part (a) are shown in Fig. 7.29, and the combined plots of δ and i_f' from all three runs are shown in Fig. 7.30.

7.10.6 Project 6: Six-phase Synchronous Machine

In Project 5 we saw how we could implement a simulation where there were coupling inductances between rotor circuits. Here in Project 6 we will look at how we can also handle coupling of the stator side windings in the qd rotor reference frame. We will use an example where the coupling on the stator side in the qd rotor reference frame remains in a synchronous machine with two three-phase sets of stator windings that are displaced by an angle. To obtain higher power capacity, the twin stator windings of a six-phase synchronous generator divides the generator's current into two paths to circumvent the limitations in current handling capacity of other equipment. Previously the limitation was from the interrupting capacity of breakers [70]; nowadays it is the current handling capacity of the power switches in the converters [71].

First, we will briefly review the equations for a six-phase synchronous machine, rearranging them into a form suitable for implementation in a simulation. Then we will proceed with the objective of this project, which is to develop and test a SIMULINK simulation of a six-phase machine. Figure 7.31 shows the windings of an idealized six-phase synchronous machine. The two sets of three-phase stator windings are labeled *abc* and *xyz* with the axes of *xyz* displaced by angle, ξ, ahead of *abc*. On the rotor are the damper *kq* and *kd* windings and the *d*-axis field *f* winding. Since parts of the distributed windings of the two sets share common stator slots, there will be mutual slot leakages. The leakage inductances of the stator and rotor may be put in the following form [72]:

$$L_l = \begin{bmatrix} [L_{l11}] & [L_{l12}] & 0 \\ [L_{l12}]^t & [L_{l22}] & 0 \\ 0 & 0 & [L_{lr}] \end{bmatrix} \tag{7.171}$$

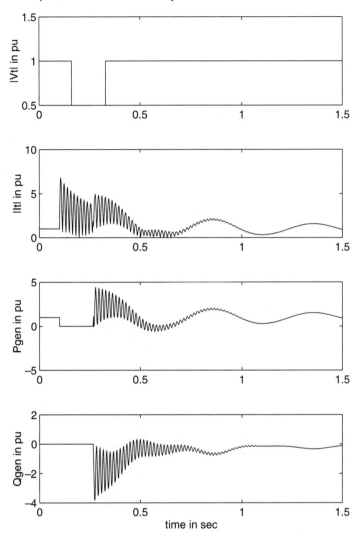

Figure 7.29   Short-circuit response from 2x3 model with non-zero x_{r1c} and x_{r2c}.

where the submatrix, $[L_{l12}]$, is a full matrix of mutual slot leakages between the two sets of three-phase stator windings, that is

$$[L_{l12}] = \begin{bmatrix} L_{lax} & L_{lay} & L_{laz} \\ L_{lbx} & L_{lby} & L_{lbz} \\ L_{lcx} & L_{lcy} & L_{lcz} \end{bmatrix} \tag{7.172}$$

and the other submatrices are diagonal in form:

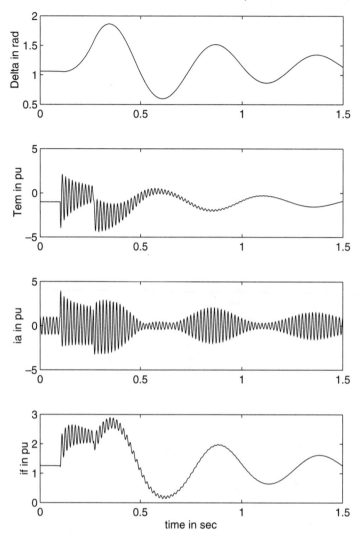

Figure 7.29   *(cont.):* Short-circuit response from 2x3 model with non-zero x_{r1c} and x_{r2c}.

$$[L_{l11}] = diag[L_{ls1}, L_{ls1}, L_{ls1}]$$

$$[L_{l22}] = diag[L_{ls2}, L_{ls2}, L_{ls2}] \qquad (7.173)$$

$$[L_{lr}] = diag[L_{lkq}, L_{lkd}, L_{lf}]$$

If the two sets of stator windings are uniformly wound, we can assume that $L_{lax} = L_{lby} = L_{lcz}$, $L_{lay} = L_{lbz} = L_{lcx}$, and $L_{laz} = L_{lbx} = L_{lcy}$, or that $[L_{l12}]$ is cyclic-symmetric. Picking the $qd0$ rotor reference as before with the angle, θ_r, measured from the stator a-phase axis, we can show that the transformation of $[L_{l12}]$ yields

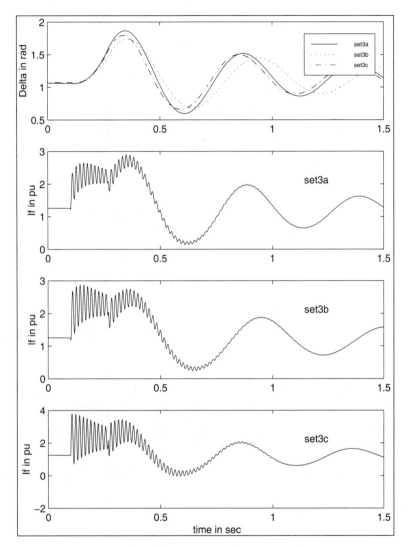

Figure 7.30   Response of δ and i'_f from three models.

$$[\mathbf{T}_{qd0}(\theta_r)]L_{l12}[\mathbf{T}_{qd0}^{-1}(\theta_r - \xi)] = \begin{bmatrix} L_{lm} & -L_{lqd} & 0 \\ L_{lqd} & L_{lm} & 0 \\ 0 & 0 & L_{lax} + L_{lay} + L_{laz} \end{bmatrix} \tag{7.174}$$

where

$$L_{lm} = L_{lax}\cos\xi + L_{lay}\cos(\xi + 2\pi/3) + L_{laz}\cos(\xi - 2\pi/3)$$
$$L_{lqd} = L_{lax}\sin\xi + L_{lay}\sin(\xi + 2\pi/3) + L_{laz}\sin(\xi - 2\pi/3) \tag{7.175}$$

The voltage equations of the stator and rotor windings of the six-phase machine shown in Fig. 7.31 using motoring notation are as follows [72]:

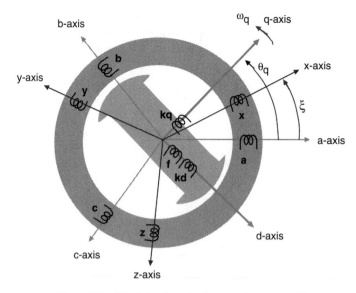

Figure 7.31 Windings of the six-phase synchronous machine.

$$v_{q1} = r_{s1}i_{q1} + \frac{d\lambda_{q1}}{dt} + \lambda_{d1}\frac{d\theta_r}{dt} \qquad V$$

$$v_{d1} = r_{s1}i_{d1} + \frac{d\lambda_{d1}}{dt} - \lambda_{q1}\frac{d\theta_r}{dt}$$

$$v_{01} = r_{s1}i_{01} + \frac{d\lambda_{01}}{dt}$$

$$v'_{q2} = r'_{s2}i'_{q2} + \frac{d\lambda'_{q2}}{dt} + \lambda'_{d2}\frac{d\theta_r}{dt} \qquad V$$

$$v'_{d2} = r'_{s2}i'_{d2} + \frac{d\lambda'_{d2}}{dt} - \lambda'_{q2}\frac{d\theta_r}{dt} \qquad (7.176)$$

$$v'_{02} = r'_{s2}i'_{02} + \frac{d\lambda'_{02}}{dt}$$

$$v'_f = r'_f i'_f + \frac{d\lambda'_f}{dt}$$

$$v'_{kd} = r'_{kd}i'_{kd} + \frac{d\lambda'_{kd}}{dt}$$

$$v'_{kq} = r'_{kq}i'_{kq} + \frac{d\lambda'_{kq}}{dt}$$

where the flux linkages are given by

$$\lambda_{q1} = L_{l1}i_{q1} + L'_{lm}(i_{q1} + i'_{q2}) - L'_{lqd}i'_{d2} + \lambda_{mq}$$

$$\lambda_{mq} = L_{mq}(i_{q1} + i'_{q2} + i'_{kq})$$

$$\lambda_{d1} = L_{l1}i_{d1} + L'_{lm}(i_{d1} + i'_{d2}) + L'_{lqd}i'_{q2} + \lambda_{md}$$

$$\lambda_{md} = L_{md}(i_{d1} + i'_{d2} + i'_{kd} + i'_f)$$

$$\lambda_{01} = L_{ls1}i_{01} + (L_{lax} + L_{lay} + L_{laz})i'_{02}$$

$$\lambda'_{q2} = L'_{l2}i'_{q2} + L'_{lm}(i_{q1} + i'_{q2}) + L'_{lqd}i'_{d1} + \lambda_{mq} \qquad (7.177)$$

$$\lambda'_{d2} = L'_{l2}i'_{d2} + L'_{lm}(i_{d1} + i'_{d2}) - L'_{lqd}i'_{q2} + \lambda_{md}$$

$$\lambda'_{02} = L'_{ls2}i'_{02} + (L_{lax} + L_{lay} + L_{laz})i_{01}$$

$$\lambda'_f = \lambda_{md} + L'_{lf}i'_f$$

$$\lambda'_{kd} = \lambda_{md} + L'_{lkd}i'_{kd}$$

$$\lambda'_{kq} = \lambda_{mq} + L'_{lkq}i'_{kq}$$

and

$$L_{l1} = L_{ls1} - L'_{lm} \qquad\qquad L'_{l2} = L'_{ls2} - L'_{lm},$$

$$L'_{lm} = \frac{N_1}{N_2}L_{lm} \qquad\qquad L'_{lqd} = \frac{N_1}{N_2}L_{lqd}, \qquad (7.178)$$

$$r'_s = \left(\frac{N_1}{N_2}\right)^2 r_2 \qquad\qquad L'_{ls2} = \left(\frac{N_1}{N_2}\right)^2 L_{ls2}$$

N_1 and N_2 are the effective number of turns of the abc and xyz stator windings, respectively; the prime on the xyz variables and parameters indicates their referred values to the abc stator windings.

The electromagnetic torque developed by the six-phase generator with P poles is given by

$$T_{em} = \frac{3}{2}\frac{P}{2}(\lambda_{d1}i_{q1} - \lambda_{q1}i_{d1} + \lambda'_{d2}i'_{q2} - \lambda'_{q2}i'_{d2})$$

$$= \frac{3}{2}\frac{P}{2}\left\{\lambda_{md}(i_{q1} + i'_{q2}) - \lambda_{md}(i_{d1} + i'_{d2})\right\} \qquad (7.179)$$

Figure 7.32 shows the $qd0$ equivalent circuits of the six-phase synchronous machine.

Equations for simulation.   As before, the equations for the six-phase synchronous machine can be arranged with ψ_{mq} and ψ_{md} maintained, so that core saturation can be included in the manner described in Section 6.8.1. The equations for the flux linkages are [72]

$$\psi_{q1} - \psi_{mq} = x_{l1}i_{q1} + x'_{lm}(i_{q1} + i'_{q2}) - x'_{ldq}i'_{d2}$$

$$\psi_{d1} - \psi_{mq} = x_{l1}i_{d1} + x'_{lm}(i_{d1} + i'_{d2}) + x'_{ldq}i'_{q2}$$

$$\psi_{q2} - \psi_{mq} = x_{l2}i'_{q2} + x'_{lm}(i_{q1} + i'_{q2}) + x'_{ldq}i_{d1}$$

$$\psi_{d2} - \psi_{mq} = x_{l2}i'_{d2} + x'_{lm}(i_{d1} + i'_{d2}) - x'_{ldq}i_{q1}$$

$$\psi_{01} = x_{ls1}i_{01} + (x_{lax} + x_{lay} + x_{laz})i'_{02} \qquad (7.180)$$

$$\psi'_{02} = x'_{ls2}i'_{02} + (x_{lax} + x_{lay} + x_{laz})i_{01}$$

$$\psi'_{kq} - \psi_{mq} = x'_{lkq}i'_{kq}$$

$$\psi'_{kd} - \psi_{md} = x'_{lkd}i'_{kd}$$

$$\psi'_f - \psi_{md} = x'_{lf}i'_f$$

When Eq. 7.181 is written in matrix form, it will be block-diagonal. For instance, the 4x4 coefficients matrix for the stator qd flux linkages may be written as

$$
\begin{bmatrix}
\psi_{q1} - \psi_{mq} \\
\psi_{d1} - \psi_{md} \\
\psi'_{q2} - \psi_{mq} \\
\psi'_{d2} - \psi_{md}
\end{bmatrix}
=
\underbrace{
\begin{bmatrix}
(x_{l1} + x'_{lm}) & 0 & x'_{lm} & -x'_{ldq} \\
0 & (x_{l1} + x'_{lm}) & x'_{ldq} & x'_{lm} \\
x'_{lm} & x'_{ldq} & (x_{l2} + x'_{lm}) & 0 \\
-x'_{ldq} & x'_{lm} & 0 & (x_{l2} + x'_{lm})
\end{bmatrix}
}_{X_{qd}}
\begin{bmatrix}
i_{q1} \\
i_{d1} \\
i'_{q2} \\
i'_{d2}
\end{bmatrix}
\qquad (7.181)
$$

The inverse relation of Eq. 7.181 with $B = X_{qd}^{-1}$ allows us to determine qd stator currents from the values of ψ_{q1}, ψ_{d1}, ψ'_{q2}, ψ'_{d2}, and ψ_{md}. The algebraic form of B is given in [72]. In general, we can write

$$
\begin{bmatrix}
i_{q1} \\
i_{d1} \\
i'_{q2} \\
i'_{d2}
\end{bmatrix}
=
\underbrace{
\begin{bmatrix}
b_{11} & b_{12} & b_{13} & b_{14} \\
b_{21} & b_{22} & b_{23} & b_{24} \\
b_{31} & b_{32} & b_{33} & b_{34} \\
b_{41} & b_{42} & b_{43} & b_{44}
\end{bmatrix}
}_{B}
\begin{bmatrix}
\psi_{q1} - \psi_{mq} \\
\psi_{d1} - \psi_{md} \\
\psi'_{q2} - \psi_{mq} \\
\psi'_{d2} - \psi_{md}
\end{bmatrix}
\qquad (7.182)
$$

The equations for the rotor circuit currents and stator zero-sequence currents can be determined in a similar way as before, that is

$$i'_{kq} = \frac{\psi'_{kq} - \psi_{mq}}{x'_{lkq}}$$

$$i'_{kd} = \frac{\psi'_{kd} - \psi_{md}}{x'_{lkd}} \qquad (7.183)$$

$$i'_f = \frac{\psi'_f - \psi_{md}}{x'_{lf}}$$

The q-axis mutual flux linkage may be expressed as

$$\psi_{mq} = x_{mq}(i_{q1} + i'_{q2} + i'_{kq})$$

$$\psi_{mq} = x_{MQ}\left\{ (b_{11} + b_{13})\psi_{q1} + (b_{12} + b_{32})\psi_{d1} \qquad (7.184) \right.$$

$$\left. + (b_{13} + b_{33})\psi'_{q2} + (b_{14} + b_{34})\psi'_{d2} + \frac{\psi'_{kq}}{x'_{lkq}} \right\}$$

where

$$\frac{1}{x_{MQ}} = \frac{1}{x_{mq}} + (b_{11} + b_{13} + b_{31} + b_{33}) + \frac{1}{x'_{lkq}} \qquad (7.185)$$

That for the d-axis flux linkage may also be expressed as

$$\psi_{md} = x_{md}(i_{d1} + i'_{d2} + i'_{kd} + i'_f)$$

$$\psi_{md} = x_{MD}\left\{ (b_{21} + b_{41})\psi_{q1} + (b_{22} + b_{42})\psi_{d1} \qquad (7.186) \right.$$

$$\left. + (b_{23} + b_{43})\psi'_{q2} + (b_{24} + b_{44})\psi'_{d2} + \frac{\psi'_{kd}}{x'_{lkd}} \right\}$$

where

$$\frac{1}{x_{MD}} = \frac{1}{x_{md}} + (b_{22} + b_{24} + b_{42} + b_{44}) + \frac{1}{x'_{lkd}} \qquad (7.187)$$

The stator and rotor circuit flux linkages are obtained by integrating their respective voltage equations, that is

$$\psi_{q1} = \omega_b \int \left\{ v_{q1} - \frac{\omega_r}{\omega_b} \psi_{d1} - r_{s1} \left(b_{11}\psi_{q1} + b_{12}\psi_{d1} + b_{13}\psi'_{q2} + b_{14}\psi'_{d2} \right) \right.$$
$$\left. -(b_{11}+b_{13})\psi_{mq} - (b_{12}+b_{14})\psi_{md} \right\} dt$$

$$\psi_{d1} = \omega_b \int \left\{ v_{d1} + \frac{\omega_r}{\omega_b} \psi_{q1} - r_{s1} \left(b_{21}\psi_{q1} + b_{22}\psi_{d1} + b_{23}\psi'_{q2} + b_{24}\psi'_{d2} \right) \right.$$
$$\left. -(b_{21}+b_{23})\psi_{mq} - (b_{22}+b_{24})\psi_{md} \right\} dt$$

$$\psi'_{q2} = \omega_b \int \left\{ v'_{q2} - \frac{\omega_r}{\omega_b} \psi_{d2} - r'_{s2} \left(b_{31}\psi_{q1} + b_{32}\psi_{d1} + b_{33}\psi'_{q2} + b_{34}\psi'_{d2} \right) \right.$$
$$\left. -(b_{31}+b_{33})\psi_{mq} - (b_{32}+b_{34})\psi_{md} \right\} dt$$

$$\psi'_{d2} = \omega_b \int \left\{ v'_{d2} + \frac{\omega_r}{\omega_b} \psi_{q2} - r'_{s2} \left(b_{41}\psi_{q1} + b_{42}\psi_{d1} + b_{43}\psi'_{q2} + b_{44}\psi'_{d2} \right) \right. \qquad (7.188)$$
$$\left. -(b_{41}+b_{43})\psi_{mq} - (b_{42}+b_{44})\psi_{md} \right\} dt$$

$$\psi_{01} = \omega_b \int \left\{ v_{01} - \frac{r_{s1}}{x_{lD}} \left(x'_{ls2}\psi_{01} - (x_{lax} + x_{lay} + x_{laz})\psi'_{02} \right) \right\} dt$$

$$\psi'_{02} = \omega_b \int \left\{ v'_{02} - \frac{r'_{s2}}{x_{lD}} \left(-(x_{lax} + x_{lay} + x_{laz})\psi_{01} + x_{ls1}\psi'_{02} \right) \right\} dt$$

$$\psi'_{kq} = -\omega_b r'_{kq} \int i'_{kq} dt$$

$$\psi'_{kd} = -\omega_b r'_{kd} \int i'_{kd} dt$$

$$\psi'_f = \frac{\omega_b r'_f}{x_{md}} \int \left\{ E_f - x_{md} i'_f \right\} dt$$

where $x_{lD} = x_{ls1}x'_{ls2} - (x_{lax} + x_{lay} + x_{laz})^2$ and $E_f = x_{md}v'_f/r'_f$.

The rotor speed, $\omega_r(t)$, is determined from the following expression of the slip speed:

$$\omega_r(t) - \omega_e = \frac{P}{2J} \int_0^t (T_{em} + T_{mech} - T_{damp})dt \qquad elect.\ rad/s \qquad (7.189)$$

where T_{em}, positive for motoring operation and negative for generating operation, is computed from

$$T_{em} = \frac{3}{2}\frac{P}{2\omega_b} \left\{ \psi_d(i_{q1} + i'_{q2}) - \psi_q(i_{d1} + i'_{d2}) \right\} \qquad N.m. \qquad (7.190)$$

Steady-state equations.   Let's consider the steady-state condition with balanced, three-phase sinusoidal variables where

$$v_a = V_{m1}\cos(\omega_e t) \qquad\qquad i_a = I_{m1}\cos(\omega_e t + \phi_1)$$

$$v_x = V_{m2}\cos(\omega_e t + \theta_2) \qquad i_x = I_{m2}\cos(\omega_e t + \theta_2 + \phi_2) \tag{7.191}$$

If the q_e-axis of the synchronously rotating reference frame is aligned with the a-axis of the *abc* stator windings at $t = 0$, we can express

$$v_{q1}^e - jv_{d1}^e = \sqrt{2}\tilde{\mathbf{V}}_1 = V_{m1}e^{j0} \qquad v_{q2}^e - jv_{d2}^e = \sqrt{2}\tilde{\mathbf{V}}_2 e^{j\xi} = V_{m2}e^{j(\xi+\theta_2)}$$

$$i_{q1}^e - ji_{d1}^e = \sqrt{2}\tilde{\mathbf{I}}_1 = I_{m1}e^{j\phi_1} \qquad i_{q2}^e - ji_{d2}^e = \sqrt{2}\tilde{\mathbf{I}}_2 e^{j\xi} = I_{m2}e^{j(\xi+\theta_2+\phi_2)} \tag{7.192}$$

Let δ be the angle by which the rotor q_r-axis leads the q_e-axis of the above synchronously rotating reference. When transformed to the rotor reference frame, the *abc* stator quantities become

$$v_{q1} - jv_{d1} = (v_{q1}^e - jv_{d1}^e)e^{-j\delta} = V_{m1}e^{-j\delta}$$

$$i_{q1} - ji_{d1} = (i_{q1}^e - ji_{d1}^e)e^{-j\delta} = I_{m1}e^{-j(\delta-\phi_1)} \tag{7.193}$$

Those of the *xyz* stator with magnitudes referred to *abc* windings are

$$v_{q2}' - jv_{d2}' = \frac{N_1}{N_2}(v_{q2}^e - jv_{d2}^e)e^{-j\delta} = \frac{N_1}{N_2}V_{m2}e^{-j(\delta-\xi-\theta_2)}$$

$$i_{q2}' - ji_{d2}' = \frac{N_2}{N_1}(i_{q2}^e - ji_{d2}^e)e^{-j\delta} = \frac{N_2}{N_1}I_{m2}e^{-j(\delta-\xi-\theta_2-\phi_2)} \tag{7.194}$$

In steady-state, with sinusoidal voltages and currents, the damper winding currents and the derivative terms in Eq. 7.176 will be zero and the voltage equations of the two stator windings simplify to

$$(v_{q1} - jv_{d1}) = \{r_1 + j\omega_e(L_{l1} + L_{lm}' + L_{mq})\}(i_{q1} - ji_{d1})$$

$$+ \{\omega_e L_{ldq}' + j\omega_e(L_{lm}' + L_{mq})\}(i_{q2}' - ji_{d2}') + E_q$$

$$(v_{q2}' - jv_{d2}') = \{r_2' + j\omega_e(L_{l2}' + L_{lm}' + L_{mq})\}(i_{q2}' - ji_{d2}') \tag{7.195}$$

$$+ \{-\omega_e L_{ldq}' + j\omega_e(L_{lm}' + L_{mq})\}(i_{q1} - ji_{d1}) + E_q$$

where

$$E_q = \omega_e(L_{md} - L_{mq})(i_{d1} + i_{d2}') + E_f \tag{7.196}$$

The voltage, E_q, has only a real component along the q_r-axis of the rotor. As described in Section 7.4.2, given the terminal operating condition or the phasor voltage and current of the stator windings, we can use the above equations to determine the rotor angle, δ, or to locate the q_r-axis.

Construct a SIMULINK simulation for a six-phase machine whose stator terminals are connected to balanced sets of three-phase sinusoidal voltage sources. Sketch the flow of variables similar to Fig. 7.10 before attempting to implement the simulation. Some of the component blocks may be established more quickly by modifying blocks of similar function in simulation *s1*. Verify your simulation using the parameters and operating conditions given in [72].

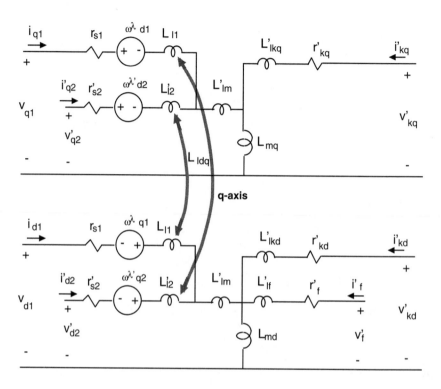

q-axis

d-axis

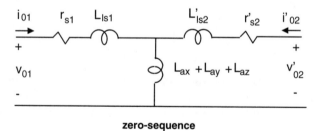

zero-sequence

Figure 7.32 $qd0$ equivalent circuits of the six-phase synchronous machine.

8

DC Machines

8.1 INTRODUCTION

As with the other rotating machines, a dc machine has a stator and a rotor. Figure 8.1 shows the basic parts of a simple dc machine. The stator typically consists of field windings wound concentrically around stator poles on the inside of a cylindrical yoke that is housed inside a metal frame. The yoke and pole irons are usually built from stamped steel laminations of about 0.5 to 1 mm thickness. At the airgap end, the pole has a pole-shoe that is used to shape the distribution of magnetic flux through the airgap.

On the rotor, the main components are the armature winding, armature core, mechanical commutator, and rotor shaft. The armature core is formed by stacking circular steel laminations with uniformly spaced slots on its outer periphery for the armature winding. The coil sides of the armature winding are distributed uniformly in the axial slots of the armature core. They are insulated from the rotor slot and held in position by insulating wedges. The two coil sides of each armature coil are located under adjacent poles of opposite polarity so that their induced emfs are additive around the coil. The coil ends are soldered to separate copper segments of the commutator, which are connected to the external circuit through the brushes riding on the commutator. The commutator segments, insulated from one another and from the clamp holding them together, have a cylindrical outer surface onto which spring-loaded brushes ride. The brushes and commutator act not only as rotary contacts between the coils of the rotating armature and stationary external

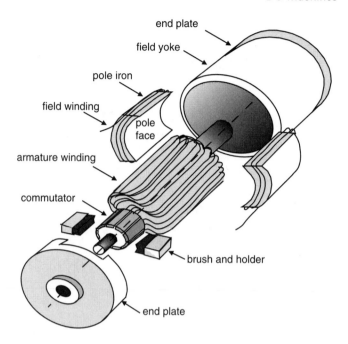

Figure 8.1   Basic parts of a dc machine.

circuit, but also as a switcher to commutate the current to the external dc circuit so that it remains unidirectional even though individual coil voltages are alternating.

The brushes are usually placed at the neutral position where the commutator segments are connected to coil sides that have no induced voltage.

8.2 ARMATURE WINDINGS

The two common winding arrangements of the armature winding are known as *lap* and *wave* windings. The coil ends of a lap winding are connected to commutator bars near each other; for example, a simplex lap winding has only one connected coil between adjacent commutator segments. The coil ends of a wave winding, on the other hand, are connected to commutator bars approximately two pole pitches apart. Though the bars of a wave-wound coil are further apart than those of a lap-wound coil, the potential between them, at approximately two pole pitches apart, are still electrically close. They cannot be exactly two pole pitches apart, because after one trip around the armature, the coil connection should not close onto the same bar but either one bar forward or backward so that the whole armature winding is traced through before the winding closes upon itself. Thus, a wave-wound armature for a P-pole machine will have $P/2$ coils connected in series between adjacent commutator segments.

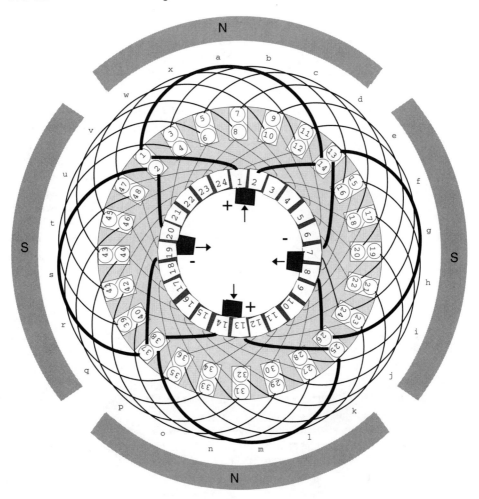

Figure 8.2   Four-pole armature lap winding.

8.2.1 Lap Windings

If the induced voltages of the two coil sides of a coil are to be additive, the two coil sides should be located under poles of opposite polarity. Furthermore, if the size and arrangement of the coils are to be uniform, the number of slots spanned by a coil, that is the slot pitch, y_s, has to be an integer value less than or equal to the total number of slots, S, divided by the number of poles.

Figure 8.2 shows a sectional view of a four-pole dc machine with a lap-wound armature winding on an armature core with 24 slots. Shown is a common two-layer winding arrangement with two coil sides per slot; top coil sides are odd-numbered and bottom coil sides are even-numbered. There are 24 coils altogether in the armature winding. The span of the coil, specifically the distance between its two coils sides around the back of the armature, is referred to as the back pitch. Using the above numbering system, the back pitch should be some odd number of coil sides given by

$$y_b = C/P \pm k \qquad (8.1)$$

where C is the total number of coil sides and k is $0, \pm 1, \pm 2, \ldots$. To ensure that the corresponding sides of a multi-element coil are in the same slot, $(y_b - 1)/\text{coil}$ sides per slot must be an integer. Using a k of 1, we obtain a y_b of 13 or 11. The former value yields a progressive lap-wound and the latter yields a retrogressive lap-wound. For the winding shown in Fig. 8.2, a y_b of 13 was used, yielding a coil span of exactly one pole pitch, or six slots. For example, the two coil sides of the highlighted coil a in Fig. 8.2 span 1 and 14, or six slot widths. With a simplex progressive lap winding, like the one shown, the return coil side of coil a is one commutator segment ahead in the clockwise direction.

At the rotor position shown in Fig. 8.2, the coil sides of the four highlighted coils, a, g, m, and s, are at the geometric midpoints between the centers of adjacent poles, also known as the geometric neutral axes of the poles. When the field distribution is perfectly symmetrical, these coil sides will be under zero field; in other words, there will be no induced emfs in them to aid or to hinder the current reversal. If the rotor rotation is counter-clockwise, a short time before the situation shown in the figure, coil sides 1 and 14 would be under the north and south pole, respectively. And a little time after, the same two coil sides would be under the south and north pole, respectively. In other words, the induced emf around coil a would flip directions when moving through the position shown. This flip is evident if we were to examine the difference in polarity of the adjacent two coils, b and x. The polarity of induced emfs in all coils are marked in Fig. 8.3, which is redrawn to show a clearer connection of the coils.

As the armature winding rotates around, the tapping is always made at the point of highest potential for the positive brush and at the lowest potential for the negative brush. It is important to note that four brush sets are required, the two positive brushes can be connected together, likewise for the negative brushes, to form four parallel armature paths. The simplex lap winding has as many parallel paths and brushes as poles. The brushes can be placed on the commutator at points known as neutral points where no voltage exists between adjacent segments. The conductors connected to those segments lie midway between the poles at zero flux density locations, that are also known as neutral axes.

As shown in Fig. 8.3, the currents in coils a, g, m, and s, bridged by the brushes, are undergoing commutation as each of these coils is moved into a different current path. In motoring convention, the current from the external circuit will enter the two brushes marked by the positive polarity of the resultant back emf of the coils. In Fig. 8.3, that current will divide among the four current paths formed by coils b, c, d, e, and f in the first quadrant, coils x, w, v, u, and t in the second quadrant, coils n, o, p, q, and r in the third quadrant, and coils l, k, j, i, and h in the fourth quadrant. A lap-wound armature, in general, has as many parallel paths as the number of poles times the *plexity* of the wound. Simplex windings having a *plexity* of one are quite common. But where the armature current is very large, multiplex windings are used to increase the number of parallel paths to keep the individual coil current low, often to obtain acceptable commutation.

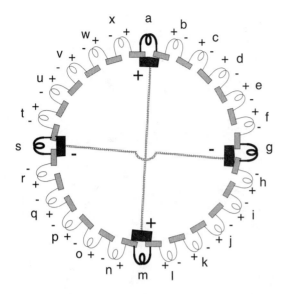

Figure 8.3 Coil connection of a four-pole, simplex lap-wound armature.

8.2.2 Wave Windings

Since the airgap field distribution of multiple pole pair machines repeats after every pole pair, the commutator connection to the two coil sides of a coil can be approximately two pole pitches away and yet the voltage between adjacent commutator segments will be nearly the same as that across a single coil in a lap winding. Going around the armature, coils under consecutive pole-pairs can be connected in series and their emfs will be additive. Coming back after going around the armature, the winding formed by the series coils must however not close on itself but progress with the series connection of other coils until the entire winding is complete. This type of winding is called a wave winding from the appearance of the end connections.

Figure 8.4 shows an example of a simplex wave winding on an armature with 21 slots in a four-pole dc machine. When the top sides of all coils are identified by odd numbers and the bottom sides by even numbers, both the back pitch, y_b, and the front pitch, y_f, will be odd since the difference between an odd and even number is always odd. The front pitch is the distance between two coil sides connected to the same commutator bar. Their sum, which has to be even for a wave winding, is to be determined from

$$y_b + y_f = \frac{C \pm 2 \times plexity}{P/2} \tag{8.2}$$

For the two-layer arrangement shown, C is twice the number of slots and the *plexity* is one; therefore, the sum of these two pitches can either be 22 or 20. The combination of $y_b = 11$ and $y_f = 11$ yields the progressive winding arrangement shown in Fig. 8.4, with a coil span that is exactly one pole pitch. Alternatively, a combination of $y_b = 11$ and $y_f = 9$ will give a retrogressive winding arrangement.

The distance between commutator segments connected to the same coil, in number of commutator segments, is given by

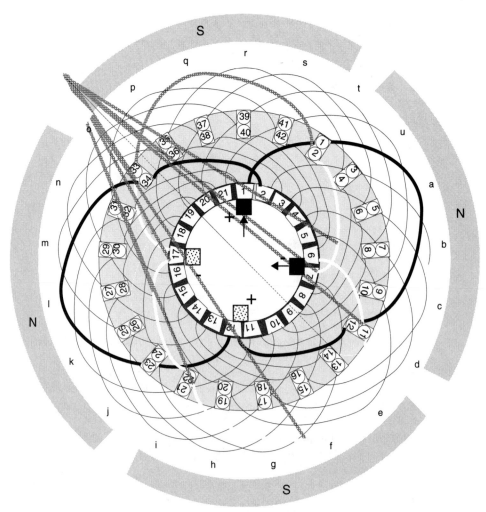

Figure 8.4   Four-pole armature wave winding.

$$y_c = \frac{y_b + y_f}{2} \tag{8.3}$$

For the winding shown, y_c is 11 and the coil span is five armature slots.

Beginning from commutator segment 1, the highlighted coils, a and l, travel one time around the armature returning to commutator segment 2, one commutator segment ahead in the clockwise direction. The number of coil elements traced through in one round of the armature is the same as the number of pole pairs, that is $P/2$. After one time around the armature, the connection must not close onto the same commutator segment, it must continue until all the other coils are similarly connected before the winding can close on itself. In the example given in Fig. 8.4, the pairs of coils in subsequent rounds are coils b

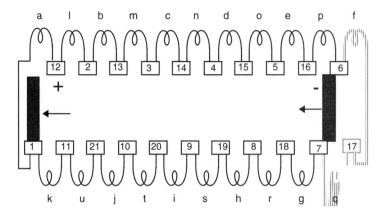

Figure 8.5   Coil connection of a four-pole, simplex, wave-wound armature.

and m, c and n, d and o, e and p, f and q, g and r, h and s, i and t, j and u, and finally the remaining coil k, which closes back onto commutator segment 1.

Tracing out the winding in the order mentioned above, we will find that the polarity of the induced coil emf does not change until half the coils are traced through. The coil connections of the wave winding in Fig. 8.4 can be redrawn as shown in Fig. 8.5, which clearly shows that the wave winding has only two parallel paths. We can also arrive at the conclusion that the coil currents can be commutated with just the two shaded brushes. However, additional brushes, shown unshaded, may be placed in other neutral axes to provide more contact area instead of using longer commutator segments. Each pair of brushes will short-circuit $P/2$ coils during commutation, because there are always $P/2$ coils between adjacent commutation segments.

8.2.3 Equipotential Connectors

Differences in induced emfs between paths may occur because of unequal pole flux, caused sometimes by small differences in airgap. With low coil resistances, even a slight difference in generated emf between parallel branches can cause a sizeable circulating current, causing higher copper losses in the winding and commutation difficulties. Equalizer connections can be used to divert some of the circulating current from the brush contacts. Such connections are made to convenient points on the armature winding that are two pole pitches apart, which normally would be at the same potential.

8.3 FIELD EXCITATION

The useful magnetic flux in the airgap under the stator poles is produced mainly by the mmf of the field windings on the stator. Figure 8.6 shows the commonly-used field winding connections. In Fig. 8.6a, the field current is from an external source that is independent of V_a. In Fig. 8.6b, the shunt field current is a function of V_a and in Fig. 8.6c, the series field current is a function of I_a. In the *compound* excitation circuits of Figs. 8.6d and 8.6e, the

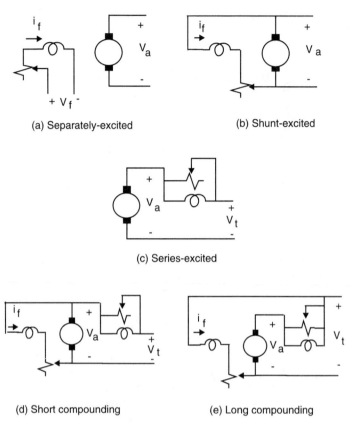

(a) Separately-excited   (b) Shunt-excited

(c) Series-excited

(d) Short compounding   (e) Long compounding

Figure 8.6   Field winding connections.

resultant field excitation is a function of both V_a and I_a. The compound excitation is referred to as *cumulative* when the mmf of the series field aids that of the main shunt field, or as *differential* when the series field mmf opposes that of the shunt field. The machine's terminal voltage vs. load current characteristic in generating and torque vs. speed characteristic in motoring are dependent on the field winding connection used.

8.4 INDUCED VOLTAGE OF THE ARMATURE WINDING

Consider the two-pole elementary machine with one armature coil shown in Fig. 8.7. The spatial distribution of the airgap flux density due to the two main poles is alternating and periodic. As drawn, the radial B field directed radially outwards from the armature under the south pole region is considered positive and that directed radially inwards under the north pole region is considered negative. With the armature rotating in the counter-clockwise direction, the direction of the induced emfs in the two coil sides can be deduced from the vector product, $\vec{v} \times \vec{B}$. In the absence of the commutator and brush connection, the induced emf of the coil as it rotates around in a steady angular velocity of ω will have a time

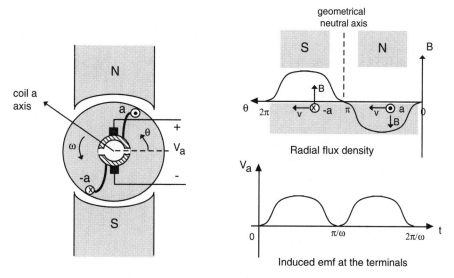

Figure 8.7   Induced emf in the coil of an elementary machine.

waveform that is similar in shape to the spatial distribution of B. With the commutator and brush, the connections to the external circuit are interchanged whenever the coil axis crosses the angular positions of $\pi/2$ and $3\pi/2$. As a result, the voltage at the external terminals is a rectified version of the actual coil voltage.

For a machine with P poles and a mean airgap diameter of D, one pole pitch, τ, will span an arc length of $\pi D/2P$. The flux per pole is given by

$$\phi = \int_0^\tau B_e(\theta)Lr d\theta = \bar{B}_e L\tau \qquad Wb \tag{8.4}$$

where $\bar{B}_e$ is the mean value of $B_e(\theta)$ over a pole, that is

$$\bar{B}_e = \frac{1}{\tau}\int_0^\tau B_e(\theta)r d\theta \qquad Wb/m^2 \tag{8.5}$$

Consider the instant when the two coil sides of the armature coil of n_c turns are located at α and $\alpha + \beta$. The flux linked by the coil at that instant is given by

$$\lambda_c = n_c L \int_\alpha^{\alpha+\beta} B_e(\theta)r d\theta \qquad Wb.\ turn \tag{8.6}$$

The voltage induced around the whole coil is given by the time derivative of the flux it links, that is

$$e_c = \frac{d\lambda_c}{dt} = n_c L \frac{d}{dt}\int_\alpha^{\alpha+\beta} B_e(\theta)r d\theta \qquad V \tag{8.7}$$

The derivative of the integral term in the above expression can be simplified using

$$\frac{d}{dt}\int_\alpha^{\alpha+\beta} f(\theta,t)d\theta = \int_\alpha^{\alpha+\beta} \frac{\partial}{\partial t} f(\theta,t)d\theta + f(\alpha+\beta,t)\frac{d(\alpha+\beta)}{dt} - f(\alpha,t)\frac{d\alpha}{dt} \tag{8.8}$$

For an airgap flux density distribution, $B_e(\theta)$, that is time-invariant, the first term is zero. With the rigid coil rotated around at an angular speed of ω, $d(\alpha + \beta)/dt = d\alpha/dt = \omega$, Eq. 8.7 simplifies to

$$e_c = n_c L r \omega \{B_e(\alpha + \beta) - B_e(\alpha)\} \quad V \quad (8.9)$$

For a complete armature winding that has a parallel paths between brushes of opposite polarity, and n_s identical coils of n_c turns each in series, the resultant voltage between brushes is the sum of the series coil voltages:

$$V_a = \sum_{i=1}^{n_s} e_{ci} = n_c r \omega \sum_{i=1}^{n_s} \{B_e(\alpha_i + \beta) - B_e(\alpha_i)\} \quad V \quad (8.10)$$

where α_i is the angular location of one of the coil sides of the ith coil in the series string.

For coils that span exactly one pole pitch, β is equal to τ. Since $B_e(\alpha_i + \tau) = -B_e(\alpha_i)$ and $\sum_{i=1}^{n_s} B_e(\alpha_i)$ can be replaced by $n_s \bar{B}_e$, where $\bar{B}_e$ is the mean value of the flux density. Substituting these into Eq. 8.10, we will obtain the following expression for the average value of the induced emf of the armature:

$$V_a = 2 n_c r \omega_m n_s \bar{B}_e \quad V \quad (8.11)$$

Alternatively, we can expressed V_a in terms of the total number of active conductors and the flux per pole, using the fact that the armature has a total of an_s coils, or a total of $an_s n_c$ turns. Since a turn is formed by two conductor sides, the total number of active conductors, Z, is $2an_s n_c$. Using the above expression for Z and those of $\phi = \tau L \bar{B}_e$ and $\tau = \pi D/2P = \pi r/P$, Eq. 8.11 can be rewritten as

$$V_a = \frac{PZ}{2\pi a} \omega \phi = k_a \omega \phi \quad V \quad (8.12)$$

where $k_a = PZ/2\pi a$. Although the above expressions for V_a and k_a have been derived for the condition where brushes are set on the neutral axis and the coils are full-pitch, the same form of V_a can still be used for other brush locations and coil pitches, but that of k_a will have to be modified.

8.5 ELECTROMAGNETIC TORQUE

Likewise, the resultant torque developed on the armature can be obtained by summing the torque on each coil. Consider motoring operation when an externally applied voltage circulates a current of I_c through the single coil in the direction against the induced emf, as shown in Fig. 8.8. The cross on coil side a in this figure denotes current flow into the paper and the dot on coil side $-a$ denotes current flow out of the paper. The force on the coil side located at position α where the flux density is $B_e(\alpha)$ is given by

$$\vec{f_a} = \int I_c \vec{dl} \times \vec{B}_e(\alpha) \quad N \quad (8.13)$$

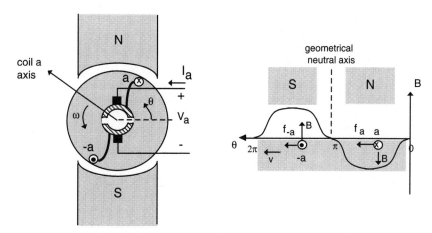

Figure 8.8   Forces on coil sides for motoring operation.

With $d\vec{l}$ perpendicular to $\vec{B}_e(\alpha)$, the force, $\vec{f}_a$, in Fig. 8.8 will be tangential. And if the coil spans one pole pitch, the force on the two sides will be equal, that is $f_a = f_{-a} = B_e(\alpha)I_cL$. Note that f_a and f_{-a} are in the direction of rotation, that is the developed tangential forces on the coil are in the direction of rotation. The torque developed may be expressed as

$$T_a = 2r B_e(\alpha) I_c L \qquad N.m \tag{8.14}$$

The resultant torque on an armature winding having a parallel current paths and n_s coils in each current path is

$$T_{em} = a \sum_{i=1}^{n_s} 2r B_e(\alpha_i) I_c L \qquad N.m \tag{8.15}$$

Using the relations $I_a = aI_c$, $\tau = 2\pi r/P$, and $Z = 2an_sn_c$, Eq. 8.15 can be rewritten as

$$T_{em} = \frac{PZ}{2\pi a}\phi I_a = k_a\phi I_a \qquad N.m \tag{8.16}$$

If we ignore saturation, ϕ may be expressed as proportional to the excitation current, I_e, that is $\phi = k_gI_e$, and the torque developed can then be written as $T_{em} = k_a'I_aI_e$, where $k_a' = k_gk_a$.

Note that under the ideal condition considered,

$$E_aI_a = k_a\phi I_a\omega = T_{em}\omega_m \qquad W \tag{8.17}$$

In other words, the electrical input power to the armature is equal to its developed mechanical output power.

8.6 ARMATURE REACTION

The current in an armature winding produces its own field, commonly referred to as the armature reaction (*AR*) field. The airgap field intensity from the armature winding current can be determined by superposition of the component field intensities from its coils. If the reluctivity of iron on both sides of the airgap is ignored, the field intensity, H_c, in the airgap due to a single coil of n_c turns is given by the line integral:

$$\oint_\Gamma H_c(x)dl = 2H_c(x)g(x) = n_c I_c \qquad A.turn \qquad (8.18)$$

where I_c is the coil current and $g(x)$ is the effective airgap length along the chosen path. By choosing a Γ that is symmetric about a coil side, we can show that the mmf drop across the airgap on either side of Γ is equal and opposite, that is

$$F(x) = H_c(x)g = \frac{n_c I_c}{2} \qquad N.m \qquad (8.19)$$

Figure 8.9 shows the spatial mmf distribution established by coils 1 and 2 of the armature winding. By superposition, the resultant spatial mmf distribution of all the coils of the armature winding is a stepped triangular wave. The resultant airgap field, H or B in the interpolar region, is lowered by the larger value of gap in that region. Note that the current distribution shown is for the case where the brushes are located at the geometric neutral. For such a case, the resulting *AR* field is in the direction of the main field under one-half pole, but is in opposition to the main field in the other half. When iron saturation is ignored, the cross-magnetizing *AR* field has no effect on the net value of the flux under a pole. But with the nonlinear effects of iron saturation, the distortion introduced by the *AR* field will result in some loss of the flux per pole.

Figure 8.10 shows the resultant B field distribution by a bold dark line; the top portions, shown lightly shaded, are flattened by iron saturation. Without iron saturation, the cross-magnetizing mmf of the armature winding results in a shift in the magnetic neutral axis, but no reduction in the flux under a pole. With saturation, however, the tops of the B field will be flattened, resulting in some reduction of the flux under each pole. The reduction in flux per pole, ϕ, is dependent on the relative position of the main and *AR* fields, and the degree of saturation.

The cross field of the armature reaction skews the resultant field distribution. Over the region where it reinforces, the increase in resultant flux is not proportional to the increase in mmf when saturation comes into effect. From the point of view of commutation, it is undesirable because it displaces the magnetic neutral from the geometric neutral axis where the brushes would normally be. Additionally, the polarity of residual resultant flux at the geometric neutral axis (g.n.a) induces an emf in the coil undergoing commutation that tends to prolong the flow of current in the same direction, in effect retarding the commutation process.

Accounting for *AR*.   With armature reaction, the net mmf in a compound machine that has a main shunt field of N_f turns carrying I_f A and a series field of N_s turns carrying the armature current, I_a, will be given by

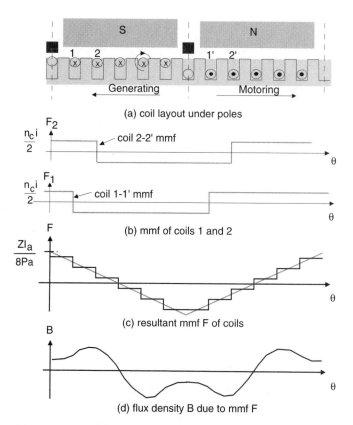

Figure 8.9   Airgap field due to armature current.

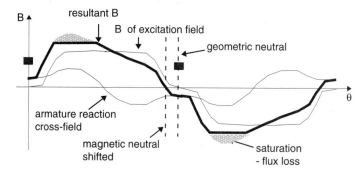

Figure 8.10   Resultant flux density with armature reaction.

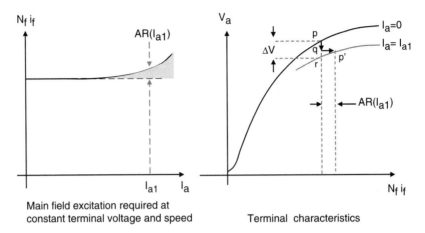

Figure 8.11 Equivalent AR drop.

$$\text{Net mmf} = N_f I_f \pm N_s I_a - AR \qquad A.turn \qquad (8.20)$$

where the $\pm$ sign is positive for cumulative compound and negative for differential compound.

In steady-state-type calculations, an equivalent armature voltage drop can be used to represent the combined effect of armature reaction and resistive drops in the brush and winding. If we have the experimental data shown on the left of Fig. 8.11, where the machine's terminal voltage is maintained constant by raising the main field excitation mmf as armature load current increases, we can obtain the loaded voltage characteristic for a certain load current from the open-circuit curve. For example, for a load current of I_{a1}, we can determine from the left figure the equivalent AR, $AR(I_{a1})$, in the main field, AT. Starting with a point, p say on the open-circuit curve, subtracting the brush and winding resistance drop at I_{a1}, represented by pq, and then adding the equivalent AR, that is $AR(I_{a1})$, we would obtain the point q on the loaded voltage curve. The rest of the loaded $V_a(I_{a1})$ curve at the same current level of I_{a1} can be constructed by repeating the procedure with other points on the open-circuit curve. Alternatively, the loaded voltage curves at various armature current values may be determined experimentally.

For steady-state calculations, we can then use pr, that is ΔV, as the equivalent voltage drop at a load current of I_{a1} when the main field excitation yields an open-circuit armature voltage corresponding to that of point p.

8.7 COMMUTATION

As the armature rotates, coils of the armature winding will occupy new positions in the various current paths. The currents in those coils that are switching from one current path to another have to be reversed in direction when going from one current path to the other by a process known as commutation. Commutation of a coil current takes place as the corresponding commutator segments of the coil move under a brush. The coils undergoing commutation are short-circuited by the brushes. During commutation, the behavior of the coil current is influenced by several factors, including the changing contact resistance under the brush, the emf due to self- and mutual inductances that react against the change of current, and the emf due to armature reaction.

Ideally, if the coil current were to change at a constant rate when the rotor was rotating at constant speed, the current density under the brush would remain uniform during commutation. As shown in Fig. 8.12, the division of the brush current between adjacent commutator segments will vary linearly when the division of the brush current is strictly proportional to the contact area. That is only so when commutation is strictly in accordance to the ohmic resistance of the contact area of the brush, The real condition,

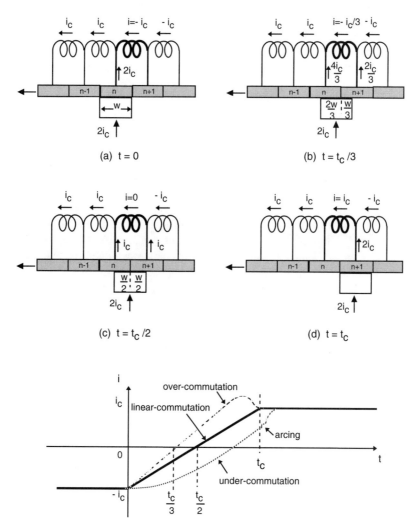

Figure 8.12   Commutation.

however, is complicated by the nonlinear nature of the film resistance between the brush and commutator segments, the induced emfs in the commutating coils. The induced voltages in the commutating coils include those from inductances, self- and mutual of the coil or coils undergoing commutation at the same time, that from eddy currents flows in surrounding conductor and iron, and that from flux density produced by armature reaction. Coil inductances will oppose the change in current, giving rise to the condition referred to as under-commutation in Fig. 8.12. On the other hand, the induced emf of the commutating coils may aid the change in current, which if in excess can contribute to the over-commutation condition shown in the same figure.

Resistive commutation, which refers to the situation where brush resistance dominates inductive effects, is used in very small machines where brush resistance can be made high. Carbon brushes work well in the current density range of 40 to $50 A/in^2$; that of copper brushes is 150-200 A/in^2.

The inductive voltage drop from the winding inductance can be compensated for by creating an induced emf of the appropriate polarity in the commutating coils. A close examination of the resultant B field distribution shown in Fig. 8.13 will show that the correct polarity of induced emf in the commutating coils can be obtained by shifting the position of the brushes riding on the commutator to advance or retard the commutation.

The required brush shift from the geometric neutral position is in the forward direction for a generator and backward for a motor if the coil undergoing commutation is to have an induced speed voltage that aids the change in current, countering that due to inductances. The magnitude of the induced speed voltage is dependent on the angular speed and flux density in the commutation zone. The latter is also a function of armature reaction, which if uncompensated, will distort the airgap flux density, shifting the magnetic neutral away from the geometric neutral axis by a varying amount. Shifting of the brushes will also alter the position of the armature reaction field, B_a, relative to the main excitation field, B_f.

Figure 8.13b shows the coils of the armature winding divided into two components: one produces an armature reaction field that is cross-magnetizing and the other, demagnetizing. The number of demagnetizing AT per pole is given by

$$AT_d = \frac{1}{2} \frac{4\alpha}{2\pi} \frac{Z}{2} \frac{I_a}{a} = \frac{1}{2} \frac{2\alpha}{\pi} \frac{Z}{P} \frac{I_a}{a} \qquad A.turn \qquad (8.21)$$

The number of cross-magnetizing AT per pole is

$$AT_c = \left(1 - \frac{2\alpha}{\pi}\right) \frac{ZI_a}{2Pa} \qquad A.turn \qquad (8.22)$$

Brush shift is mainly limited to small, inexpensive machines for several reasons. As shown in Fig. 8.13b, it produces a demagnetizing component to the main field, the direction of shift is dependent on armature current direction (motoring or generating), and the amount of shift is dependent on the magnitude of the armature current or loading.

8.7.1 Interpoles

In larger machines, interpoles that are excited by the armature current can be used to induce a voltage of proper polarity and magnitude in the commutating coils that will counteract

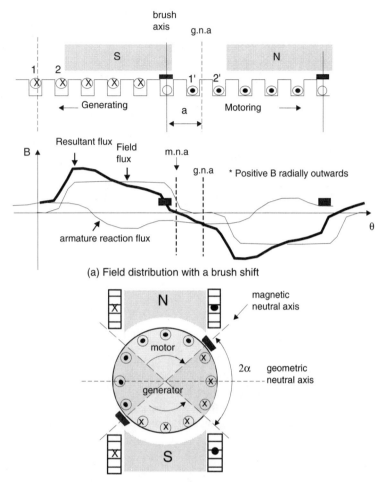

(a) Field distribution with a brush shift

(b) Demagnetizing sector of armature mmf shown shaded

Figure 8.13   Effects of brush shift on field.

the delay in current change from the coil inductance. Interpoles are positioned in between main poles, usually centered over the neutral axes where coil sides of the coil undergoing commutation are located. As shown in Fig. 8.14, the magnitude of the interpole excitation, which is proportional to the armature current being commutated, creates a resultant field for the proper polarity of induced emf in the armature winding that will aid commutation. The polarity of the induced emf changes with the direction of armature current flow; as such, it will function as desired in both motoring and generating operations.

If an interpole is to be positioned over each and every coil side undergoing commutation, the number of interpoles for a lap-wound armature will be equal to the number of main poles. In practice, half that number is sufficient because the required voltage can be induced in just one of the coil sides of the coil undergoing commutation. Similarly,

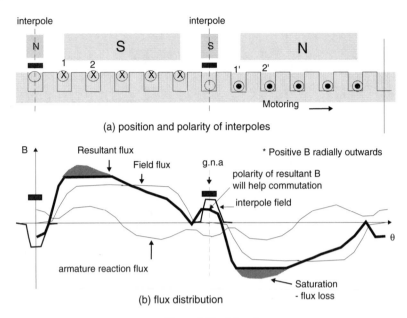

Figure 8.14   Interpole.

we see in Fig. 8.4 that for a wave-wound armature, the brushes short-circuit $P/2$ coils in series. With two brushes, only two interpoles, each to provide sufficient induced emf to commutate the current of the $P/2$ coils in series, will suffice.

Excitation of each interpole must be sufficient to balance out the armature reaction, $ZI_a/2Pa$, and to drive interpole flux across the reluctances of its path (gap + iron in pole + iron in armature). Neglecting the reluctances of iron, the estimated mmf from an interpole is

$$F_{interpole} = N_{ip}I_a = \frac{ZI_a}{2Pa} + k_c\frac{B_{ip}g_{ip}}{\mu_0} \qquad A.turn \tag{8.23}$$

where N_{ip} is the number of turns of the interpole winding, B_{ip}, the mean flux density, and k_c, the product of two factors: a Carter's coefficient to allow for slotting and a peak-to-mean form factor of the flux distribution under the interpole. In practice, excess excitation can be trimmed by using a non-inductive shunt across the interpole winding.

8.7.2 Compensating Windings

Large and rapid changes in armature current can in some instances cause large voltage differences between coils connected to adjacent commutator segments. When the voltage between commutator segments is in excess of 30V, there is a danger of flashover, which could ultimately lead to flashover between brush arms. An expensive but effective remedy is to provide a compensating winding with a magnetic axis that is along that of the armature and also excited by the armature current. The compensating winding is normally embedded in pole face slots; its mmf is supposed to cancel the armature reaction mmf under the

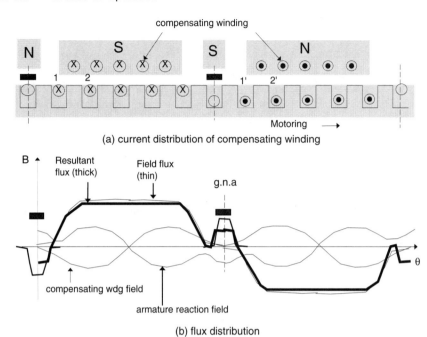

(a) current distribution of compensating winding

(b) flux distribution

Figure 8.15   Field distribution with interpole and compensating windings.

main pole arc. Figure 8.15 shows the layout and field distribution with interpole and compensating windings.

The required mmf of the compensating winding to cancel the armature reaction can be estimated from

$$F_{comp} = \frac{I_a Z}{2Pa} \times \frac{\text{pole arc}}{\text{pole pitch}} \qquad A.turn \qquad (8.24)$$

8.8 MODES OF OPERATION

In terms of energy flow, an operating machine has not only an input and an output, but also energy storage capability in the form of magnetic energy in the fields and kinetic energy in the rotating components. Based on the direction of energy flow, we can identify two common modes of operation. The machine is said to be motoring when energy drawn from the external electrical source connected to its armature winding terminals is converted to perform mechanical work and/or increase the kinetic energy of the rotor. It is said to be generating when energy from a mechanical prime mover driving the rotor is converted to electrical energy flowing out of the armature terminals into the external circuit.

8.8.1 Motoring

In the motoring mode of operation shown in Fig. 8.16, the direct voltage from an external source applied to the terminals of the armature winding injects an armature current, I_a,

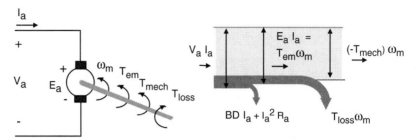

Figure 8.16 Motoring.

against the armature induced emf, E_a. The current, I_a, interacts with the airgap flux to produce an electromagnetic torque, T_{em}. In motoring, the rotor will accelerate in the direction of T_{em}, reaching a steady-state speed at which T_{em} is balanced by the friction and load torques, T_{loss} and T_{mech}.

With the induced emf, E_a, of the armature winding opposing the flow of current, I_a, in motoring, the voltage equation of the armature winding is

$$V_a = E_a + R_a I_a + V_{brush} + L_{aq} \frac{dI_a}{dt} \qquad V \qquad (8.25)$$

where R_a is the resistance of the armature winding, V_{brush} is the total brush drop, and L_{aq} is the armature winding inductance. For motoring, positive I_a is to flow into the armature winding at the positive terminal of V_a, that is, net power flows from the external voltage source into the armature winding. The actual brush drop is a complex function of brush material, brush pressure, and current density and temperature; but, for simple calculations, an approximation of one volt drop per brush contact is considered reasonable.

Equating the acceleration torque of the rotor to its inertia torque, we have

$$T_{em} - T_{loss} + T_{mech} = J \frac{d\omega_m}{dt} \qquad N.m \qquad (8.26)$$

where T_{mech} is the externally applied mechanical torque on the shaft in the direction of rotation, T_{em}, the electromagnetic torque developed by the machine, and T_{loss}, the equivalent torque representing friction and windage and stray load losses. The value of T_{em} in the direction of rotation shown is positive for motoring and negative for generating, that of T_{mech} is negative for motoring and positive for generating.

Multiplying the stator voltage equation by I_a, we obtain

$$V_a I_a = E_a I_a + R_a I_a^2 + V_{brush} I_a + \frac{d(L_{aq} I_a^2/2)}{dt} \qquad W \qquad (8.27)$$

Similarly, multiplying through the rotor equation by ω_m, we have

$$P_{em} + P_{mech} - P_{losses} = \frac{d(J\omega_m^2/2)}{dt} \qquad W \qquad (8.28)$$

Equating P_{em} with $E_a I_a$, Eqs. 8.27 and 8.28 can be combined to give the following relation of the power flow through the machine:

$$\underbrace{V_a I_a + P_{mech}}_{input\ power} = \underbrace{\frac{d(J\omega_m^2/2 + L_{aq}^2 I_a^2/2)}{dt}}_{change\ in\ stored\ energy} + \underbrace{I_a^2 R_a + V_{brush} I_a + P_{losses}}_{losses} \qquad W \qquad (8.29)$$

In steady-state, the net rate of change in stored energy will become zero and the flow of power through the motor may then be presented as shown on the right side of Fig. 8.16.

8.8.2 Generating

In the generating mode of operation, a mechanical prime mover provides the externally applied torque, T_{mech}, to rotate the rotor forward. In the presence of a non-zero flux density in the airgap from either residual magnetism or external excitation, speed voltages will be induced in the rotating coils of the armature winding. An armature current, I_{ag}, as shown in Fig. 8.17, will flow in the armature winding if the external circuit connected to the armature winding has a closed path, delivering power from the induced emf to the rest of the armature circuit.

The torque and power equations of rotor motion can be the same as those given by Eqs. 8.26 and 8.28 for motoring, with I_a replaced by $-I_{ag}$. In generating, the value of T_{mech} and P_{mech} will be positive and those of T_{em}, in the direction of rotation, and P_{em} will be negative. Thus, T_{mech} is being opposed by T_{em} and T_{loss}, in the direction of rotation.

The voltage equation of the armature winding for generating in terms of I_{ag} is

$$E_a = V_a + R_a I_{ag} + V_{brush} + L_{aq}\frac{dI_{ag}}{dt} \qquad V \qquad (8.30)$$

The above voltage equation is the same as Eq. 8.25 when I_a is replaced by $-I_{ag}$. In generating, net power flows from the armature winding to the external circuit connected across the armature terminals. Multiplying Eq. 8.30 by I_{ag}, we obtain

$$E_a I_{ag} = V_a I_{ag} + R_a I_{ag}^2 + V_{brush} I_{ag} + \frac{d(L_{aq} I_{ag}^2/2)}{dt} \qquad W \qquad (8.31)$$

Equating $-P_{em}$ in Eq. 8.28 with $E_a I_{ag}$ in Eq. 8.31, we obtain the following expression for the power flow through the machine for generating:

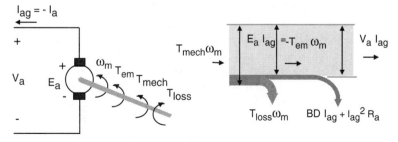

Figure 8.17   Generating.

$$\underbrace{P_{mech}}_{input\ power} - \underbrace{R_a I_{ag}^2 + V_{brush} I_{ag} + P_{losses}}_{losses} - \underbrace{\frac{d(J\omega_m^2/2 + L_{aq} I_{ag}^2/2)}{dt}}_{change\ in\ stored\ energy} = \underbrace{V_a I_{ag}}_{output\ power} \quad (8.32)$$

In steady-state, the net or average change in stored energy is zero and the flow of power through the generator can be depicted by the diagram on the right side of Figure 8.17.

8.9 FEASIBLE TORQUE-SPEED REGION

Using the motoring convention, the relations between internal emf and the armature circuit component voltages of a dc machine operating in steady-state at a rotor speed of ω_m are

$$E_a = k_a\phi\omega_m = V_a - R_a I_a - V_{brush} \quad V \quad (8.33)$$

Ignoring brush drops and dividing through by $k_a\phi$, and using the relation $T_{em} = k_a\phi I_a$ to replace I_a on the right-hand side, we can express the rotor speed in terms of the developed torque as

$$\begin{aligned}
\omega_m &= \frac{V_a}{k_a\phi} - \frac{R_a T_{em}}{k_a^2\phi^2} \\
&= \omega_{mo} - \frac{R_a T_{em}}{k_a^2\phi^2} \quad rad/s
\end{aligned} \quad (8.34)$$

In the above expression, ω_{mo} is the no-load speed when $T_{em} = 0$. When V_a, R_a, and ϕ are held constant, the above expression is a linear relationship between ω_m and T_{em}. The intercept on the speed axis corresponds to ω_{mo} and the slope of the line is proportional to R_a. The figures on the left-hand side of Fig. 8.18 show the direction and polarity of the main variables of interest using motoring convention with counter-clockwise rotation of ω_m considered positive. On the $T_{em}\omega_m$ plane, forward motoring is in the first quadrant, forward generating is in the second quadrant, reverse motoring is in the third quadrant, and reverse generating is in the fourth quadrant. The relations of $E_a = k_a\phi\omega_m$ and $T_{em} = k_a\phi I_a$ hold true in algebraic terms. For example, a change in sign of ω_m will be reflected by a corresponding change in E_a.

The dark line in the diagram on the right side of Fig. 8.18 is the steady-state torque-speed characteristic of the machine with V_a and $k_a\phi$ held fixed. A family of parallel characteristics like the ones shown in dashed lines for $V_{a1} > V_{a2} > \ldots > V_{an}$ can be generated by varying the value of the applied voltage, V_a. In steady-state motoring, T_{em} is balanced by the sum of T_{mech} and T_{loss}. When V_a and $k_a\phi$ are held fixed and the load torque varies, the steady-state operating point slides along the linear operating characteristic.

For points on the operating characteristics in the second quadrant, their T_{em} are negative. The direction of the developed torque is opposite to that of the rotor's rotation. Energy from the rotor's momentum or the external applied torque is being converted by the machine to electrical energy. With ϕ unchanged, I_a will reverse along with T_{em}; thus, the direction of power flow in the armature circuit will reverse too. A temporary venture into the second quadrant, however, is usually referred to as regenerative braking since the motor

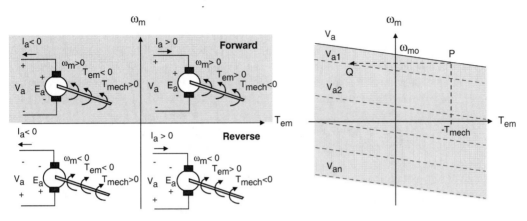

Figure 8.18   Speed-torque region.

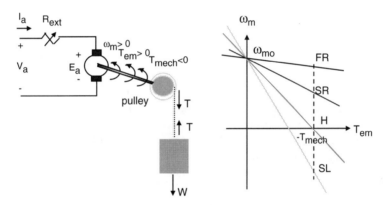

Figure 8.19   Operating a hoist.

develops a torque that opposes a further increase in speed, as in the case of the rotor being temporarily overhauled by the load torque. Sustained operation in the second quadrant is simply referred to as generating.

With flux ϕ held constant, the vertical intercept and slope of the operating characteristic can be varied by changing the applied voltage, V_a, and the armature circuit resistance respectively. For example, Fig. 8.19 shows an operating sequence of a hoist raising and lowering a load. Control of the machine in this case is assumed to be confined to changing the armature circuit resistance by adjusting a variable external resistance, R_{ext}, in the armature circuit. Increasing R_{ext} increases the slope of speed vs. torque line. Point *FR* corresponds to the initial stage of rapidly raising the load with a R_{ext} of zero. As the load approaches a certain height, the speed can be reduced by gradually increasing the value of the external circuit resistance, R_{ext}, for example, to point SR. At H, the load has reached the desired height and the dc machine is providing a braking torque to hold the load steady at that position against the gravitational force. Lowering of the load is accomplished by unwinding in the opposite direction of rotation, that is $(-\omega_m)$, in a controlled manner. For example, at point SL, obtained here by using an even larger R_{ext}, the machine works as a brake where the potential energy of the load is absorbed by the machine while at the same time energy is also fed into the machine from the electrical supply – all this energy is dissipated as heat in the resistance of the armature and braking resistance. The machine is, in fact, regenerating in the reverse direction of rotation.

Alternatively, to go from motoring to generating momentarily, we could lower V_a to V_{a1} to obtain the new operating characteristic shown in Fig. 8.18 by the dashed line. Assuming that the rotor inertia is finite and neglecting the electromechanical transients, the new operating point, Q, immediately following the change in V_a will have the same operating speed as the initial motoring point, P. However, T_{em} at Q in Fig. 8.18 is negative, that is, of opposite sign to that at P. The reversal of T_{em} will slow the rotor down unless the rotor speed is being sustained by T_{mech} from a mechanical prime mover, as in the case of prolonged generating operation.

8.10 BRAKING

Here, braking refers to the operating condition where the rotor is deliberately being decelerated by electrical means. Electrical braking is usually used to achieve rapid stops in emergency situations, accurate stops, as in lift and machine tool control, or to counteract overhauling by gravitational forces, as in hoist and traction. Electrical braking complements mechanical braking by reducing the wear and size of the mechanical brakes, by providing smoother and more reliable braking, and by boosting overall efficiency using a regenerative method that recovers a portion of the change in kinetic energy.

The changeover from forward motoring operation to braking can be achieved in several ways: Rapid braking can be obtained by reversing the polarity of the applied voltage while simultaneously inserting an external braking resistance to the armature circuit to limit the armature current. Since the direction of rotation and flux are the same as motoring, the polarity of the induced armature voltage will remain the same as in motoring. However, the direction of I_a will be opposite to that in motoring with the reversal of the applied voltage, V_a; as such, the direction of T_{em} too will be opposite to that in motoring, that is, T_{em} in braking will be in a direction opposite to ω_m.

The voltage equation of the armature winding in braking with the reversal of V_a and the insertion of R_{ext} is

$$-V_a = E_a - (R_{ext} + R_a)I_a - V_{brush} - L_{aq}\frac{dI_a}{dt} \qquad V \qquad (8.35)$$

Using the same torque equation as given by Eq. 8.26 for motoring, the flow of power through the machine can be shown to be

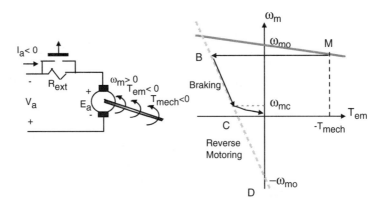

Figure 8.20   Braking by reversing V_a and inserting R_{ext}.

$$\underbrace{\frac{d(J\omega_m^2/2 + L_{aq}^2 I_a^2/2)}{dt}}_{change\ in\ stored\ energy} = \underbrace{P_{mech}}_{mech\ power} + \underbrace{-V_a I_a}_{\substack{electrical \\ input\ power}} \underbrace{-I_a^2 R_a + V_{brush} I_a + P_{losses}}_{losses} \qquad W \qquad (8.36)$$

In Fig. 8.20, the solid line is the operating characteristic of the machine that is initially motoring at point M, and the dashed line is that for braking with V_a flipped and R_{ext} inserted. The latter has a no-load speed, $-\omega_{mo}$ or $-V_a/k_a\phi$, that is equal and opposite to that when the machine is motoring. Point B is the operating point immediately following the reversal of V_a, the speed being continuous with a finite inertia rotor must be equal immediately before and after the changeover. Along BCD, where both I_a and V_a are reversed, electrical power is still supplied to the machine. In BC, ω_m and therefore E_a remain in the same direction as in M. However, T_{em} which is reversed along with I_a, now acts in the opposite direction to the inertia torque, slowing down the rotor. Kinetic energy from the rotor and electrical input energy from the armature input are dissipated as heat in R_a and R_{ext}. The dc machine functions as a brake until its speed becomes zero at C. Further along CD, the direction of rotor rotation reverses and the machine motors up in the reverse direction. With shunt self-excitation, however, there is a critical speed, ω_{mc}, below which the machine can no longer maintain self-excitation with a given field resistance; effective electrical braking will cease and the motor will coast to a stop. With series self-excitation, to maintain self-excitation, the reversed I_a must continue to aid the residual flux. This is satisfied by either reversing the field or the armature connection.

Dynamic braking.   Dynamic braking refers to braking in the generating mode where V_a is zero. It is widely used in electric traction drives. The dynamic braking characteristic for a motor with fixed excitation on the speed-torque plane is a line through the origin since V_a is zero. The negative slope of the characteristic is proportional to $R_a + R_{ext}$. A changeover from motoring to dynamic braking can simply be effected by

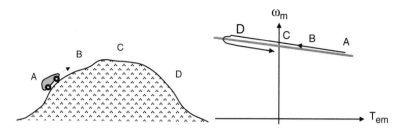

Figure 8.21   Regenerative braking.

disconnecting the supply voltage and replacing it with an external braking resistor, R_{ext}, across the armature terminals. With the rotor speed continuing in the same direction, the sign of E_a will be the same. With V_a removed, I_a will now be circulated by E_a, or the machine will operate as a generator. The recovered kinetic energy from the rotor assembly will be dissipated in R_a and R_{ext}.

Regenerative braking.   Regenerative braking refers to braking with energy recuperation above the no-load speed, ω_{mo}. For example, to illustrate when such an operating mode may be advantageous, let's consider an electric vehicle moving over a hilly terrain as shown on the left side of Fig. 8.21. On the uphill climb, at point A, the machine must produce torque to overcome friction, windage, and gravitation. On reaching B, the gravitational force component becomes smaller because the incline is less. With a fixed terminal voltage, I_a falls off and ω_m rises, and the operating point shifts from A to B along the torque-speed line of the machine shown on the right side of Fig. 8.21.

On the downhill run, the gravitational force component helps to overcome the frictional forces. When these are equal, $I_a = 0$ (no-load operation). As the downhill slope becomes steeper, the gravitational force component can be greater than the frictional forces, overhauling the vehicle forward. The value of induced emf, E_a, rises above that of the terminal voltage, V_a, reversing the flow of I_a and causing a reversal in power flow from the armature to the external circuit. The torque, T_{em}, reverses in direction along with I_a, retarding the acceleration in the forward direction. Thus, part of the electrical energy that was used to raise the potential energy of the train in the uphill climb can be recovered electrically during the downhill run by regenerative braking.

8.11 SPEED CONTROL

Within limits, as can be seen from the equation below, the operating speed of a dc motor can be controlled in a straightforward manner by varying V_a, ϕ, or R_a:

$$\omega_m = \frac{V_a - R_a I_a - V_{brush}}{k_a \phi} \qquad rad/s \qquad (8.37)$$

8.11.1 Speed Control by Varying Circuit Resistance

The effects of external resistance in the armature circuit on the operating speed of a separately excited motor can be determined from the following equation that is a modified form of Eq. 8.37:

$$\omega_m = \underbrace{\frac{V_a}{k_a \phi}}_{no-load\ speed} - \frac{(R_a + R_{ext})T_{em}}{k_a^2 \phi^2} \qquad rad/s. \tag{8.38}$$

When the parameters in the first term are held constant, the operating speed can only be adjusted downwards by varying the external resistance, R_{ext}. With V_a, ϕ and T_{em} constant, the output power will decrease along with speed though the input power, $V_a I_a$, remains constant; as such, the efficiency will decrease with speed. Keeping the machine within an allowable temperature rise can be a problem at low speeds as draft cooling decreases with speed, unless separate forced cooling is used. Besides operational efficiency, the resistor, R_{ext}, for continuous operation of large motors can be bulky and expensive. Nevertheless, its simplicity and low initial cost can be attractive, especially for intermittent-duty drives where the decrease in efficiency and cooling at low speed for intermittent operation may not be of as much concern.

That for a series excited motor with $\phi = k_s I_a$ is of the form:

$$\omega_m = \frac{V_a}{\sqrt{k_a k_s T_{em}}} - (R_a + R_s + R_{ext})\sqrt{\frac{T_{em}}{k_a k_s}} \qquad rad/s. \tag{8.39}$$

R_{ext} and R_s are the resistances of the external resistor and the series field winding, respectively.

8.11.2 Speed Control by Varying Excitation Flux

The effect of a change in excitation flux on the operating speed of a dc machine can be deduced from the ratio of the induced emfs at two different operating conditions, that is

$$\frac{\omega_{m1}}{\omega_{m2}} = \frac{E_{a1}/\phi_1}{E_{a2}/\phi_2} = \frac{\phi_2}{\phi_1}\frac{V_a - R_a I_{a1} - V_{brush}}{V_a - R_a I_{a2} - V_{brush}} \approx \frac{\phi_2}{\phi_1} \tag{8.40}$$

If the load torque remains constant, both input and output power increase with motor speed; thus, motor efficiency is maintained when ϕ is varied. Raising the speed by reducing flux and maintaining the torque constant will require a higher I_a, which, in continuous operation, must stay within some allowable value. Commutation will also become more difficult because I_a and ω_m will increase simultaneously. At very low excitation, the armature reaction of I_a required to develop a certain torque may reduce the ϕ by such a proportion as to cause the speed to increase with I_a, giving rise to an ascending speed-torque characteristic that is unstable for motor operation. To avoid this situation, the range of speed control by reducing excitation is usually limited to one to three times the base speed of the machine (base speed being the speed where rated excitation yields the rated voltage).

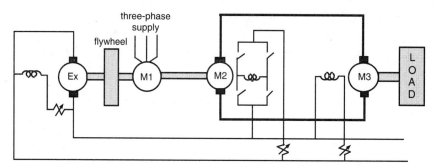

Figure 8.22   Ward-Leonard system.

Decreasing speed by increasing excitation, on the other hand, has a rather limited range because excitation is usually kept close to its rated value to fully utilize the iron. Speed control by changing excitation is economical because the power losses in the field rheostat are small and the rheostat required is compact and inexpensive.

8.11.3 Speed Control by Varying Applied Voltage

As evident from Eq. 8.37, the operating speed can also be varied by adjusting V_a. An adjustable dc supply of V_a can be obtained from another dc machine or from a static converter. Figure 8.22 shows a Ward-Leonard system, where the main dc motor, M3, is fed with an adjustable dc supply generated by another dc machine, M2, that is driven by an ac motor, M1. In low-power applications, M1 can be a squirrel-cage induction motor, but for higher power applications, a wound-rotor induction motor, synchronous motor, or a diesel engine would be needed. The exciter, Ex, provides a constant excitation voltage. Speed reversal of the dc motor, M3, is obtained by reversing the polarity of the excitation to the dc generator, M2.

Some of the advantages of such a system are:

1. Wide range of speed control, typically 25:1, often limited by the useful range of M2's output voltage of about 10:1 because of residual magnetic flux and the steepness of the speed-torque characteristic of motor M1.
2. Fast acceleration of high inertia loads possible with I_a set at the maximum allowable.
3. Regenerative load braking can be effected simply by lowering generator M2's voltage.
4. Dispensation of lossy and bulky resistance and switches in the armature circuit for starting and braking.

The disadvantages are in the number of full-size machines in the chain, which raises the cost and lowers the overall efficiency.

Electronic control.   Figure 8.23 shows an electronic speed control dc machine drive with outer speed and excitation regulation loops operating through a faster current

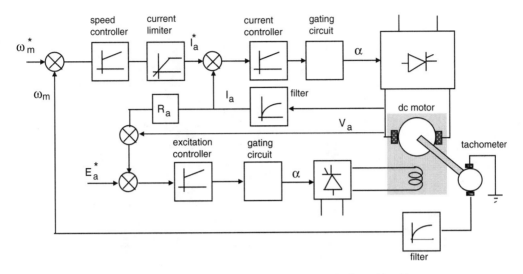

Figure 8.23   Current and speed control loops of a dc machine drive.

regulation inner loop. The current control via α is discrete (six times per cycle for six pulse units). The speed response of the drive is mainly governed by the mechanical time constant. The inner current loop response is determined mainly by the armature circuit time constant, which is usually much smaller than the mechanical time constant. The feedback current value can be obtained from a Hall effect sensor, a non-inductive resistive shunt and isolation amplifier, or by measuring the ac-side current using a combination of current transformers (C.T.s), rectifier, and filter.

Speed sensing based on the back emf by measuring $(V_a - I_a R_a)$ may not be as accurate as that from a tachogenerator with a good linear voltage to speed characteristic. Speed and position sensing can also be accomplished with a digital encoder.

Full utilization of both machine and converter ratings over a wide speed range can be achieved by a combination of voltage and flux control. For this discussion, base speed is the operating speed at which rated open-circuit voltage is obtained with rated field current excitation. For a converter that is appropriately sized to handle the dc machine's voltage and current ratings, maximum torque at speeds below base speed can be obtained when armature and field currents are regulated at their respective maximum values. Lower values of developed torque can be obtained by regulating the armature current accordingly, and by keeping field excitation at the maximum. Since E_a will increase with speed if flux remains constant, beyond base speed, field excitation must be lowered to keep the terminal voltage within the allowable operating voltage of the converter. Thus, operation beyond base speed is also referred to as the field weakening mode, in which the output power will be constant if E_a and I_a are held at their respective limits. The profile of I_a, I_f, E_a, output power, and torque over the speed range are shown in Fig. 8.24.

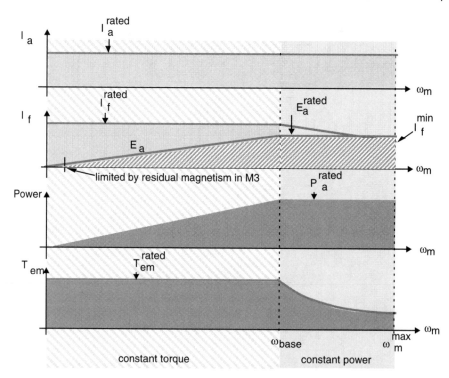

Figure 8.24   Speed range with torque and power control.

8.12 FOUR-QUADRANT OPERATION

Most of the newer dc drives are of the electronic type - supplied by controlled rectifiers.
Figure 8.25a shows a dc drive with an electronic bridge converter supply to both the
armature and field circuit. The converters can be single-phase or three-phase controlled
bridges. With a fully-controlled bridge, the terminal voltage, V_a, can be varied by
controlling the firing angle of the switching devices in the bridge. For example, if the ac
source inductance is neglected, the average values of the dc output voltage of a two- and
six- pulse rectifier fed from an ac supply of voltage magnitude V_m with the firing angle, α,
are $(2V_m/\pi)\cos\alpha$ and $(3\sqrt{3}V_m/\pi)\cos\alpha$, respectively. Theoretically, V_a can be adjusted to
any desired value between the full-positive and full-negative limits by changing the firing
angle, α. In practice, some margin angle in the inversion mode would be needed to allow
for the finite turn-off of real switches.

A motor usually has a large thermal constant that will allow it to withstand
transient overcurrent for a short while with no serious consequence. Unlike motors, the
semiconductor switches in a converter are less forgiving insofar as their capability to safely
withstand overcurrent or overvoltage. Better controlled performance may be achieved with
the converter's current rating higher than that of the motor, if the current control of the
converter can be properly coordinated to protect the motor against sustained overloading.

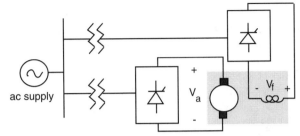

(a) dc drive with electronically-controlled armature and field voltages

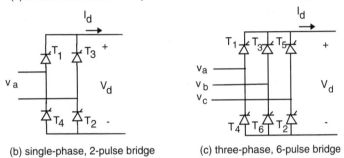

(b) single-phase, 2-pulse bridge

(c) three-phase, 6-pulse bridge

Figure 8.25   Electronic control of armature and field voltage.

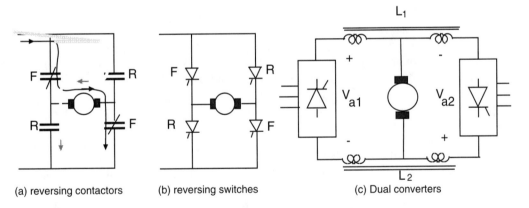

(a) reversing contactors

(b) reversing switches

(c) Dual converters

Figure 8.26   Connections for bi-directional current operation of dc machine.

The usual bridge circuit of semiconductor switches will conduct current in one direction. Thus, a dc machine with only a single bridge supply will only be capable of operation in the first and second quadrants of the torque-speed plot shown in Fig. 8.18. For the dc machine to operate in all four quadrants, its armature current must be reversible. This can be accomplished using reversing switches or contactors, or two converters connected back-to-back, such as those shown in Fig. 8.26.

With reversing switches or contactors, the changeover needed to handle a reverse flow of I_a of the dc machine should be performed at a current zero, as chopping the current can produce damaging voltage spikes. The switching time with mechanical contactors is long compared to that with electronic switches. Faster current reversal can be obtained by simultaneously operating two fully-controlled converters in anti-parallel as shown in Fig. 8.26c. To minimize the dc component of the circulatory current flowing around the two back-to-back converters, the converters are controlled such that $V_{a1} + V_{a2} = 0$, equivalent to $\cos\alpha_1 + \cos\alpha_2 = 0$, or $\alpha_1 + \alpha_2 = \pi$. Even then, harmonic components of the circulating current will be created by the difference in instantaneous dc output voltages. Inductors L_1 and L_2 are introduced to absorb the difference in instantaneous output voltages of the two bridges and to limit the harmonic currents. Another strategy of operating the two back-to-back converters is to not have both simultaneously active. In non-simultaneous control, only one of the converters is operating at any given time and the other is blocked. Furthermore, a dead time of 2 to 10 ms after current zero is to be provided in a changeover.

8.13 STARTING A SHUNT DC GENERATOR

For a voltage to be induced in the armature winding, the armature winding must be rotating relative to some flux. In a self-excited generator, the initial flux is from residual magnetism in the core. The additional conditions required for the internal voltage to build up under self-excitation are that the field current circulated by induced emf from the residual magnetism reinforces the initial flux, and that the field circuit resistance is less than the critical value for that rotor speed. Let's examine the two additional conditions with the help of the drawings in Fig. 8.27. The circuit diagram in Fig. 8.27a indicates the positive polarity and direction of the key variables. A positive I_f flowing in the direction shown produces positive core flux, ϕ, the sign of both the flux per pole and the speed in turn determines the sign of the induced emf, E_a, for the polarity marked. Recall that $E_a = k_a\phi\omega_m$. Figure 8.27c is an enlargement of the shaded area about the origin of the plot in Figure 8.27b.

Let's assume that the armature is rotated at a fixed speed of ω_m in the positive direction by a prime mover and that the core has an initial residual flux, ϕ_R, in the positive direction as shown in Figure 8.27c. The magnitude of induced emf, E_a, from the residual magnetism is given by OA in Figure 8.27c. With zero I_f, this initial positive E_a across the field winding circuit will all be absorbed initially by the field circuit inductance, that is $L_{ff} dI_f/dt$ equal to E_a; in other words, dI_f/dt will be positive. But when I_f increases from zero, the induced emf generated will be absorbed by a combination of $I_f R_f$ drop and $L_{ff} dI_f/dt$. The voltage, E_a, will continue to build up along with the increase in excitation if the field resistance line is below the magnetization curve of the generator at the driven speed to some useful magnitude at point F in Fig. 8.27b, when all of E_a will then be absorbed by resistive drop and dI_f/dt becomes zero. Point F will be the steady-state operating point for that rotor speed, assuming that the field circuit resistance is kept fixed. However, if the field circuit resistance line is above that of the critical field resistance line shown, the induced emf will build up to a value that is not much higher than the initial value of OA, as determined by the new crossover point. To obtain a higher operating voltage, either the speed will have to be increased, raising the magnetization curve relative to the field resistance line, thus shifting the crossover point upwards, or the field circuit resistance will have to be decreased to give the desired cross-over point.

On the other hand, if the field winding connection is such that the initial induced emf OA from the residual magnetism creates a field mmf that drives down ϕ_R, instead of reinforcing it, in effect a negative I_f, OA will still cause I_f to increase from zero, but in the negative direction. Point B in the second quadrant will be the steady-state operating point.

The conditions shown in the fourth and third quadrants of Fig. 8.27 are similar to what have been discussed for the first and second quadrants, respectively, but for a rotation opposite to that shown.

8.14 STARTING A SHUNT DC MOTOR

Some of the methods used to start a dc motors are:

- Direct connection to supply voltage (direct-on-line, DOL starting).
- Inserting starting resistors (resistance starting).
- Using an adjustable supply voltage ("soft" start).

At standstill, the back emf of the armature is zero, but as the rotor speed increases, the back emf will increase along with it. Applying the full-rated voltage to an armature with low resistance at standstill can cause the starting current to reach 20 or more times its rated value. To minimize the risk of damage to the motor by the large starting current or torque, the voltage applied to the armature could be lowered during starting and progressively raised as the motor speed increases to keep the starting period short.

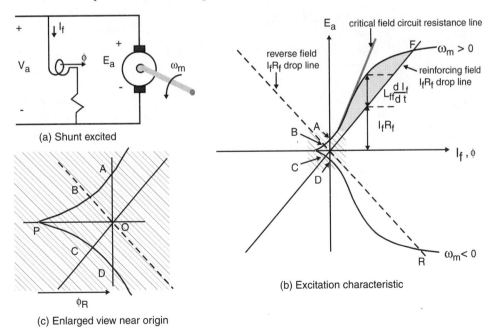

(a) Shunt excited

(b) Excitation characteristic

(c) Enlarged view near origin

Figure 8.27   Startup of a shunt generator.

8.14.1 Direct-on-line Starting

The advantages of direct-on-line (DOL) starting are low cost and simplicity. On the other hand, the large starting current can cause dangerous sparking, overheating of the armature winding, increased complexity in operation of protection equipment, a large voltage drop in supply network, and a large transient torque which can be damaging to the mechanical drive train. As such, DOL starting is limited to low-power motors with inherently high armature resistance drop.

When frictional and windage losses are ignored, the governing differential equations for the armature circuit and rotor speed are

$$V_a = I_a R_a + L_{aq}\frac{dI_a}{dt} + \underbrace{k_a \omega_m \phi}_{E_a} \qquad V$$

$$J\frac{d\omega_m}{dt} = T_{mech} + \underbrace{k_a \phi I_a}_{T_{em}} \qquad N.m$$

(8.41)

In DOL starting, the full-rated value of V_a is applied during the startup of the motor from rest. With full voltage applied and no internal voltage at standstill, the starting current rises quickly, limited only by the small resistance and inductance of the armature winding. As the motor accelerates up, the internal emf, E_a, increases along with speed, modifying the behavior of I_a, which eventually will decrease to its steady-state value.

8.14.2 Resistance Starting

External resistors may be temporarily inserted into the armature circuit during starting to reduce the starting current. These resistors can be manually or automatically shorted out as the motor accelerates. Figure 8.28 shows a simple starting circuit with m external starting resistor segments that are successively shorted out by contactors as the machine accelerates up to speed. It also shows the waveforms of the armature current and rotor speed during starting. The value of the starting resistors for a shunt motor can be determined as follows:

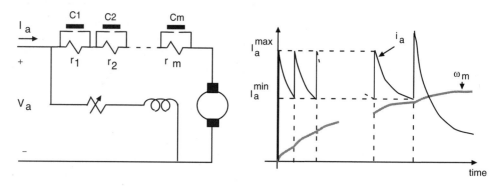

Figure 8.28   Starting circuit and waveforms of i_a and ω_m.

First, we need to establish the allowable maximum and minimum values of the current during starting.

The maximum transient current that a motor can safely carry is usually limited to what it can commutate without excessive sparking. Typically, this is less than twice the rated value for a motor without compensating winding, or even as high as 3.5 times the rated value for special cases where the machine is well-compensated. The minimum value is usually based on maintaining a high average starting torque to minimize the run-up time with a reasonable small number of switchings. A typical range for I_a^{min} is between 1.1 to 1.3 of I_a^{rated}.

Consider the situation when the flux, ϕ, is held constant and the small armature inductance can be ignored. Let $R_{s0}, R_{s1}, \ldots, R_{sm}$ be the total resistances of the armature circuit for the initial condition where all the contactors are open, and for the subsequent stages when the contactors are progressively closed. At the initial position, with R_{s0} limiting the current to I_a^{max}, $I_a^{max} = V_a/R_{s0}$. As the motor speeds up, the increase of E_a with ω_m reduces I_a. When I_a drops to I_a^{min}, contactor 1 closes and the resistance of the armature circuit changes from R_{s0} to R_{s1}, that is

$$E_a = V_a - I_a^{min} R_{s0} = V_a - I_a^{max} R_{s1} \qquad V \qquad (8.42)$$

or

$$R_{s1} = \frac{I_a^{min}}{I_a^{max}} R_{s0} = \beta R_{s0} \qquad \Omega \qquad (8.43)$$

where β denotes the current ratio. Repeating the same procedure for the closing of the other contactors, we will obtain

$$R_{s2} = \beta R_{s1} = \beta^2 R_{s0}$$

$$R_{sm} = \beta^m R_{s0} \qquad (8.44)$$

Finally, when all m contactors are closed, the remaining resistance will be that of the armature winding, that is $R_{sm} = R_a$. Substituting this into Eq. 8.44, the required number of segments for a given β can then be determined using

$$m = \frac{\ln R_a/R_{s0}}{\ln \beta} \qquad (8.45)$$

Alternatively, the I_a^{min}/I_a^{max} ratio with a fixed number of segments, m, is given by

$$\beta = \sqrt[m]{\frac{R_a}{R_{s0}}} \qquad (8.46)$$

The values of the m external resistor segments can be computed from

$$r_1 = R_{s0} - R_{s1} = R_{s0}(1 - \beta) \qquad \Omega$$

$$r_2 = R_{s1} - R_{s2} = R_{s1}(1 - \beta) = \beta r_1 \qquad (8.47)$$

$$r_m = R_{sm-1} - R_{sm} = \beta^{m-1} r_1$$

8.15 PROJECTS

8.15.1 Project 1: Startup and Loading of a Shunt dc Generator

In this project, we shall implement the SIMULINK simulation *s1* of a dc shunt generator and use it to study the required condition of residual flux and direction of rotation for the successful build-up of self-excitation and load characteristic of a dc shunt generator.

The equation of the field winding is

$$v_f = I_f(R_f + R_{rh}) + L_f \frac{dI_f}{dt} \qquad V \qquad (8.48)$$

where R_f and R_{rh} are the resistances of the main field winding and field circuit rheostat, and L_f is the field winding inductance.

The armature winding is represented by an equivalent winding in quadrature with the field winding. Using generator notation, the average value of the terminal voltage of the armature winding can be expressed as

$$E_a = I_a R_a + L_{aq} \frac{dI_a}{dt} + V_a + V_{brush}, \qquad E_a = k_a \omega_m \phi \qquad V \qquad (8.49)$$

where V_a is the average value of the armature terminal voltage, I_a is the average current flowing out of the armature winding at the positive terminal of V_a, E_a is the average value of the internal voltage, R_a is the armature winding resistance, and L_{aq} is the armature winding inductance. In simulation, the terminal voltage, V_a, can be developed using the expression of the load resistor, whose value can also be set very large to approximate an open-circuit condition.

In this project, we will be dealing with fixed-speed operation; as such, the equation of motion of the rotor will not be needed. However, if speed variation is to be simulated, the equation of motion of the rotor, using the convention of positive torque in the direction of rotor rotation, is

$$T_{em} + T_{mech} = J \frac{d\omega_m}{dt} + D\omega_m, \qquad T_{em} = k_a\phi(-I_a) = E_{ao}(-I_a)/\omega_{mo} \qquad N.m \qquad (8.50)$$

The negative sign in front of I_a in T_{em} is consistent with the generating convention used for the circuit equation given in Eq. 8.49. In the generating mode, I_a here will be positive, whereas T_{em} will remain negative.

Due to magnetic saturation, the relation between the flux per pole, ϕ, and the effective field excitation, I_f, is nonlinear. The effects of magnetic saturation and rotor speed on E_a can be accounted for using the open-circuit magnetization curve at some known speed and speed scaling, that is

$$\frac{E_a}{\omega_m} = k_a\phi = \frac{E_{a0}}{\omega_{mo}} \qquad (8.51)$$

Where magnetic saturation is negligible, ϕ is proportional to I_f; therefore, $k_a\phi$ can be replaced by $k_f I_f$. In which case, $E_{a0} = k_f \omega_{mo} I_f = k_g I_f$. The value of k_g at the speed of

ω_{mo} can be determined from the slope of the airgap line of the open-circuit characteristic of the machine.

In the first part of this project we shall examine the condition for successful self-excitation of a shunt generator with a simple resistive load. Figure 8.29 shows the SIMULINK simulation in *s1* of a dc generator with a series RL load connected to its armature. The corresponding MATLAB M-file, *m1*, may be used to initialize your simulation with the following machine parameters of a 120-V, 2-kW, 1750 rev/min dc shunt generator:

Rated power is 2 kW	Rated armature voltage is 125 V
Rated armature current is 16 A	Rated speed is 1750 rev/min
Armature winding resistance, $R_a = 0.24\Omega$	Armature winding inductance, $L_{aq} = 18$ mH
Shunt field winding resistance, $R_f = 111\Omega$	Shunt field winding inductance, $L_f = 10$H

Figure 8.30 is the magnetization characteristic of the machine taken at a constant speed of 2000 rev/min. For the shunt generator to self-excite, the direction of rotation and residual flux must be such that the polarity of the induced voltage, E_a, circulates a field current that further increases the airgap flux. In the above SIMULINK simulation, the residual flux is represented indirectly by the open-circuit voltage on the magnetization curve at zero field current, and the direction of rotation is represented by the sign of the speed of rotation. The field circuit connection can be reversed in the simulation by changing the sign of the *Field Polarity* from $+1$ to -1.

If we were to neglect the passive element voltage drops of the armature winding, brush, and field circuit resistances, the final voltage to which the generator will self-excite would be determined by the intersection of the field circuit line and the open-circuit curve of the armature. The slope of the field circuit line corresponds to the total field circuit resistance, R_f and R_{rh}. When the field circuit resistance exceeds the initial slope of the magnetization curve at that speed, the terminal voltage of the shunt generator will not self-excite to a useful level.

Use speed scaling to determine the open-circuit magnetization curve for the rated speed of 1750 rev/min from that of Fig. 8.30 given for 2000 rev/min. Next, determine the critical value of field circuit resistance that corresponds to the initial slope of the magnetization curve at the rated speed of 1750 rev/min. Also determine the value of external field rheostat, R_{rh}, for the shunt generator to self-excite to the rated terminal voltage of 125 V. Approximate an open-circuit condition on the armature terminals with an R_{load} of $1e^6$ Ω, and set ω_m to the rated speed in radian/sec. With these parameters, the resulting system equations are quite stiff. Try running your simulation using the *ode*15*s* or Adams/Gear numerical method with a minimum step size and tolerance set to $5e^{-5}$ and $1e^{-6}$, respectively. For the first set of runs, a stop time of two seconds is sufficient. Check the final value of V_a to see if it is close to the rated voltage.

Reverse the sign of ω_m and rerun the simulation with the other parameters kept the same. Note the final value of V_a. Change the sign of the field polarity connection and rerun the simulation. What is the final value of V_a? Explain the difference in results caused by a change in direction of rotation and field circuit connection.

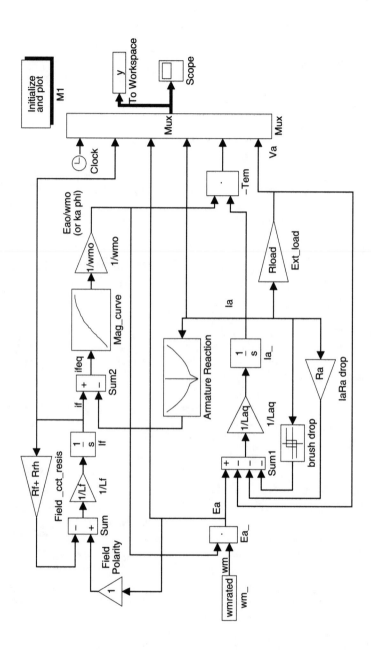

Figure 8.29 Simulation *s1* of shunt generator.

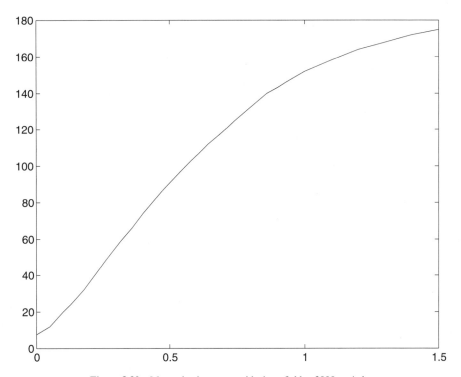

Figure 8.30   Magnetization curve with shunt field at 2000 rev/min.

The next set of runs examines the droop in terminal voltage of the armature as loading increases. The droop is caused by armature reaction and resistive drop in the armature winding, as well as brush drop. For this study, brush drop may be assumed to be a constant one volt per brush, and armature reaction may be approximated by an equivalent main field demagnetizing current of the form:

$$I_{ar} = \frac{AR}{N_f} = 0.04 * abs(\arctan(I_a)) + 0.0001 * I_a^2 \qquad A \qquad (8.52)$$

In reality the armature reaction is also dependent on the level of net excitation, that is also of I_f, and will occur only when the iron of the flux path becomes saturated.

Reset the stop time to about 12 seconds. Simulate the loading sequence where R_{load} is being decreased in five steps of $1e^6$, 20, 10, 5, and 4 Ω. Make the changes to R_{load} during the simulation on a fly and allow some time after each change for the terminal voltage to settle. Obtain a plot of the armature current, I_a, the terminal voltage, V_a, and the electromagnetic torque developed by the machine, T_{em}, over the entire loading sequence.

8.15.2 Project 2: Resistance Starting of a dc Shunt Motor

In this project, we shall examine the starting transients of a separately excited dc motor. The objectives are to implement the SIMULINK simulation, *s2*, of a dc motor and to use it to study the starting transients when the motor is started DOL, when it is started with a resistance starter that is switched at fixed armature current level, and when it is started with a resistance starter that is switched at fixed time intervals.

For this project, we will assume that the field excitation of the dc motor has reached the desired steady-state value before energizing the armature circuit to start the motor. The internal emf, E_a, of the armature is assumed to be proportional to the product of the field flux, ϕ, and the motor speed, ω_m, that is

$$E_a = k_a \phi \omega_m \quad V \tag{8.53}$$

The electromagnetic torque developed by the motor is given by

$$T_{em} = k_a \phi I_a \quad N.m \tag{8.54}$$

where I_a is the current of the armature.

When the rotor is at or near standstill, ω_m and also E_a will be zero. In resistance starting, the motor is started with a fixed dc supply voltage, and starting resistors are inserted during the starting period to keep the armature current within some safe limits. The upper current limit is often decided by the ability to commutate properly, and the lower limit is to maintain acceptable acceleration or run-up time.

A simple representation of the separately excited dc motor with its starting circuit is shown in Fig. 8.31a. Figure 8.31b shows the SIMULINK simulation of the circuit in s2.

The KVL equation of the armature circuit is

$$V_a = I_a R_t + L_{aq}\frac{dI_a}{dt} + E_a + V_{brush} \quad V \tag{8.55}$$

where R_t, the total resistance in the armature circuit circuit, is the sum of the starter resistance and the armature winding resistance, R_a, and V_{brush} is the voltage drop across the brushes on the commutator. The equation of motion of the rotor is obtained by equating the net accelerating torque to the inertia torque on the rotor, that is

$$T_{em} + T_{mech} - D_\omega \omega_m = J\frac{d\omega_m}{dt} \quad N.m \tag{8.56}$$

where T_{mech} is the externally applied torque in the direction of rotation, D_ω, the damping coefficient, and J, the inertia of the rotor. The above two equations can be rewritten into integral form, that is

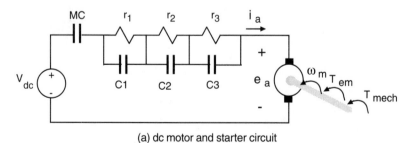

(a) dc motor and starter circuit

Figure 8.31   Starting of a dc motor under fixed excitation.

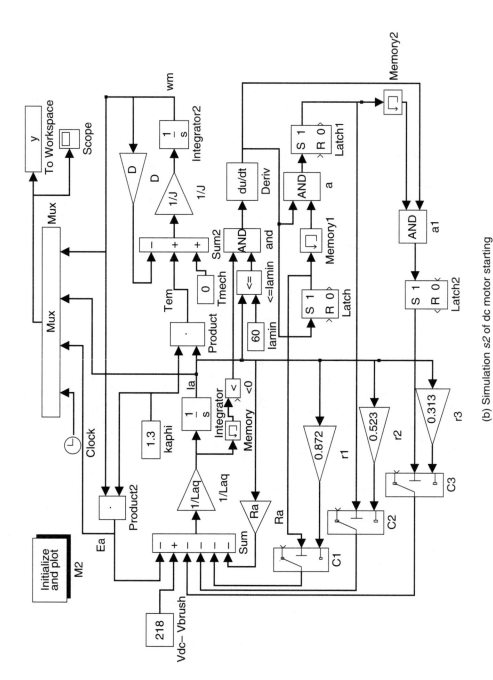

(b) Simulation *s2* of dc motor starting

Figure 8.31 (cont.): Starting of a dc motor under fixed excitation.

$$I_a(t) = \frac{1}{L_{aq}} \int_0^t (V_{dc} - V_{brush} - I_a R_t - E_a)dt + I_a(0) \qquad A$$

$$(8.57)$$

$$\omega_m(t) = \frac{1}{J} \int_0^t (T_{em} + T_{mech} - D_\omega \omega_m) + \omega_m(0) \qquad rad/s$$

During starting, the switching of the three resistor segments in $s2$ is being triggered by the crossing of I_a below the threshold value of I_a^{min}. The contacts, $C1$, $C2$, and $C3$, are initially open. When the main contact, MC, is closed to start the motor, the initial starting current drawn will be limited to its upper value, I_a^{max}, by all of the starting resistors. As the rotor increases in speed, the back emf, E_a, will correspondingly increase with speed, causing the armature current to decrease. When the starting current decreases to the lower limit, I_a^{min}, contact $C1$ closes to short out the first starting resistance segment, r_1. When r_1 is shorted, the armature current rises again and the torque, T_{em}, increases along with it, accelerating the rotor to a higher speed. As the back emf, E_a, increases with speed, the armature current will again decrease. When the armature current decreases again to the lower limit, the next starting resistor segment, r_2, will be shorted out by the contact, $C2$. In this way, the rotor accelerates further and the same process is repeated until the last starting resistance segment, r_3, is shorted out, leaving only the armature winding resistance as the rotor accelerates on to full speed.

The sequential logic for starting is implemented in Fig. 8.31b by detecting the condition when I_a decreases down to I_a^{min}. The logic signal drives a derivative module, producing a pulse input to a set of SR latches connected to operate like a counter. The outputs of the three latches are used to switch off the resistive drops of the three switchable starting resistor segments. The SR latches can be obtained from the **extras/Flip Flops** block library. The logic and relational operator modules are taken from the **Nonlinear** block library. Memory modules, from the **Nonlinear** block library, are used for different purposes: the module placed before the relational operator module is used to break an algebraic loop, whereas the modules placed between latches are used to create a short delay to get the latches to operate as desired. The switches, C1, C2, and C3, will pass input 1 through if their input 2 is greater or equal to an adjustable threshold, otherwise they pass input 3 through. For these switches to be operated by the standard zero one logic levels, their threshold level must be set at 0.5.

Set up the simulation, $s2$, to simulate the starting of a separately excited dc motor. The M-file, $m2$, has been programmed to set up the parameters of a 10-kW, 220-V, 1490 rev/min, separately-excited motor that has an armature resistance, R_a, of 0.3 ohm, an armature winding inductance, L_{aq}, of 12 mH, a brush drop of two volts, and a rated current of 50 A. First, we will use $s2$ to simulate the motor starting with no mechanical load under full flux condition, with $k_a\phi = 1.3$. The inertia of the rotor and load assembly, J, is 2.5 kgm^2, and the damping coefficient, D_ω, is zero.

As an illustration, we will determine the three starting resistor sections for a requirement that $I_a^{max} = 100A$ and $I_a^{min} = 60A$. Ignoring the armature inductance and assuming finite inertia, at the closing of the main contact, MC, we have

$$E_a(0) = 0 = V_{dc} - V_{brush} - I_a^{max}(R_a + r_1 + r_2 + r_3) \qquad V \qquad (8.58)$$

At the instant of shorting out the first section of the starting resistor, we have

$$E_a(t_1) = V_{dc} - V_{brush} - I_a^{min}(R_a + r_1 + r_2 + r_3)$$
$$= V_{dc} - V_{brush} - I_a^{max}(R_a + r_2 + r_3) \tag{8.59}$$

At the instant of shorting out the next section of the starting resistor, we have

$$E_a(t_2) = V_{dc} - V_{brush} - I_a^{min}(R_a + r_2 + r_3)$$
$$= V_{dc} - V_{brush} - I_a^{max}(R_a + r_3) \tag{8.60}$$

At the instant of shorting out the last section of the starting resistor, we have

$$E_a(t_2) = V_{dc} - V_{brush} - I_a^{min}(R_a + r_3) = V_{dc} - V_{brush} - I_a^{max} R_a \tag{8.61}$$

From Eq. 8.58, we obtain

$$(R_a + r_1 + r_2 + r_3) = \frac{V_{dc} - V_{brush}}{I_a^{max}} \quad \Omega \tag{8.62}$$

And from Eqs. 8.59 to 8.61, we obtain

$$(R_a + r_2 + r_3) = \frac{I_a^{min}}{I_a^{max}}(R_a + r_1 + r_2 + r_3)$$

$$(R_a + r_3) = \frac{I_a^{min}}{I_a^{max}}(R_a + r_2 + r_3) \tag{8.63}$$

$$(R_a) = \frac{I_a^{min}}{I_a^{max}}(R_a + r_3)$$

With $I_a^{max} = 100A$ and $I_a^{min} = 60A$, the values of the three starting resistance segments are

$$r_1 = 0.872 \ \Omega$$
$$r_2 = 0.523 \ \Omega \tag{8.64}$$
$$r_3 = 0.313 \ \Omega$$

A sample set of results showing the armature current, I_a, the back emf, E_a, and the rotor speed, ω_m is given in Fig. 8.32. It was obtained using the Adams/Gear numerical integration method on MATLAB4 with a start time of zero, a stop time of 7 seconds, a minimum step size of 0.5 msec, a maximum step size of 5 msec, and an error tolerance of $1e^{-6}$.
    Conduct a simulation of the following cases:

(a) Start up the same motor with a load torque of 50 N.m

(b) Resize the starting resistor sections for an I_a^{max} of 120 A and an I_a^{min} of 75 A and run a case of the same motor starting from standstill with no mechanical load using these starting resistances.

(c) Modify a copy of the simulation given in s2 to simulate an open-loop starter where the contacts, $C1, C2$, and $C3$, are sequentially closed by timers. Simulate a case of the same motor starting with no load and time-delays of 30, 20, and 10 seconds between the closings of MC and $C1$, $C1$ and $C2$, and $C2$ and $C3$, respectively.

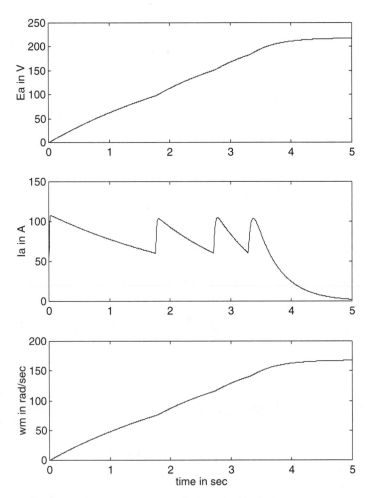

Figure 8.32 Starting of a dc motor: E_a (top), I_a (middle), and ω_m (bottom).

8.15.3 Project 3: Methods of Braking

For dc machines, a changeover from motoring to braking can be accomplished by reversing the flow of the motoring armature current, I_a. Although a reversal of field flux will also result in a reversal of T_{em}, it is seldom employed because the time constant for ϕ is much longer than that for I_a.

In this project, we shall examine braking methods on the second quadrant of the speed torque domain in which I_a and T_{em} are reversed:

1. *Plugging*: When V_a is reversed and an external resistor is also inserted simultaneously to limit I_a to within some allowable value.

2. *Dynamic braking*: When V_a is switched off and an external resistor is simultaneously inserted across the armature terminals to limit I_a.

3. *Regenerative braking*: When V_a becomes smaller than E_a, reversing the flow of I_a. This condition can be brought about by an adjustable V_a supply or by an increase in E_a from an increase in ω_m, as in the case of the load overhauling the rotor.

For this study, we will again use the parameters of the 2-kW, 125-V, 1750 rev/min dc machine given earlier in Project 1. The machine, however, will be connected to operate as a separately-excited motor, with its field excitation held fixed at the rated condition. The inertia of the rotor assembly, J, is 0.5 kgm^2. For simplicity, we will ignore brush drop and armature reaction. In each case, we will begin with the machine motoring at a speed of 1750 rev/min, drawing rated armature current of 16 A from a supply voltage, V_a, of 125V.

Resistor value for plugging.   Assuming that the armature current is not to exceed 250% of its rated value, or 40 A, we first estimate the limiting resistors needed for the plugging and dynamic braking methods. With armature reaction ignored, we can assume that flux is maintained at rated motoring condition during braking. The flux at rated motoring condition is given by

$$k_a\phi = \frac{V_a - R_a I_a}{\omega_m} = 60\frac{125 - 0.1416}{1750(2\pi)} = 0.6699 \qquad N.m/A \qquad (8.65)$$

Thus the developed torque , T_{em}, given by $k_a\phi I_a$ is 10.72 N.m. And the internal voltage, E_a, given by $k_a\phi\omega_m$ is 122.76 V.

A changeover from motoring to braking using the plugging method is effected by simultaneously reversing the polarity of the applied voltage across the armature and inserting an external resistor, R_{ext}, into the armature circuit. With flux maintained constant and the rotor speed immediately after the changeover the same as before, the value of R_{ext} that will keep I_a within 250% of its rated value can be determined from

$$E_a - (-V_a) = I_a^{max}(R_{ext} + R_a) \qquad V \qquad (8.66)$$

Rearranging and substituting in the known values, we have

$$R_{ext} = \frac{122.76 + 125}{2.5(16)} - 0.14 = 6.054 \qquad \Omega \qquad (8.67)$$

Resistor value for dynamic braking.   Similar equations can be used to compute the limiting resistor for the dynamic braking method, except that V_a during braking will be zero. Thus, for the dynamic braking method, the value of the limiting resistor is

$$R_{ext} = \frac{122.76}{2.5(16)} - 0.14 = 2.929 \qquad \Omega \qquad (8.68)$$

Simulation of plugging and dynamic braking.   Let's examine the braking performance of the plugging and dynamic braking methods assuming that the load torque behaves in the manner described by

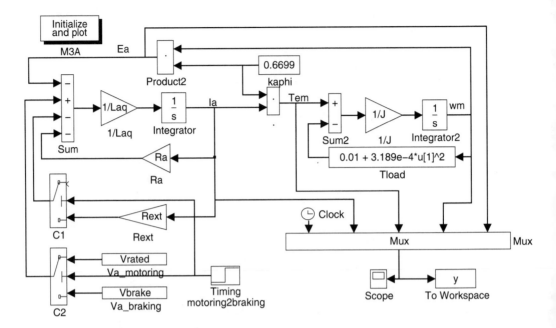

Figure 8.33   Simulation *s3a* for plugging or dynamic braking operation.

$$T_{load} = -T_{mech} = 0.01 + 3.189e^{-4}\omega_m^2 \qquad N.m \qquad (8.69)$$

Figure 8.33 shows the SIMULINK simulation, *s3a*, for this project. In this simulation, a step generator, *Timing motoring2generating*, is used to trigger the changeover from motoring to braking. The two switches, $C1$ and $C2$, are to be loaded with a threshold level of 0.5 when the step generator is set to have an initial output of 1 and a final output of zero. Steady-state values of the armature current and rotor speed can be used as initial values for the corresponding integrators to minimize the startup transients, thus saving simulation time. With the given system parameters, the simulation will run satisfactorily using the *ode15s* or Adams/Gear method with tolerance set to $1e^{-6}$ and minimum step size of $1e^{-4}$. Implement and run your simulation of the above system. Obtain the plots of T_{em}, I_a, and ω_m. Based on the observed results, verify the changeover from motoring to braking and check the agreement between the values of I_a and E_a used to determine the R_{ext} with those obtained from your simulation.

Sample results of the armature current and rotor speed using the plugging and dynamic braking methods are shown in Fig. 8.34.

Regenerative braking in the second quadrant.   As shown earlier in Fig. 8.21, regenerative braking will occur when the rotor speed rises above the no-load speed as in the case of the rotor being overhauled by its load torque. We can examine the dynamics of this braking method using the same simulation shown in Fig. 8.33 by replacing

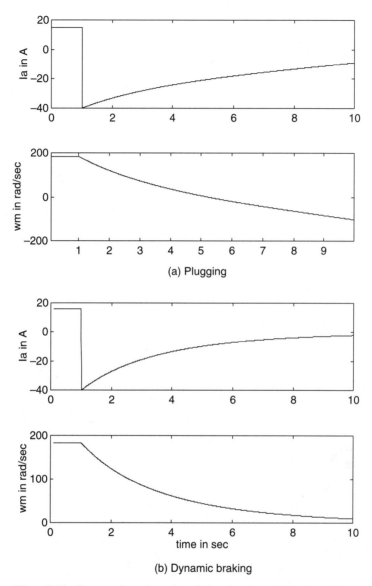

Figure 8.34 Current and speed transients during plugging and dynamic braking.

the passive load with an active load, one that will reverse in sign to overhaul the rotor of the motor. Replace the passive load torque simulation with a sine generator and set the peak of the sinusoidal loading to be the same as the rated torque of $10.72 N.m$, the initial phase of the sine wave to $\pi/2$ so that it will start off motoring at rated torque. Set the frequency of variation of the sinusoidal loading to $\pi/2$ rad/sec. Set the step time of the step generator, *Timing motoring2generating*, to some value larger than the run-time so that the switches,

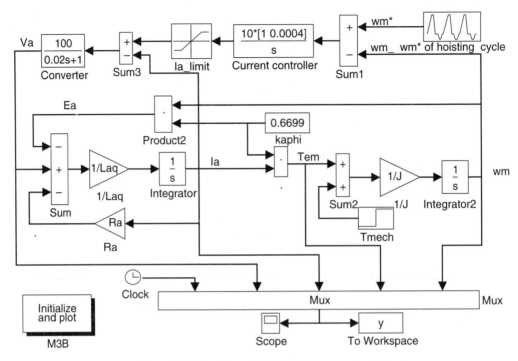

Figure 8.35   Simulation *s3b* of a closed loop speed control system.

$C1$ and $C2$, will not be activated during the run. Run the simulation and obtain plots of T_{em}, I_a, and ω_m.

Regenerative braking in the fourth quadrant.   Braking in the reverse direction of rotation against an active load torque that is overhauling the motor, as in the lowering operation of a hoist, is also used in practice. This corresponds to operation in the fourth quadrant of the speed-torque domain. In this part of the project, we shall examine the operation of a separately-excited dc motor operating with adjustable armature supply voltage that is controlled electronically to perform the raising and lowering of a constant load torque, as in a routine hoisting operation. For raising the load, the dc machine motors in the first quadrant where both speed and torque are in the positive direction. At the end of the raising, the load is held steady for a while before it is lowered at controlled speed, countering the gravitational force on the load. Figure 8.35 shows the SIMULINK simulation, *s3b*, of a simple closed loop speed system that can be used to demonstrate the raising, holding, and lowering operations.

In *s3b*, the reference speed, ω_m^*, of a hypothetical hoisting operation is represented by a *repeating sequence source*. The load torque is represented by a *delay step* source with the step change in load torque from zero to T_{rated} occurring at $t = 0$, synchronized with the first ramp-up of the speed reference cycle. The time values of the repeating sequence are [0 30 40 50 65 75 85 95 100] and the corresponding output speed values are [0 wraise

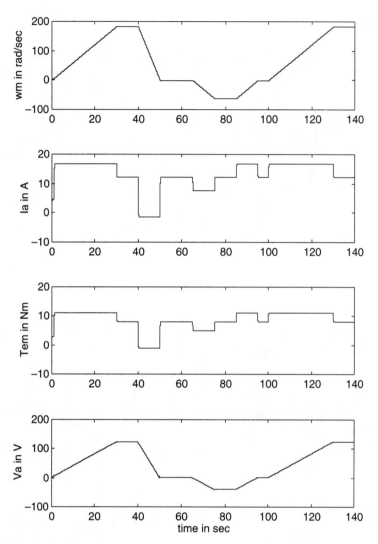

Figure 8.36   Controlled raising, holding, and lowering of a load.

wraise 0 0 wlower wlower 0 0]. In the simulation, *s3b*, the dynamics of the ac/dc bridge and its controller that provides the adjustable V_a to the machine, is represented by a simple transfer function with a first-order delay.

The sample result of the hoisting operation given in Fig. 8.36 has been obtained using the values of the transfer function of the current controller shown in Fig. 8.35. It is obtained with *wraise* set to rated speed, *wlower* to minus one-third rated speed, and the upper and lower limits of the current limiter set at ± 120 percent of the rated armature current. Repeat the simulation of the hoisting operation for a load torque that is half the rated value and comment on the profiles of the armature current and voltage.

8.15.4 Project 4: Universal Motor

A Universal motor is a small dc series motor that is designed to operate from an ac supply. The characteristics of Universal motors are high no-load speed and high starting torque. Because of the high operating speed, the size of these motors for a given hp rating is typically smaller than other fractional hp ac machines, making it ideal for hand-held tools and appliances where weight, compactness, and speed are important factors.

If the total armature current, I_a, is $I_m \cos \omega t$, the current in the coils of an armature with a parallel identical paths will be given by

$$I_c = \frac{1}{a} I_m \cos \omega t \qquad A \tag{8.70}$$

In a series-excited machine, the flux produced by the ac current, I_a, will also be alternating in time. Thus, the flux per pole may be expressed as

$$\sum_{i=1}^{n_s} B(\theta, t) L \tau = \phi_m \cos \omega t \qquad Wb \tag{8.71}$$

Substituting these into Eq. 8.16, we obtain

$$T_{em} = \frac{PZ}{2\pi a} \phi_m I_m \cos^2 \omega t \qquad N.m \tag{8.72}$$

The average value of the above torque is

$$\bar{T}_{em} = \frac{PZ}{2\pi a} \frac{\phi_m I_m}{2} \qquad N.m \tag{8.73}$$

It is only half that of the case where the flux and current are at a steady level of ϕ_m and I_m, respectively.

Figure 8.37 shows the SIMULINK simulation, *s4*, of a Universal motor. The corresponding MATLAB M-file, *m4*, can be used to set up the MATLAB workspace to simulate a motor having the following parameters and magnetization curve:

Rated power is 325 W   Rated frequency of supply voltage is 60 Hz
Rated terminal voltage is 120 Vrms   Rated armature current is 3.5 Arms
Rated speed is 2800 rev/min   Armature winding resistance, $R_a = 0.6\Omega$
Armature winding inductance, $L_{aq} = 10mH$ Series field winding resistance, $R_{se} = 0.1\Omega$
Series field winding inductance, $L_{se} = 26mH$ Rotor inertia, $J = 0.015\ kgm^2$.

The magnetization characteristic of the machine at a constant speed of 1500 rev/min is shown in Fig. 8.38. In the simulation, the switch, *Sw4AC*, controls whether dc or ac voltage supply is fed to the armature circuit. A delay step function generator allows the timing of the step change in load torque and the two levels of T_{mech} used in here.

We shall first examine the startup transients of the series motor with a terminal voltage V_a of $120\sqrt{(2)} \sin \omega t$ V and a load torque equal to the rated value of 1.1084 $N.m$. Obtain plots of the armature current, I_a, internal voltage, E_a, developed torque, T_{em}, and rotor speed, ω_m, for the startup. Take an expanded time scale plot of the same variables when

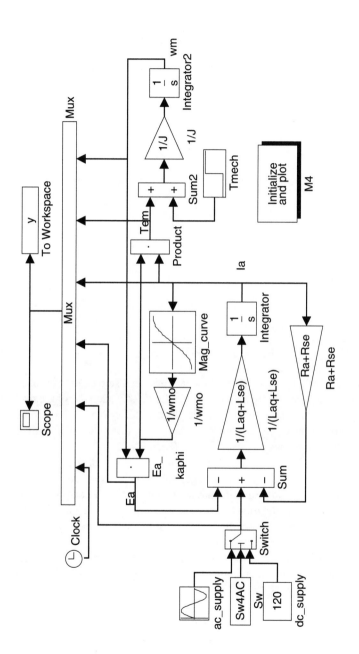

Figure 8.37 Simulation *s4* of dc series motor.

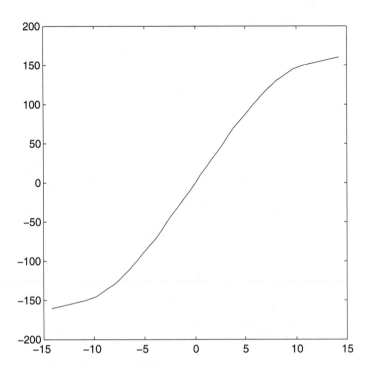

Figure 8.38 Magnetization curve with series field at 1500 rev/min.

the machine has attained steady-state. Perform a step change in load torque from full-rated to half-rated and obtain the plots of V_a, I_a, E_a, T_{em}, and ω_m of the machine's response to the step change in load torque. Note the phase difference between V_a and I_a, and the frequency of the pulsating component in T_{em}. Explain the difference in magnitude between V_a and E_a, aside from the resistive drops across R_a and R_{se}.

Replace the ac source for V_a with a dc source of the same rms value, that is a V_a of 120 V. Using the dc supply voltage, repeat the startup run with rated load torque and the step change in load torque from rated to half-rated. Obtain plots of V_a, I_a, E_a, T_{em}, and ω_m of the startup and the transient response to the step load change. Comment on the difference in magnitude of I_a and in the speed of startup and response to a step change in load torque of the motor with the two kinds of voltage supply.

Sample results of the startup with ac supply are shown in Fig. 8.39, and those of the machine's response to a step reduction in load torque at $t = 0.5s$ with ac and dc supplies are shown in Fig. 8.40.

8.15.5 Project 5: Series dc Machine Hoist

In this project, we shall examine the routine operation of a dc series motor drive for a hoist. We will consider the case where the load on the hoist is constant. In the motoring mode, the dc machine of the hoist will operate in the first quadrant, motoring to raise the load. The rotor speed will increase in the positive direction until steady-state is reached or until a

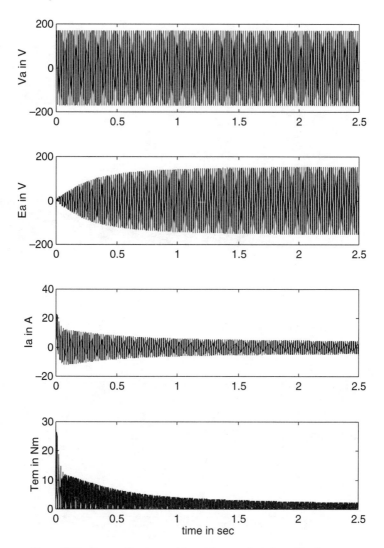

Figure 8.39   Startup of series machine with ac supply and rated load torque.

change in operating mode to hold or lower is detected. In the braking or lowering mode of operation, the dc machine will be overhauled by the active load torque that remains at rated value in the opposite direction. Its rotor speed will decrease, finally reversing in direction. With the proper value of external resistor inserted, the motor's speed in the reverse direction should not exceed the specified lowering speed that is used to determine the value of the external braking resistor.

Figure 8.41 shows the steady-state speed torque curves of the test series motor when motoring with rated voltage, when braking with rated voltage applied and a series braking resistor of 13.236 Ω inserted, and when braking with supply voltage replaced by a braking

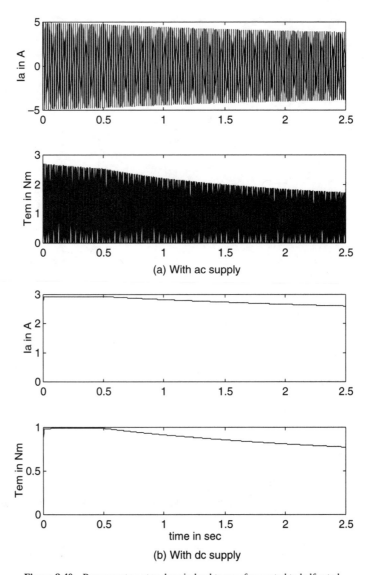

Figure 8.40   Response to a step drop in load torque from rated to half-rated.

resistor of 2.667 Ω. Figures 8.42 and 8.43 show the plots of the speed vs. torque and the speed vs. power curves of the test motor for six equally spaced values of the terminal voltage between rated and $-$ Vrated/5.

Steady-state stability of operating point.   From Fig. 8.42 we can deduce that the speed-torque characteristic for negative V_a is not suitable for braking an active load because the slope of the curve has a positive $\partial T_{em}/\partial \omega_m$ slope. A linearized

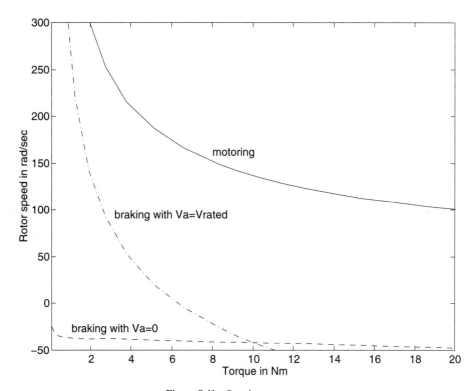

Figure 8.41   Speed-torque curves.

analysis of the equation of motion of the rotor without the damping term yields

$$J\frac{d\Delta\omega_m}{dt} = \Delta T_{em} + \Delta T_{mech} = \Delta T_{em} - \Delta T_{load} \tag{8.74}$$

For small perturbations, the changes in T_{em} and T_{load} with respect to a change in speed, $\Delta\omega_m$, can be approximated by the corresponding slopes of the torque vs. speed curves of the machine and load, respectively. That is $\Delta T_{em} = (\partial T_{em}/\partial\omega_m)\Delta\omega_m$ and $\Delta T_{load} = (\partial T_{load}/\partial\omega_m)\Delta\omega_m$, at the operating point defined by their intersection. With these substituted into Eq. 8.74 and solving, we obtain

$$\Delta\omega_m(t) = \Delta\omega_m(0)e^{\frac{1}{J}(\frac{\partial T_{em}}{\partial\omega_m} - \frac{\partial T_{load}}{\partial\omega_m})t} \tag{8.75}$$

For the operating point to be stable, the exponent in Eq. 8.75 cannot be positive or

$$\frac{\partial T_{em}}{\partial\omega_m} \le \frac{\partial T_{load}}{\partial\omega_m} \tag{8.76}$$

With a constant load torque, $\partial T_{load}/\partial\omega_m$ is zero. From Fig. 8.42, we can discern that the value of $\partial T_{em}/\partial\omega_m$ is slightly positive when operating in the fourth quadrant with a negative V_a, in which case the machine driving a constant load torque will not have a stable operating point. Verify this observation by conducting a simulation of such a condition using your simulation.

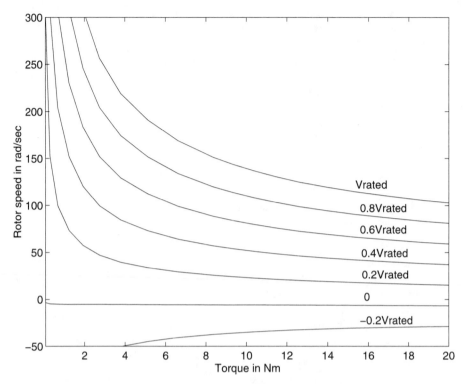

Figure 8.42 Speed-torque curves with various supply voltages.

First we will use the simulation shown in Fig. 8.44 to study the raising of a fixed load and the use of two methods of braking to lower the same load. The magnetization curve of the series motor shown in Fig. 8.45 must have some residual flux, without which the machine will not self-excite. The values of E_a and I_a vectors for the *2-D Lookup table* representing the machine's magnetization curve are

$$E_a = -160\ -155\ -150\ -145\ -140\ -135\ -130\ -125\ -120\ -115\ -110\ -105$$
$$-100\,-90\ -80\ -70\ -60\ -50\ -40\ -30\ -20\ -10\ 0\ 10\ 20\ 30\ 40\ 50\ 60\ 70\ 80$$
$$90\ 100\ 105\ 110\ 115\ 120\ 125\ 130\ 135\ 140\ 145\ 150\ 155\ 160$$

$$I_a = -27.22\ -23.55\ -20.62\ -18.74\ -17.56\ -16.62\ -15.63\ -14.83\ -13.98$$
$$-13.32\ -12.64\ -11.97\ -11.29\ -10.04\ -8.89\ -7.61\ -6.58\ -5.65$$
$$-4.64\ -3.58\ -2.52\ -1.42\ -0.40\ 0.62\ 1.72\ 2.78\ 3.84\ 4.85\ 5.78$$
$$6.81\ 8.09\ 9.24\ 10.49\ 11.17\ 11.84\ 12.52\ 13.18\ 14.03\ 14.83\ 15.82\ 16.76$$
$$17.94\ 19.82\ 22.75\ 26.42$$

The MATLAB M-file given in *m5* will perform the following tasks:

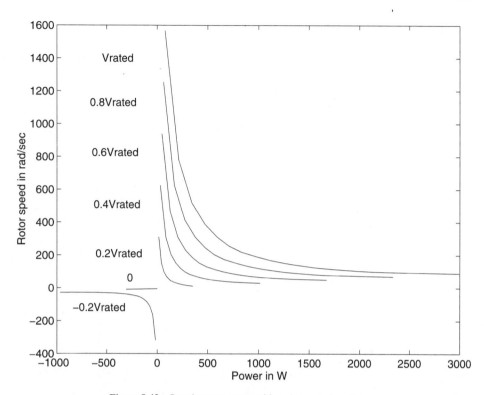

Figure 8.43   Speed-power curves with various supply voltages.

1. Load the machine parameters and magnetization curve data into the MATLAB workspace.

2. Determine certain steady-state characteristics of the dc motor supplied with rated voltage.

3. Determine the value of the braking resistor to be inserted that will limit the lowering speed to 500 rev/min when the dc series machine is braking in the fourth quadrant with V_a maintained at the rated voltage.

4. Determine the value of the braking resistor to be inserted that will limit the lowering speed to 500 rev/min when the dc series machine is operating with V_a set to zero, under dynamic braking, in the fourth quadrant.

5. Plot the steady-state speed vs. torque characteristics corresponding to the above three operating conditions.

```
% M-file m5  is for Project 5 on series dc machine hoist in Chapter 8

% m5 loads the following dc machine parameters
% and plots results of simulation

Prated = 1500;
```

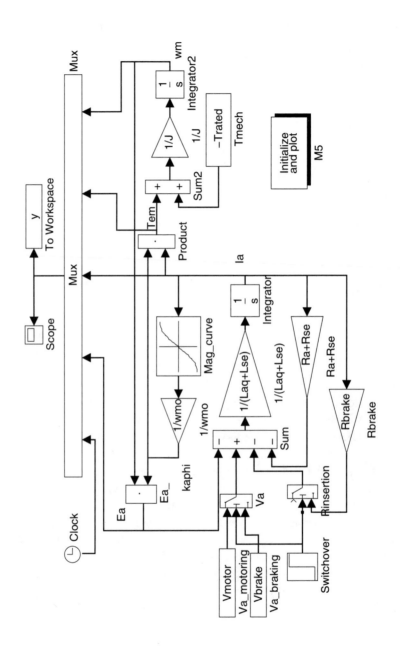

Figure 8.44 Simulation s5 of dc series motor.

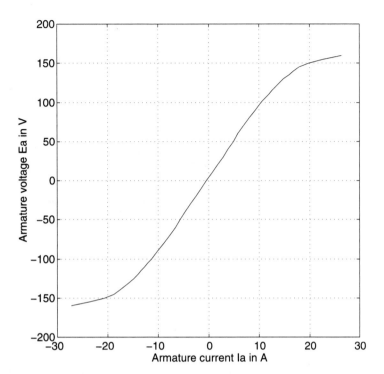

Figure 8.45   Magnetization curve of dc series motor for Project 5.

```
Vrated = 125;
Iarated = 13.2;
wmrated = 1425*(2*pi)/60;
Trated = Prated/wmrated;
Ra = 0.24;
Rse = 0.2;
Laq =  0.018;
Lse = 0.044;
J = 0.5; % rotor inertia in kgm2

% Enter magnetization curve data
wmo = 1200*(2*pi)/60; % speed at which mag. curve data was taken
% load voltage values of mag. curve
SEVP5 = [-160 -155 -150 -145 -140 -135 -130 -125 -120 -115 ...
-110 -105 -100 -90 -80 -70 -60 -50 -40 -30 -20 -10 0 ...
10 20 30 40 50 60 70 80 90 100 105 110 115 120 125 130 135 ...
140 145 150 155 160 ];

% load main field current values of mag. curve
SEIP5 = [-27.224 -23.547 -20.624 -18.739 -17.560 -16.617 ...
 -15.627 -14.826 -13.977 -13.317 -12.643 -11.969 -11.290 ...
-10.041 -8.886 -7.613 -6.576 -5.647 -4.643 -3.582 -2.521 ...
-1.423 -0.400  0.623  1.721  2.782  3.843  4.847  5.776  ...
```

```
6.813  8.086  9.241 10.490 11.169 11.843 12.517 13.177 14.026 ...
14.827 15.817 16.760 17.939 19.824 22.747 26.424 ];
% plotting of machine characteristics
plot_option = menu('Plot machine characteristics?', 'Yes' ,'No');
if plot_option == 1,
clf;
plot(SEIP5,SEVP5); % plot mag curve measured at wmo
xlabel('Armature current Ia in A')
ylabel('Armature voltage Ea in V')
axis('square')
grid on
% Transfer to keyboard for actions (e.g. printing) on above plot
disp('Displaying Mag characteristics, return for simulation');
keyboard
end % if plot_option

hseriesV = SEVP5(24:1:45); % obtain positive half of E_awithout
% zero value
hseriesI = SEIP5(24:1:45); % obtain positive half of I_a
kaphi = (hseriesV)/(wmo);
tem = (kaphi.*hseriesI);
%plot(tem,kaphi,'+') % check smoothness of kaphi curve
%keyboard

% Compute speed-torque characteristic with rated supply voltage
% and no external resistor

Vmotor = Vrated-2; % provide for 2 volt of brush drop
Rtotal = Ra + Rse;
E = Vmotor - hseriesI.*Rtotal;
wm = E./kaphi;
pem = E.*hseriesI; % tem.*wm;

% Case of Vbrake = Vmotor % only insert Rbrake
Vbrake = Vmotor
% Compute braking resistor to limit speed to wbrake with Tload=Trated
wbrake = -400*2*pi/60 % braking speed to rad/sec
Iabrake = interp1(tem,hseriesI,Trated) % interpolate for Trated
kaphibrake = interp1(hseriesI,kaphi,Iabrake)
Rbrake =  (Vbrake - kaphibrake*wbrake)/Iabrake - Ra - Rse
Rtotal = Ra + Rse + Rbrake;
E1 = Vbrake - hseriesI.*Rtotal;
wm1 = E1./kaphi;

% Transferring to keyboard to simulate the case of braking
% with Va = Vrated
disp('Run simulation of braking with Va=Vrated, before returning')
keyboard
% Plotting results. Save and return
clf;
subplot(4,1,1)
plot(y(:,1),y(:,5))
title('Rotor speed')
```

```
ylabel('wm in rad/sec')
subplot(4,1,2)
plot(y(:,1),y(:,4))
axis([-inf inf -200 200])
title('Braking Torque Tem with Va=Vrated')
ylabel('Tem in Nm')
subplot(4,1,3)
plot(y(:,1),y(:,3))
title('Armature current')
ylabel('Ia in A')
subplot(4,1,4)
plot(y(:,1),y(:,2))
title('Internal voltage Ea')
ylabel('Ea in V')
xlabel('time in sec')
disp('Displaying results of simulation in Fig. 1')
disp('Return next for plot of braking characteristics');
keyboard
% Case of Vbrake = 0
Vbrake = 0
Rbrake =  (Vbrake - kaphibrake*wbrake)/Iabrake - Ra - Rse
Rtotal = Ra + Rse + Rbrake;
E2 = Vbrake - hseriesI.*Rtotal;
wm2 = E2./kaphi;
% Plot braking performance with Va = 0 and Vrated along with
% motoring char. with rated voltage
clf;
plot(tem,wm,'-',tem,wm1,'-.',tem,wm2,'--')
axis([-inf 20 -50 300])
axis('square')
xlabel('Torque in Nm')
ylabel('Rotor speed in rad/sec')
% Transfer to keyboard for simulating the braking with Va = 0
disp('Displaying braking characteristics in Fig. 1');
disp('Run simulation of braking with Va=0 before returning')
keyboard
clf;
subplot(4,1,1)
plot(y(:,1),y(:,5))
title('Rotor speed')
ylabel('wm in rad/sec')
subplot(4,1,2)
plot(y(:,1),y(:,4))
axis([-inf inf -200 200])
title('Braking Torque Tem with Va=0')
ylabel('Tem in Nm')
subplot(4,1,3)
plot(y(:,1),y(:,3))
title('Armature current')
ylabel('Ia in A')
subplot(4,1,4)
plot(y(:,1),y(:,2))
title('Internal voltage Ea')
```

```
ylabel('Ea in V')
xlabel('time in sec')
```

Implement the SIMULINK simulation, *s5*, and set the threshold level of the switches, *Va* and *Rinsertion*, to 0.5, and the delay time of the *Switchover* switch to eight seconds to allow time for the motor to raise the rated load torque to almost the steady-state speed. For a changeover from raising to lowering after the delay time, the initial output level from the *Switchover* switch is 1 and the final output level is zero.

Execute the MATLAB M-file, *m5*, to set up the appropriate operating condition for you to simulate and observe the performance of the two braking methods in lowering operations. Use the *ode15s* or Adams/Gear method of integration, a stop time of 35 secs, a minimum step size of 0.5 msec, and a tolerance of $1e^{-6}$ to obtain the plots of E_a, I_a, T_{em}, and ω_m vs. time of the raising and lowering of the active load of rated value for braking with $V_a = 0$ and $V_a = V_{rated}$.

Sample results showing the braking performance of the series motor with these two values of V_a are given in Fig. 8.46.

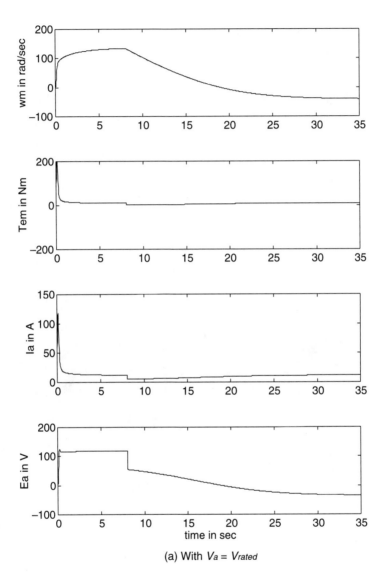

(a) With $V_a = V_{rated}$

Figure 8.46   Braking performance of series motor.

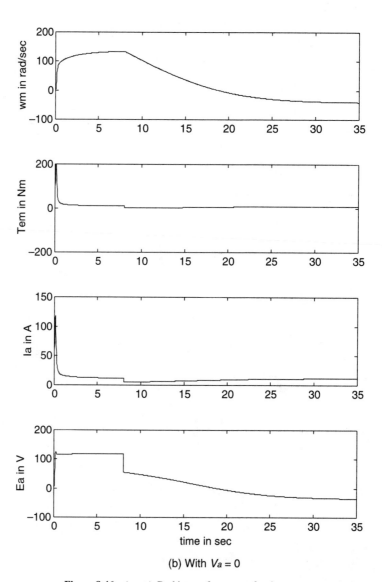

(b) With $V_a = 0$

Figure 8.46 *(cont.):* Braking performance of series motor.

9

Control

of

Induction Machines

9.1 INTRODUCTION

With the availability of faster and less expensive processors and solid-state switches, ac induction motor drives now compare favorably to dc motor drives on considerations such as power to weight ratio, acceleration performance, maintenance, operating environment, and higher operating speed without the mechanical commutator. Costs and robustness of the machine, and perhaps control flexibility, are often reasons for choosing induction machine drives in small to medium power range applications. In this chapter, we will first review the basics of multi-quadrant drive operation, after which we will examine some common methods by which the speed or torque of an induction machine is controlled. Even without simulating the switching waveforms in detail, that is with only the fundamental component of inverter outputs simulated, we can still gain useful insights into the dynamic performance attainable by common control strategies used in today's adjustable-speed induction motor drives.

9.1.1 Basics of Multi-quadrant Operation

For drive applications, the primary requirement is the ability to handle the load torque over a range of speeds with good dynamic performance. Figure 9.1 depicts the relation between torque and speed in all four quadrants of drive operation. In quadrant one, both

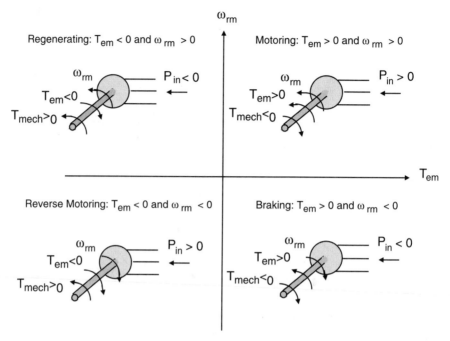

Figure 9.1 Four quadrants of drive operation.

the developed torque and speed of the machine are positive, as shown, counter-clockwise rotation is taken as positive. A machine operating in the first quadrant is said to be motoring, when the electrical power flowing into the machine terminals is converted into electromechanical power to turn the rotor against an externally applied mechanical torque, T_{mech}. In motoring, the externally applied mechanical torque is usually referred to as the load torque. As shown in quadrant one for motoring operation, T_{em} accelerates the rotor in the positive direction against a negative value, T_{mech}.

In quadrant two, the developed torque is negative, opposite to the positive rotor rotation. When the net electromechanical power output, given by $T_{em} \times \omega_{rm}$, is negative, the machine is said to be generating. Sustained operation in the second quadrant will require a positive T_{mech} to drive ω_{rm} in the forward direction. But for momentary excursion into second quadrant, as in the case of braking in the forward direction, both T_{mech} and T_{em} can be acting in a direction opposite to ω_{rm} in the forward direction. Thus, electrical power could still be flowing into the machine's stator winding when it is producing forward braking torque. In forward braking, the kinetic energy of the rotating rotor assembly is being converted to mechanical work and electrical power.

Operations in the third quadrant may be regarded as motoring in the reverse direction, since both developed torque and speed of machine are negative. In this quadrant, T_{mech} in the counter-clockwise direction is still considered positive. Likewise, operations in the fourth quadrant may be regarded as regeneration or braking in the reverse direction; T_{em} being counter-clockwise is positive, opposing the rotation clockwise. Again, for sustained

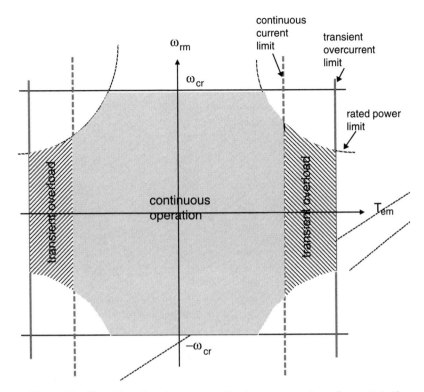

Figure 9.2 Allowable region of operation, taking into account ratings of converter and motor.

operation in the fourth quadrant, T_{mech} must balance T_{em} and the clockwise acting T_{mech} will be negative.

9.1.2 Operating Region Limited by Equipment Ratings

Since the induced voltage is proportional to flux and speed, and the developed torque is proportional to flux and current, the upper limit of operating speed can be determined by the maximum allowable speed of the rotor assembly or the voltage rating of the motor and converter. Because the construction of the motor is usually more robust than the power electronic devices of the converter, often a higher torque/current dynamic performance is obtained by allowing the motor to operate with a reasonable margin of transient overloading. But not so for the power electronics devices, as they are less forgiving in that exceeding the voltage and current limits of the power switches can often lead to catastrophic failure. Figure 9.2 shows the allowable region of operation for continuous and transient operation of such a drive where the converter has higher current and voltage ratings than those of the motor.

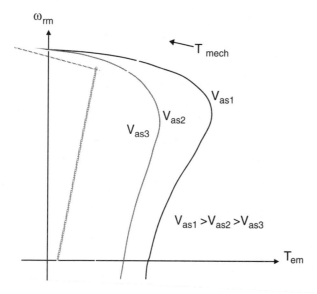

Figure 9.3 Speed control by reducing applied voltage.

9.2 SPEED CONTROL

9.2.1 Reducing Applied Voltage Magnitude

A simple means by which a limited range of speed control below the nominal motoring speed of the motor can be obtained is to reduce the applied voltage below its rated value. It is apparent from the intersections between the motor and load characteristics shown in Fig. 9.3 that the range of speed control is dependent not only on how much the torque-speed curve of the motor is changed by a reduction in applied voltage, but also on the load's torque-speed curve. Since the developed torque is proportional to the square of the applied voltage and the rotor current is proportional to the applied voltage, the ratio of torque developed to rotor current is proportional to the applied voltage. If the thermal loading is affected mainly by the rotor current, the developed torque for a given level of thermal loading will decrease as the applied voltage is reduced. In which case, low-speed operations without overheating are possible only if the load torque decreases with speed. Furthermore, since $P_{em} = (1-s)P_{ag}$, efficiency will decrease as slip increases.

9.2.2 Adjusting Rotor Circuit Resistance

With wound-rotor machines, external rotor circuit resistances can be introduced to limit the starting current drawn. Figure 9.4 shows that a limited range of speed control below the nominal motoring speed can also be obtained by varying the amount of external rotor circuit resistance. Unlike the previous method of reducing the applied voltage, where the breakdown torque also decreases with applied voltage, this method of speed control can provide constant torque operation at a high torque to current ratio.

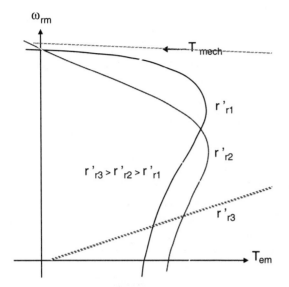

T_{em} vs. w_{rm} for three different values of rotor circuit resistance

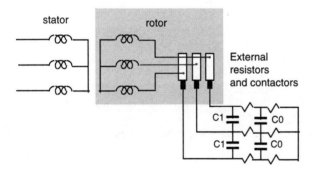

Figure 9.4 Speed control by adjusting rotor resistance.

9.2.3 Adjusting Stator Voltage and Frequency

With an inverter supply, the magnitude, frequency, and phase of the voltages applied to the motor can all be varied electronically. If the inverter can handle bi-directional power flow, the motor can be made to operate in all four quadrants. But, if the inverter can only handle current flow in one direction, from supply to motor, the operation will be limited to one or two quadrants.

When the excitation frequency, ω_e, is zero, the value of slip s given by $(\omega_e - \omega_r)/\omega_e$ becomes undefined. This can be a problem with the conventional form of the steady-state equations given in Eq. 6.68. For variable frequency operations that include operation with dc stator excitation, the earlier form of the steady-state equations given in Eq. 6.66, repeated below, is more useful:

$$\tilde{V}_{as} = (r_s + j\omega_e L_{ls})\tilde{I}_{as} + j\omega_e L_m(\tilde{I}_{as} + \tilde{I}'_{ar}) \qquad V$$

$$\tilde{V}'_{ar} = (r'_r + j(\omega_e - \omega_r)L'_{lr})\tilde{I}'_{ar} + j(\omega_e - \omega_r)L_m(\tilde{I}_{as} + \tilde{I}'_{ar}) \qquad V \qquad (9.1)$$

The corresponding expression to use for the average value of the electromagnetic torque developed by a P-pole machine is

$$T_{em} = \frac{3I'^2_{ar}r'_r}{\omega_{sm} - \omega_{rm}} = \frac{3P}{2(\omega_e - \omega_r)}I'^2_{ar}r'_r \qquad N.m \qquad (9.2)$$

9.3 CONSTANT AIRGAP FLUX OPERATION

In this section, we will show that the torque-speed curve of an induction machine that is operated with its mutual flux held constant is translated along the speed axis with no changes to its shape as the supply's excitation frequency changes.

The airgap, or mutual flux of the induction machine, may be expressed as

$$|\lambda_m| = L_m|\tilde{I}_{as} + \tilde{I}'_{ar}| = \frac{E_m}{\omega_e} \qquad Wb.\ turn \qquad (9.3)$$

Thus, holding the airgap or mutual flux constant is equivalent to holding the ratio E_m/ω_e constant, that is

$$E_m = \omega_e \frac{E_{m,rated}}{\omega_b} \qquad V \qquad (9.4)$$

where $E_{m,rated}$ is the value of E_m at rated or base frequency. The maximum continuous value of E_m should not be higher than its value at rated frequency if excessive core saturation is to be avoided.

When the machine is singly-excited, that is V'_{ar} is zero, we can deduce from Fig. 9.5 that the rotor current is given by

$$\tilde{I}'_{ar} = -\frac{\tilde{E}_m}{\frac{\omega_e r'_r}{\omega_e - \omega_r} + j\frac{\omega_e}{\omega_b}x'_{lr}} \qquad A \qquad (9.5)$$

Using Eq. 9.4 to replace E_m in the above rotor current expression and squaring, we obtain

$$I'^2_{ar} = \frac{E_m^2}{\left(\frac{\omega_e r'_r}{\omega_e - \omega_r}\right)^2 + \left(\frac{\omega_e x'_{lr}}{\omega_b}\right)^2} = \frac{E_{m,rated}^2}{\left(\frac{\omega_b r'_r}{\omega_e - \omega_r}\right)^2 + x'^2_{lr}} \qquad (9.6)$$

Substituting the square of the rotor current in Eq. 9.2 with Eq. 9.6, the expression for the developed torque with constant airgap flux becomes

$$T_{em} = \frac{3P}{2(\omega_e - \omega_r)}\left(\frac{E_{m,rated}^2}{\left(\frac{\omega_b r'_r}{\omega_e - \omega_r}\right)^2 + x'^2_{lr}}\right)r'_r \qquad N.m \qquad (9.7)$$

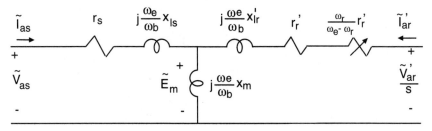

Figure 9.5   Equivalent circuit of induction machine.

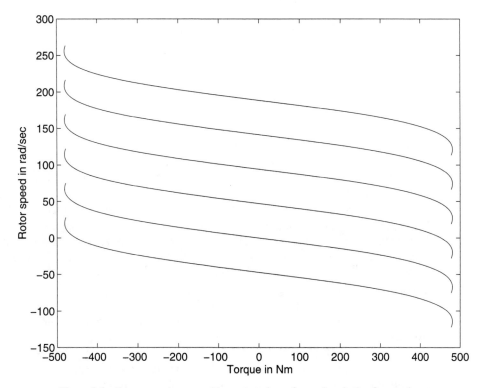

Figure 9.6   Torque speed curves with constant airgap flux and excitation frequencies at 60, 45, 30, 15, 0, and −15 Hz.

An examination of the above expression of the developed torque for constant airgap flux operation will reveal that the value of torque remains the same at a given value of the slip speed, $\omega_e - \omega_r$, for any value of ω_e. Graphically, this is equivalent to a vertical translation of the speed vs. torque curve for rated condition along the speed axis as frequency changes; the shape of the curve is unaffected. Figure 9.6 shows the torque-speed curves for several excitation frequencies of the 20-hp, 60-Hz, 220-V three-phase induction motor of Project 2 in Chapter 6 operating with its mutual or airgap flux maintained constant at the rated condition value for all values of slip. These curves indicate that with constant airgap flux operation, we will obtain the same value of torque at the same slip speed of $\omega_e - \omega_r$. From Eq. 9.6, we see that the same remark also applies to the magnitude of $\tilde{I}'_{ar}$.

With E_m held constant, maximum power for a given excitation frequency will be delivered across the airgap when

$$\frac{\omega_e r'_r}{\omega_e - \omega_{rmaxt}} = \frac{\omega_e}{\omega_b} x'_{lr} \tag{9.8}$$

In other words, the slip speed at maximum or pull-out torque is

$$\omega_e - \omega_{rmaxt} = \pm \frac{\omega_b r'_r}{x'_{lr}} \qquad elect. \ rad/s \tag{9.9}$$

The positive value on the right side of Eq. 9.9 corresponds to a motoring slip and the negative value corresponds to a generating slip. The value of maximum torque developed, obtained by substituting Eq. 9.9 into Eq. 9.7, is

$$T_{em}^{max} = \frac{3P}{4\omega_b} \frac{E_{m,rated}^2}{x_{lr}'^2} \qquad N.m \tag{9.10}$$

The above value of maximum torque is independent of the excitation frequency, ω_e; it is, in fact, the same as that obtained with rated angular frequency, ω_b.

It can also be seen from Eq. 9.7 that the torque can be controlled by adjusting stator flux, slip speed or both, via the control of the motor's stator voltage magnitude and frequency. With a voltage-fed inverter, the slip speed is usually kept well within the maximum slip value given by Eq. 9.9 to keep the input power factor high, and the stator current and losses low. Within the voltage rating of the inverter, the magnitude of the stator voltage can be adjusted to keep stator flux constant. However, keeping both stator flux and slip speed constant will result in higher slip and therefore higher rotor losses at the lower operating frequencies. Higher than needed excitation flux will be accompanied by higher core losses, but the higher excitation will provide a ready reserve capacity to handle potentially higher torque in another part of the loading cycle.

Rotor excitation flux may be considered as stator excitation flux times slip. Thus, over the constant torque range, maintaining rotor flux at a level capable of providing the maximum expected torque is preferred. Above base speed, in the constant horsepower range, slip speed is held constant by allowing slip to increase with speed until it reaches its maximum, s_{maxt}. With the rated slip often about half the value of s_{maxt}, the upper end of the constant horsepower range is usually about twice rated or base speed. A higher upper limit of the speed range for constant horsepower operation can be obtained by using an oversized motor.

9.4 CONSTANT VOLTS/HERTZ OPERATION

Although the regulation of airgap flux may be accomplished with direct feedback of the measured airgap flux, in practice, the use of terminal voltage measurement is preferred because airgap flux measurement with Hall sensors or search coils, besides having filtering

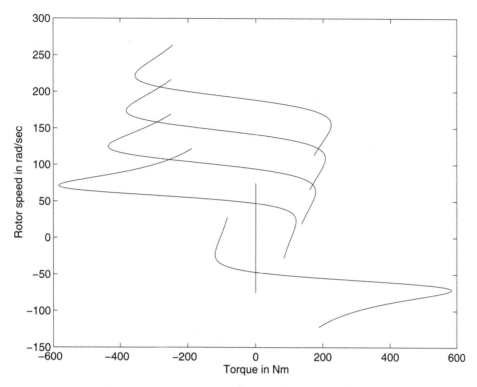

Figure 9.7   Torque speed curves with $|v_{as}| = V_{rated} * f/f_{rated}$ and excitation frequencies at 60, 45, 30, 15, 0, and -15 Hz.

and mounting problems, must be customized to the machine, precluding the use of off-the-shelf replacement motors in the field. Indirect control of the airgap voltage with terminal voltage feedback is simpler. However, as shown in Fig. 9.7, the torque speed curves of the same motor operated with a straightforward constant volts per hertz control are not the same as those of Fig. 9.6 obtained with airgap flux held constant for all operating conditions. For non-zero excitation frequencies, the distortion and skew in Fig. 9.7 may be attributed to the change in mutual flux caused by the series drop across the $r_s + j(\omega_e/\omega_b)x_{ls}$, especially at low excitation frequencies where the resistive drop component becomes dominant.

Figure 9.8a shows the stator volts/hertz curves for the same 20-hp motor for three values of stator current. Each curve is obtained by keeping the magnitude of both the stator current and airgap flux constant at their respective values at rated operating condition as excitation frequency is being changed. These curves indicate that the magnitude of stator terminal voltage required will be a function of both frequency and loading. In general, the value of terminal voltage needed for motoring is higher than that for no-load, which in turn is higher than that for generating. In motoring operations, the series drop reduces the airgap voltage, but for generating operations, the power flow and the series drop reverses. With the exception of the curve for $\omega_e = 0$, the impact of the series voltage at the low excitation

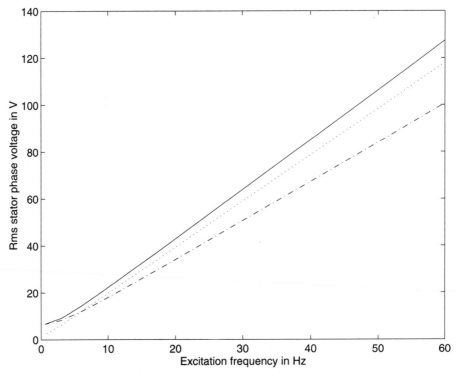

(a) Stator volts/hertz with $e_m = E_{mrated}^* f/f_{rated}$ for three loadings:
solid line for rated motoring current, dotted line for no-load,
and dash-dot line for rated generating current

Figure 9.8   Effects of loading on operating characteristics.

frequency range can be compensated for by adding a small voltage boost to the constant V/f characteristic at the lower excitation frequency end for motoring operations.

Figure 9.8b shows the torque-speed curves obtained with the V/f curves of Fig. 9.8a. They are different from those of Fig. 9.6; the new curves have lower pull-out torque. By examining the currents of both sets, we will be able to see the reason for the reduced torque. Those of Fig. 9.6 have higher stator and rotor currents, especially near the pull-out torque condition.

Figure 9.9a shows the stator volts/hertz curves of the same 20-hp motor when its airgap flux is held constant at rated value and rotor currents give the same pull-out torque given by Eq. 9.10. The torque-speed curves using excitation voltages based on these curves are shown in Fig. 9.9b. These torque-speed curves compare favorably with those of Fig. 9.6, with the exception of that for $\omega_e = 0$. The curve for $\omega_e = 0$ has not only a much higher value of pull-out torque, but also a different shape.

Let's examine the case where ω_e is zero further to determine the difference in shape of the curve. With $\omega_e = 0$, the magnetizing voltage, E_m, is always zero and the stator current will be determined by stator terminal voltage and the stator winding resistance. When the

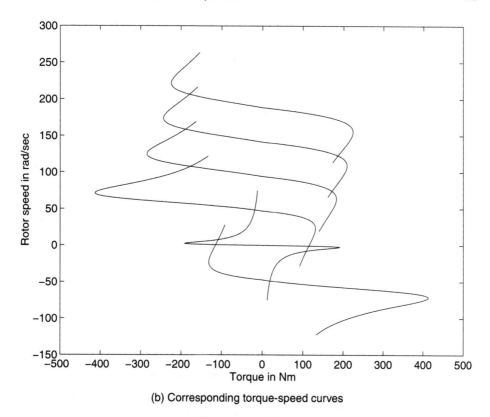

(b) Corresponding torque-speed curves

Figure 9.8   *(cont.):* Effects of loading on operating characteristics.

stator terminal voltage is fixed, the stator current remains unchanged at all slip speeds. Consequently, as determined, the sharper torque-speed curve in Fig. 9.9b is one of constant current rather than constant flux excitation.

The large pull-out torque of the torque-speed curve for the case of $\omega_e = 0$ can be reduced by using a lower stator voltage. Figure 9.10a shows the modified stator volts/hertz curves for the pull-out torque at $\omega_e = 0$ to be same as those of other excitation frequencies.

The case of dc stator excitation, that is $\omega_e = 0$, occurs not only in an adjustable-frequency drive, but also in the use of dc injection to obtain a braking torque. When the voltage equations of the machine are written in the form shown in Eq. 9.1, the solution of these equations can be handled numerically for any ω_e, including zero. With ω_e set to zero, Eq. 9.1 becomes

$$\tilde{V}_{as} = r_s \tilde{I}_{as} \qquad V$$

$$\tilde{V}'_{ar} = (r'_r - j\omega_r L'_{lr})\tilde{I}'_{ar} - j\omega_r L_m (\tilde{I}_{as} + \tilde{I}'_{ar}) \qquad (9.11)$$

With only stator excitation, that is $\tilde{V}'_{ar}$ is zero, the rotor phasor current is given by

$$\tilde{I}'_{ar} = -\frac{j\omega_r L_m}{-r'_r + j\omega_r L'_r}\tilde{I}_{as} \qquad A \qquad (9.12)$$

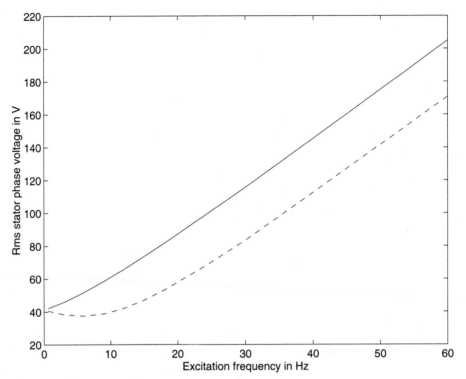

(a) Stator volts/hertz: solid line for motoring and dash-dot line for generating

Figure 9.9   Stator volts/hertz with $e_m = E_{mrated} * f/f_{rated}$ and rotor current to produce the pull-out torque as given by Eq. 9.10.

The negative equivalent rotor resistance acts like a source of power driving the rotor current, regenerating power back through the stator. The motoring torque is

$$T_{em} = -\frac{3P}{2}\frac{\omega_r L_m^2 r_r'}{r_r'^2 + \omega_r^2 L_r'^2} I_{as}^2 \qquad N.m \tag{9.13}$$

In this situation, the torque developed is negative or a braking torque for positive value of ω_r.

9.5 INDUCTION MACHINE DRIVES

The availability of efficient power switches and fast processors has facilitated the growth in development and use of induction motor drives. In a typical induction motor drive, the power converter serves to convert the supply energy into a form suited for the operation of the motor. The output characteristic of the power converter can be controlled to appear like an adjustable magnitude and frequency current or voltage source to the motor. Figure 9.11 shows a basic block diagram of an induction motor drive. The power converter conditions

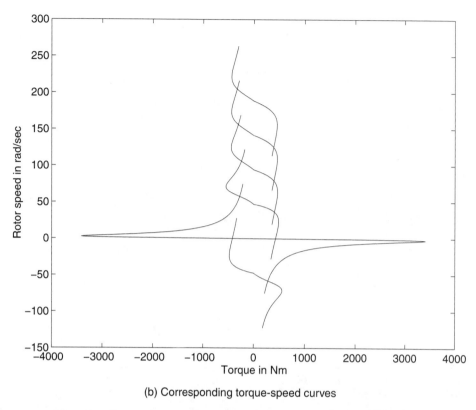

(b) Corresponding torque-speed curves

Figure 9.9 *(cont.):* Stator volts/hertz with $e_m = E_{mrated} * f/f_{rated}$ and rotor current to produce the pull-out torque as given by Eq. 9.10.

the power delivered to the motor in some desired manner as directed by the controller. Usually, the conditioning involves generating the proper voltage or current waveforms with the right timing. The primary function of the controller is to translate the command value and feedback signals to control signals for operating the converter. It may also have other functions, such as condition monitoring and protection.

The subject of simulating power electronic circuits is by itself a major topic, which is beyond the scope of this text. Nevertheless, the electromechanical dynamics of induction motor drives under different control strategies can be studied with simulation using a sinusoidal voltage or current source of variable magnitude and frequency to represent the fundamental component of actual inverter waveform. Since power is mainly carried by the fundamental component, the primary control interaction will be portrayed. Such a simulation can provide some understanding of the control principles involved without going into the details of the power electronics of the inverter, which can be difficult and time-consuming to simulate because of the frequent discontinuities caused by the inverter switchings. It, of course, cannot be used to address problems which are associated with the harmonics of the actual inverter waveforms.

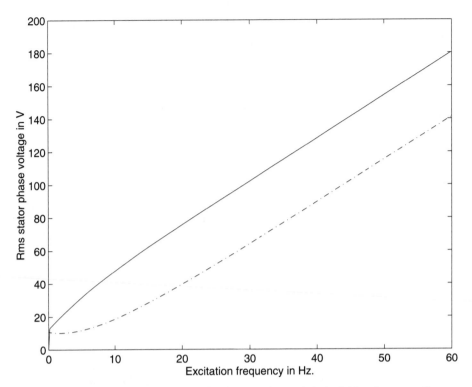

(a) Modified stator volts/hertz: solid line for motoring and dash-dot line for generating

Figure 9.10   Stator volts/hertz with $e_m = E_{mrated} * f/f_{rated}$ and all pull-out torque the same as in Fig. 9.6 at 60, 45, 30, 15, 0, and -15 Hz.

Inverter types can be divided into two broad categories by the source characteristic of their dc links: voltage- or current-fed inverters. Voltage-fed inverters have supply sources that behave like dc voltage sources, often rectifiers with shunt capacitor in the dc links. The supply to a current-fed inverter usually is a controlled rectifier with a series large inductor in the dc link. Most of today's small induction motor drives use a voltage-fed pulse-width-modulation (pwm) inverter that allows both magnitude and frequency of its ac output voltages to be changed electronically.

The function of a pulse-width modulator is to translate the modulation waveforms of variable amplitude and frequency into a train of switching pulses for the inverter. In a classical sinusoidal pulse-width modulator, the crossovers between a sinusoidal modulation wave and a triangular carrier wave determine the points of commutations of the pulse trains. Figure 9.12 shows the output voltage waveforms of a three-phase sinusoidal pwm inverter. In a sinusoidal pwm inverter, the modulation index is defined as the ratio of the amplitude of the modulating wave to the amplitude of the triangular carrier wave. The magnitude of the fundamental component of the pwm output is proportional to the modulation index, when the modulation index is less than unity. With a modulation index of one, the magnitude

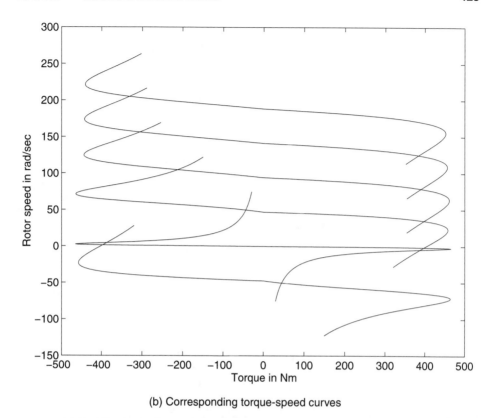

(b) Corresponding torque-speed curves

Figure 9.10   *(cont.):* Stator volts/hertz with $e_m = E_{mrated} * f / f_{rated}$ and all pull-out torque the same as in Fig. 9.6 at 60, 45, 30, 15, 0, and -15 Hz.

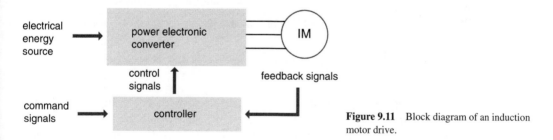

Figure 9.11   Block diagram of an induction motor drive.

of the fundamental component is about 79 percent of that of a square wave of the same amplitude. As the modulation index approaches unity, the pulse width becomes too narrow, in that there will not be sufficient time for the semiconductor switches to turn off and recover their voltage-blocking capability. In inverters where a dc bus short-circuit through two complementary switched devices of an inverter leg is possible, pulse duration that is insufficient for complete recovery of the device during turn off can be prevented by locking out further switching until after a specified minimum time.

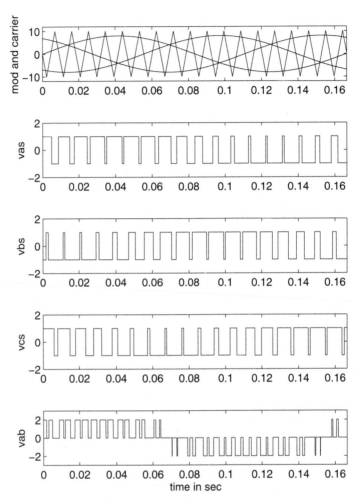

Figure 9.12   Pwm and square wave ac output waveforms.

Above unity modulation index, the relationship between output fundamental magnitude and modulation index becomes nonlinear. In approaching the upper limit of the
ac output voltage, many voltage-fed inverters will switch from pulse modulated to some
form of block modulated and finally to quasi-square wave output. The techniques for a
gradual transition from sine pwm to square are discussed in References [89] - [91]. Block
modulation tends to produce relatively high content of lower harmonics, which are usually
not a problem at the higher speeds. It is a modulation technique that is favored in a current-
fed inverter, especially at higher power.

There are other pwm schemes besides the basic sinusoidal pwm scheme mentioned
here, each with something to offer. For example, the harmonic elimination technique
described in [84] allows selective elimination of undesirable harmonics staircase [82, 83] or

sub-harmonic injection [86] pwm schemes that employ non-sinusoidal modulating waves, pwm schemes using voltage space vector to reduce certain current harmonics [88], and optimal pwm precalculated off-line to meet some specific criteria.

9.5.1 Operating Strategies

Figure 9.13 shows the operating strategies often used for motoring to obtain a wide speed range. For motoring, we can identify the following three modes:

> **Mode 1:** Holds slip speed constant and regulates stator current to obtain constant torque.
>
> **Mode 2:** Holds stator voltage at its rated value and regulates stator current to obtain constant power.
>
> **Mode 3:** Holds stator voltage at its rated value and regulates slip speed just below its pull-out torque value.

In Mode 1, the ratio of magnitude of inverter output voltage to inverter frequency is adjusted to maintain airgap flux approximately constant. The maximum available torque in the constant torque region is usually set by limiting the inverter current to be under that corresponding to the pull-out torque. The transition from Mode 1 to Mode 2 occurs when the maximum available voltage of the inverter is reached. In Mode 2, the drive operates with maximum available inverter voltage; often its output waveform will be quasi-square. As frequency continues to increase in this mode, the machine will operate with reduced airgap flux. The envelope of this mode is obtained by increasing slip to maintain stator current at its limit. Transition from Mode 2 to Mode 3 occurs when slip finally approaches a value corresponding to the pull-out torque condition. Thereafter, slip will be kept just under its pull-out torque value and regulation of stator current ceases. The upper limit of speed may be determined by considerations, such as unacceptably low pull-out, excessive iron and windage losses, mechanical clearance, and bearing loading.

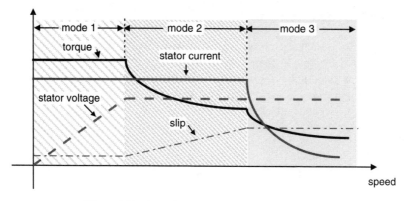

Figure 9.13   Operating modes over a wide speed range.

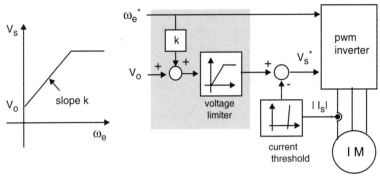

(a) Linearly compensated volts/hertz control

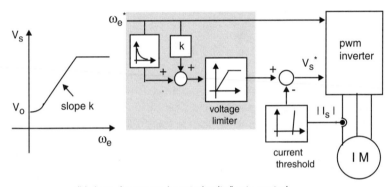

(b) Low frequency boosted volts/hertz control

Figure 9.14   Volts/hertz controls.

Figure 9.13 shows the operating strategies for only motoring; those for braking are about the same. The upper limit of Mode 1 is at a higher speed than the case for motoring, because in braking, the voltage drops in the machine are reversed.

9.5.2 Scalar Control

For many variable-speed applications where a small variation of motor speed with loading is tolerable, a simple open-loop system using a volt per hertz control with low-frequency compensation, such as those shown in Fig. 9.14, may be satisfactory. Figure 9.14 shows only two simple ways of generating a compensated volts per hertz characteristic. Other ways with load-dependent features are discussed in reference [92].

As shown, a slip speed command is added to the measured rotor speed to produce the desired inverter frequency. The command value of slip speed can be negative, in which case, the machine will be generating instead of motoring; however, its value must be limited to some safe margin below the slip speed corresponding to the pull-out torque point. Since the slip speed is normally small relative to the rotor speed, such a scheme requires rather precise measurement of the rotor speed. Operating with negative slip speed will cause the

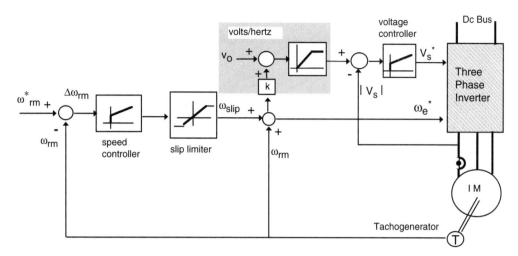

Figure 9.15   Closed-loop speed control with volts/hertz and slip regulation.

motor to regenerate power back into the dc link. The regenerated power has to be dissipated in a braking resistor or regenerated back into the ac mains to prevent excessive rise in dc voltage due to the over-charging of the dc link smoothing capacitor.

The controlled slip strategy is widely used because the induction machine's input power factor and torque to stator current ratio can both be kept high, resulting in better utilization of the available inverter current. When airgap flux and slip speed are both held constant, the torque developed will be the same, but the efficiency is not as good as that obtained with airgap flux and slip held constant. When slip is held constant, the slip speed will vary linearly with excitation frequency and the slope of the torque-speed curve on the synchronous speed side will decrease with excitation frequency. Figure 9.15 shows a closed-loop speed control scheme which uses volts/hertz and slip regulation.

9.6 FIELD-ORIENTED CONTROL

In a dc machine, the axes of the armature and field winding are usually orthogonal to one another. The mmfs established by currents in these windings will also be orthogonal. If iron saturation is ignored, the orthogonal fields produce no net interaction effect on one another. The developed torque may be expressed as

$$T_{em} = k_a \phi(I_f) I_a \qquad N.m \qquad (9.14)$$

where k_a is a constant coefficient, $\phi(I_f)$, the field flux, and I_a, the armature current. Here, the torque angle is naturally 90°, flux may be controlled by adjusting the field current, I_f, and torque may be controlled independently of flux by adjusting the armature current, I_a. Since the time constant of the armature circuit is usually much smaller than that of the field winding, controlling torque by changing armature current is quicker than changing I_f, or both.

In general, torque control of a three-phase induction machine is not as straightforward as that of a dc machine because of the interactions between the stator and rotor fields whose orientation are not held spatially at 90° but vary with operating condition. The field of the rotor winding in an induction machine may be likened to that of the field winding in a dc machine, except that it being induced is not independently controllable. With sinusoidal excitation, the rotor field rotates at synchronous speed. If we were to select a synchronously rotating $qd0$ frame whose d-axis is aligned with the rotor field [95], the q component of the rotor field, $\lambda_{qr}^{\prime e}$, in the chosen reference frame would be zero, that is

$$\lambda_{qr}^{\prime e} = L_m i_{qs}^e + L_r^\prime i_{qr}^{\prime e} = 0 \qquad Wb. \ turn \tag{9.15}$$

$$i_{qr}^{\prime e} = -\frac{L_m}{L_r^\prime} i_{qs}^e \qquad A \tag{9.16}$$

With $\lambda_{qr}^{\prime e}$ zero, the first equation in Eq. 6.38 for the developed torque reduces to

$$T_{em} = -\frac{3}{2}\frac{P}{2}\lambda_{dr}^{\prime e} i_{qr}^{\prime e} \qquad N.m \tag{9.17}$$

Substituting for $i_{qr}^{\prime e}$ using Eq. 9.16, Eq. 9.17 can be written in the desired form of

$$T_{em} = \frac{3}{2}\frac{P}{2}\frac{L_m}{L_r^\prime}\lambda_{dr}^{\prime e} i_{qs}^e \tag{9.18}$$

which shows that if the rotor flux linkage, $\lambda_{dr}^{\prime e}$, is not disturbed, the torque can be independently controlled by adjusting the stator q component current, i_{qs}^e.

For $\lambda_{qr}^{\prime e}$ to remain unchanged at zero, $p\lambda_{qr}^{\prime e}$ must be zero, in which case, the q-axis voltage equation of the rotor winding with no applied rotor voltages reduces to

$$v_{qr}^{\prime e} = \underbrace{r_r^\prime i_{qr}^{\prime e}}_{=0} + \underbrace{p\lambda_{qr}^{\prime e}}_{=0} + (\omega_e - \omega_r)\lambda_{dr}^{\prime e} \qquad V \tag{9.19}$$

In other words, the slip speed must satisfy

$$\omega_e - \omega_r = -\frac{r_r^\prime i_{qr}^{\prime e}}{\lambda_{dr}^{\prime e}} \qquad elect. \ rad/s \tag{9.20}$$

Also, if $\lambda_{dr}^{\prime e}$ is to remain unchanged, $p\lambda_{dr}^{\prime e}$ must be zero too. Using this condition and that of $\lambda_{qr}^{\prime e}$ being zero in the d-axis rotor voltage equation, we will obtain the condition that $i_{dr}^{\prime e}$ must be zero, that is

$$v_{dr}^{\prime e} = \underbrace{r_r^\prime i_{dr}^{\prime e}}_{=0} + \underbrace{p\lambda_{dr}^{\prime e}}_{=0} - (\omega_e - \omega_r)\underbrace{\lambda_{qr}^{\prime e}}_{=0} \qquad V \tag{9.21}$$

And, when $i_{dr}^{\prime e}$ is zero, $\lambda_{dr}^{\prime e} = L_m i_{ds}^e$. Substituting this into Eq. 9.20 and using Eq. 9.16, we obtain the following relationship between slip speed and the ratio of the stator qd current components for the d-axis of the synchronously rotating frame to be aligned with the rotor field:

$$\omega_e - \omega_r = \frac{r_r^\prime}{L_r^\prime}\frac{i_{qs}^e}{i_{ds}^e} \qquad elect. \ rad/s \tag{9.22}$$

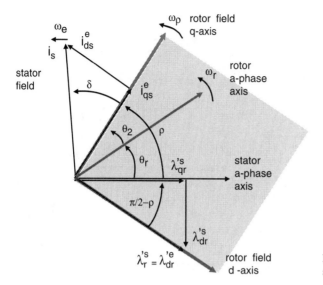

Figure 9.16   Properly oriented qd synchronously rotating reference.

In practice, the magnitude of rotor flux can be adjusted by controlling i_{ds}^e, and the orientation of the d-axis to the rotor field can be maintained by keeping either slip speed or i_{qs}^e in accordance to Eq. 9.22. With proper field orientation, the dynamic of $\lambda_{dr}^{\prime e}$ will be confined to the d-axis and is determined by the rotor circuit time constant. This can be seen from Eq. 9.21 with $i_{dr}^{\prime e}$ replaced by $(\lambda_{dr}^{\prime e} - L_m i_{ds}^e)/L_r'$, that is

$$\lambda_{dr}^{\prime e} = \frac{r_r' L_m}{r_r' + L_r' p} i_{ds}^e \qquad Wb.\ turn \qquad (9.23)$$

Field-oriented control schemes for the induction machine are referred to as the *direct* type when the angle, ρ, shown in Fig. 9.16, is being determined directly as with the case of direct airgap flux measurement, or as the *indirect* type when the rotor angle is being determined from surrogate measures, such as slip speed.

9.6.1 Direct Field-oriented Current Control

Figure 9.17 shows a direct field orientation control scheme for torque control using a current-regulated pwm inverter. For field orientation, controlling stator current is more direct than controlling stator voltage; the latter approach must allow for the additional effects of stator transient inductances. With adequate dc bus voltage and fast switching devices, direct control of the stator current can be readily achieved. The direct method relies on the sensing of airgap flux, using specially fitted search coils or Hall-effect devices. The drift in the integrator associated with the search coil is especially problematic at very low frequencies. Hall devices are also temperature-sensitive and fragile.

The measured flux in the airgap is the resultant or mutual flux. It is not the same as the flux linking the rotor winding, whose angle, ρ, is the desired angle for field orientation. But, as shown by the expressions below, in conjunction with the measured stator current,

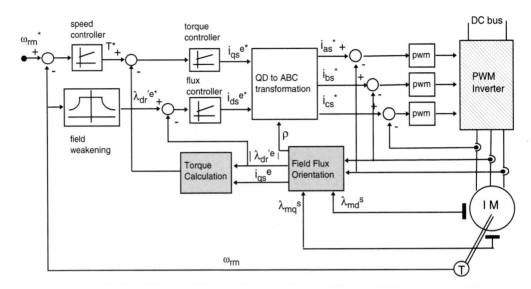

Figure 9.17   Direct field-oriented control of a current-regulated pwm inverter induction motor drive.

we can determine the value of ρ and the magnitude of the rotor flux. The measured *abc* stator currents are first transformed to the stationary qd currents using

$$i_{qs}^s = \frac{2}{3}i_{as} - \frac{1}{3}i_{bs} - \frac{1}{3}i_{cs} \qquad A$$

$$i_{ds}^s = \frac{1}{\sqrt{3}}(i_{cs} - i_{bs}) \tag{9.24}$$

Adding and subtracting a $L_{lr}'i_{qs}^s$ term to the right-hand side, the rotor q-axis flux linkage in the stationary reference frame may be expressed as

$$\lambda_{qr}'^s = (L_m + L_{lr}' - L_{lr}')i_{qs}^s + (L_m + L_{lr}')i_{qr}'^s \qquad Wb.\ turn \tag{9.25}$$

Since λ_{mq}^s is equal to $L_m(i_{qs}^s + i_{qr}'^s)$, we can determine $\lambda_{qr}'^s$ from the measured quantities, that is

$$\lambda_{qr}'^s = \frac{L_r'}{L_m}\lambda_{mq}^s - L_{lr}'i_{qs}^s \tag{9.26}$$

Similarly, $\lambda_{dr}'^s$ can be determined from

$$\lambda_{dr}'^s = \frac{L_r'}{L_m}\lambda_{md}^s - L_{lr}'i_{ds}^s \qquad Wb.\ turn \tag{9.27}$$

Using $\lambda_{qr}'^s$ and $\lambda_{dr}'^s$ computed from Eqs. 9.26 and 9.27, we can determine the cosine and sine of ρ by the following geometrical relations that are deducible from Fig. 9.16:

$$\sin\left(\frac{\pi}{2} - \rho\right) = \cos\rho = \frac{\lambda_{dr}'^s}{|\lambda_r'^s|}$$

$$\cos\left(\frac{\pi}{2} - \rho\right) = \sin\rho = \frac{\lambda_{qr}'^s}{|\lambda_r'^s|} \tag{9.28}$$

where

$$|\lambda_r'^e| = |\lambda_r'^s| = \sqrt{\lambda_{dr}'^{s\,2} + \lambda_{qr}'^{s\,2}} \tag{9.29}$$

The above computations, from Eqs. 9.24 to 9.29, are performed inside the field orientation block shown in the middle of Fig. 9.17. The calculated value of $|\lambda_r'^e|$ is fed back to the input of the flux controller regulating airgap flux. Inside the torque calculation block, the calculated values of $\lambda_r'^e$ and i_{qs}^e are used in Eq. 9.18 to estimate the value of torque developed by the machine, and the estimated torque is fed back to the input of the torque controller.

The respective outputs of the torque and flux controllers are the command values, i_{qs}^{e*} and i_{ds}^{e*}, in the field-oriented rotor reference frame. Inside the qd to abc transformation block are the following transformations from qde to qds and qds to balanced abc:

$$
\begin{aligned}
i_{qs}^{s*} &= i_{qs}^{e*} \cos\rho + i_{ds}^{e*} \sin\rho \qquad A \\
i_{ds}^{s*} &= -i_{qs}^{e*} \sin\rho + i_{ds}^{e*} \cos\rho
\end{aligned}
\tag{9.30}
$$

$$
\begin{aligned}
i_{as}^* &= i_{qs}^{s*} \qquad A \\
i_{bs}^* &= -\frac{1}{2} i_{qs}^{s*} - \frac{\sqrt{3}}{2} i_{ds}^{s*} \\
i_{cs}^* &= -\frac{1}{2} i_{qs}^{s*} + \frac{\sqrt{3}}{2} i_{ds}^{s*}
\end{aligned}
\tag{9.31}
$$

9.6.2 Direct Field-oriented Voltage Control

Field orientation of the stator currents can also be achieved by applying the proper stator voltages. Since the strategy in a field-oriented scheme is to avoid perturbing the rotor flux linkage as much as possible when responding to a change in load torque, we can use the transient model derived in Section 6.7 in conjunction with the properly-oriented qd stator currents to determine the stator voltages to apply. The field-oriented qd stator currents are determined by converting the measured abc currents to qd stationary using Eq. 9.24 and the value of ρ as determined by Eq. 9.28 in the transformation below:

$$
\begin{aligned}
i_{qs}^e &= i_{qs}^s \cos\rho - i_{ds}^s \sin\rho \qquad A \\
i_{ds}^e &= i_{qs}^s \sin\rho + i_{ds}^s \cos\rho
\end{aligned}
\tag{9.32}
$$

In the transient model of Section 6.7, we have shown that for a situation where the rotor flux linkages may be assumed to remain constant, the machine can be represented by constant voltages behind the stator transient inductance. The response of the machine to stator side transients can be represented by the set of equations given in Eq. 6.88. The stator flux linkages can be expressed in terms of only the stator currents and rotor flux linkages, that is

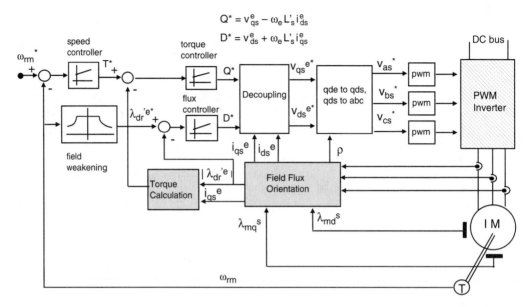

Figure 9.18   Direct field-oriented control of a voltage-regulated pwm inverter induction motor drive.

$$\lambda_{qs}^e = L_s' i_{qs}^e + \frac{L_m}{L_r'} \lambda_{qr}'^e \qquad Wb.\ turn$$

$$\lambda_{ds}^e = L_s' i_{ds}^e + \frac{L_m}{L_r'} \lambda_{dr}'^e$$

(9.33)

Substituting the above expressions for the stator flux linkages into the derivative term on the left side of Eq. 6.89, we obtain

$$L_s' \frac{di_{qs}^e}{dt} + \frac{L_m}{L_r'} \frac{d\lambda_{qr}'^e}{dt} = v_{qs}^e - r_s i_{qs}^e - E_{qs}' - \omega_e L_s' i_{ds}^e \qquad V$$

$$L_s' \frac{di_{ds}^e}{dt} + \frac{L_m}{L_r'} \frac{d\lambda_{dr}'^e}{dt} = v_{ds}^e - r_s i_{ds}^e - E_{ds}' + \omega_e L_s' i_{qs}^e$$

(9.34)

Setting the time derivative terms of the rotor flux linkages to zero and rearranging so that the left-hand side has the sum of the voltage behind the transient impedance and the voltage drop across the stator transient impedance, we obtain

$$r_s i_{qs}^e + L_s' \frac{di_{qs}^e}{dt} + E_{qs}' = v_{qs}^e - \omega_e L_s' i_{ds}^e$$

$$r_s i_{ds}^e + L_s' \frac{di_{ds}^e}{dt} + E_{ds}' = v_{ds}^e + \omega_e L_s' i_{qs}^e$$

(9.35)

Figure 9.18 shows a direct field orientation scheme using stator voltage regulation. In this scheme, the two left-hand side values of Eq. 9.35 are assumed to be produced by the torque and flux controllers. Adjusting these controller outputs for the cross-coupling current terms on the right-hand side of Eq. 9.35, we will obtain the desired command values for v_{qs}^e and v_{ds}^e. The command values for the abc stator voltages can then be computed as follows:

$$v_{qs}^{s*} = v_{qs}^e \cos \rho + v_{ds}^e \sin \rho \qquad V$$
$$v_{ds}^{s*} = -v_{qs}^e \sin \rho + v_{ds}^e \cos \rho \tag{9.36}$$

$$v_{as}^* = v_{qs}^{s*}$$

$$v_{bs}^* = -\frac{1}{2} v_{qs}^{s*} - \frac{\sqrt{3}}{2} v_{ds}^{s*} \tag{9.37}$$

$$v_{cs}^* = -\frac{1}{2} v_{qs}^{s*} + \frac{\sqrt{3}}{2} v_{ds}^{s*}$$

9.7 INDIRECT FIELD ORIENTATION METHODS

For very low-speed operations and for position type control, the use of flux sensing that relies on integration which has a tendency to drift may not be acceptable. A commonly-used alternative is indirect field orientation, which does not rely on the measurement of the airgap flux, but uses the condition in Eqs. 9.18, 9.22, and 9.23 to satisfy the condition for proper orientation. Torque can be controlled by regulating i_{qs}^e and slip speed, $\omega_e - \omega_r$. Rotor flux can be controlled by regulating i_{ds}^e. Given some desired level of rotor flux, $\lambda_r'^*$, the desired value of i_{ds}^{e*} may be obtained from

$$\lambda_{dr}'^* = \frac{r_r' L_m}{r_r' + L_r' p} i_{ds}^{e*} \qquad Wb. \ turn \tag{9.38}$$

For the desired torque of T_{em}^* at the given level of rotor flux, the desired value of i_{qs}^{e*} in accordance with Eq. 9.18 is

$$T_{em}^* = \frac{3}{2} \frac{P}{2} \frac{L_m}{L_r'} \lambda_{dr}'^{e*} i_{qs}^{e*} \qquad N.m \tag{9.39}$$

It has been shown that when properly oriented, $i_{dr}'^e$ is zero and $\lambda_{dr}'^e = L_m i_{ds}^e$; thus, the slip speed relation of Eq. 9.22 can also be written as

$$\omega_2^* = \omega_e - \omega_r = \frac{r_r'}{L_r'} \frac{i_{qs}^{e*}}{i_{ds}^{e*}} \qquad elect. \ rad/s \tag{9.40}$$

The above conditions, if satisfied, ensure the decoupling of the rotor voltage equations. To what extent this decoupling is actually achieved will depend on the accuracy of motor parameters used. Since the values of rotor resistance and magnetizing inductance are known to vary somewhat more than the other parameters, on-line parameter adaptive techniques are often employed to tune the value of these parameters used in an indirect field-oriented controller to ensure proper operation.

Figure 9.19 shows an indirect field-oriented control scheme for a current controlled pwm induction motor drive. The field orientation, ρ, is the sum of the rotor angle from the

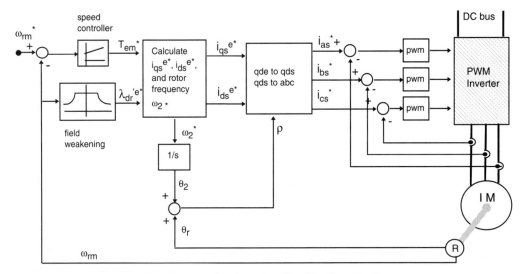

Figure 9.19   Indirect field-oriented control of a current regulated pwm inverter induction motor drive.

position sensor, θ_r, and the angle, θ_2, from integrating the slip speed. If orthogonal outputs of the form $\cos\theta_r$ and $\sin\theta_r$ are available from the shaft encoder, the values of $\cos\rho$ and $\sin\rho$ can be generated from the following trigonometric identities:

$$\cos\rho = \cos(\theta_r + \theta_2) = \cos\theta_r \cos\theta_2 - \sin\theta_r \sin\theta_2$$
$$\sin\rho = \sin(\theta_r + \theta_2) = \sin\theta_r \cos\theta_2 + \cos\theta_r \sin\theta_2$$

$$(9.41)$$

In simulation, the value of $\cos\theta_2$ and $\sin\theta_2$ may be generated from a variable-frequency oscillator of the kind shown earlier in Section 2.8.

9.8 PROJECTS

The objectives of this group of projects on closed-loop control of induction motors are to experiment firsthand the implementation of the control methods discussed earlier through simulation and to gain insight into some elements of these drives, their wave shapes, and transient performances.

9.8.1 Project 1: Closed-loop Speed Using Scalar Control

In this project, we will experiment with the implementation of the closed-loop speed scalar volts/hertz control induction machine drive shown in Fig. 9.15. We will examine the startup transients of the motor with a preset ramp rate of the speed reference, the dynamic response of the drive to cyclic changes in speed command, and load torque. For purposes of comparison, we will use the transient behavior of the same motor operated with fixed supply voltage and no feedback control.

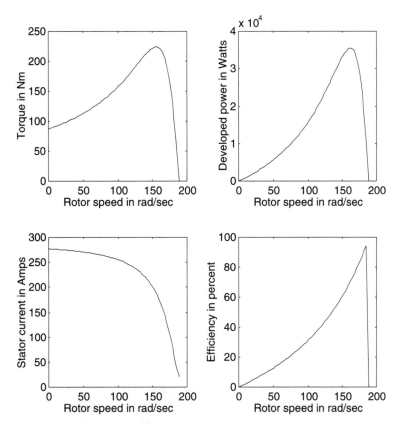

Figure 9.20   Steady-state characteristics of 20-hp, 220-V, four-pole motor.

For comparison purposes, we will use the same 20-hp motor used in Project 2 of Chapter 6. The 20-hp, 220-V, 60-Hz, four-pole, squirrel-cage induction motor has the following parameters.

$$r_s = 0.1062 \ \Omega \qquad\qquad x_{ls} = 0.2145 \ \Omega$$

$$r_r' = 0.0764 \ \Omega \qquad\qquad x_{lr}' = 0.2145 \ \Omega$$

$$x_m = 5.834 \ \Omega \qquad\qquad J_{rotor} = 2.5 \ kgm^2$$

Rated stator current is 49.68 A, rated speed, 1748.3 rev/min, and rated slip, 0.0287. Plots of some steady-state characteristics of the motor vs. rotor speed are shown in Fig. 9.20. Operating from a rated frequency and sinusoidal voltage supply, the rated torque is 81.49 N.m and the slip at which maximum torque occurs is 0.1758.

Fixed voltage open-loop operation.   Figure 9.21a shows an overall diagram of the SIMULINK simulation, *s1o,* for the first part of this project on open-loop operations. The SIMULINK simulation is set up for simulating the dynamic behavior of

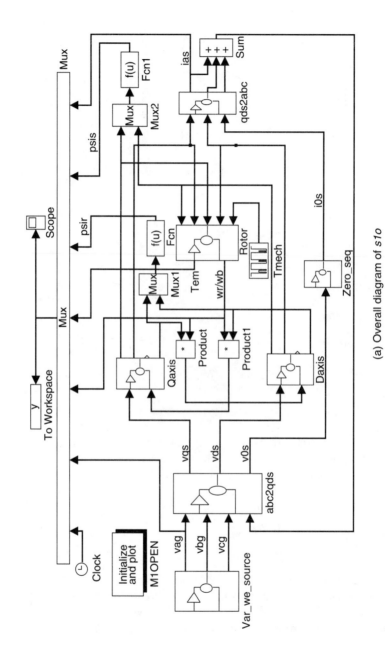

(a) Overall diagram of *s1o*

Figure 9.21 Simulation *s1o* of an open-loop induction motor drive.

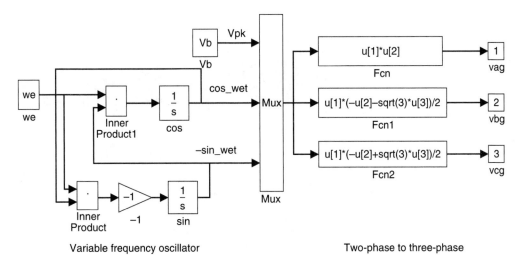

(b) Inside the *Var_we_source block of s10*

Figure 9.21   *(cont.):* Simulation *s10* of an open-loop induction motor drive.

the motor when started DOL with the rated voltage and when subjected to cyclic changes in torque loading. The results from these open-loop operations will later be used as a benchmark to compare of the performance of the same motor operated with closed-loop feedback.

The MATLAB M-files, *m1o* and *m1c*, can be used in conjunction with the SIMULINK simulations of the open-loop and closed-loop system, *s1o* and *s1c*, respectively to set up the machine parameters, the run conditions, and to plot the results of the simulation. The value of the externally applied mechanical torque is generated by a *repeating sequence* source with the time and and output values scheduled as follows:

Time array: $time\_tmech = [0\ 0.75\ 0.75\ 1.0\ 1.0\ 1.25\ 1.25\ 1.5\ 1.5\ 2]$
Output array: $tmech\_tmech = [0\ 0$ *-Trated -Trated -Trated/2 -Trated/2 -Trated -Trated* $0\ 0$ $]$

The two function blocks, *Fcn* and *Fcn1*, in the main diagram take the square root of the sum of the square of the rotor and stator flux linkages. The inside of the *Var_we_source* block is shown in Fig. 9.21b. The details inside the other simulation blocks of the main diagram are the same as those given in Project 1 of Chapter 6.

Sample results showing the starting of the motor with full stator voltage and the response of the motor to the programmed sequence of loading are shown in Fig. 9.22. Note the large starting current and fluctuations in the magnitude of the stator and rotor flux linkages as the loading changes.

Closed-loop volts/hertz speed control.   Figure 9.23 shows the SIMULINK simulation, *s1c*, that is to be used for the second part of this project on closed-loop operations. The SIMULINK simulation is set up to simulate closed-loop control of the rotor speed. In this project, we will not simulate the switching of the inverter. To save simulation

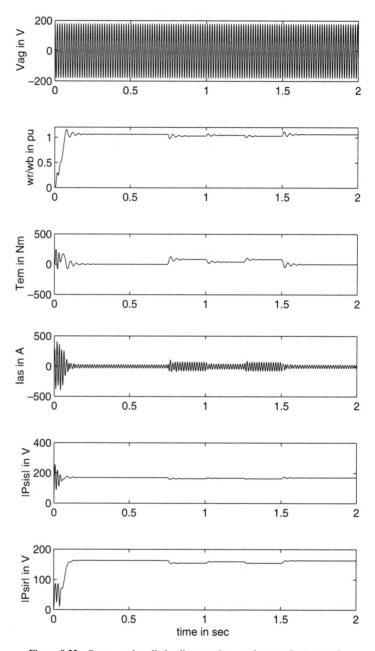

Figure 9.22   Startup and cyclic loading transients under open-loop control.

time, only the fundamental components of the inverter output voltages are represented by variable frequency and magnitude sinusoidal voltage sources. The generation of these voltages are shown in Fig. 9.23b.

Often the neutral point of the stator winding is not brought out or connected to the neutral point, g, of the sources. As such, the source voltages, v_{ag}, v_{bg}, and v_{cg}, are, in general, not equal to the voltages, v_{as}, v_{bs}, and v_{cs}, of the stator windings. As shown in Fig. 9.23a, in $s1c$ we have chosen the neutral point, g, of the three-phase sinusoidal voltage sources as the system reference. The neutral point, s, of the stator windings of the induction motor is assumed to be free-floating. To obtain the applied voltages across the phase windings of the stator, we must first determine the voltage, v_{sg}. In the $abc2qds$ block of $s1c$, the neutral voltage of the stator winding is determined from the voltage drop of three stator phase currents flowing in a fictitious capacitor, C_{sg}, connecting the points s and g. The value of C_{sg} should be kept relatively small so as not to compromise the accuracy of the simulation of the actual open-circuit condition. However, if the value of C_{sg} used is too small, relative to the other parameters, the resulting system equations can become artificially stiff, which can cause convergence difficulties. On the other hand, too large a value of C_{sg} will not only result in inaccuracy, but also spikes on the v_{sg} waveform. When the actual system condition is not marginally stable, using the proper algorithm and a capacitance reactance of about 50 per unit on the machine base will not usually give any numerical problems.

The voltage, v_{sg}, between the neutral points, s and g, can be determined from

$$v_{sg} = \frac{1}{C_{sg}} \int (i_{as} + i_{bs} + i_{cs})dt \qquad V \tag{9.42}$$

Given the phase voltages of the three-phase source, v_{ag}, v_{bg}, and v_{cg}, those of the stator phase windings of the induction machine in the simulation can then be determined from

$$v_{as} = v_{ag} - v_{sg} \qquad V$$
$$v_{bs} = v_{bg} - v_{sg} \tag{9.43}$$
$$v_{cs} = v_{cg} - v_{sg}$$

A *look-up* table is used to produce the desired volts per hertz characteristic shown in Fig. 9.24 for the drive. Satisfactory transient performance can be obtained using the peak value of the rms voltage, V, and the operating frequency, ω_e, based on the following relationships:

$$\tilde{E}_{mb} = jx_m I_{rated} \qquad V$$
$$\tilde{E}_m = \frac{\tilde{E}_{mb}}{\omega_b}\omega_e \qquad V$$
$$\mathbf{Z_s} = r_s + j\frac{\omega_e}{\omega_b}x_{ls} \qquad \Omega \tag{9.44}$$
$$V = |\tilde{E}_m + I_{rated}\mathbf{Z_s}| \qquad V$$

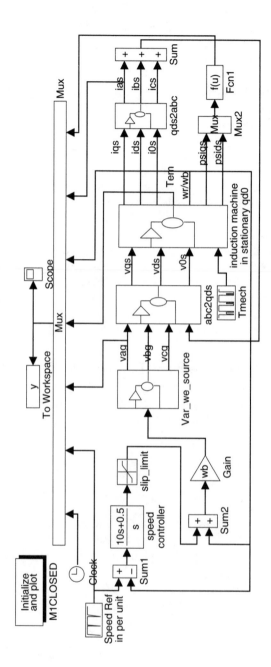

Figure 9.23 Simulation *s1c* of a closed-loop volts/hertz speed control drive.

(a) Overall diagram of *s1c*

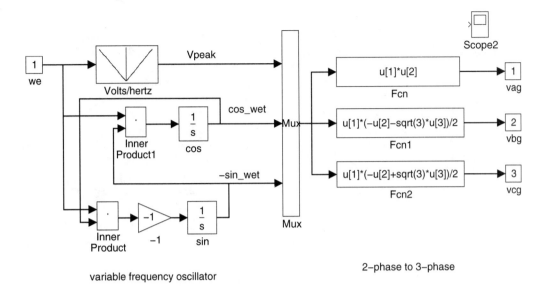

(b) Inside the *Var_we_source* block

Figure 9.23   *(cont.):* Simulation *s1c* of a closed-loop volts/hertz speed control drive.

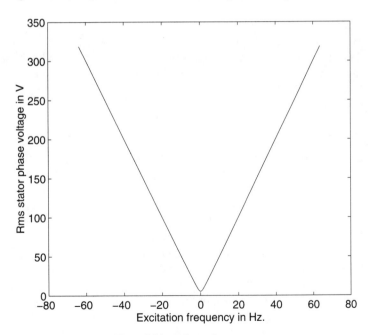

Figure 9.24   Volts per hertz curve.

Startup and loading transients.   Set up the SIMULINK simulation, *s1c*, shown in Fig. 9.23. Use *m1c* to set up the parameters of the 20-hp induction machine in the MATLAB workspace, as well as the operating conditions for the run. Set the upper and lower limits on the output slip of the slip controller to $\pm 0.7 s_{maxt}$. Repeat the startup and load cycling run as in the open-loop case, this time, however, ramp the stator excitation frequency from zero to 60 Hz in 0.5 second. Aside from the initial speed ramping to start the machine, the loading cycle is the same as that used in the open-loop case. The values of the time and output arrays of the *repeating sequence* source for speed reference are *time_wref* = [0 0.5 4] and *speed_wref* =[0 1 1], respectively. Those of the *repeating sequence* source for the externally applied mechanical torque are *time_tmech* = [0 0.75 0.75 1.0 1.0 1.25 1.25 1.5 1.5 2] and *tmech_tmech* = [0 0 -Trated -Trated -Trated/2 -Trated/2 -Trated -Trated 0 0]. Check your results against the sample result shown in Fig. 9.25. Comment on differences in the startup and loading transients when compared to the case with no control.

Effect of changing the limits on slip.   Repeat the above startup and loading run but with the upper and lower limits on the output slip of the slip controller changed from $\pm 0.7 s_{maxt}$ to $\pm 0.5 s_{maxt}$. Compare the startup transients obtained with those obtained previously with the higher limits on the slip speed. Explain the differences.

Response to changes in speed reference.   Next, we will use the simulation, *s1c*, to examine the dynamic response of the motor to a cyclic change in speed reference beginning with the motor unexcited. For this run, the values of the time and output arrays of the *repeating sequence* source for speed reference are to be changed to *time_wref* = [0 0.25 0.5 1.0 1.25 1.5] and *speed_wref* = [0 0.5 0.5 -0.5 -0.5 0], respectively. Also, those for the externally applied mechanical torque are to be changed to *time_tmech* = [0 4] and *tmech_tmech* = [0 0]. Reset the upper and lower limits on the output slip of the slip controller to $\pm 0.7 s_{maxt}$. Sample results of the unloaded motor's response to the cyclic change in speed reference are shown in Fig. 9.26.

9.8.2 Project 2: Six-step Inverter/Induction Motor Drive

In this second project on the closed-loop speed control induction motor drive, we will represent the inverter output by a set of three-phase, six-step voltages of variable frequency and magnitude, instead of just their fundamental components used in Project 1. Each of the phase voltages is displaced 120° from the other.

Some of the things that will become apparent from doing these two projects are that the dynamic response of the system with just the fundamental components represented is almost the same as that obtained with higher harmonics of the inverter switchings included, and that a simulation with step changes in variables not only will have to be done with greater care, but will also take longer to run than the case when the variables are changing smoothly. In this project, we will also have the opportunity to examine the harmonics in the line current and voltage of the motor when supplied from a six-step voltage-fed inverter.

The SIMULINK simulations of a six-step voltage-fed inverter/motor drive with open-loop and closed-loop scalar control are given in *s2o* and *s2c*, respectively. The

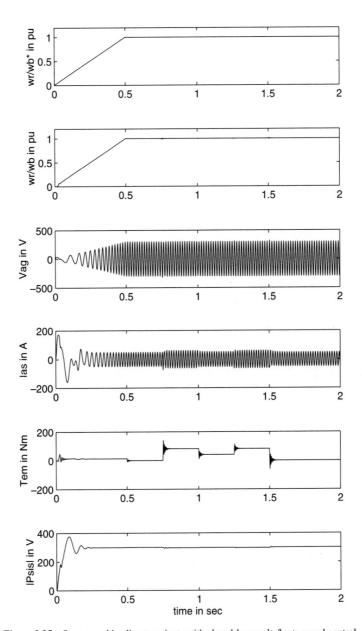

Figure 9.25   Startup and loading transients with closed-loop volts/hertz speed control.

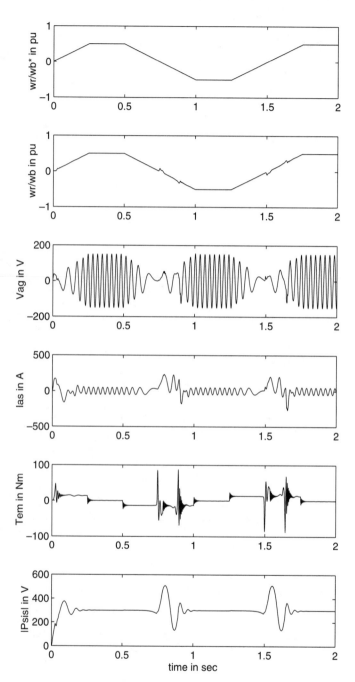

Figure 9.26   Response to speed reference changes with closed-loop volts/hertz speed control.

corresponding MATLAB script files for setting up the parameters of these two systems, the operating conditions, and the plotting of the simulated results are the same as those used in Project 1, that is *m1o* may be used for *s2o* and *m1c* for *s2c*. Figure 9.27 shows the SIMULINK simulation, *s2c*, for the closed-loop speed control induction motor drive with a six-step inverter waveform representation. Compared to the simulation used in Project 1 of this chapter, the modifications to change the inverter voltage output waveforms from sinusoidal to six-step is in the variable frequency source block shown in Fig. 9.27b. The additional *sign* functions after the three-phase sinusoidal voltages will produce a set of square waves, phase-displaced 120° from one another.

Since a square wave has a fundamental component that is of magnitude $4/\pi$ of its amplitude, a compensation factor of $\pi/4$ must be introduced to the slope of the volt/hertz gain, so that the amplitude of the square waves will produce the same fundamental component as the rated sinusoidal voltage at the rated frequency. Note also that the input to the volts/hertz *look-up* table is taken directly from the speed reference, instead of from the *we* output of the amplifier summing per unit slip and rotor speed. If it were taken from the latter *we* output, the amplitude of the step wave output would be modulated by the pulsations in the rotor speed. In practice, the magnitude of the step wave output is equal to the dc link voltage at the inverter input. Adjustment of the dc link voltage can be accomplished using a controllable rectifier.

Wave shapes.   Show that the stationary $qd0$ components of the stator winding phase voltages, that is v_{as}, v_{bs}, and v_{cs} are given by

$$v_q^s = \frac{2}{3}v_{as} - \frac{1}{3}(v_{bs} + v_{cs}) = \frac{2}{3}v_{ag} - \frac{1}{3}(v_{bg} + v_{cg}) - v_{sg} \qquad V$$

$$v_d^s = \frac{1}{\sqrt{3}}(v_{cs} - v_{bs}) = \frac{1}{\sqrt{3}}(v_{cg} - v_{bg}) \qquad (9.45)$$

$$v_0^s = \frac{1}{3}(v_{as} + v_{bs} + v_{cs}) = \frac{1}{3}(v_{ag} + v_{bg} + v_{cg}) - v_{sg}$$

Examine the wave shape of v_{sg}, v_{ag}, v_{as}, v_{ab}, i_{as}, i_{0s}, and T_{em}. The term six-step refers to the steps on the output line-to-line voltages, as in v_{ab}. Satisfy yourself that the level of the fundamental component of v_{as} is the same as v_{ag}, in spite of the difference in their appearance.

Startup and loading transients.   Repeat the startup and load cycling run, ramping the stator excitation frequency from zero to 60 Hz in 0.5 second as in the closed-loop control case of Project 1. Set the values of the time and output arrays of the *repeating sequence* source for the speed reference to correspond to a *time_wref* = [0 0.5 4], and *speed_wref* = [0 1 1], respectively. Similarly, set those of the *repeating sequence* source for the externally applied mechanical torque to *time_tmech* = [0 0.75 0.75 1.0 1.0 1.25 1.25 1.5 1.5 2] and *tmech_tmech* = [0 0 -Trated -Trated -Trated/2 -Trated/2 -Trated -Trated 0 0].

Check your results against the sample result shown in Fig. 9.28. Comment on the differences in the startup and loading transients when compared to the case with no control.

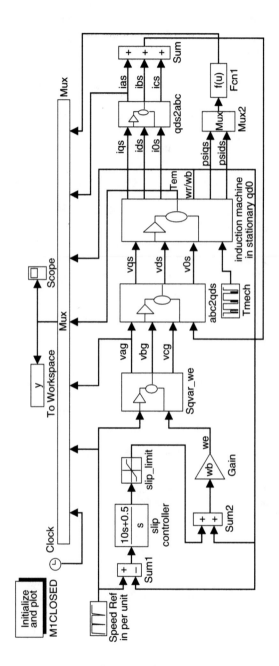

(a) Overall diagram of *s2c*

Figure 9.27 Closed-loop volts/hertz speed control with six-step inverter waveforms.

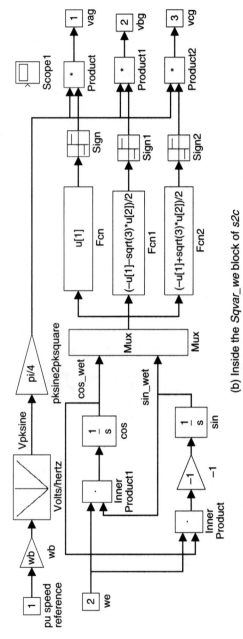

(b) Inside the *Sqvar_we* block of *s2c*

Figure 9.27 (cont.): Closed-loop volts/hertz speed control with six-step inverter waveforms.

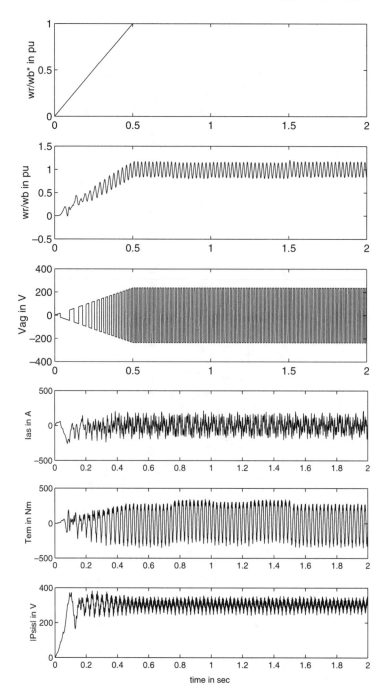

Figure 9.28   Startup and loading transients with volts/hertz speed control.

Response to changes in speed reference.   Next, we will examine the dynamic response of the motor to a cyclic change in speed reference beginning with the motor unexcited. The values of the time and output arrays of the *repeating sequence* source for speed reference are to be changed to $time\_wref = [0\ 0.25\ 0.5\ 1.0\ 1.25\ 1.5]$ and $speed\_wref = [0\ 0.5\ 0.5\ -0.5\ -0.5\ 0]$, respectively. Those for the externally applied mechanical torque are correspondingly changed to $time\_tmech = [0\ 4]$ and $tmech\_tmech = [0\ 0]$. Sample results of the unloaded motor's response to the cyclic change in speed reference are shown in Fig. 9.29.

9.8.3 Project 3: Field-oriented Control

In this project, we will implement the SIMULINK simulation, *s3*, of a current regulated pwm induction motor drive with indirect field-oriented control. The objectives of the project are to become familiar with the implementation of a type of field-oriented control, examine how well such a control keeps the rotor flux constant during changes in load torque, and to observe the improvement in dynamic response with this method of vector control to that with the scalar control in Project 1 of this chapter. For comparison purposes, we will use the same 20-hp, 220-V, four-pole induction machine of Project 1.

Figure 9.30a shows the overall diagram of the SIMULINK simulation, *s3*, for a current regulated pwm induction motor drive with indirect field-oriented control. To avoid lengthy simulation time, the switchings of the pwm converter are omitted and only the fundamental components of the pwm output voltages are portrayed. On the left side of the diagram, a proportional-integral (PI) torque controller converts the speed error to a reference torque, T_{em}^*. Going into the field orientation block are the reference torque, T_{em}^*, the d-axis rotor flux, $\lambda_{dr}^{'e*}$, and the rotor angle, θ_r.

Figure 9.30b shows the inside of the *Field_Orient* block. Inside it, Eqs. 9.38, 9.39, and 9.40 are used to compute the values of i_{ds}^{e*}, i_{qs}^{e*}, and ω_2^*. The angle, ρ, is the sum of the slip angle, θ_2, and the rotor angle, θ_r. In the *qde2abc* block, the transformations given in Eqs. 9.30 and 9.32 are used to generate the *abc* reference currents. In the SIMULINK simulation of Fig. 9.30, three large shunt resistors, each of a value that is 500 times the base impedance of the machine, are connected across the stator phase terminals to the star point, *g*, of the source. They are used to generate the input terminal *abc* phase voltages to the stator windings of the induction motor. The star point of the stator windings, *s*, is floating. The simulation of the motor and its rotor equation is the same as that described earlier in Project 1 of Chapter 6.

The *look-up* table for *field-weakening* matches the desired value of the rotor d-axis flux, $\lambda_{dr}^{'e}$, to that of the mechanical speed of the rotor, ω_{rm}. For speed less than the base or rated speed, $\lambda_{dr}^{'e}$ is set equal to its no-load value with rated supply voltage, as determined by Eqs. 6.90 and 6.93. Beyond the base speed, the flux speed product is held constant at the base speed value. The values of $\lambda_{dr}^{'e}$ and mechanical speed are generated by the *m3* file, which is also used to set up the parameters and other run conditions of the simulation in the MATLAB workspace. The values for $\lambda_{dr}^{'e}$ in the *field-weakening look-up* table are

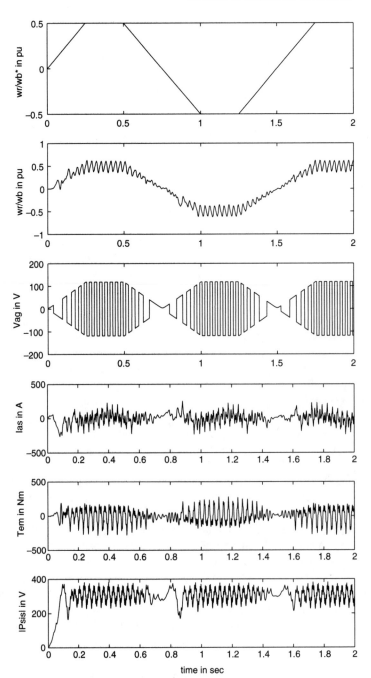

Figure 9.29   Response to cyclic change in speed reference with volts/hertz speed control.

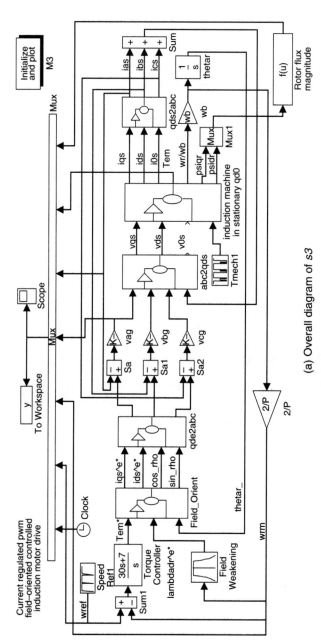

(a) Overall diagram of *s3*

Figure 9.30 Current regulated induction motor drive with indirect field-oriented control.

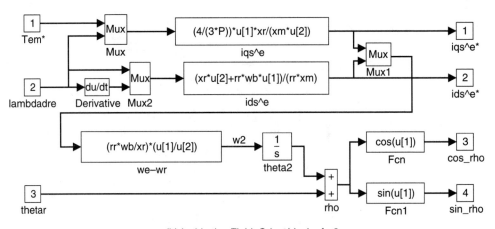

(b) Inside the *Field_Orient* block of *s3*

Figure 9.30 *(cont.)*: Current regulated induction motor drive with indirect field-oriented control.

0.1624	0.1710	0.1805	0.1911	0.2030	0.2166	0.2320	0.2499
0.2707	0.2953	0.3248	0.3248	0.3248	0.3248	0.3248	0.3248
0.3248	0.3248	0.3248	0.3248	0.3248	0.3248	0.3248	0.3248
0.3248	0.3248	0.3248	0.3248	0.3248	0.3248	0.3248	0.2953
0.2707	0.2499	0.2320	0.2166	0.2030	0.1911	0.1805	0.1710
0.1624							

The corresponding values of the mechanical speed are

−376.9911	−358.1416	−339.2920	−320.4425	−301.5929	−282.7433	−263.8938	−245.0442
−226.1947	−207.3451	−188.4956	−169.6460	−150.7964	−131.9469	−113.0973	−94.2478
−75.3982	−56.5487	−37.6991	−18.8496	0	18.8496	37.6991	56.5487
75.3982	94.2478	113.0973	131.9469	150.7964	169.6460	188.4956	207.3451
226.1947	245.0442	263.8938	282.7433	301.5929	320.4425	339.2920	358.1416
376.9911							

Set up the SIMULINK simulation, *s3*, of the current regulated induction motor drive with indirect field-oriented control shown in Fig. 9.30. Use the MATLAB script file, *m3*, to set up the machine and control parameters, the reference signal, and the load disturbance for the two studies described below.

Step changes in torque at a fixed reference speed. In this study, the machine is ramped up to speed using the speed reference after which it is subjected to a sequence of step changes in load torque. Use a text editor to check that *m3* sets the following values

tstop: The study time to two seconds.

time_wref: The time array of the speed reference *repeating sequence* signal source to [*0 0.5 tstop*].

speed_wref: The value array of the speed reference *repeating sequence* signal source to [*0 wbm wbm*].

time_tmech: The time array of the T_{mech} *repeating sequence* source to [*0 0.75 0.75 1. 1. 1.25 1.25 1.5 1.5 2*].

tmech_tmech: The value array of the T_{mech} *repeating sequence* source to [*0 0 -Trated -Trated -Trated/2 -Trated/2 -Trated -Trated 0 0*].

Run the simulation using the *ode15s* or Adams/Gear method, a stop time of two seconds, a minimum step size of $1e^{-4}$ second, a maximum step size of 0.01 second, and a tolerance of $1e^{-5}$. These parameters are picked, along with the size of shunt capacitors used to generate the stator phase voltages, to give reasonable accuracy and simulation time. Reducing the size of the shunt capacitors will give a closer approximation to the actual of no physical capacitance; it will however, require finer time steps and a longer run-time.

Plot the values of the reference speed, $wref$, the rotor mechanical speed, wrm, the stator a-phase voltage, vag, the stator a-phase current, ias, the electromagnetic torque, Tem, and the magnitude of the rotor flux, $\sqrt{\psi_{qr}^2 + \psi_{dr}^2}$. Sample results from such a run are shown in Fig. 9.31. Observe how quickly and smoothly the speed of the motor responds to the programmed speed reference with the field-oriented control. Observe also how well your field-oriented control maintains the magnitude of the rotor flux over the simulated condition.

Compare your results with the sample results of such a run shown in Fig. 9.31. Discuss the response of the motor to the loading sequence and check the values of phase current and voltage with the steady-state curves shown in Fig. 9.20 for the same machine.

Cyclic change of speed reference.   For this study, check to see that *m3* sets the following values:

tstop: The study time to two seconds.

time_wref: The time array of the speed reference *repeating sequence* signal source to [*0 0.25 0.5 1.0 1.25 1.5*].

speed_wref: The value array of the speed reference *repeating sequence* signal source to [*0 wbm/2 wbm/2 -wbm/2 -wbm/2 0*].

time_tmech: The time array of the T_{mech} *repeating sequence* source to [*0 tstop*].

tmech_tmech: The value array of the T_{mech} *repeating sequence* source to [*0 0*].

Repeat the above run with the motor carrying an externally applied mechanical torque equal to the rated load torque, that is $T_{mech} = -Trated$. Sample results for this run are given in Fig. 9.32. Note the slightly larger transient of the rotor flux at the initial startup of the motor from standstill. Examine the acceleration torque, that is $T_{mech} + T_{em}$, and the rotor angle, θ_r, during the initial startup transients. Explain why the same kind of transient in the rotor flux is not present in the beginning of the second cycle of the speed ramp-up.

Controller parameter detuning.   In the field, detuning is often due to changes in machine parameters with operating condition. For example, as rotor resistance

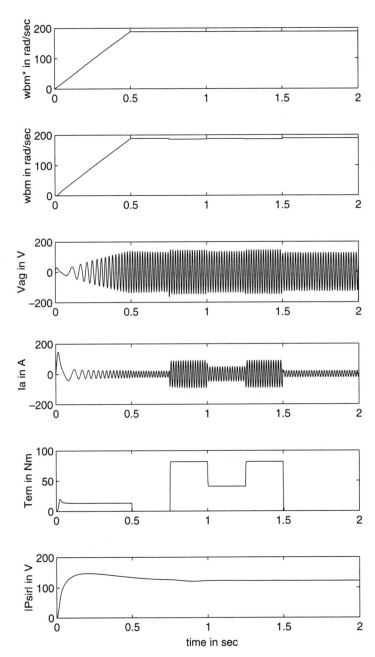

Figure 9.31   Startup and loading transients with field-oriented control.

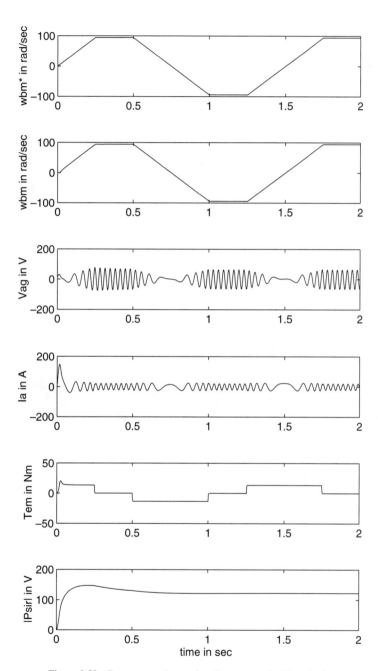

Figure 9.32   Response to changes in reference speed with no-load.

increases, the response time of the rotor flux decreases, but the ensuing transient is more oscillatory. With simulation, we can isolate the effect of detuning by changing the machine parameter used in the controller while keeping those in the machine simulation unchanged. In this manner, observed changes in simulated behavior of the drive may then be attributed to the off-tuned parameters in the controller and not to the influence of a change in machine parameters.

Let's examine the effect of detuning in the value of rotor resistance by introducing an estimation factor, k_r, to all the r'_r terms of the *Field_Orient* block. As set up, perfect tuning is when $k_r = 1$. Repeat the previous runs of step changes in torque at a fixed reference speed and of cyclic change of speed reference for a k_r of 1.5 and 0.5, with no-load and with rated load. Comment on observed differences in dynamic response and in rotor flux level as compared to the case where k_r is 1. Explain why differences in behavior with the motor loaded are greater than those when it is unloaded.

10

Synchronous Machines
in Power Systems
and Drives

10.1 INTRODUCTION

Most of today's electrical power generators in power systems are three-phase synchronous generators. As motors, they are not as competitive in initial costs as induction motors at low and medium horsepowers, but at the higher power ranges, synchronous motors are competitive because their higher efficiencies can lower operating costs. Other forms of synchronous motors, such as the reluctance and permanent magnet motors are, nevertheless, quite popular at the low power ranges. In this chapter, we will be looking into methods of modeling and simulating synchronous generators and ancillary equipment in system-type studies. We will also study synchronous motor drives.

10.2 SYNCHRONOUS GENERATORS IN POWER SYSTEMS

In modern power systems, a number of synchronous generators are operated in parallel. Studies are routinely conducted to ensure that the generators will operate properly in the event of probable faults or changes in system conditions. Studies concerned with the dynamic behavior of synchronous generators are often divided into three classes:

1. Transient stability studies examine the ability of the generators to maintain synchronism from large oscillations created by a severe transient disturbance. Since

463

the oscillations are large, the machine models used should portray the essential nonlinearity in the frequency range from 0.1 to 5 Hz. The dynamic behavior of such synchronizing oscillations is known to be influenced by system parameters and the type of control. Of interest for some time have been the possible destabilizing effects of the voltage regulator, techniques to improve stability via excitation control, and the phenomenon of subsynchronous resonance, in which the synchronous complement of the natural frequency of the electrical network is close to one of the natural torsional frequencies of the rotor shaft.

2. Dynamic stability studies examine the small-signal behavior and stability about some operating point. Such studies often use a linearized representation derived from perturbating the nonlinear model.

3. Long-term dynamic energy balance studies are concerned with system behavior beyond the synchronizing oscillations, usually over an extended period. For such long-term studies, the nonlinear generator models may not need to have as high a frequency fidelity as those used for studying synchronizing oscillations; instead, the dynamics of slower acting components will be important.

When dealing with a large power system, it is not practical to represent each and every component in full detail. The electromechanical oscillation frequency between synchronous generators in a power system typically lies between 0.5 to 3 Hz. The sub-transient time constant of most machines is between 0.03 to 0.04 seconds, which is short compared to the typical period of the electromechanical oscillations of machines. But, the transient time constant ranging from 0.5 to 10 seconds is usually longer than the period of the electromechanical oscillations. In practice, depending on the range of response frequency considered to be important for the problem at hand, appropriate models of the required fidelity would be selected.

To minimize the effort spent on setting up the models and on computation, we also take advantage of simplifications that are offered by the separation in time scales of the different dynamic behaviors and the fact that the severity of a disturbance is usually attenuated as it propagates through the system. For example, the duration of the electrical transients of the network is very short relative to the electromechanical dynamics of the generator; as such, a static representation of the network can be used where longer electromechanical oscillations are primarily of interest. On the other hand, the duration of interest may not be long enough to require the inclusion of the dynamics of slower-acting components, such as those of boilers and automatic generation control. However, where separate groups of machines are interconnected by weak ties and the stability of the generators cannot be determined from the first few swings, the dynamics of slower-acting components should be represented. For extended time studies, the choice of models would be a compromise between better fidelity offered by a full model and the savings in computational effort offered by simpler models.

The models chosen for the generators on the network need not be the same; they only have to be compatible with the network representation in the context of the solution algorithm used. If we know the type and location of the disturbances, we can selectively

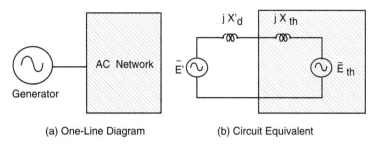

(a) One-Line Diagram   (b) Circuit Equivalent

Figure 10.1   Equivalent circuit representation of generator and network.

employ a detailed model for generators electrically close to the fault locations, and then choose progressively simpler models for generators further away.

Some understanding of the basic dynamic behavior of synchronous generators in transient situations can be obtained using a simple model of a generator and network. Figure 10.1 shows such a simplified representation of a system in which the generator is represented by a *voltage behind the transient reactance* model and the rest of the ac network to which the generator is connected is represented by a single-phase Thevenin's equivalent. The KVL equation of the equivalent circuit representation of the system is

$$\tilde{\mathbf{E}}' = \tilde{\mathbf{E}}_{th} + jX_t\tilde{\mathbf{I}}, \qquad X_t = X'_d + X_{th} \qquad (10.1)$$

If the Thevenin's voltage is used as a reference phasor, that is $\tilde{\mathbf{E}}_{th} = E_{th}\angle 0$ and $\tilde{\mathbf{E}}' = E'\angle\delta$, the electrical output power of the generator is given by

$$P_{gen} = \Re(\tilde{\mathbf{E}}'\tilde{\mathbf{I}}^*) = \frac{E'E_{th}}{X_t}\sin\delta \qquad pu \qquad (10.2)$$

The above expression describes the power vs. angle relationship of the simple system. It indicates that the power transfer characteristic for the system is a sine curve with a maximum value of $E'E_{th}/X_t$. If the small losses within the generator are neglected, its electrical output power, P_{gen}, will be equal to the input mechanical power, P_{mech}, in steady-state.

Where the deviation of the generator's rotor speed from the synchronous value is small, the per unit torque can be approximated reasonably by the per unit power, that is $T_{em}\omega_b \approx -P_{gen}$ and $T_{mech}\omega_b \approx P_{mech}$. The equation of motion of the rotor for generator operation, without damping, becomes

$$P_{mech} - P_{gen} = \frac{2H}{\omega_b}\frac{d\omega}{dt} \qquad pu \qquad (10.3)$$

Using Eqs. 7.84 and 7.85, we can replace $d\omega/dt$ in the above equation by $d^2\delta/dt^2$ and obtain what is commonly known as the *swing equation*:

$$P_{mech} - P_{gen} = \frac{2H}{\omega_b}\frac{d^2\delta}{dt^2} \qquad pu \qquad (10.4)$$

Multiplying both sides of the swing equation by $d\delta/dt$ and integrating, we obtain

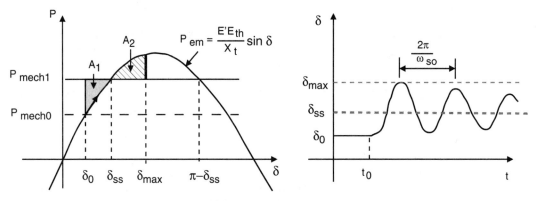

Figure 10.2   Transient power angle characteristic.

$$\left(\frac{d\delta}{dt}\right)^2 = \frac{\omega_b}{H}\int (P_{mech} - P_{gen})d\delta \tag{10.5}$$

If the machine were to maintain synchronism, the excursion of δ would be bounded and $d\delta/dt$ would have to return to zero at the new steady-state operating point. If the new steady-state value of δ is δ_{ss} and the maximum and minimum values of δ following the disturbance are δ_{max} and δ_{min}, respectively, the right-hand side of Eq. 10.5 can be expressed as

$$\underbrace{\int_{\delta_{min}}^{\delta_{max}} (P_{mech} - P_{gen})d\delta}_{} = \underbrace{\int_{\delta_{min}}^{\delta_{ss}} (P_{mech} - P_{gen})d\delta}_{A_1} + \underbrace{\int_{\delta_{ss}}^{\delta_{max}} (P_{mech} - P_{gen})\,d\delta}_{A_2} \tag{10.6}$$

where $\delta_{max} > \delta_{ss} > \delta_{min}$. Figure 10.2 depicts the areas, A_1 and A_2, of the above stability criterion when applied to a stable case where the disturbance is a step change in value of P_{mech}, from P_{mech0} to P_{mech1}. The step increase in P_{mech} initially causes the generator's rotor to accelerate above synchronous speed, advancing δ. Since the output power, P_{gen}, increases along with δ, the accelerating power decreases to zero when δ first reaches its new steady-state value of δ_{ss}; however, the gain in momentum will carry the rotor beyond that angle. But, as δ moves beyond δ_{ss}, P_{gen} becomes larger than P_{mech}, the difference in power is drawn from the kinetic energy of the rotor, and the rotor starts to decelerate. However, δ will continue to increase until all of the momentum gained by the rotor during the acceleration is returned to the electrical network at the point where $\delta = \delta_{max}$, after which δ reverses in direction. If damping losses in the system are neglected, the rotor will oscillate about the new steady-state value of δ_{ss}, between $\delta_{min} = \delta_0$ and δ_{max}. With damping losses, the oscillation of δ will eventually be damped out and δ will settle to its new steady-state value of δ_{ss}, where $d\delta/dt$ will again be zero. By Eqs. 10.5 and 10.6, the areas, A_1 and A_2, must be equal and of opposite sign if the machine is to return to some steady-state operation; the above condition of stability is known as the *equal-area criterion*.

     If however, the gain in rotor momentum is so large as to carry δ beyond the *critical angle* of $\pi - \delta_{ss}$, beyond which $P_{mech1} > P_{gen}(\delta)$ and the rotor accelerates, δ would continue to increase beyond the critical angle, causing the machine to lose synchronism. Thus, for purposes of determining whether a machine subjected to some disturbance will remain in synchronism or not according to the equal-area criterion, we need only establish that the available margin area, A_2^{max}, is larger than the area A_1, where A_2^{max} is the maximum area of A_2 up to the critical angle. For instance, as shown in Fig. 10.2, A_2^{max} is the area of A_2 from δ_{ss} to $\pi - \delta_{ss}$.

The difference between A_2^{max} and the required A_2 may be considered as a margin of the transient stability. The transient power angle curve may be raised by excitation control of E'. This observation and the concern over sluggish voltage regulation led to the introduction of high-speed excitation systems. But high-speed excitation systems could introduce negative damping, which might adversely affect the dynamic stability. An excellent analysis showing how negative damping can come from the excitation system and a discussion on stabilizing signal design to improve the damping of the dynamic oscillations are given in [103]. This has led to many more studies on power system stabilizers with the objective of obtaining better transient performance of the generator through its excitation system. A related problem is the adverse interaction between power system stabilizers and the torsional modes of the turbine shaft, which gives rise to subsynchronous oscillations.

In this chapter, we will derive two more machine models that are used in transient stability studies: the transient model for the transient period and a full model that is good even for the short sub-transient period. The difference between the second model and that described earlier in Chapter 7 is basically one of form. Here, the equations of the machine are manipulated into a form to directly use as many of the machine parameters obtainable from standard tests, usually in the form of reactances and time constants. The model derivations in this chapter can help shed some light onto certain assumptions used in determining the parameters experimentally. For example, for the open-circuit field time, the coupling effect of the kd damper winding is assumed to be over with, that is, i_{kd} has already reached its steady-state value of zero. Whereas in the determination of the open-circuit sub-transient time constant, changes in both the field and damper winding currents are included since these changes initially maintain their respective winding flux linkages as constant. Such approximations are reasonable when $r_{kd} \gg r_f$, but experience has shown that considerable differences in parameters can be obtained when different methods are used to calculate the parameters. The fit between a model and its parameters is important to the final accuracy, as the advantage of sophisticated mathematical models can be negated by poorly fitted parameters.

As transient stability studies are mainly concerned with synchronous generators, these models are often written using the generator notation in which the assumed direction of stator currents flow out of the stator windings. To switch from the motor notation used in Chapter 7 to the generator notation in this chapter, all that is required is to invert the sign of all the stator currents in the voltage, flux linkage, and torque expressions.

10.3 TRANSIENT MODEL WITH D AND Q FIELD WINDINGS

We shall begin the derivation of the transient model by deriving the relations between stator current and flux linkage in the d-axis for the transient period where the damper windings may be assumed to be no longer active. The d-axis winding flux linkages without the contribution of the kd damper winding, obtained by setting i_{kd} to zero, are

$$\lambda_d = -L_d i_d + L_{md} i_f' \qquad Wb. \; turn$$

$$\lambda_f' = -L_{md} i_d + L_{ff}' i_f' \tag{10.7}$$

Multiplying λ_f' by $\omega_r L_{md}/L_{ff}'$, we have

$$\frac{\omega_r L_{md}}{L_{ff}'} \lambda_f' = -\frac{\omega_r L_{md}^2}{L_{ff}'} i_d + \omega_r L_{md} i_f' \tag{10.8}$$

Using the following relationships:

$$E_q' = \omega_r \lambda_d' = \omega_r \frac{L_{md}}{L_{ff}'} \lambda_f', \qquad E_q = \omega_r L_{md} i_f', \qquad \frac{L_{md}^2}{L_{ff}'} = L_d - L_d'. \tag{10.9}$$

Equation 10.8 can be written as

$$E_q' = E_q - \omega_r (L_d - L_d') i_d \qquad V \tag{10.10}$$

Eliminating the field current, i_f' between the two flux linkage equations in Eq. 10.7, and using some of the relationships given in Eq. 10.9, we can obtain the following expressions for λ_d:

$$\lambda_d = \frac{L_{md}}{L_{ff}'} \lambda_f' - L_d i_d + \frac{L_{md}^2 i_d}{L_{ff}'} = \frac{L_{md}}{L_{ff}'} \lambda_f' - L_d' i_d \qquad Wb. \; turn$$

$$= \frac{E_q'}{\omega_r} - L_d' i_d \tag{10.11}$$

Rearranging Eq. 10.11, the expression for i_d is

$$i_d = \frac{1}{L_d'} \left(\frac{E_q'}{\omega_r} - \lambda_d \right) \qquad A \tag{10.12}$$

Similar expressions for the q-axis windings can be obtained as follows: Multiplying

$$\lambda_g' = -L_{mq} i_q + L_{gg}' i_g' \qquad Wb. \; turn \tag{10.13}$$

by $\omega_r L_{mq}/L_{gg}'$, we obtain

$$\frac{\omega_r L_{mq}}{L_{gg}'} \lambda_g' = -\frac{\omega_r L_{mq}^2}{L_{gg}'} i_q + \omega_r L_{mq} i_g' \tag{10.14}$$

Using the following relationships:

$$E_d' = -\omega_r \lambda_q' = -\omega_r \frac{L_{mq}}{L_{gg}'} \lambda_g', \qquad E_d = -\omega_r L_{mq} i_g' \qquad \frac{L_{mq}^2}{L_{gg}'} = L_q - L_q', \tag{10.15}$$

Equation 10.14 can be rewritten as

$$-E_d' = -E_d - \omega_r (L_q - L_q') i_q \qquad V \tag{10.16}$$

Again, eliminating i'_g between the expressions of λ'_g and λ_q and using the relationships in Eq. 10.15, we can obtain the following expressions for λ_q and i_q:

$$\lambda_q = \frac{L_{mq}\lambda'_g}{L'_{gg}} - L_q i_q + \frac{L^2_{mq}}{L'_{gg}} i_q$$

$$= -\frac{E'_d}{\omega_r} - L_q i_q + (L_q - L'_q)i_q \qquad (10.17)$$

$$= -\frac{E'_d}{\omega_r} - L'_q i_q$$

or

$$i_q = \frac{1}{L'_q}\left(\frac{-E'_d}{\omega_r} - \lambda_q\right) \qquad (10.18)$$

Note the similarity in form between the expressions of corresponding quantities of the d- and q-axes. The above expressions of i_d and i_q can also be obtained from Eq. 7.41. For example, the expression for i_d given in Eq. 7.41, without the damper windings, can be written as

$$-i_d = (1 - \frac{L_{MD}}{L_{ls}})\frac{\lambda_d}{L_{ls}} - \frac{L_{MD}}{L_{ls}}\frac{\lambda'_f}{L'_{lf}} \qquad A \qquad (10.19)$$

and that for L_{MD} from Eq. 7.40 is

$$\frac{1}{L_{MD}} = \frac{1}{L_{ls}} + \frac{1}{L'_{lf}} + \frac{1}{L_{md}} = \frac{1}{L_{ls}} + \frac{1}{L'_d - L_{ls}}$$

$$= \frac{L'_d}{L_{ls}(L'_d - L_{ls})} \qquad (10.20)$$

With Eq. 10.20, we can replace the coefficient of the first term in Eq. 10.19 by

$$(1 - \frac{L_{MD}}{L_{ls}}) = \left(1 - \frac{L'_d - L_{ls}}{L'_d}\right) = \frac{L_{ls}}{L'_d} \qquad (10.21)$$

Using the expressions for E'_q given in Eq. 10.9, we can express the coefficient of the second term in Eq. 10.19 as

$$\frac{\lambda'_f}{L'_{lf}} = \frac{1}{L'_d - L_{ls}}\frac{E'_q}{\omega_r} \qquad (10.22)$$

Back-substituting Eqs. 10.21 and 10.22 into Eq. 10.19, we will obtain the same expression for i_d as given in Eq. 10.12. Equations 10.12 and 10.19 are alternate forms of expressing i_d.

Voltage equations of the stator windings.   The qd voltage equations of the stator winding in terms of the stator currents, i_q and i_d that are flowing out of the stator windings, become

$$\frac{d\lambda_q}{dt} = v_q - r_s(-i_q) - \omega_r \lambda_d \qquad Wb.\ turn/s\ or\ V$$

$$\frac{d\lambda_d}{dt} = v_d - r_s(-i_d) + \omega_r \lambda_q$$

(10.23)

If the flux linkages are to be the state variables in the model, Eqs. 10.12 and 10.18 can be used to replace i_d and i_q in Eq. 10.23, yielding

$$\frac{d\lambda_q}{dt} = v_q + \frac{r_s}{L'_q}\left(\frac{-E'_d}{\omega_r} - \lambda_q\right) - \omega_r \lambda_d$$

$$\frac{d\lambda_d}{dt} = v_d + \frac{r_s}{L'_d}\left(\frac{E'_q}{\omega_r} - \lambda_d\right) + \omega_r \lambda_q$$

(10.24)

Field voltage equation. The voltage equations of the f and g field windings are to be manipulated into a form containing their respective time constants. Beginning with the d-axis, the f field voltage equation is given by

$$v'_f = r'_f i'_f + \frac{d\lambda'_f}{dt} \qquad V$$

(10.25)

Multiplying through by $\omega_r L_{md}/r'_f$, we have

$$\frac{\omega_r L_{md} v'_f}{r'_f} = \omega_r L_{md} i'_f + \frac{\omega_r L_{md}}{r'_f}\frac{d\lambda'_f}{dt}$$

(10.26)

From Eq. 7.96, $E_f = \omega_e L_{md} v'_f/r'_f$, the left-hand side of Eq. 10.26 is the internal voltage, E_f. Using the relationships given in Eqs. 7.102 and 7.113, the second term on the right side of Eq. 10.26 can be expressed in terms of E'_q and T'_{do}, that is

$$\frac{\omega_r L_{md}}{r'_f}\lambda'_f = \frac{L'_{ff}}{r'_f}\left(\frac{\omega_r L_{md}}{L'_{ff}}\lambda'_f\right) = T'_{do}E'_q$$

(10.27)

Thus, Eq. 10.26 can be written as

$$E_f = E_q + T_{do}\frac{dE'_q}{dt} \qquad V$$

(10.28)

or

$$T'_{do}\frac{dE'_q}{dt} = E_f - E_q$$

(10.29)

Replacing E_q by $E'_q - \omega_r(L_d - L'_d)i_d$, Eq. 10.29 can also be expressed in the following forms:

$$T'_{do}\frac{dE'_q}{dt} + E'_q = E_f - \omega_r(L_d - L'_d)i_d$$

(10.30)

or

$$T'_{do}\frac{dE'_q}{dt} + \frac{L_d}{L'_d}E'_q = E_f + \left(\frac{L_d - L'_d}{L'_d}\right)\omega_r \lambda_d \tag{10.31}$$

q-axis field winding voltage equation.   For the sake of symmetry in the equations, we will keep a g winding on the q-axis. The symmetry makes it easy to deduce or to verify corresponding expressions of the q-axis windings once we have derived the expressions for the d-axis windings. The g winding can also be used as a fictitious winding to represent the effects of induced q-axis currents in a solid iron rotor.

Again, when the g winding's voltage equation,

$$v'_g = i'_g r'_g + \frac{d\lambda'_g}{dt} \tag{10.32}$$

is multiplied through by $\omega_r L_{mq}/r'_g$, we obtain

$$\frac{\omega_r L_{mq} v'_g}{r'_g} = \omega_r L_{mq} i'_g + \frac{\omega_r L_{mq}}{r'_g}\frac{d\lambda'_g}{dt} \tag{10.33}$$

Making the following substitutions:

$$\frac{\omega_r L_{mq}}{r'_g}\lambda'_g = \frac{L'_{gg}}{r'_g}\left(\frac{\omega_r L_{mq}}{L'_{gg}}\lambda'_g\right) = -T'_{qo}E'_d$$

$$E_g = \omega_r L_{mq}\left(\frac{v'_g}{r'_g}\right), \qquad E_d = -\omega_r L_{mq} i'_g \tag{10.34}$$

Equation 10.33 becomes

$$T'_{qo}\frac{dE'_d}{dt} = -E_g - E_d \tag{10.35}$$

Replacing E_d by $E'_d - \omega_r(L_q - L'_q)i_q$, Eq. 10.35 becomes

$$T'_{qo}\frac{dE'_d}{dt} + E'_d = -E_g + \omega_r(L_q - L'_q)i_q \tag{10.36}$$

We can also use Eq. 10.18 to replace the stator current, i_q in Eq. 10.36, to obtain

$$T'_{qo}\frac{dE'_d}{dt} + \frac{L_q}{L'_q}E'_d = -E_g - \left(\frac{L_q - L'_q}{L'_q}\right)\omega_r \lambda_q \tag{10.37}$$

Torque equation.   The expression for the electromagnetic torque developed in the direction of rotation, that is positive for motoring, in terms of the outgoing stator currents, i_q and i_d, is

$$T_{em} = \frac{3}{2}\frac{P}{2}\{\lambda_d(-i_q) - \lambda_q(-i_d)\} \qquad N.m$$

$$= \lambda_d(-i_q) - \lambda_q(-i_d) \qquad pu \tag{10.38}$$

Note that as written, T_{em} will still be positive for motoring and negative for generating since the currents, $-i_q$ and $-i_d$, are into the stator windings. Replacing λ_q and λ_d in Eq. 10.38 by $\lambda'_q + L'_q(-i_q)$ and $\lambda'_d + L'_d(-i_d)$, respectively, we have

$$T_{em} = \frac{3}{2}\frac{P}{2}\left\{\lambda'_d(-i_q) - \lambda'_q(-i_d) + (L'_d - L'_q)i_d i_q\right\} \tag{10.39}$$

Furthermore, replacing λ'_q by $-E'_d/\omega_r$ and λ'_d by E'_q/ω_r, Eq. 10.39 can be expressed as

$$T_{em} = -\frac{3}{2}\frac{P}{2}\left\{\frac{E'_q i_q + E'_d i_d}{\omega_r} - (L'_d - L'_q)i_d i_q\right\} \qquad N.m$$

$$= -\left\{\frac{E'_q i_q + E'_d i_d}{\omega_r/\omega_b} - \omega_b(L'_d - L'_q)i_d i_q\right\} \qquad pu \tag{10.40}$$

Alternatively, using the expressions given in Eqs. 10.12 and 10.18, Eq. 10.38 can also be written as

$$T_{em} = -\frac{3}{2}\frac{P}{2}\left\{\frac{\lambda_d}{L'_q}\left(\frac{-E'_d}{\omega_r} - \lambda_q\right) - \frac{\lambda_q}{L'_d}\left(\frac{E'_q}{\omega_r} - \lambda_d\right)\right\}$$

$$= \frac{3}{2}\frac{P}{2}\left\{\frac{\lambda_d E'_d}{\omega_r L'_q} + \frac{\lambda_q E'_q}{\omega_r L'_d} - \left(\frac{1}{L'_d} - \frac{1}{L'_q}\right)\lambda_d \lambda_q\right\} \qquad N.m \tag{10.41}$$

The equations for the transient model are summarized in Table 10.1.

TABLE 10.1   TRANSIENT MODEL WITHOUT DAMPER WINDINGS

Stator winding equations:

$$v_q = -\frac{r_s}{L'_q}\left(\frac{-E'_d}{\omega} - \lambda_q\right) + \frac{d\lambda_q}{dt} + \omega_r \lambda_d \qquad V \ or \ pu$$

$$v_d = -\frac{r_s}{L'_d}\left(\frac{E'_q}{\omega_r} - \lambda_d\right) + \frac{d\lambda_d}{dt} - \omega_r \lambda_q \tag{10.42}$$

Rotor winding equations:

$$T'_{do}\frac{dE'_q}{dt} + \frac{L_d}{L'_d}E'_q = E_f + \left(\frac{L_d - L'_d}{L'_d}\right)\omega_r \lambda_d$$

$$T'_{qo}\frac{dE'_d}{dt} + \frac{L_q}{L'_q}E'_d = -E_g - \left(\frac{L_q - L'_q}{L'_q}\right)\omega_r \lambda_q \tag{10.43}$$

or, in terms of the stator currents,

$$T'_{do}\frac{dE'_q}{dt} + E'_q = E_f - \omega_r(L_d - L'_d)i_d$$

$$T'_{qo}\frac{dE'_d}{dt} + E'_d = -E_g + \omega_r(L_q - L'_q)i_q \tag{10.44}$$

Torque equation:

$$T_{em} = \frac{3}{2}\frac{P}{2}\left\{ \frac{\lambda_d E_d'}{\omega_r L_q'} + \frac{\lambda_q E_q'}{\omega_r L_d'} - (\frac{1}{L_d'} - \frac{1}{L_q'})\lambda_d\lambda_q \right\} \qquad N.m \qquad (10.45)$$

or

$$T_{em} = -\frac{3}{2}\frac{P}{2}\left\{ \frac{E_q' i_q + E_d' i_d}{\omega_r} - (L_d' - L_q')i_d i_q \right\} \qquad N.m$$

$$= -\left\{ \frac{E_q' i_q + E_d' i_d}{\omega_r/\omega_b} - \omega_b(L_d' - L_q')i_d i_q \right\} \qquad pu \qquad (10.46)$$

Rotor equation:

$$J\frac{d\omega_{rm}}{dt} = T_{em} + T_{mech} - T_{damp} \qquad N.m$$

$$2H\frac{d\{(\omega_r - \omega_e)/\omega_b\}}{dt} = T_{em(pu)} + T_{mech(pu)} - T_{damp(pu)} \qquad pu \qquad (10.47)$$

$$\frac{d\delta_e}{dt} = \omega_r - \omega_e, \qquad \omega_r = \frac{P}{2}\omega_{rm} \qquad (10.48)$$

When the g winding equation is used to represent the effects of induced currents in the rotor iron, its external input voltage, v_g', and correspondly E_g, would be zero. For completeness, we have retained the derivative terms of the stator flux linkages in the above transient model. In many instances, these derivative terms in the transient model may be neglected. It has been observed that the rotor winding transients are dominant when such a model is used for transient stability prediction. Over the interval of interest, which often is the first swing of the rotor, the rotor transients vary at the rate of T_{do}' and T_{qo}'. Their impact through the speed voltage terms, $\omega_r\lambda_d$ and $\omega_r\lambda_q$, is usually observed to be greater than that of the $d\lambda_q/dt$ and $d\lambda_d/dt$ terms. Consequently, the order of the model can be further reduced by two by neglecting the $d\lambda_q/dt$ and $d\lambda_d/dt$ terms in the stator voltage equations. To be consistent with that simplification, ω_r in the rest of the equations (except that of the rotor dynamics), should be replaced with the excitation frequency, ω_e. Assuming $\omega_e = \omega_b$ and dropping the changes in stator qd flux linkage terms, the equations of the simplified transient model will reduce to those shown in Table 10.2

TABLE 10.2   TRANSIENT MODEL WITH CHANGES IN STATOR qd FLUX LINKAGES NEGLECTED

Stator winding equations:

$$v_q = -r_s i_q - x_d' i_d + E_q' \qquad V \text{ or } pu$$

$$v_d = -r_s i_d + x_q' i_q + E_d' \qquad (10.49)$$

Rotor winding equations:

$$T'_{do}\frac{dE'_q}{dt} + E'_q = E_f - (x_d - x'_d)i_d$$

$$T'_{qo}\frac{dE'_d}{dt} + E'_d = -E_g + (x_q - x'_q)i_q$$

$$(10.50)$$

where

$$\lambda'_q = \lambda_q - L'_q(-i_q), \qquad\qquad E'_d = -\omega_e\lambda'_q$$

$$\lambda'_d = \lambda_d - L'_d(-i_d), \qquad\qquad E'_q = \omega_e\lambda'_d$$

$$(10.51)$$

Torque equation:

$$T_{em} = -\frac{3}{2}\frac{P}{2\omega_e}\left\{E'_q i_q + E'_d i_d + (x'_q - x'_d)i_d i_q\right\} \qquad N.m$$

$$= -\left\{E'_q i_q + E'_d i_d + (x'_q - x'_d)i_d i_q\right\} \qquad pu$$

$$(10.52)$$

Rotor equation:

$$J\frac{d\omega_{rm}}{dt} = T_{em} + T_{mech} - T_{damp} \qquad N.m$$

$$2H\frac{d\{(\omega_r - \omega_e)/\omega_b\}}{dt} = T_{em(pu)} + T_{mech(pu)} - T_{damp(pu)} \qquad pu \qquad (10.53)$$

$$\frac{d\delta_e}{dt} = \omega_r - \omega_e, \qquad \omega_r = \frac{P}{2}\omega_{rm} \qquad (10.54)$$

10.4 SUB-TRANSIENT MODEL WITH FIELD AND DAMPER WINDINGS

In this section, we will derive the equations for a sub-transient model that has f and g field windings and kd and kq damper windings. During the sub-transient period, the coupling effects of both field and damper windings are active in that the currents in these windings are free to change. When a transient disturbance occurs on the stator side and the stator currents change, the corresponding change in currents flowing in the field and damper windings on the rotor would initially maintain the flux linkages of these rotor windings as constant. We have discussed the effective inductances to the stator currents and sub-transient flux linkages in Section 7.6. The sub-transient model that we are deriving in this section will make use of some of the relations mentioned earlier in Section 7.6.

Stator voltage equations.   With the stator currents, i_q and i_d, assumed to be flowing out of the stator windings and replacing λ_q and λ_d by $\lambda''_q + L''_q(-i_q)$ and $\lambda''_d + L''_d(-i_d)$, respectively, the qd voltage equations of the stator windings become

$$v_q = r_s(-i_q) + \omega_r \left\{ \lambda_d'' + L_d''(-i_d) \right\} + \frac{d\lambda_q}{dt} \quad \text{V}$$

$$= -ri_q + E_q'' - \omega_r L_d'' i_d + \frac{d\lambda_q}{dt}$$

$$v_d = r_s(-i_d) - \omega_r \left\{ \lambda_q'' + L_q''(-i_q) \right\} + \frac{d\lambda_d}{dt} \tag{10.55}$$

$$= -ri_d + E_d'' + \omega_r L_q'' i_q + \frac{d\lambda_d}{dt}$$

The next step is to express the sub-transient flux linkages and voltages in terms of the damper flux linkages. We will show only the derivation of the relations for the d-axis in detail; those for the q-axis can be obtained similarly. Using Eq. 7.41, the d-axis stator current flowing out of the stator winding, expressed in terms of the flux linkages, is given by

$$(-i_d) = \left(1 - \frac{L_{MD}}{L_{ls}} \right) \frac{\lambda_d}{L_{ls}} - \frac{L_{MD}}{L_{ls}} \frac{\lambda_f'}{L_{lf}'} - \frac{L_{MD}}{L_{ls}} \frac{\lambda_{kd}'}{L_{lkd}'} \tag{10.56}$$

Substituting the above expression for $-i_d$ in $\lambda_d'' = \lambda_d - L_d''(-i_d)$, we obtain

$$\lambda_d'' = \lambda_d - L_d'' \left\{ \left(1 - \frac{L_{MD}}{L_{ls}} \right) \frac{\lambda_d}{L_{ls}} - \frac{L_{MD}}{L_{ls}} \frac{\lambda_f'}{L_{lf}'} - \frac{L_{MD}}{L_{ls}} \frac{\lambda_{kd}'}{L_{lkd}'} \right\} \tag{10.57}$$

Since

$$\frac{1}{L_d''} = \frac{1}{L_{ls}} \left(1 - \frac{L_{MD}}{L_{ls}} \right)$$

Eq. 10.57 simplifies to

$$\lambda_d'' = \frac{L_d'' L_{MD}}{L_{ls}} \left(\frac{\lambda_f'}{L_{lf}'} + \frac{\lambda_{kd}'}{L_{lkd}'} \right) \quad \text{Wb. turn} \tag{10.58}$$

Using the following relationships:

$$\lambda_f' = \frac{L_{ff}' E_q'}{\omega_r L_{md}}$$

$$L_d'' L_{MD} = (L_d'' - L_{ls}) L_{ls}$$

$$\frac{L_{md} L_{lf}'}{L_{ff}'} = L_d' - L_{ls} \tag{10.59}$$

$$\frac{L_d'' - L_{ls}}{L_{lkd}'} = \frac{L_d' - L_d''}{L_d' - L_{ls}}$$

Equation 10.58 can be rewritten as

$$\lambda_d'' = \left(\frac{L_d'' - L_{ls}}{L_d' - L_{ls}} \right) \left(\frac{E_q'}{\omega_r} - \lambda_{kd}' \right) + \lambda_{kd}' \quad \text{Wb.turn} \tag{10.60}$$

Using Eq. 10.60 to substitute for λ_d'' in Eq. 7.110, we can express the sub-transient voltage, E_q'', as

$$
\begin{aligned}
E_q'' = \omega_r \lambda_d'' &= \left(\frac{L_d'' - L_{ls}}{L_d' - L_{ls}}\right)(E_q' - \omega_r \lambda_{kd}') + \omega_r \lambda_{kd}' \qquad \text{V} \\
&= \left(\frac{L_d'' - L_{ls}}{L_d' - L_{ls}}\right)E_q' + \left(\frac{L_d' - L_d''}{L_d' - L_{ls}}\right)\omega_r \lambda_{kd}'
\end{aligned}
\tag{10.61}
$$

By symmetry, the corresponding expressions of the q-axis quantities should be similar in form to those of the d-axis quantities. For example, we can show that the q-axis sub-transient flux is given by

$$
\begin{aligned}
\lambda_q'' = \lambda_q - L_q'' &\left\{\left(1 - \frac{L_{MQ}}{L_{ls}}\right)\frac{\lambda_q}{L_{ls}} - \frac{L_{MQ}}{L_{ls}}\frac{\lambda_g'}{L_{lg}'} - \frac{L_{MD}}{L_{ls}}\frac{\lambda_{kq}'}{L_{lkq}'}\right\} \qquad \text{Wb. turn} \\
&= \frac{L_q'' L_{MQ}}{L_{ls}}\left(\frac{\lambda_g'}{L_{lg}'} + \frac{\lambda_{kq}'}{L_{lkq}'}\right)
\end{aligned}
\tag{10.62}
$$

Again, using the following substitutions:

$$
E_d' = -\frac{\omega_r L_{mq}\lambda_g'}{L_{gg}'}, \qquad L_q'' L_{MQ} = (L_q'' - L_{ls})L_{ls}, \qquad \text{and} \qquad \frac{L_{mq}L_{lg}'}{L_{gg}'} = L_q' - L_{ls}
\tag{10.63}
$$

Equation 10.62 can be rewritten as

$$
\lambda_q'' = -\left(\frac{L_q'' - L_{ls}}{L_q' - L_{ls}}\right)\frac{E_d'}{\omega_r} + \left(\frac{L_q'' - L_{ls}}{L_{lkq}'}\right)\lambda_{kq}' \qquad \text{Wb.turn}
\tag{10.64}
$$

Replacing L_{lkq}' in Eq. 10.64 by the transient and sub-transient inductances, that is $L_{lkq}' = (L_d'' - L_{ls})(L_d' - L_{ls})/(L_d' - L_d'')$, we obtain

$$
\lambda_q'' = \left(\frac{L_q'' - L_{ls}}{L_q' - L_{ls}}\right)\left(\frac{-E_d'}{\omega_r} - \lambda_{kq}'\right) + \lambda_{kq}'
\tag{10.65}
$$

The d-axis sub-transient voltage defined earlier in Eq. 7.111 can now be expressed as

$$
\begin{aligned}
E_d'' = -\omega_r \lambda_q'' &= \left(\frac{L_q'' - L_{ls}}{L_q' - L_{ls}}\right)\left(E_d' + \omega_r \lambda_{kq}'\right) - \omega_r \lambda_{kq}' \\
&= \left(\frac{L_q'' - L_{ls}}{L_q' - L_{ls}}\right)E_d' - \left(\frac{L_q' - L_q''}{L_q' - L_{ls}}\right)\omega_r \lambda_{kq}'
\end{aligned}
\tag{10.66}
$$

Now that we have the stator voltage equations expressed in the desired rotor flux linkages and sub-transient voltages, we will next manipulate the voltage equations of the rotor windings into compatible forms.

Voltage equations of the rotor windings.   We will begin with the following d-axis rotor flux linkage expressions, manipulating them to express i'_{kd} in terms of the flux linkage, λ'_{kd}, and the voltage, E'_q:

$$\lambda'_{kd} = -L_{md}i_d + L_{md}i'_f + L'_{kdkd}i'_{kd} \qquad Wb.\ turn$$

$$\lambda'_f = -L_{md}i_d + L'_{ff}i'_f + L_{md}i'_{kd}$$

(10.67)

Eliminating the field current by pre-multiplying λ'_f by (L_{md}/L'_{ff}) and subtracting from it λ'_{kd}, we obtain

$$\frac{L_{md}}{L'_{ff}}\lambda'_f - \lambda'_{kd} = \left(-\frac{L^2_{md}}{L'_{ff}} + L_{md}\right)i_d + \left(\frac{L^2_{md}}{L'_{ff}} - L'_{kdkd}\right)i'_{kd}$$

(10.68)

Equation 10.68 can be rewritten as

$$\frac{E'_q}{\omega_r} - \lambda'_{kd} = -(L_d - L'_d - L_{md})i_d - \frac{(L'_d - L_{ls})^2}{L'_d - L''_d}i'_{kd}$$

(10.69)

from which we obtain the desired expression for i'_{kd}, that is

$$i'_{kd} = \frac{L'_d - L''_d}{(L'_d - L_{ls})^2}\left\{\lambda'_{kd} - \frac{E'_q}{\omega_r} + (L'_d - L_{ls})i_d\right\}$$

(10.70)

d-axis field voltage equation.   The voltage equation of the f field winding, given by

$$v'_f = i'_f r'_f + \frac{d\lambda'_f}{dt} \qquad V$$

(10.71)

when multiplied by $\omega_r L_{md}/r'_f$, yields

$$\frac{\omega_r L_{md}v'_f}{r'_f} = \omega_r L_{md}i'_f + \frac{L'_{ff}}{r'_f}\frac{d}{dt}\left(\frac{\omega_r L_{md}}{L'_{ff}}\lambda'_f\right)$$

(10.72)

Replacing $\omega_r L_{md}v'_f/r'_f$ by E_f and substituting E_q for $\omega_r L_{md}i'_f$, T'_{do} for L'_{ff}/r'_f, and E'_q for $\omega_r L_{md}\lambda'_f/L'_{ff}$, Eq. 10.72 becomes

$$E_f = E_q + T'_{do}\frac{dE'_q}{dt} \qquad V$$

(10.73)

Also, when the equation for λ'_f in Eq. 10.57 is multiplied through by $\omega_r L_{md}/L'_{ff}$, we obtain

$$\frac{\omega_r L_{md}\lambda'_f}{L'_{ff}} = \omega_r L_{md}i'_f - \frac{\omega_r L^2_{md}}{L'_{ff}}i_d + \frac{\omega_r L^2_{md}}{L'_{ff}}i'_{kd}$$

(10.74)

which, in turn, can be expressed in the form

$$E'_q = E_q - \omega_r(L_d - L'_d)(i_d - i'_{kd})$$

(10.75)

Using Eq. 10.75 to replace E_q in Eq. 10.73, we obtain the following expressions for the d-axis field winding:

$$T'_{do}\frac{dE'_q}{dt} + E'_q = E_f - \omega_r(L_d - L'_d)(i_d - i'_{kd}) \tag{10.76}$$

Substituting for i'_{kd} with Eq. 10.70, Eq. 10.76 becomes

$$T'_{do}\frac{dE'_q}{dt} + E'_q = E_f - \omega_r(L_d - L'_d)i_d$$
$$+ \omega_r\left\{\frac{(L'_d - L''_d)(L_d - L'_d)}{(L'_d - L_{ls})^2}\right\}\left\{\lambda'_{kd} - \frac{E'_q}{\omega_r} + (L'_d - L_{ls})i_d\right\} \tag{10.77}$$

or

$$T'_{do}\frac{dE'_q}{dt} = E_f - \omega_r\left\{\frac{(L_d - L'_d)(L''_d - L_{ls})}{L'_d - L_{ls}}\right\}i_d - \left\{1 + \frac{(L'_d - L''_d)(L_d - L'_d)}{(L'_d - L_{ls})^2}\right\}E'_q$$
$$+ \left\{\frac{(L'_d - L''_d)(L_d - L'_d)}{(L'_d - L_{ls})^2}\right\}\omega_r\lambda'_{kd} \tag{10.78}$$

q-axis field voltage equation.   Similarly, the voltage equation of the g field winding, given by

$$v'_g = r'_g i'_g + \frac{d\lambda'_g}{dt} \qquad V \tag{10.79}$$

can also be manipulated using the relations, $E_g = \omega_r L_{mq} v'_g / r'_g$, $E_d = \omega_r L_{mq} i'_g$, $E'_d = -\omega_r L_{mq}\lambda'_g / L'_g$, and $T'_{qo} = L'_{gg}/r'_g$ into the form:

$$E_g = E_d - T'_{qo}\frac{dE'_d}{dt} \tag{10.80}$$

Multiplying both sides of $\lambda'_g = -L_{mq}i_q + L'_{gg}i'_g + L_{mq}i'_{kq}$ by $\omega_r L_{mq}/L'_{gg}$, we obtain

$$\frac{\omega_r L_{mq}\lambda'_g}{L'_{gg}} = \omega_r L_{mq}i'_g - \frac{\omega_r L^2_{mq}}{L'_{gg}}i_q + \frac{\omega_r L^2_{mq}}{L'_{gg}}i'_{kq} \tag{10.81}$$

This can be expressed as

$$-E'_d = E_d - \omega_r(L_q - L'_q)(i_q - i'_{kq}) \tag{10.82}$$

Using Eq. 10.82 to substitute for E_d in Eq. 10.80, we obtain

$$T'_{qo}\frac{dE'_d}{dt} + E'_d = -E_g + \omega_r(L_q - L'_q)(i_q - i'_{kq}) \tag{10.83}$$

As in the case of the d-axis damper winding, we can obtain the following expression for i'_{kq} by eliminating the field current i'_g between the expressions of λ'_{kq} and λ'_g, that is

$$i'_{kq} = \frac{L'_q - L''_q}{(L'_q - L_{ls})^2}\left\{\lambda'_{kq} + \frac{E'_d}{\omega_r} + (L'_q - L_{ls})i_q\right\} \qquad A \tag{10.84}$$

Using Eq. 10.84 to substitute for i'_{kq}, Eq. 10.83 can be expressed in the following forms:

$$T'_{qo}\frac{dE'_d}{dt} + E'_d = -E_g + \omega_r(L_q - L'_q)i_q$$

$$-\omega_r\left\{\frac{(L_q - L'_q)(L'_g - L''_q)}{(L'_q - L_{ls})^2}\right\}\left\{\lambda'_{kq} + \frac{E'_d}{\omega_r} + (L'_q - L_{ls})i_q\right\}$$

$$T'_{qo}\frac{dE'_d}{dt} = -E_g + \omega_r\left\{\frac{(L_q - L'_q)(L''_q - L_{ls})}{L'_q - L_{ls}}\right\} - E'_d\left\{1 + \frac{(L'_q - L''_q)(L_q - L'_q)}{(L'_q - L_{ls})^2}\right\}$$

$$-\left\{\frac{(L_q - L'_q)(L'_q - L''_q)}{(L'_q - L_{ls})^2}\right\}\omega_r\lambda'_{kq} \qquad (10.85)$$

d-axis damper voltage equation.   The voltage equation of the d-axis damper winding is

$$\frac{d\lambda'_{kd}}{dt} + r'_{kd}i'_{kd} = 0 \qquad (10.86)$$

Using the following relation to replace r'_{kd} by T''_{do}:

$$T''_{do} = \frac{(L'_d - L_{ls})^2}{(L'_d - L''_d)r'_{kd}} \qquad (10.87)$$

Equation 10.86 can be rewritten as

$$T''_{do}\frac{d\lambda_{kd}}{dt} + \frac{(L'_d - L_{ls})^2}{L'_d - L''_d}i'_{kd} = 0 \qquad (10.88)$$

Using Eq. 10.70 to replace i'_{kd} in Eq. 10.88, we obtain the desired expression for the d-axis damper winding, that is

$$T''_{do}\frac{d\lambda'_{kd}}{dt} + \lambda'_{kd} - \frac{E'_q}{\omega_r} + (L'_d - L_{ls})i_d = 0 \qquad (10.89)$$

q-axis damper voltage equation.   As in the case of the d-axis damper winding, the voltage equation of the q-axis damper winding can first be manipulated into the form:

$$T''_{qo}\frac{d\lambda'_{kq}}{dt} + \frac{(L'_q - L_{ls})^2}{L'_q - L''_q}i'_{kq} = 0 \qquad (10.90)$$

Using Eq. 10.84 to replace i'_{kq}, we can obtain the following form of the q-axis damper winding voltage equation:

$$T''_{qo}\frac{d\lambda'_{kq}}{dt} + \lambda'_{kq} + \frac{E'_d}{\omega_r} + (L'_q - L_{ls})i_q = 0 \qquad (10.91)$$

Electromagnetic torque.   Replacing λ_q and λ_d respectively by $\lambda_q'' + \lambda_q''(-i_q)$ and $\lambda_d'' + \lambda_d''(-i_d)$, the expression of the developed electromagnetic torque in terms of the sub-transient quantities, with assumed direction of the stator currents, i_q and i_d, out of the stator winding, is

$$T_{em} = -\frac{3}{2}\frac{P}{2}\left\{\lambda_d'' i_q - \lambda_q'' i_d + (L_q'' - L_d'')i_d i_q\right\} \qquad N.m \qquad (10.92)$$

In terms of the voltages behind the sub-transient inductances, $E_d'' = -\omega_r \lambda_q''$ and $E_q'' = \omega_r \lambda_d''$, the expression of the developed torque can be rewritten as

$$T_{em} = -\frac{3}{2}\frac{P}{2}\left\{\left(\frac{E_q'' i_q + E_d'' i_d}{\omega_r}\right) + (L_q'' - L_d'')i_d i_q\right\} \qquad N.m$$

$$\qquad\qquad (10.93)$$

$$= -\left\{\left(\frac{E_q'' i_q + E_d'' i_d}{\omega_r/\omega_b}\right) + \omega_b(L_q'' - L_d'')i_d i_q\right\} \qquad pu$$

The equations for the stator and rotor windings, along with the torque equation of the sub-transient model, are summarized in Table 10.3.

TABLE 10.3   SUB-TRANSIENT MODEL

Stator winding equations:

$$v_q = -ri_q + E_q'' - \omega_r L_d'' i_d + \frac{d\lambda_q}{dt}$$

$$\qquad\qquad (10.94)$$

$$v_d = -ri_d + E_d'' + \omega_r L_q'' i_q + \frac{d\lambda_d}{dt}$$

where

$$E_q'' = \left(\frac{L_d'' - L_{ls}}{L_d' - L_{ls}}\right)E_q' + \left(\frac{L_d' - L_d''}{L_d' - L_{ls}}\right)\omega_r \lambda_{kd}'$$

$$\qquad\qquad (10.95)$$

$$E_d'' = \left(\frac{L_q'' - L_{ls}}{L_q' - L_{ls}}\right)E_d' - \left(\frac{L_q' - L_q''}{L_q' - L_{ls}}\right)\omega_r \lambda_{kq}'$$

Rotor winding equations:

$$T_{do}'\frac{dE_q'}{dt} = E_f - \omega_r\left\{\frac{(L_d - L_d')(L_d'' - L_{ls})}{L_d' - L_{ls}}\right\}i_d - \left\{1 + \frac{(L_d' - L_d'')(L_d - L_d')}{(L_d' - L_{ls})^2}\right\}E_q'$$

$$\qquad + \left\{\frac{(L_d' - L_d'')(L_d - L_d')}{(L_d' - L_{ls})^2}\right\}\omega_r \lambda_{kd}' \qquad\qquad (10.96)$$

$$T_{qo}'\frac{dE_d'}{dt} = -E_g + \omega\left\{\frac{(L_q - L_q')(L_q'' - L_{ls})}{L_q' - L_{ls}}\right\}i_q - \left\{1 + \frac{(L_q' - L_q'')(L_q - L_q')}{(L_q' - L_{ls})^2}\right\}E_d'$$

$$\qquad - \left\{\frac{(L_q - L_q')(L_q' - L_q'')}{(L_q' - L_{ls})^2}\right\}\omega_r \lambda_{kq}' \qquad\qquad (10.97)$$

$$T_{do}'' \frac{d\lambda_{kd}'}{dt} = \frac{E_q'}{\omega_r} - \lambda_{kd}' - (L_d' - L_{ls})i_d$$

$$T_{qo}'' \frac{d\lambda_{kq}'}{dt} = \frac{-E_d'}{\omega_r} - \lambda_{kq}' - (L_q' - L_{ls})i_q \tag{10.98}$$

Torque equation:

$$T_{em} = -\frac{3}{2}\frac{P}{2}\left\{\left(\frac{E_q''i_q + E_d''i_d}{\omega_r}\right) + (L_q'' - L_d'')i_d i_q\right\} \qquad N.m$$

$$= -\left\{\left(\frac{E_q''i_q + E_d''i_d}{\omega_r/\omega_b}\right) + \omega_b(L_q'' - L_d'')i_d i_q\right\} \qquad pu \tag{10.99}$$

Rotor equation:

$$J\frac{d\omega_{rm}}{dt} = T_{em} + T_{mech} - T_{damp} \qquad N.m$$

$$2H\frac{d\{(\omega_r - \omega_e)/\omega_b\}}{dt} = T_{em(pu)} + T_{mech(pu)} - T_{damp(pu)} \qquad pu \tag{10.100}$$

$$\frac{d\delta_e}{dt} = \omega_r - \omega_e, \qquad \omega_r = \frac{P}{2}\omega_{rm} \tag{10.101}$$

As given in Eq. 10.95, the expressions of the sub-transient voltages, E_q'' and E_d'', consisted of a linear combination of the transient voltages, E_q' and E_d', and the speed voltages associated with the damper winding flux linkages, λ_{kq}' and λ_{kd}'. Since the sub-transient time constants, T_{do}'' and T_{qo}'', are shorter than the transient time constants, initial changes in E_q'' and E_d'' will be mainly from those components associated with λ_{kq}' and λ_{kd}'.

10.5 SPECIFIC CASES

Here we will consider specific cases in modeling which are fairly common in power system studies.

10.5.1 Salient-pole Machines

The sub-transient model given in Table 10.3, with two rotor windings in the q-axis, is often used for round-rotor machines. For salient-pole machines, $L_q' \approx L_q$; as such,

$$T_{qo}' = \left(\frac{L_q - L_{ls}}{L_q - L_q'}\right)r_g' \qquad \text{is large.} \tag{10.102}$$

Applying these conditions to

$$T'_{qo}\frac{dE'_d}{dt} + E'_d = \omega_r(L_q - L'_q)(i_q - i'_{kq}) \tag{10.103}$$

we obtain

$$\frac{dE'_d}{dt} = 0 \quad \text{and} \quad E'_d = 0 \tag{10.104}$$

Furthermore, for salient-pole machines, $L''_q \ll L'_q$, the expression for E''_d reduces to

$$E''_d = -\omega_r\lambda'_{kq}\left(\frac{L'_q - L''_q}{L'_q - L_{ls}}\right) \tag{10.105}$$

The voltage equation for the kq winding becomes

$$T''_{qo}\frac{dE''_d}{dt} + E''_d = -\omega_r(L'_q - L''_q)i_q$$

10.5.2 Neglecting the $d\lambda_q/dt$ and $d\lambda_d/dt$ Terms

The $d\lambda_q/dt$ and $d\lambda_d/dt$ terms in the stator voltage equations of Eq. 10.55 are usually small in comparison to the voltages behind the sub-transient reactances, E''_q and E''_d. The reason being that speed voltage components of E''_q and E''_d are proportional to ω_r, whereas the maximum time rates of change of λ_d and λ_q are of the order of $1/T''_{do}$ and $1/T''_{qo}$, respectively, which is typically an order smaller than ω_r. As such, the $d\lambda_q/dt$ and $d\lambda_d/dt$ terms may be omitted, in which case the ω_r of the speed voltage terms should also be replaced by the excitation frequency, ω_e.

The modified equations of such a simplified sub-transient model are given in Table 10.4.

TABLE 10.4   SUB-TRANSIENT MODEL WITH CHANGES IN STATOR qd FLUX LINKAGES NEGLECTED

Stator winding equations:

$$v_q = -ri_q + E''_q - \omega_e L''_d i_d \quad \text{V}$$
$$v_d = -ri_d + E''_d + \omega_e L''_q i_q \tag{10.106}$$

where

$$E''_q = \left(\frac{L''_d - L_{ls}}{L'_d - L_{ls}}\right)E'_q + \left(\frac{L'_d - L''_d}{L'_d - L_{ls}}\right)\omega_e\lambda'_{kd}$$
$$E''_d = \left(\frac{L''_q - L_{ls}}{L'_q - L_{ls}}\right)E'_d - \left(\frac{L'_q - L''_q}{L'_q - L_{ls}}\right)\omega_e\lambda'_{kq} \tag{10.107}$$

Rotor winding equations:

$$T'_{do}\frac{dE'_q}{dt} = E_f - \omega_e\left\{\frac{(L_d-L'_d)(L''_d-L_{ls})}{L'_d-L_{ls}}\right\}i_d - \left\{1+\frac{(L'_d-L''_d)(L_d-L'_d)}{(L'_d-L_{ls})^2}\right\}E'_q$$

$$+\left\{\frac{(L_d-L'_d)(L'_d-L''_d)}{(L'_d-L_{ls})^2}\right\}\omega_e\lambda'_{kd}$$

$$T'_{qo}\frac{dE'_d}{dt} = -E_g + \omega_e\left\{\frac{(L_q-L'_q)(L''_q-L_{ls})}{L'_q-L_{ls}}\right\}i_q - \left\{1+\frac{(L'_q-L''_q)(L_q-L'_q)}{(L'_q-L_{ls})^2}\right\}E'_d \quad (10.108)$$

$$-\left\{\frac{(L_q-L'_q)(L'_q-L''_q)}{(L'_q-L_{ls})^2}\right\}\omega_e\lambda'_{kq}$$

$$T''_{do}\frac{d\lambda'_{kd}}{dt} = \frac{E'_q}{\omega_e} - \lambda'_{kd} - (L'_d-L_{ls})i_d$$

$$T''_{qo}\frac{d\lambda'_{kq}}{dt} = \frac{-E'_d}{\omega_e} - \lambda'_{kq} - (L'_q-L_{ls})i_q \quad (10.109)$$

Torque equation:

$$T_{em} = -\frac{3}{2}\frac{P}{2}\left\{\left(\frac{E''_q i_q + E''_d i_d}{\omega_e}\right)+(L''_q-L''_d)i_d i_q\right\} \quad N.m$$

$$= -\left\{\left(\frac{E''_q i_q + E''_d i_d}{\omega_e/\omega_b}\right)+\omega_b(L''_q-L''_d)i_d i_q\right\} \quad pu \quad (10.110)$$

Rotor equation:

$$J\frac{d\omega_{rm}}{dt} = T_{em} + T_{mech} - T_{damp} \quad N.m$$

$$2H\frac{d\{(\omega_r-\omega_e)/\omega_b\}}{dt} = T_{em(pu)} + T_{mech(pu)} - T_{damp(pu)} \quad pu \quad (10.111)$$

$$\frac{d\delta_e}{dt} = \omega_r - \omega_e, \qquad \omega_r = \frac{P}{2}\omega_{rm} \quad (10.112)$$

10.5.3 Neglecting Sub-transient and Transient Saliency

Most practical power systems have a fairly extensive network linking the generators. For transient stability studies, the focus is on the ability of the generators to remain in synchronism in the ensuing electromechanical oscillations. Usually, the much faster electromagnetic transients of the network can be neglected. Representing the large

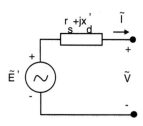

Figure 10.3   The voltage behind a transient impedance model.

network by a static rather than dynamic representation results in considerable savings in computational time in transient stability studies.

Further simplification, at the generator level, can be made by ignoring the changes in stator qd flux linkage terms of the stator voltage equations. To avoid algebraic loops involving the static network equations and the static equations of the stator, these equations ought to be properly combined. As illustrated in greater detail later in Project 2, the transformation of the stator equations in the rotor reference frame to the synchronously rotating frame of the network involves the time-varying rotor angle. If the effective stator impedances in the q- and the d-axis are the same, in other words, there is no rotor saliency, the resulting impedance in the synchronously rotating reference will not be a function of the rotor angle. If not, the resulting impedance is a function of the rotor angle and its value will have to be updated as the rotor angle changes. The computational advantage of a constant effective stator impedance which can be absorbed into the network's admittance or impedance network can be significant.

Fortunately, the sub-transient saliency of many machines is usually small, that is $L_d'' \approx L_q''$. Where simpler models are desirable, the transient saliency can also be neglected by setting $L_d' = L_q'$. For example, the model given in Table 10.2 can be further simplified by neglecting transient saliency. Using the relationship between time phasor and space vector given earlier in Eqs. 7.62 and 7.63, we can rewrite the stator qd voltage equations of the model given in Table 10.2 as

$$\underbrace{v_q - jv_d}_{\tilde{V}} = -(r_s + jx_d')\underbrace{(i_q - ji_d)}_{\tilde{I}} + \underbrace{(E_q' - jE)}_{\tilde{E}'} \tag{10.113}$$

The dynamics of E_q' and E_d' are governed by Eq. 10.50. With $x_d' = x_q'$, the per unit torque equation in Eq. 10.52 becomes

$$T_{em} = -\tilde{E}'\tilde{I}* \qquad pu \tag{10.114}$$

If the dynamics of the transient voltages, E_q' and E_d' in Eq. 10.50, are ignored, $\tilde{E}'$ will be a constant phasor and we will obtain the classical *voltage behind transient impedance* model consisting of only Eqs. 10.113 and 10.114. An equivalent circuit representation of Eq. 10.113 is given in Fig. 10.3.

10.6 EXCITATION SYSTEMS

Excitation systems may be classified according to the primary source of excitation power: dc, ac rotary, or ac static. In many of the older dc-type excitation systems, the primary

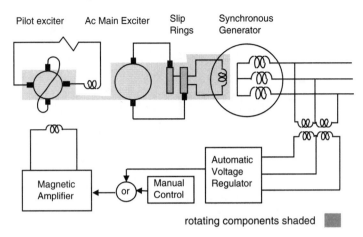

Figure 10.4   dc-type excitation system.

excitation power is from a dc generator whose field winding is mounted on the same shaft as the rotor of the synchronous machine. The dc generator serves as the main exciter; it, in turn, may be separately excited by another exciter. Figure 10.4 shows an example of such a dc excitation system with an amplidyne exciter.

Most modern excitation systems are of the ac rotary- or ac static-type. The ac rotary-type uses the output of a rotary ac alternator as the main exciter to supply the dc excitation to the synchronous generator. The two main arrangements are shown in Fig. 10.5. In Fig. 10.5a, the field winding of the ac alternator is on the same shaft as the rotor of the synchronous generator. Its stator and the rectifier are stationary. As shown, the rectifier is a thyristor-controlled bridge whose dc output voltage is electronically controlled by adjusting the phase delay in turning on the thyristors of the bridge. In this arrangement, the dc excitation current can only flow in one direction. However, bi-directional dc excitation current can be obtained by using two such bridges connected in anti-parallel. The rectified dc output of the bridge is connected to the main field winding of the synchronous generator via a pair of slip rings. The slip rings can be eliminated by interchanging the position of the armature and field windings of the ac alternator. As shown in Fig. 10.5b, a brushless ac excitation system has the armature of the ac exciter and the rectifier bridge rotating with the rotor, and the ac exciter field is stationary.

Most ac static-type exciters draw their primary power from a local ac bus and use controlled rectification to provide an adjustable dc excitation to the field winding of the synchronous generator. Figure 10.6 shows an example of an ac static-type excitation system. Such bus-fed systems are dependent on the availability of the ac voltage, which could be adversely affected by nearby faults. Some incorporate fault current compounding to offset the reduction in ac voltage during certain kinds of faults. Compared to rotary-type exciters, the ac static excitation systems are more compact, less expensive, and have much quicker response time. For the definitions of various terms used in describing the performances of excitation systems, the reader is referred to References [109, 111, 115].

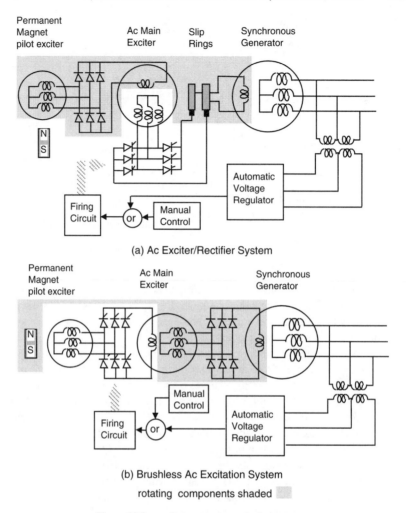

(a) Ac Exciter/Rectifier System

(b) Brushless Ac Excitation System

rotating components shaded

Figure 10.5   ac alternator-type excitation systems.

As mentioned earlier, there are a number of stability issues in which the excitation system has an important role. Thus, it is not surprising that much has been written about these systems and the modeling of them in transient and stability studies. For details on modeling, the reader is referred to References [108, 111]. The wide variety of design arrangements makes the task of defining generic models difficult. However, the mathematical representations of some principal parts commonly found in such systems will be discussed in the following subsections.

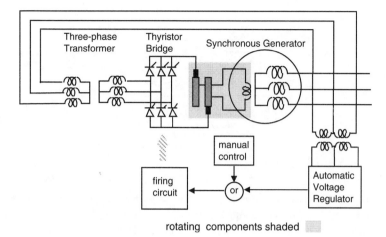

Figure 10.6   ac static-type excitation system.

10.6.1 Voltage Transducer and Load Compensation Circuit

The error signal to the excitation system is usually obtained by comparing the desired or reference value to the corresponding rectified value of the controlled ac quantity. The voltage transducer and rectifier are modeled simply by a single time constant with unity gain as shown in Fig. 10.7. Any compensation of the voltage droop caused by the load current using a compensating impedance, $R_c + jX_c$, is modeled by the corresponding voltage magnitude expression, as shown in Fig. 10.7.

10.6.2 Regulator

The regulator section typically consists of an error amplifier with limiters. Its gain vs. frequency characteristic usually can be approximated quite well by the transfer function blocks shown in Fig. 10.8. Some degree of transient gain reduction can be achieved using a compensator that has a $T_C < T_B$. Also shown in Fig. 10.8 at the input of the regulator are the stabilizer feedback signal, v_f, and the supplementary signal, v_{supp}, from a power system stabilizer.

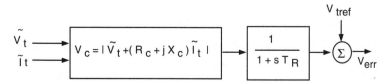

Figure 10.7   Voltage transducer and load compensation circuit.

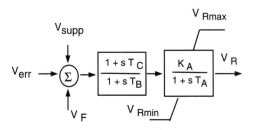

Figure 10.8 Regulator amplifier.

10.6.3 dc Exciter

The output signal from the regulator usually must be amplified by the exciter before it has the necessary power and range to excite the field winding of a large synchronous generator. The average-value model of a dc exciter includes the field winding, magnetic nonlinearity of the exciter's main field path, and the armature. The armature winding usually has a small number of turns compared to the field winding; as such, the small resistance and inductance of the armature winding are often neglected. Since the intended application of these computer models include large signal variations, limits and nonlinearities are considered important characteristics to be modeled.

The voltage equation of the field winding and terminal voltage vs. field current curve of the armature may be expressed as

$$v_f = i_f r_f + \frac{d\lambda_f(i_f)}{dt}$$

$$v_x = f(i_f, i_x)$$

$$(10.115)$$

The subscripts, f and x in this section, denote the quantities of the field and armature windings of the exciter. The output voltage of the exciter is typically a nonlinear function of the field and armature currents. At a given i_x, the field current may be expressed in terms of the armature output voltage, v_x, and the exciter saturation function, S_e, that is

$$i_f = \frac{v_x}{R_{ag}} + S_e v_x \qquad (10.116)$$

Typically, two values of the saturation function, S_e, are given: one at 75 percent and the other at the full maximum value of v_x. Based on these two values, a useful region of the S_e curve about the normal operating condition can then be approximated by an exponential function of the form:

$$S_e = A_{ex} \exp^{B_{ex} v_x} \qquad (10.117)$$

Figure 10.9 shows the magnetization curve of the dc exciter loaded with a fixed resistance corresponding to some desired steady-state value of i_x. It is evident from Fig. 10.9 that the value of S_e changes with the operating point. Using Eq. 10.116 to replace i_f in the field voltage equation of Eq. 10.115, we obtain

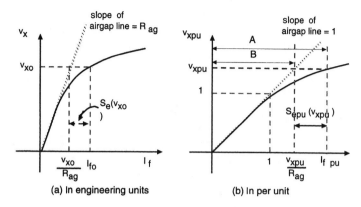

Figure 10.9 Magnetization curve of dc exciter.

$$v_f = \left\{ \frac{r_f}{R_{ag}} + S_e(i_f) r_f \right\} v_x + \frac{d\lambda_f(v_x)}{dt}$$

$$= \frac{r_f}{R_{ag}} \{1 + R_{ag} S_e(i_f)\} v_x + \frac{d\lambda_f(v_x)}{dv_x} \frac{dv_x}{dt} \tag{10.118}$$

There is more than one base system in use for per unitizing the parameters of the exciter. For example, in Appendix A of [111], the value of exciter output voltage that produces rated open-circuit stator voltage on its airgap line is chosen as the base, v_{xbase}, for both v_x and v_f. That of v_{xbase}/R_{ag} is the base value for the exciter field current. With this per unit system, the exciter's field current and voltage equations become

$$i_{fpu} = v_{xpu} + R_{ag} S_e(i_f) v_{xpu}$$

$$v_{fpu} = \frac{r_f}{R_{ag}} \{1 + R_{ag} S_e(i_f)\} v_{xpu} + \frac{d\lambda_f(v_x)}{dv_x} \frac{dv_{xpu}}{dt}$$

$$= \frac{r_f}{R_{ag}} \{1 + S_{epu}(v_{xpu})\} v_{xpu} + \tau_E \frac{dv_{xpu}}{dt} \tag{10.119}$$

$$= \left\{ K_E + \frac{r_f}{R_{ag}} S_{epu}(v_{xpu}) \right\} v_{xpu} + \tau_E \frac{dv_{xpu}}{dt}$$

where

$$S_{epu} = \frac{i_{fpu} - v_{xpu}/R_{ag}}{v_{xpu}/R_{ag}} = \frac{A - B}{B}$$

$$\tau_E = \frac{d\lambda_f(v_x)}{dv_x} = \frac{d\lambda_f(v_{xpu})}{dv_{xpu}}$$

$$K_E = \frac{r_f}{R_{ag}} \tag{10.120}$$

$$S_E = \frac{r_f}{R_{ag}} S_{epu}(v_{xpu})$$

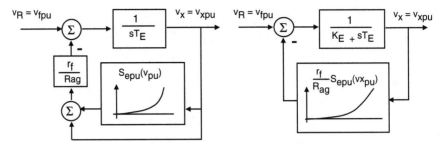

Figure 10.10   Transfer function of dc exciter.

Figure 10.10 shows two possible block diagrams of the transfer function between v_x and v_f, based on the two different forms of the field voltage equation given in Eq. 10.119. When using these representations to simulate operating conditions where the exciter field circuit resistance may be varied, it is important to note that the feedback gain is a function of r_f; however, τ_E will not be affected by changes in r_f.

10.6.4 Stabilizer

The role of the stabilizer shown in Fig. 10.11 is to provide the needed phase advance to achieve the proper gain and phase margins in the open-loop frequency response of the regulator/exciter loop [113]. Stabilizers are used in two situations: one is to enable a higher regulator gain in off-line operations by compensating for the large time constant of the exciter, and the other is to counter the negative damping introduced by a high initial response excitation system in on-line operation. A similar reduction in gain of the excitation system over the transient period can also be achieved using a lag-lead circuit with appropriately chosen values of T_B and T_C in the forward path shown in Fig. 10.11.

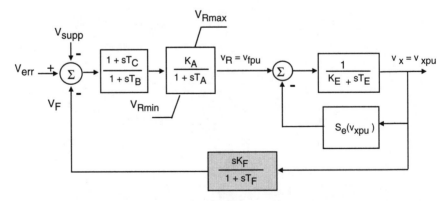

Figure 10.11   Stabilizer for the regulator/exciter loop.

10.6.5 Diode Bridge Operating Characteristic

The full operational characteristic of a bridge rectifier fed from a three-phase ac voltage supply with a finite ac inductance may be divided into three modes:

In Mode 1, two or three diodes conduct alternatively, three when the diodes are commutating. The delay angle, α, is zero and the commutation angle, u, increases from 0 to 60°. It can be shown that the range of the dc output current, I_d, in Mode 1 is from 0 to I_{s2}, where I_{s2} is the peak value of the prospective short-circuit current during commutation. Over that range, the average value of the output dc voltage decreases linearly from an open-circuit value of V_{do} to $0.75V_{do}$. When the peak value of ac phase voltage is V_s, the open-circuit dc voltage, V_{do}, is

$$V_{do} = \frac{3\sqrt{3}}{\pi} V_s \tag{10.121}$$

The output characteristic of the diode bridge in Mode 1 is given by

$$V_d = V_{do} - R_c I_d \tag{10.122}$$

where the commutation resistance, R_c, of a single six-pulse bridge is equal to $3\omega_e L_c/\pi$, L_c being the ac-side inductance on each phase of ac supply to the bridge.

As I_d increases, the longer commutation time may extend beyond the natural commutation point of the next diode coming on. The so-called Mode 2 is characterized by three diodes always conducting, as the delay angle, α, increases with I_d from 0 to 30°, while the commutation angle, u, remains at 60°. The range of the dc output current is from $I_{s2}/2$ to $\sqrt{3}I_{s2}/2$. Over that range of dc output current, the average value of the output dc voltage decreases on a curve from $0.75V_{do}$ to $\sqrt{3}V_{do}/4$ in a nonlinear manner given by

$$V_d = \frac{\sqrt{3}V_{do}}{2}\sqrt{1 - \left(\frac{2X_c I_d}{\sqrt{3}V_s}\right)^2} \tag{10.123}$$

In Mode 3, three and four diodes of the bridge are conducting alternatively. The delay angle, α, remains at 30°, while the commutation angle increases with I_d from 60° to 120°. The range of I_d in Mode 3 is from $\sqrt{3}I_{s2}/2$ to $2I_{s2}/\sqrt{3}$. As I_d increases, the average value of the dc output voltage decreases linearly from $\sqrt{3}V_{do}/4$ to zero. At the end of Mode 3, u is 120° and four diodes will be conducting, shorting all three phases of the ac supply. Note that at that point, $I_d = 2I_{s2}/\sqrt{3} = V_s/\omega_e L_c$; its value is equal to the peak three-phase short-circuit current. Figure 10.12 shows the range of the three modes of a diode bridge rectifier with finite ac inductance.

10.6.6 Power System Stabilizer

When it became apparent that the action of some voltage regulators could result in negative damping of the electromechanical oscillations below the full-power transfer capability, power system stabilizers (pss) were introduced as a means to enhance damping through the modulation of the generator's excitation so as to extend the power transfer limit. In power

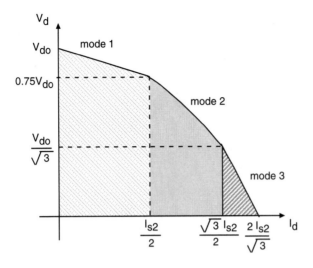

Figure 10.12   Output characteristic of diode bridge.

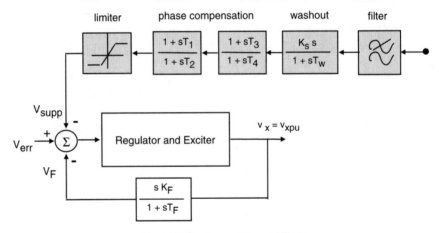

Figure 10.13   Power system stabilizer.

system applications, the oscillation frequency may be as low as 0.1 Hertz between areas, to perhaps as high as 5 Hertz for smaller units oscillating in the local mode.

Figure 10.13 shows the main circuit components of a power system stabilizer; it consists of the following parts:

1. A wash-out circuit for reset action to eliminate steady offset. The value of the T_w is usually not critical, as long as the frequency response contribution from this part does not interfere with the phase compensation over the critical frequency range. It can range from 0.5 to 10 seconds.

2. The two stages of phase compensation have compensation center frequencies of $1/2\pi \sqrt{T_1 T_2}$ and $1/2\pi \sqrt{T_3 T_4}$.

3. A filter section may be added to suppress frequency components in the input signal of the pss that could excite undesirable interactions.

4. Limits are included to prevent the output signal of the power system stabilizer from driving the excitation into heavy saturation.

The output signal of the power system stabilizer is fed as a supplementary input signal, v_{supp}, to the regulator of the excitation system. Although the preliminary circuit design may be based on a small-signal model of the system using phase compensation or root locus techniques, the nonlinear action of the system and pss should not be ignored.

10.7 SHAFT TORSIONAL

The instantaneous torque of an electric machine may contain components which interact with the torsional modes of the mechanical shaft. Since the synchronous torque or power is a function of rotor position, a transient twist of the generator rotor translates to torque or power transients which, if reinforced, could lead to disastrous consequence. Examples of known torsional interactions include subsynchronous oscillations between a generator and series capacitor-compensated lines [116] or the firing control of a high-voltage dc converter nearby [107]. Besides control interactions, material fatigue from accumulated torsional oscillations are also of concern.

    The shaft of a turbine generator is a complex mechanical assembly of several shaft sections with a large number of torsional modes of vibration. Although the complex shaft assembly can be represented by a continuum model in which the shaft is subdivided into minute cylindrical sections, such detailed modeling is often seldom necessary. When the torsional modes of concern are below the synchronous frequency, it has been shown that a simpler representation of the shaft using a lumped-mass model is often adequate. The classical lumped-mass model of a turbine shaft represents the generator's rotor, turbine blades, and the exciter's rotor as separate masses connected by weightless springs. The torsional modes or natural frequencies below the system electrical frequency from such a model are quite close to those obtained from the more sophisticated continuum model.

    Figure 10.14 shows a typical lumped-mass representation of a generator shaft. The equation of motion of the n-disk torsional system of Fig. 10.14, expressed in the same per unit torque as that used for the generator, is

$$\frac{2}{\omega_b}\mathbf{H}\ddot{\theta} + \mathbf{D}\dot{\theta} + \mathbf{K}\theta = \mathbf{T} \qquad pu \qquad (10.124)$$

where θ is the vector of angular displacements or twists of the individual disk to a common reference, in electrical degrees. Here, θ is positive in the direction of rotation of the shaft, and for convenience of expressing the terms in $\mathbf{K}\theta$, θ is measured with respect to the first disk in the order that Eq. 10.124 is written, that is $\theta_1 = 0$. The base angular frequency in electrical radians per second is ω_b. $\mathbf{T}$ is the vector of externally applied torques on the disks. $\mathbf{H}$ is a diagonal matrix whose diagonal elements are the inertia constants of the disks in per unit torque electrical radians to per unit speed, that is $\mathbf{H} = diag(H_1, H_2, \ldots H_n)$. The inertia constants of the various masses are with respect to the same volt-amp base as the generator.

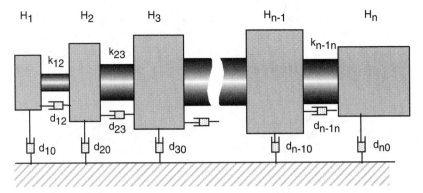

Figure 10.14 Flexible shaft representation of a generator shaft.

Using the angle of the first disk as the common reference for the angular displacements of the other disks, the stiffness matrix, **K**, for the serial assembly is a tri-diagonal matrix:

$$\mathbf{K} = \begin{bmatrix} k_{12} & -k_{12} \\ -k_{21} & k_{12}+k_{23} & -k_{23} \\ & -k_{32} & k_{23}+k_{34} & -k_{34} \\ & & \cdots & \cdots \\ & & & \cdots & \ddots \\ & & & & \cdots \\ & & & & & k_{(n-1)(n-1)} & k_{(n-1)n} \\ & & & & & k_{n(n-1)} & k_{nn} \end{bmatrix} \qquad (10.125)$$

$$\mathbf{K}_{(1,1)} = k_{12}$$

$$\mathbf{K}_{(i,i)} = k_{(i-1)i} + k_{i(i+1)}$$

$$\mathbf{K}_{(i,i+1)} = \mathbf{K}_{(i+1,i)} = k_{(i+1)i} = -k_{i(i+1)} \qquad i = 2,3,\ldots,n-1 \qquad (10.126)$$

$$\mathbf{K}_{(n-n)} = k_{(n-1)n}$$

where k_{ij} is the torsional stiffness or spring constant of the portion of the shaft connecting masses i and j in units of per unit torque to electrical radians. It is a function of the modulus of rigidity of the material, the section modulus, and the length of that portion of the shaft.

The matrix, **D**, is the damping coefficient matrix, representing torsional damping between the masses. In modeling, it is usually assumed to be tri-diagonal, like **K**:

$$\mathbf{D} = \begin{bmatrix} d_{12}+d_{10} & -d_{21} \\ -d_{21} & d_{12}+d_{23}+d_{20} & -d_{23} & \cdots \\ & \cdots & \cdots & \ddots \\ & & \cdots & \ddots \\ & & & \cdots \\ & & \ddots & d_{(n-1)(n-1)} & d_{n(n-1)n} \\ & & & d_{(n-1)n} & d_{nn} \end{bmatrix} \qquad (10.127)$$

$$\mathbf{D}_{(1,1)} = d_{12} + d_{10}$$

$$\mathbf{D}_{(i,i)} = d_{(i-1)i} + d_{i(i+1)} + d_{i0}$$

$$\mathbf{D}_{(i,i+1)} = \mathbf{D}_{(i+1,i)} = d_{(i+1)i} = -d_{i(i+1)} \qquad i = 2, 3, \ldots, n-1 \qquad (10.128)$$

$$\mathbf{D}_{(n,n)} = d_{(n-1)n}$$

where d_{i0} is the viscous damping of the ith disk to the synchronously rotating reference frame, d_{ij} is the viscous damping between the ith and jth disks, both in units of per unit torque to per unit speed. In practice, the damping coefficients are difficult to determine accurately. Their values are usually determined in the modal domain where the damping of individual modal components is more easily measured.

For simplicity, we will begin the modal analysis without the damping term, in which case, Eq. 10.124 of the shaft assembly reduces to

$$\frac{2\mathbf{H}}{\omega_b}\ddot{\theta} + \mathbf{K}\theta = \mathbf{T} \tag{10.129}$$

or

$$\ddot{\theta} + \frac{\omega_b}{2}\mathbf{H}^{-1}\mathbf{K}\theta = \frac{\omega_b}{2}\mathbf{H}^{-1}\mathbf{T} \tag{10.130}$$

The matrix, $\omega_b\mathbf{H}^{-1}\mathbf{K}/2$, is referred to as the torsional system matrix, which in general, is a non-symmetric tri-diagonal matrix. Ignoring the excitation term on the right side of Eq. 10.130 and assuming harmonic motion, that is $\ddot{\theta} = -\lambda\theta$, the natural (unforced) system equation becomes

$$\left[\frac{\omega_b}{2}\mathbf{H}^{-1}\mathbf{K} - \lambda\mathbf{I}\right]\theta = 0 \tag{10.131}$$

The characteristic equation of the torsional system is the determinant of the matrix, $[\frac{\omega_b}{2}\mathbf{H}^{-1}\mathbf{K} - \lambda\mathbf{I}]$, equated to zero. The roots, λ, of the characteristic equation are known as the eigenvalues. The eigenvalues are the square of the natural or modal frequencies, ω_m, of the torsional system, that is

$$\lambda_i = \omega_{mi}^2 \qquad i = 1, 2, \ldots, n \tag{10.132}$$

Since the entire shaft assembly rotates as a solid body, it belongs to the class called semidefinite systems, which are characterized by one or more of shaft's natural frequencies being equal to zero. In this instance, the mode with zero natural frequency is referred to as the zero mode of the torsional system. It is the mode that we have modeled in previous equations of motion of the rotor where we have implicitly assumed that the entire rotor assembly rotates as a single body. When the n disks of the shaft assembly are connected by $(n-1)$ flexible shaft sections, there will be $(n-1)$ other modes of torsional oscillations besides the zero mode.

By substituting λ_i into the natural system equation, we can solve for the right eigenvector or mode shape vector, θ_{mi}.

$$\left[\frac{\omega_b}{2}\mathbf{H}^{-1}\mathbf{K} - \lambda_i\mathbf{I}\right]\theta_{mi} = 0 \tag{10.133}$$

or

$$\frac{2\lambda_i}{\omega_b} \mathbf{H}\theta_{mi} = \mathbf{K}\theta_{mi} \tag{10.134}$$

Pre-multiplying Eq. 10.134 by the transpose of the mode shape vector, θ_{mj}, we have

$$\frac{2\lambda_i}{\omega_b} \theta_{mj}^t \mathbf{H}\theta_{mi} = \theta_{mj}^t \mathbf{K}\theta_{mi} \tag{10.135}$$

Using a similar procedure, but with the ith and jth modes interchanged, we obtain

$$\frac{2\lambda_j}{\omega_b} \theta_{mi}^t \mathbf{H}\theta_{mj} = \theta_{mi}^t \mathbf{K}\theta_{mj} \tag{10.136}$$

Since the matrices, $\mathbf{H}$ and $\mathbf{K}$, are symmetric,

$$\theta_{mi}^t \mathbf{H}\theta_{mj} = \theta_{mj}^t \mathbf{H}\theta_{mi}$$

$$\theta_{mi}^t \mathbf{K}\theta_{mj} = \theta_{mj}^t \mathbf{K}\theta_{mi} \tag{10.137}$$

Subtracting Eq. 10.135 from Eq. 10.136 and using the relations given in Eq. 10.137, we obtain

$$\frac{2}{\omega_b}(\lambda_i - \lambda_j)\theta_{mj}^t \mathbf{H}\theta_{mi} = 0 \tag{10.138}$$

Since the eigenvectors are orthonormal, for $\lambda_i \neq \lambda_j$,

$$\theta_{mj}^t \mathbf{H}\theta_{mi} = 0 \tag{10.139}$$

We can see from Eq. 10.135 that

$$\theta_{mj}^t \mathbf{K}\theta_{mi} = 0 \tag{10.140}$$

However, for $i = j$,

$$\theta_{mi}^t \mathbf{H}\theta_{mi} = H_{mi}$$

$$\theta_{mi}^t \mathbf{K}\theta_{mi} = K_{mi} \tag{10.141}$$

where H_{mi} and K_{mi} are known as the generalized (or modal) inertia constants and stiffness, respectively.

The orthonormal eigenvectors form a set of principal coordinates in which Eq. 10.124 can be expressed in a decoupled form, in which each component can be solved independently of the other. Let $\mathbf{Q}$ be the matrix whose columns are the right eigenvectors of the matrix $\mathbf{H}^{-1}\mathbf{K}$. The matrix, $\mathbf{Q}$, is not unique, it can be multiplied by an arbitrary constant, that is

$$\bar{\mathbf{Q}} = \mathbf{Q}\mathbf{R} \tag{10.142}$$

where the scaling matrix, $\mathbf{R}$, is a diagonal matrix of the same order as $\mathbf{Q}$. $\bar{\mathbf{Q}}$ or $\mathbf{Q}$ is often referred to as the modal matrix, because its n columns represent the modes of vibration or degree of freedom of the n-disk system. For example, that which corresponds to a natural frequency of zero is referred to as Mode 0. And that which corresponds to the next higher natural frequency is referred to as Mode 1, and so on.

The transformation from actual to modal angles can be performed by substituting $\theta = \bar{\mathbf{Q}}\theta_m$. Pre-multiplying Eq. 10.124 through by $\bar{\mathbf{Q}}^t$ and making the above-mentioned substitution, we obtain

$$\frac{2}{\omega_b}\underbrace{\bar{\mathbf{Q}}^t\mathbf{H}\bar{\mathbf{Q}}}_{\mathbf{H_m}}\ddot{\theta}_m + \underbrace{\bar{\mathbf{Q}}^t\mathbf{D}\bar{\mathbf{Q}}}_{\mathbf{D_m}}\dot{\theta}_m + \underbrace{\bar{\mathbf{Q}}^t\mathbf{K}\bar{\mathbf{Q}}}_{\mathbf{K_m}}\theta_m = \underbrace{\bar{\mathbf{Q}}^t T}_{\mathbf{T_m}} \tag{10.143}$$

A subscript, m, is used here to denote the corresponding modal quantities. For the shaft assembly, the modal inertia and stiffness matrices, $\mathbf{H_m}$ and $\mathbf{K_m}$, are diagonal matrices. Often, the damping coefficient matrix, $\mathbf{D}$, is casted into the same form as $\mathbf{H}$ and $\mathbf{K}$, in which case, the matrix, $\bar{\mathbf{Q}}^t\mathbf{D}\bar{\mathbf{Q}}$, will also be diagonal. Denoting the modal inertia, stiffness, and damping coefficient matrices by $\mathbf{H}_m$, $\mathbf{K}_m$, and $\mathbf{D}_m$, respectively, Eq. 10.143 can be written as

$$\frac{2}{\omega_b}\mathbf{H}_m\ddot{\theta}_m + \mathbf{D}_m\dot{\theta}_m + \mathbf{K}_m\theta_m = \bar{\mathbf{Q}}^t T \tag{10.144}$$

or

$$\ddot{\theta}_m + \frac{\omega_b}{2}\mathbf{H}_m^{-1}\mathbf{D}_m\dot{\theta}_m + \frac{\omega_b}{2}\mathbf{H}_m^{-1}\mathbf{K}_m\theta_m = \frac{\omega_b}{2}\mathbf{H}_m^{-1}\bar{\mathbf{Q}}^t T \tag{10.145}$$

Since $\mathbf{H}_m$, $\mathbf{D}_m$ and $\mathbf{K}_m$ are diagonal matrices, the equations for the modes are decoupled from one another and have a form similar to Eq. 10.145. For example, the torsional equation of the mith mode is

$$\ddot{\theta}_{mi} + \frac{\omega_b D_{mi}}{2H_{mi}}\dot{\theta}_{mi} + \omega_{mi}^2\theta_{mi} = \frac{\omega_b}{2H_{mi}}T_{mi} \qquad i = 1,2,\ldots,n \tag{10.146}$$

where $m1$ is referred to as Mode 0, $m2$ is Mode 1, and so on. Comparing the form of the above modal equation with that of the normal second-order equation, that is

$$\ddot{\theta}_{mi} + 2\zeta_i\omega_{mi}\dot{\theta}_{mi} + \omega_{mi}^2\theta_{mi} = \frac{\omega_b}{2H_{mi}}T_{mi} \tag{10.147}$$

we can identify that

$$2\zeta_i\omega_{mi} = \frac{\omega_b}{2}\frac{D_{mi}}{H_{mi}} \tag{10.148}$$

The damping factor, ζ_i, is a dimensionless quantity. It is defined as the ratio of the actual damping to the critical damping. When $\zeta_i \ll 1$, the response will be under-damped and may be expressed in the form:

$$\theta_{mi}(t) = \theta_{mi}(0)e^{-\zeta_i\omega_{mi}t}\sin(\omega_{mi}\sqrt{1-\zeta_i^2}\,t + \phi_i) \tag{10.149}$$

In practice, the damping values are often determined from measurements of the modal component's decay in torsional oscillations. One such measure is the logarithmic decrement. It is defined as

$$\hat{\delta}_i = \frac{1}{n_c}\ln\frac{\theta_{mi}(0)}{\theta_{mi}\left(\dfrac{2\pi n_c}{\omega_{mi}}\right)} \tag{10.150}$$

where $\theta_{mi}(0)$ and $\theta_{mi}\left(\frac{2\pi n_c}{\omega_{mi}}\right)$ are the peak values of θ_{mi} at the first and n_cth cycle. Eq. 10.150 may also be expressed as

$$\hat{\delta}_i = \frac{1}{n_c} \ln \frac{\theta_{mi}(0)}{\theta_{mi}(0)e^{-\zeta_i\omega_{mi}\left(\frac{2\pi n_c}{\omega_{mi}}\right)}} \qquad (10.151)$$

which simplifies to

$$\hat{\delta}_i = 2\pi\,\zeta_i \qquad (10.152)$$

In terms of the logarithmic decrement, the damping coefficient in Eq. 10.148 is

$$2\zeta_i\omega_{mi} = \frac{\omega_{mi}}{\pi}\hat{\delta}_i \qquad (10.153)$$

Another measure is the decrement factor. It is defined as the inverse of the time constant for the envelope of the measured oscillations of a modal component to decay to 1/e of its original amplitude. The relationship between the decrement factor and logarithmic decrement is

$$\sigma_i = f_{mi}\hat{\delta}_i = \frac{\omega_{mi}}{2\pi}\hat{\delta}_i \qquad sec^{-1} \qquad (10.154)$$

From the relationships given in Eqs. 10.148, 10.152, and 10.154, the modal damping can be determined from the measured decrement values using

$$D_{mi} = \frac{2}{\pi}\frac{\omega_{mi}}{\omega_b}H_{mi}\hat{\delta}_i$$

$$= \frac{4}{\omega_b}H_{mi}\sigma_i \qquad (10.155)$$

10.7.1 Simulation

As mentioned earlier, Mode 0 of the rotor corresponds to rigid body rotation; the elements of the Mode 0 eigenvector are all of the same value. In the following discussion, we will assume that the eigenvector of Mode 0 is the first column of $\bar{\mathbf{Q}}$. Since the modal frequency of Mode 0, ω_{m1}, is zero, Eq. 10.146 for Mode 0 may be rewritten into a form similar to Eq. 7.91, and simulated as before. From $\mathbf{T_m} = \bar{\mathbf{Q}}^t\mathbf{T}$, the input modal torque of Mode 0 may be expressed as

$$T_{m1} = \sum_{j=1}^{n}\bar{q}_{j1}T_j \qquad (10.156)$$

where $\bar{q}_{j1}$ is the jth element of the first column of $\bar{Q}$. Using the above expression for T_{m1} and a ω_{m1} of zero, Eq. 10.146 for Mode 0 may be expressed in the form

$$\frac{2H_{m1}}{\omega_b}\frac{d^2\theta_{m1}}{dt^2} = \sum_{j=1}^{n}\bar{q}_{j1}T_j - D_{m1}\frac{d\theta_{m1}}{dt} \qquad (10.157)$$

From the relation, $\mathbf{H}_m = \bar{\mathbf{Q}}^t\mathbf{H}\bar{\mathbf{Q}}$, we can deduce that the value of modal inertia for Mode 0, H_{m1}, is given by

$$H_{m1} = \sum_{j=1}^{n} \bar{q}_{j1}^2 H_j \tag{10.158}$$

where H_j's are the inertia constants of the disks on the rotor assembly. In general, the values of the modal inertia and damping will depend on the scaling applied to $\mathbf{Q}$. The right-hand side of Eq. 10.157 corresponds to some scaled magnitude of the net acceleration torque on the entire rotor. For generating operations, the turbine torques will be positive in the direction of rotation. They are opposed by the electromagnetic torque developed by the generator and the damping torque.

If the angle of the generator disk corresponds to the gth row in the transformation $\theta = \bar{\mathbf{Q}}\theta_m$, the generator's rotor angle, δ, measured with respect to a synchronously rotating reference, is

$$\delta = \underbrace{q_{g1}\theta_{m1}}_{\text{Mode 0}} + \underbrace{q_{g2}\theta_{m2} + \ldots + q_{gn}\theta_{mn}}_{\text{other modes}} \tag{10.159}$$

The modal angles of the other torsional modes can be obtained by integrating their respective equations in Eq. 10.146, rearranged in the form:

$$\ddot{\theta}_{mi} = \frac{\omega_b}{2H_{mi}} T_{mi} - 2\zeta_i \omega_{mi}\dot{\theta}_{mi} - \omega_{mi}^2 \theta_{mi} \qquad i = 2, 3, \ldots, n \tag{10.160}$$

The input modal torques, T_{mi}, in Eq. 10.160 can be derived from the externally applied torques on the disks using $\mathbf{T}_m = \bar{\mathbf{Q}}^t \mathbf{T}$, that is

$$T_{mi} = \sum_{j=1}^{n} \bar{q}_{ij}^t T_j \qquad i = 2, 3, \ldots, n \tag{10.161}$$

where $\bar{q}_{ij}^t$ is the (i, j) element of the transposed eigenvector matrix, $\bar{\mathbf{Q}}^t$, and T_j is the jth element of the vector of externally applied torque, $\mathbf{T}$. The solution of Eq. 10.160 yields the rest of the modal angles in θ_m.

The angles of twist with respect those of disk 1 can be determined from $\theta = \bar{\mathbf{Q}}\theta_{\mathbf{m}}$. More frequently, the shaft torques between disks are of interest. Shaft torque can be computed from the product of the shaft stiffness and the relative angle of twist. For example, the torque carried by the shaft section between the ith and jth disks can be computed using

$$T_{ij} = k_{ij}(\theta_i - \theta_j)$$
$$= k_{ij} \sum_{l=1}^{n} (\bar{q}_{il} - \bar{q}_{jl})\theta_{ml} \tag{10.162}$$

10.7.2 Computing Q and Scaling

When computing $\bar{\mathbf{Q}}$, there may be advantages in noting that the computation of the eigenvectors of a symmetric matrix is much easier than that of a non-symmetric matrix. We can avoid computing $\mathbf{Q}$ directly from the matrix, $\mathbf{H}^{-1}\mathbf{K}$, by introducing an intermediate transformation, $\theta = \mathbf{H}^{-\frac{1}{2}}\delta$. Substituting this into Eq. 10.130 yields

$$\frac{1}{\omega_b}\mathbf{H}^{\frac{1}{2}}\ddot{\delta} + \mathbf{K}\mathbf{H}^{-\frac{1}{2}}\delta = \mathbf{T}$$

or

$$\ddot{\delta} + \frac{\omega_b}{2}\mathbf{H}^{-\frac{1}{2}}\mathbf{K}\mathbf{H}^{-\frac{1}{2}}\delta = \frac{\omega_b}{2}\mathbf{H}^{-\frac{1}{2}}\mathbf{T} \qquad (10.163)$$

Since $\mathbf{K}$ is tri-diagonal symmetric and $\mathbf{H}^{-\frac{1}{2}}$ is diagonal, the matrix, $\mathbf{H}^{-\frac{1}{2}}\mathbf{K}\mathbf{H}^{-\frac{1}{2}}$, is a tri-diagonal symmetric matrix.

Let $\mathbf{S}$ be the matrix whose columns are the eigenvectors of the tri-diagonal symmetric matrix, $\mathbf{H}^{-\frac{1}{2}}\mathbf{K}\mathbf{H}^{-\frac{1}{2}}$. The eigenvectors of the symmetric real matrix are real and orthonormal. If we were to reintroduce the damping term to Eq. 10.163 and apply the transformation, $\delta = \mathbf{S}\theta_m$, and the scaling transformation of Eq. 10.142, we could obtain the modal equation given in Eq. 10.143. However, by using this two-stage transformation, we have simplified the computation of the eigenvectors. The computation of the matrix, $\mathbf{S}$, is less time-consuming than that of computing $\mathbf{Q}$ directly because the matrix $\mathbf{H}^{-\frac{1}{2}}\mathbf{K}\mathbf{H}^{-\frac{1}{2}}$ is symmetric. The matrix, $\mathbf{Q}$, can then be determined using

$$\mathbf{Q} = \mathbf{H}^{-\frac{1}{2}}\mathbf{S} \qquad (10.164)$$

and that of $\bar{\mathbf{Q}}$ from

$$\bar{\mathbf{Q}} = \mathbf{Q}\mathbf{R} \qquad (10.165)$$

Since both $\mathbf{H}^{-\frac{1}{2}}$ and $\mathbf{R}$ are diagonal matrices, the computational effort involved in Eqs. 10.164 and 10.165 will also be small.

The MATLAB **eig** function yields eigenvectors that are already scaled to unit length vectors. If you are using a routine that doesn't and you want the eigenvectors to be scaled to unit length, the scaling matrix to use will be of the form:

$$\mathbf{R} = diag\left(\frac{1}{\sqrt{\sum_i q_{i1}^2}} \quad \frac{1}{\sqrt{\sum_i q_{i2}^2}} \quad \cdots \quad \frac{1}{\sqrt{\sum_i q_{in}^2}}\right) \qquad (10.166)$$

If the generator simulation is to handle either rigid- or flexible-shaft-type studies, there may be an advantage in choosing a scaling matrix such that the eigenvector corresponding to Mode 0 is all ones, that is $q_{j1} = 1$ for all j. It is equivalent to specifying that the Mode 0 shape vector is all ones. This condition is obviously satisfied using the scaling matrix given below:

$$\mathbf{R} = diag\left(\frac{1}{q_{g1}} \quad \frac{1}{q_{g2}} \quad \cdots \frac{1}{q_{gn}}\right) \qquad (10.167)$$

Note that scaling the largest element of each eigenvector to unity will also produce a Mode 0 shape vector of all ones, since the elements of the Mode 0 vector are all equal. With an all-one Mode 0 eigenvector, the resultant Mode 0 inertia and acceleration torque in Eqs. 10.156 and 10.158 will reduce to simply the sum of the actual inertias and the algebraic sum of the torques on the rotor assembly, respectively.

In addition to an all-one Mode 0 eigenvector, the $\mathbf{R}$ of Eq. 10.167 also produces a $\bar{\mathbf{Q}}$ matrix that has a row vector of all ones in the row corresponding to the generator mass. Since $\mathbf{T_m} = \bar{\mathbf{Q}}^t \mathbf{T}$, using such a transformation will produce in each modal torque a unit of the generator's electromagnetic torque, T_g. This may be advantageous when the objective of the study is the torsional interactions of the rotor shaft assembly from only the electromagnetic torque, T_g, of the generator. For such a study, the other excitation torques of the turbines and exciter may be ignored, and the resulting input modal torques will all be equal to T_g.

10.8 SYNCHRONOUS MOTOR DRIVES

Earlier, in Section 5.6, we showed that a set of three-phase windings excited with a balanced set of three-phase currents will produce a rotating mmf, sinusoidally distributed in airgap and of constant amplitude. Specifically, if $i_a(t) = I_m \cos \omega t$ and i_b and i_c lag i_a to form a balanced set, the resultant mmf at an airgap point θ_a from the axis of the a-phase winding may be expressed as

$$F(\theta_a, t) = \underbrace{\frac{3}{2} \frac{4}{\pi} \left(\frac{N_{ph} k_w}{P} \right)}_{F_s} I_m \cos(\theta_a - \omega t) \tag{10.168}$$

where N_{ph} and k_w are the number of turns per phase and the winding factor of the phase windings, respectively. If the excitation source on the rotor produces an mmf in the airgap that is also sinusoidally distributed about the rotor axis with a peak value of F_r, say displaced at an angle δ_{sr} ahead of F_s as shown in Fig. 5.14, the excitation torque from the interaction of the stator and rotor mmfs given in Eq. 5.55 may be written as

$$T = -K I_m F_r \sin \delta_{sr} \tag{10.169}$$

where K is a constant factor that includes machine dimensions and winding factor. Equation 7.67 also shows that even when the rotor has no mmf source of its own, as in the case of a reluctance motor, the magnetic saliency of the rotor will still result in a steady alignment or reluctance torque from the rotating stator's mmf.

With a power electronic inverter supplying power to the stator windings, both the frequency and phase of the stator currents can be electronically controlled to produce a stator mmf that will rotate in synchronism with the rotor, that is, δ_{sr} can be maintained constant. Equation 10.169 indicates that the torque per stator ampere, that is T/I_m, is maximum when δ_{sr} is held at 90°. Such an operating condition is analogous to the operation of a dc machine, where the brushes and commutator switch the currents flowing in its rotating armature windings to produce an armature mmf that is at right angles to the stationary field winding's mmf. Thus, a synchronous motor with an inverter supply controlled in this manner may be viewed as a dc machine with stationary armature and the inverter functioning much like the brushes and commutator of a dc machine to keep the stator mmf in synchronism with the rotor mmf. This similarity has resulted in the use of the term *brushless dc motor*, usually in reference to a permanent magnet synchronous motor with an inverter that is switched in accordance with its rotor position. The use of rotor position information to direct the switching of the inverter is generally referred to as *self-controlled*.

Synchronous motors can operate in parallel from an adjustable frequency inverter in an open-loop manner that is not self-controlled; their speed will be determined by the common inverter frequency. Their operating characteristic in open-loop control will be just the typical synchronous machine characteristic of load angle varying with output power, rotor swing, and the possibility of losing synchronism with the inverter supply. Most modern drives are of the self-controlled type with an additional speed or torque regulation loop. Excluding cycloconverter drives, most are dc link converter arrangements with either voltage or current source characteristic to the motor. A line-commutated current source inverter often is a practical choice for a large synchronous motor drive from the standpoint of commutation requirements, freedom of the motor voltage, and protection against overcurrent.

Medium to large horsepower synchronous motor drives are mostly line-commutated, mainly because forced commutation with auxiliary commutation circuits and associated commutation losses are not practical at high power. Besides, line-commutated converters with thyristor switches have a proven record in high-voltage dc transmission systems. Figure 10.15 shows a simplified diagram of a line-commutated synchronous motor drive with self-control. Such a drive is also referred to as a load-commutated synchronous motor drive. Power from the ac supply source is first converted to dc by the source-side converter, which normally operates in the rectifier mode to control the dc current while the motor-side converter operates normally in the inverter mode to power the motor. Adjustment of the output torque, and hence also the speed of the motor, is through the control of the dc

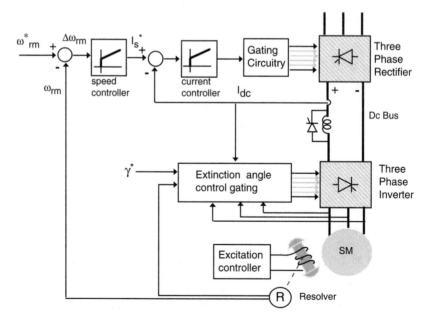

Figure 10.15   Line-commutated synchronous motor drive.

current. In the self-controlled mode, the switching frequency of the inverter is decided by the rotor speed.

The line-commutated inverter relies on the back emf of the motor to naturally commutate the current from one path to another. The line voltage provides a forward bias to the thyristor that is to be turned on. The same voltage will commutate the current from the thyristor that is to be turned off to that which has just turned on. Although the theoretical range of the phase delay for inversion is from 90° to 180°, in practice, a lower limit of around 150° may be used to allow for the commutation delay and the turn-off time for the thyristor to recover its blocking voltage capability. The commutation delay is a function of the magnitude of the dc current and the available commutating voltage at that point of the cycle.

That the phase delay must be less than 180° implies a motor phase current that is in advance of or leading the phase voltage. In other words, the motor should operate with a leading power factor, or overexcited. At very low speeds, typically under 10% rated, the back emf of the motor may not be of sufficient magnitude to commutate the thyristors reliably. Schemes that can be used to start the motor include using a pony motor, adding a fourth leg to the inverter and connecting a commutating capacitor between the center of the fourth leg to the neutral of the motor to provide forced commutation capability at low speed, or using the rectifier to modulate the dc link current to zero after each pulse of current to allow the conducting inverter thyristors to regain their voltage blocking capability. The last approach is often preferred as it requires no additional components, just additional coordination between the controls of the rectifier and inverter.

The drive is inherently capable of four-quadrant operation. Reversal of power flow from motor to ac source can be easily accomplished by a reversal in polarity of the dc link voltage, and not the direction of dc current, by switching the mode of operation of the ac source side and motor converters, the former to inverting and the latter to rectifying. With six pulse bridges, low order line harmonics and harmonic torque pulsations can be of concern. Another disadvantage is that the ac-side power factor will decrease with the dc link voltage when the motor is not operating close to rated speed.

The control of output torque has to be coordinated with the control of the field excitation, as the increase of output torque with stator current for fixed excitation has a certain ceiling beyond which torque can actually decrease with a further increase in stator current. A measure of the degree of iron utilization is the volt/hertz. To keep the stator's volt/hertz constant when excitation is held fixed, the stator's reaction along the d-axis, that is $I_d x_{ad}$, will have to vary with output torque.

At the lower voltage and current ranges, forced-commutation by auxiliary resonant circuits or devices with inherent turn-off capability becomes practical. One obvious operating advantage with forced commutation is that the motor no longer has to operate with a leading terminal power factor. At the low power range, fast switching field-effect and insulated gate bipolar transistors can easily control the waveforms of the stator currents over a wide frequency range. For a wound-field synchronous motor that has almost sinusoidal airgap field distribution, supplying it with the proper sinusoidal stator currents will result in a smooth average torque. But, for a permanent magnet motor with surface mount magnets that has trapezoidal rather than sinusoidal airgap field distribution, a simple block modulation of the stator currents can be used to obtain a steady and reasonably smooth torque. Whereas true sinusoidal waveform control of the stator currents will require continuous and more precise information about the rotor's position, such as that from a resolver, that for simple block current modulation can utilize Hall-effect devices as rotor position detectors to perform the necessary switchings at discrete positions of the rotor.

10.9 PROJECTS

10.9.1 Project 1: Transient Models

The purpose of this project is to compare the transient and small-signal results of the transient model with stator transients neglected with those obtained using the full model described in Chapter 7. To facilitate the comparison we will again perform these tests using the parameters of the 9375-kVA, 12.5-kV, 60-Hz, two-pole synchronous generator of Projects 1 and 3 of Chapter 7 in which the full model was used.

The system of our study is the one machine to infinite bus system shown in Figure 10.16. The external line parameters, r_e and x_e, are to be varied to change the electrical strength of the connection between generator and infinite bus, which in this case supposedly represents a large system to which the generator is connected.

The following remarks are applicable to the transient and sub-transient models given in Tables 10.2 and 10.4, respectively, where changes in stator qd flux linkages in the qd rotor reference frame have been neglected. First we take note that the variables in these machine models are in the rotor reference frame. For the test system of Fig. 10.16, where we have only one machine and a rather simple external network, we can easily transform the variables of the external network to the rotor reference frame of that one machine. In phasor form, the voltage drop across the external line is

$$\tilde{V}_z = (r_e + jx_e)\tilde{I} \tag{10.170}$$

Using the relation given in Eq. 7.62, the phasor quantities can be expressed in qd components of a synchronously rotating reference frame, that is $\tilde{V}_z = v_{qz}^e - jv_{dz}^e$ and $\tilde{I} = i_q^e - ji_d^e$. In terms of the qd rotor reference frame of the machine,

$$v_{qz} - jv_{dz} = e^{-j\delta}v_{qz}^e - jv_{dz}^e, \qquad i_q - ji_d = e^{-j\delta}(i_q^e - ji_d^e) \tag{10.171}$$

Using the above-mentioned relationships between phasors and synchronously rotating quantities and between synchronously rotating quantities and rotor qd quantities, we can convert Eq. 10.170 to an equivalent voltage drop in the rotor qd reference frame:

infinite bus   generator   **Figure 10.16** One machine to infinite bus system.

$$\tilde{\mathbf{V}}_z = v_{qz}^e - jv_{dz}^e = (r_e + jx_e)(i_q^e - ji_d^e)$$

$$v_q - jv_d = e^{-j\delta}(r_e + jx_e)e^{j\delta}(i_q - ji_d) \tag{10.172}$$

$$= (r_e + jx_e)(i_q - ji_d)$$

Expressed in terms of the same i_q and i_d of the stator, the external line drop of $r_e + jx_e$ can be readily added to the stator winding voltage equations. The stator voltage equations with the external impedance included become

$$v_q = -(r_s + r_e)i_q - (x_d' + x_e)i_d + E_q'$$

$$v_d = -(r_s + r_e)i_d + (x_q' + x_e)i_q + E_d' \tag{10.173}$$

Likewise, when using the sub-transient model given in Table 10.4, the external impedance can be included directly into Eq. 10.106 of the stator qd voltages.

Figure 10.17 shows the SIMULINK simulation given in *s1eig* for a synchronous generator connected by a series RL line to an infinite bus. Inside the *tmodel* block, the generator is represented by the transient model given in Table 10.2. The components of the infinite bus voltages, expressed in the synchronously rotating reference frame, are transformed into the rotor reference frame of the generator in the *qde2qdr* block using

$$v_{qi} = v_{qi}^e \cos\delta - v_{di}^e \sin\delta$$

$$v_{di} = v_{qi}^e \sin\delta + v_{di}^e \cos\delta \tag{10.174}$$

where δ is the rotor angle measured with respect to the q-axis of a synchronously rotating reference frame. By definition, the frequency of the infinite bus is constant at the synchronous value. With only one machine, it will be convenient for calculation purposes to select the phasor of the infinite bus voltage as the reference phasor and also the q-axis of the synchronously rotating reference frame. With this choice of synchronously rotating reference frame, v_{di} in the above equation will be identically zero.

Inside the *stator_wdg* block, the stator qd currents are determined using the stator voltage equations that included the external series RL line parameters in series with the

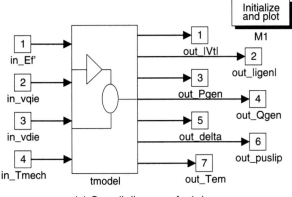

(a) Overall diagram of *s1eig*

Figure 10.17   Simulation *s1eig* for linearized analysis.

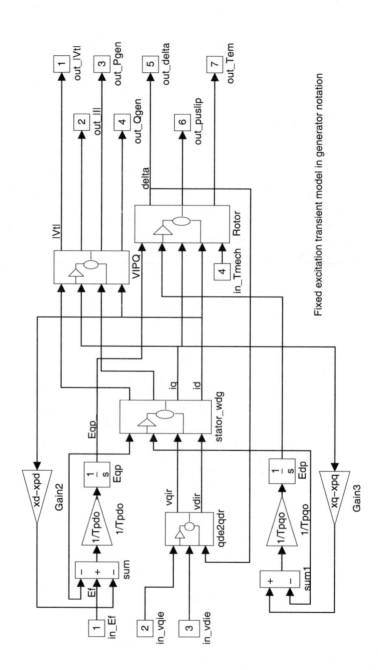

(b) Inside the *tmodel* block of *s1eig*

Figure 10.17 *(cont.)*: Simulation *s1eig* for linearized analysis.

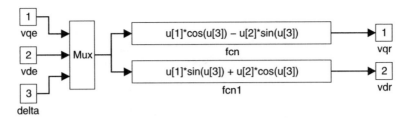

(c) Inside the *qde2qdr* block

Figure 10.17   *(cont.):* Simulation *s1eig* for linearized analysis.

stator resistance and leakage reactance, as given in Eq. 10.173. The same stator voltage equations, without the external RL line parameters, are used to compute the terminal voltage of the generator within the *stator_wdg* block.

The two rotor circuits in this model are represented by the differential equations in E'_q and E'_d. Only the physical field winding on the d-axis has an external excitation of E_f; that for the q-axis circuit, E_g, is zero.

Linearized analysis of a synchronous generator.   As in Project 3 of Chapter 7, we will determine the linear models of a synchronous machine that is generating directly into an infinite bus at two different operating conditions, this time with a simpler transient model of the generator. The linear models obtained will be used to determine the transfer functions, $\Delta P_{gen}/\Delta T_{mech}$ and $\Delta Q_{gen}/\Delta E_f$. We will then compare these transfer functions with those obtained earlier in Project 3 of Chapter 7 for the same operating condition so that we can identify what part of the characteristics are retained by the simpler model.

First set up the SIMULINK simulation, *s1eig*, and use *m1* to set up the parameters of the 9375 kVA generator of Project 1, Chapter 7, given in file *set1.m*. Note that in *s1eig*, the inputs and outputs are defined by sequentially numbered input and output ports as required by the SIMULINK **trim** and MATLAB **linmod** functions.

Use the MATLAB M-file, *m1*, to perform the following tasks:

1. Load the parameters and rating of the synchronous machine given in Project 1 onto the MATLAB workspace. Make sure that the units used are consistent with those employed in your SIMULINK representation of the system.

2. Have in mind the operating conditions for which you would like to obtain the linear models of the system. For the first part of this project, set the values of the external line resistance and reactance to zero for the generator to be directly connected to the infinite bus and obtain the transfer functions for the following two operating conditions:

   a. The complex power delivered by the machine to its terminal, $P_{gen} + jQ_{gen}$, is 0, $v_{qi}^e = 1$ *pu*, $v_{di}^e = 0$, and $D_\omega = 2$ *pu*.

   b. The complex power delivered by the machine to its terminal, $P_{gen} + jQ_{gen}$, is $0.8 + j0.6$ *pu*, $v_{qi}^e = 1$ *pu*, $v_{di}^e = 0$, and $D_\omega = 2$ *pu*.

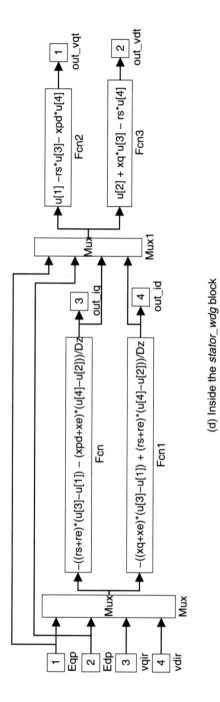

(d) Inside the *stator_wdg* block

Figure 10.17 (cont.): Simulation *sleig* for linearized analysis.

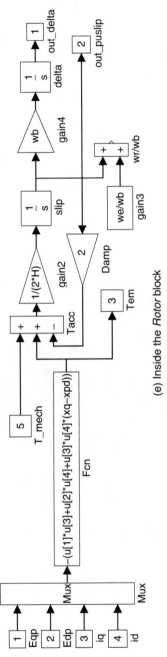

(e) Inside the *Rotor* block

Figure 10.17 (cont.): Simulation *sleig* for linearized analysis.

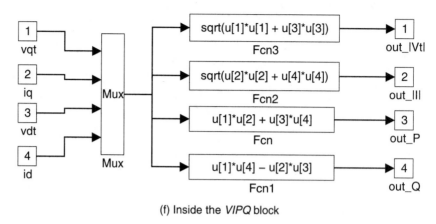

(f) Inside the *VIPQ* block

Figure 10.17   *(cont.)*: Simulation *sleig* for linearized analysis.

3. Set up a `for` loop to perform the following subtasks for each of the above operating conditions:

   a. Use the SIMULINK **trim** function to determine steady-state of a desired operating point of the SIMULINK subsystem, *tmodel* in *sleig*.

   b. Use the MATLAB **linmod** function to determine the state variable model, that is the A,B,C, and D matrices of the linear model. Obtain the system eigenvalues using the MATLAB **eig** function on the system matrix A, that is from eig(A).

   c. Use the MATLAB **ss2tf** to determine the transfer functions of $\Delta Q_{gen}/\Delta E_f$ and $\Delta P_{gen}/\Delta T_{mech}$ at each of the above operating points.

   d. Use the MATLAB **tf2zp** to determine the poles and zeros of the two sets of transfer functions.

   e. Plot the root-locus of the transfer functions, $\Delta Q_{gen}/\Delta E_f$ and $\Delta P_{gen}/\Delta T_{mech}$, at the operating point where $P_{gen} = 0.8$, $Q_{gen} = 0.6$, $v_{qs}^e = 1$, and $v_{ds}^e = 0$.

Repeat all of tasks 2 and 3 above for the condition when the external line has a resistance and reactance of $0.01 + j0.1$ pu.

For comparison purposes, the results of the transfer functions given in Tables 10.5 and 10.6 are for the same condition as in Project 3 of Chapter 7, where the generator is directly connected to the infinite bus; in other words, the external line resistance and reactance are both zero.

The sample root-locus plot in Fig. 10.18 is for $\Delta Q_{gen}/\Delta E_f$ at the operating point where $P_{gen} + jQ_{gen} = 0.8 + j0.6$ *pu*, $v_{qi}^e - jv_{di}^e = 1 + j0$ *pu*, and $D_\omega = 2$ *pu*.

Operating characteristics and parameter sensitivity studies.
Next, we will use the simulation, *s1*, to examine the dynamic response of the system to a step change of the input mechanical torque about the operating point to see if the response is as predicted by the eigenvalues of the linearized model. Figure 10.19 shows the content of the SIMULINK file, *s1*; it contains a modified copy of *sleig*. The input ports are replaced with

TABLE 10.5 $\Delta P_{gen}/\Delta T_{mech}$

Output is $P_{gen}+jQ_{gen}=0+j0$ Gain is 2.66e−15 Zeros are:	Output is $P_{gen}+jQ_{gen}=0.8+j0.6$ Gain is −4.44e−15 Zeros are:
−1.130e+16 0.000e+00i −2.439e+00 0.000e+00i −6.383e−01 0.000e+00i Poles are: −7.870e−01 5.126e+00i −7.870e−01 −5.126e+00i −2.732e+00 0.000e+00i −6.383e−01 0.000e+00i	2.368e+16 0.000e+00i −6.383e−01 0.000e+00i −3.374e−01 0.000e+00i Poles are: −3.951e−01 1.025e+01i −3.951e−01 −1.025e+01i −3.818e+00 0.000e+00i −3.358e−01 0.000e+00i

TABLE 10.6 $\Delta Q_{gen}/\Delta E_f$

Output is $P_{gen}+jQ_{gen}=0+j0$ Gain is 3.57e−01 Zeros are:	Output is $P_{gen}+jQ_{gen}=0.8+j0.6$ Gain is 2.74e−01 Zeros are:
−7.870e−01 5.126e+00i −7.870e−01 −5.126e+00i −2.732e+00 0.000e+00i Poles are: −7.870e−01 5.126e+00i −7.870e−01 −5.126e+00i −2.732e+00 0.000e+00i −6.383e−01 0.000e+00i	−3.643e−01 9.620e+00i −3.643e−01 −9.620e+00i −3.581e+00 0.000e+00i Poles are: −3.951e−01 1.025e+01i −3.951e−01 −1.025e+01i −3.818e+00 0.000e+00i −3.358e−01 0.000e+00i

constant sources whose values are set to correspond to the steady-state values determined by the SIMULINK **trim** function for the desired operating condition.

(a) Set $v_{qi}^e - jv_{di}^e = 1+j0$, $P_{gen}+jQ_{gen}=0.8+j0.6$, $D_\omega=2$, and the values of the external line resistance and reactance to zero. Program the *repeating sequence* source representing the externally applied mechanical torque, T_{mech}, with *time* values of [0 7.5 7.5 15 15 22.5 22.5 30] and *output* values of [0.8 0.8 0.9 0.9 0.7 0.7 0.8 0.8]. This sequence of input allows an initial two seconds for any simulation startup transients to settle before applying the step changes. Use any of the RK methods of integration with a minimum step size of 0.1 msec, and a tolerance of $1e^{-6}$. Record the response of $|Vt|$, $|Igen|$, T_{em}, P, Q, and δ. Compare the simulated transient behavior with that predicted from the eigenvalues of the corresponding transfer function.

(b) Change the value of H, the inertia constant of the machine, from the base value of three seconds to five seconds and repeat Part (a). Compare the oscillation frequency of δ obtained in Parts (a) and (b) with the corresponding runs in Project 1 of Chapter 7.

(c) Repeat the simulation runs of Parts (a) and (b) with the value of the external line resistance and reactance changed from zero to $0.01 + j0.1$ pu. Comment on the effects that an increase in line impedance has on the frequency and damping of the electromechanical oscillations.

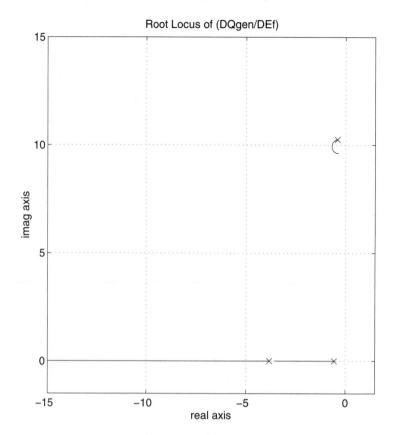

Figure 10.18   Root-locus plot of $\Delta Q_{gen}/\Delta E_f$.

10.9.2 Project 2: Multi-machines System

The objectives of this project are to implement the SIMULINK simulation, *s2*, of the simple multi-machines system in Fig. 10.20 using the transient model for the generators, and to use the simulation to examine the interactions between generators.

In stability studies, the main interest is the transients of the electromechanical oscillations and the concern is whether the generators will maintain synchronism. To keep simulation effort to a reasonable level, the fast electromagnetic transients are usually neglected. Furthermore, the static representation of the network retains only a subset of the network nodes where the generation and load behavior are of interest.

In this project, we will examine the modeling issues arising from the use of the transient model given in Table 10.2 in conjunction with a static network representation. In Fig. 10.20, the infinite bus represents the rest of the system and generators *gen*1 and *gen*2 are local generators, whose transient behavior are of interest in this instance. Each of these two generators will be represented by a transient model. With an infinite bus in the network, its voltage phasor can be conveniently chosen as the reference phasor for the angle

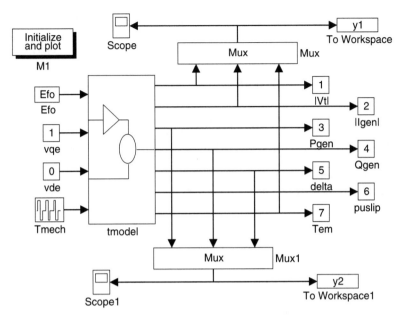

Figure 10.19   Simulation *s1eig* of generator connected to infinite bus by RL line.

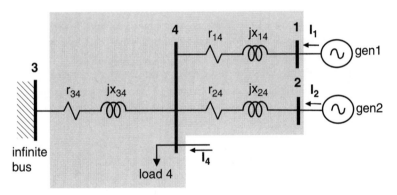

Figure 10.20   Four-bus test system.

of other bus voltages and rotor angles. If there is no infinite bus, the rotor q-axis of one of the generators may be chosen as the reference axis instead. In a simulation, the rotor speed of the reference machine can be held steady by setting its inertia to some very large value.

We will take this opportunity to elaborate on the conditions for proper interfacing of the dynamic equations of the generator to the algebraic equations of the static network representation. When the stator equations of the generator model are a set of dynamic equations, interfacing them with the static equations of the network is not unlike that of interfacing the full model to an external set of resistive loads that has no dynamics. However, when the time derivative terms of the stator qd flux linkages are dropped from the stator voltage equations, as in the stator voltage equations of the transient and sub-transient models presented in Tables 10.2 and 10.4, the resulting stator voltage equations will become algebraic. Algebraic loops will be formed if these algebraic stator equations and those of the static network representation are to be solved separately. To avoid algebraic loops, we will have to combine them. As shown in the previous project, with one machine and a simple radial line, we can easily incorporate the network equations into the stator qd equations. But, for a system with more than one machine or where the dimension of the network equations is large, the stator voltage equations are best incorporated into those of the network. Since the stator voltage equations of the models given in Tables 10.2 and 10.4 are in terms of qd variables in the machine's own rotor reference frame, to incorporate these stator voltage equations into those of the network equations which are usually expressed in phasor form, we will need to express the stator voltage equations in the synchronously rotating reference frame whose q-axis is aligned with the reference phasor.

For example, those of Table 10.2 with variables in their respective rotor reference frames can be expressed in the complex or space vector form as

$$(v_q - jv_d) = -r_s(i_q - ji_d) - x'_d i_d - jx'_q i_q + (E'_q - jE'_d) \tag{10.175}$$

Substituting $(v_q - jv_d) = e^{-j\delta}(v_q^e - jv_d^e)$ and $(i_q - ji_d) = e^{-j\delta}(i_q^e - ji_d^e)$, and rearranging, we obtain

$$(v_q^e - jv_d^e) = -r_s(i_q^e - ji_d^e) - x'_d i_d^e - jx'_q i_q^e + e^{j\delta}(E'_q - jE'_d) \tag{10.176}$$

With saliency, the reactances to the real and imaginary parts of the current are unequal. On the other hand, if transient saliency, $x'_d \neq x'_q$ in the transient model, or sub-transient saliency, $x''_d \neq x''_q$ in the case of the sub-transient model, is ignored, incorporating the stator voltage equations into the network equations will be straightforward.

Let's continue with the equations of the transient model to illustrate the procedure involved. When transient saliency is ignored, Eq. 10.176 reduces to

$$\underbrace{(v_q^e - jv_d^e)}_{\tilde{V}} = -(r_s + x'_d)\underbrace{(i_q^e - ji_d^e)}_{\tilde{I}} + \underbrace{e^{j\delta}(E'_q - jE'_d)}_{\tilde{E}'} \tag{10.177}$$

The above simplified stator voltage equation in phasor form can be modeled by a Thevenin's or Norton's equivalent, as shown in Fig. 10.21, whose fixed impedance of $(r_s + x'_d)$ can now be easily added to the Z_{bus} or Y_{bus} of the network.

If we were to use the Thevenin's equivalent shown in Fig. 10.21a, the value of the internal voltage, $\tilde{E}'$, could be obtained from the simulation of the rotor's field winding equation. When the transient impedances of the machines are incorporated in the network admittance matrix, the injected current, $\tilde{I}$, can be determined by multiplying the resultant admittance matrix with the vector of internal voltages. The value of $\tilde{I}$ or $i_q^e - ji_d^e$ is then transformed back to the rotor reference frame of the machine to obtain $i_q - ji_d$ that is needed for the rest of the machine simulation, including that of the terminal bus voltage of the machine.

If, on the other hand, we were to use the Norton's equivalent in Fig. 10.21b, the equivalent transient source current, $\tilde{I}'$, could be computed from $\tilde{E}'/(r_s + x'_d)$ and the value of the external bus voltage determined from the bus impedance matrix equation of the network.

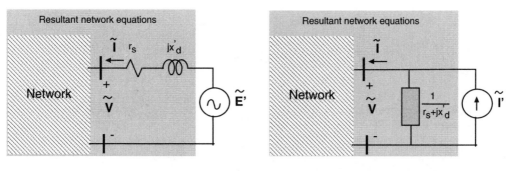

(a) Thevenin's equivalent   (b) Norton's equivalent

Figure 10.21   Equivalent network representations.

In the test system of Fig. 10.20, load bus 4 is purposely retained so that we can use the injected current, $\mathbf{I}_4$, to introduce a fault at that bus. The bus admittance matrix equation of the network, including the transient impedances of the two generators, is of the form:

$$
\begin{bmatrix} i_{q1}^{e} - ji_{d1}^{e} \\ i_{q2}^{e} - ji_{d2}^{e} \\ i_{q3}^{e} - ji_{d3}^{e} \\ i_{q4}^{e} - ji_{d4}^{e} \end{bmatrix} = \begin{bmatrix} Y_{11} & Y_{12} & Y_{13} & Y_{14} \\ Y_{21} & Y_{22} & Y_{23} & Y_{24} \\ Y_{31} & Y_{32} & Y_{33} & Y_{34} \\ Y_{41} & Y_{42} & Y_{43} & Y_{44} \end{bmatrix} \begin{bmatrix} E_{q1}^{\prime e} - jE_{d1}^{\prime e} \\ E_{q2}^{\prime e} - jE_{d2}^{\prime e} \\ v_{q3}^{e} - jv_{d3}^{e} \\ v_{q4}^{e} - jv_{d4}^{e} \end{bmatrix} \qquad (10.178)
$$

where the Y's are the elements of the admittance matrix of the network with ground as the reference. Since we plan to use the injected current at bus 4 as the load or fault current, the corresponding row and column should be gyrated, that is

$$
\begin{bmatrix} i_{q1}^{e} - ji_{d1}^{e} \\ i_{q2}^{e} - ji_{d2}^{e} \\ i_{q3}^{e} - ji_{d3}^{e} \\ v_{q4}^{e} - jv_{d4}^{e} \end{bmatrix} =
$$

$$
\begin{bmatrix} Y_{11} - \frac{Y_{14}Y_{41}}{Y_{44}} & Y_{12} - \frac{Y_{14}Y_{42}}{Y_{44}} & Y_{13} - \frac{Y_{14}Y_{43}}{Y_{44}} & \frac{Y_{14}}{Y_{44}} \\ Y_{21} - \frac{Y_{24}Y_{41}}{Y_{44}} & Y_{22} - \frac{Y_{24}Y_{42}}{Y_{44}} & Y_{23} - \frac{Y_{24}Y_{43}}{Y_{44}} & \frac{Y_{24}}{Y_{44}} \\ Y_{31} - \frac{Y_{34}Y_{41}}{Y_{44}} & Y_{32} - \frac{Y_{34}Y_{42}}{Y_{44}} & Y_{33} - \frac{Y_{34}Y_{43}}{Y_{44}} & \frac{Y_{34}}{Y_{44}} \\ \frac{-Y_{41}}{Y_{44}} & \frac{-Y_{42}}{Y_{44}} & \frac{-Y_{43}}{Y_{44}} & \frac{1}{Y_{44}} \end{bmatrix} \begin{bmatrix} E_{q1}^{\prime e} - jE_{d1}^{\prime e} \\ E_{q2}^{\prime e} - jE_{d2}^{\prime e} \\ v_{q3}^{e} - jv_{d3}^{e} \\ i_{q4}^{e} - ji_{d4}^{e} \end{bmatrix} \qquad (10.179)
$$

The above system of equations in complex variables may be rewritten as a system of equations of real variables with twice the order. For example, the complex equation:

$$
(x - jy) = (G + jB)(p - jq) \qquad (10.180)
$$

where x, y, G, B, p, and q are of order n, is equivalent to

$$
\begin{bmatrix} x \\ y \end{bmatrix} = \begin{bmatrix} G & B \\ -B & G \end{bmatrix} \begin{bmatrix} p \\ q \end{bmatrix} \qquad (10.181)
$$

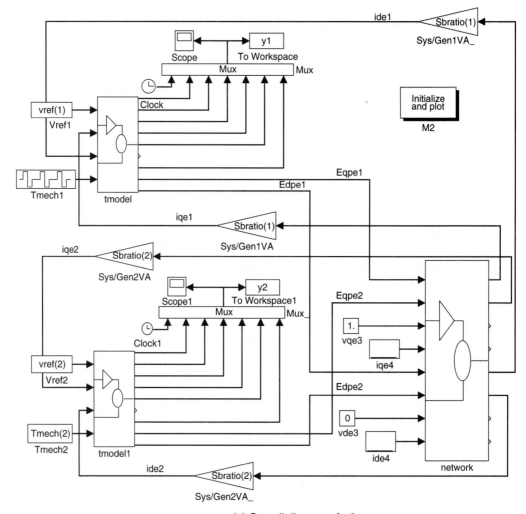

(a) Overall diagram of *s2*

Figure 10.22   Simulation *s2* of two-machine, four-bus system.

In the form of Eq. 10.181, the components of the injected currents of buses 1, 2 and 3, and the voltage of load bus 4 can be computed from a simple matrix multiplication of the network matrix and the input vector consisting of the internal transient voltages of the two generators, the infinite bus voltage, and the injected current at bus 4.

Figure 10.22a shows the overall diagram of the SIMULINK simulation, *s2*, for the four-bus system based on the above-mentioned method. Figure 10.22b shows the inside of the *tmodel* block of *s2*. Note the change in the inputs as compared to the earlier implementation shown in Figs. 10.17a through 10.17e. Instead of the stator *qd* voltages, stator *qd* currents in the synchronously rotating reference frame injected by the machine

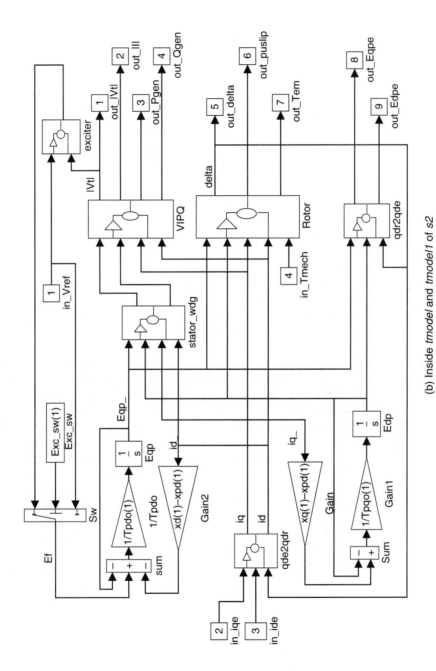

(b) Inside *tmodel* and *tmodel1* of *s2*

Figure 10.22 (cont.): Simulation *s2* of two-machine, four-bus system.

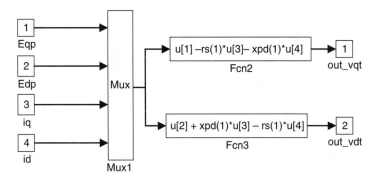

(c) Inside the *stator_wdg* block of *s2*

Figure 10.22   *(cont.):* Simulation *s2* of two-machine, four-bus system.

into the network are now used as input. Among its outputs are the internal transient voltages, E'_q and E'_d, of the machine in the synchronously rotating reference frame.

Figure 10.22c shows the modification to the inside of the *stator_wdg* block of *tmodel* to accommodate the change in input stator qd currents. Note that the machine parameters in the expressions used are relabeled to correspond to those of the machine. In Fig. 10.22b, we have also added an excitation system block. As shown in Fig. 10.22d, the inside of the excitation system block has an IEEE Type 1 excitation system representation [108]. Integrators instead of SIMULINK **Transfer Fcn** blocks are used to build up the components of the excitation system because we like to be able to initialize the states of the excitation system with the outputs from the **trim** routine to keep the startup transients of the simulation to a minimum. Note that the input to the field excitation field can be switched by the control signal, *Exc_sw*, to use either dc field voltage or ac terminal voltage regulation. Using a common input for the reference voltage is convenient, but we must be aware that when operating with dc field voltage regulation, the ac regulator section must be deactivated, otherwise the ac voltage input error will drive the ac exciter to its limit. This can be avoided by reducing the gain of the regulator. Also, with dc field voltage regulation, the negative value of K_E in the exciter model can result in an unstable eigenvalue from the representation of the exciter not in use. This can be avoided by resetting the value of K_E to 1 when the exciter is not being used.

The network equations expressed in the form shown in Eq. 10.181 are implemented using four matrix gain SIMULINK blocks. Figure 10.22e shows the inside of the *network* block of Fig. 10.22. Among the outputs of the *network* block are injected currents from the generators. For power systems with many transformers connecting different voltage levels, the representation of the network is simplified when using a common per unit system. The parameters of the generators can also be expressed in the common system base. In the simulation shown in Fig. 10.22, the parameters of the generators in the *tmodel* blocks remain in their own respective bases. As such, their currents from the *network* must be scaled by the ratio of the system to generator base *VA*, which is equivalent to the ratio of system base current to generator base current when the base voltages of the system and generator per unit system are the same at the generator buses. In addition to the scaling of the generator

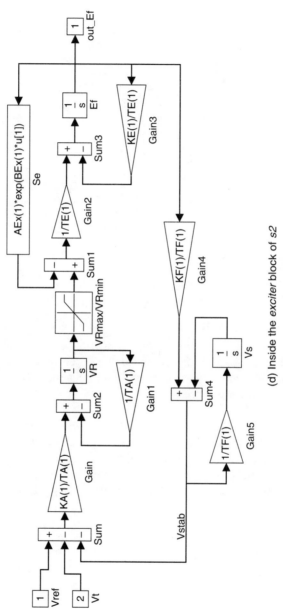

(d) Inside the *exciter* block of *s2*

Figure 10.22 (cont.): Simulation *s2* of two-machine, four-bus system.

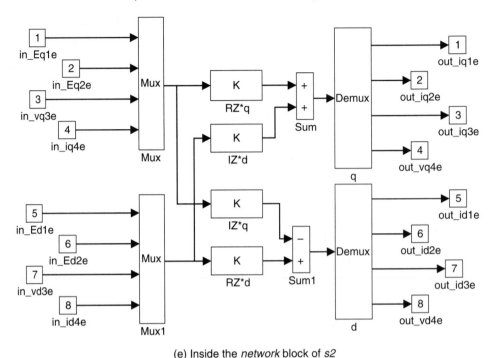

(e) Inside the *network* block of *s2*

Figure 10.22   *(cont.):* Simulation *s2* of two-machine, four-bus system.

current, to account for the difference in per unit system between the generator and network, the stator transient impedance must also be appropriately scaled when it is incorporated into the network admittance matrix.

Interactions and effects of network strength, excitation, and inertia on rotor oscillations.   Having discussed the method of putting together the four-bus system using the transient model and a static network, we can now proceed to implement and use the simulation to investigate the interactions between the two machines and the effects that network strength, excitation method, and rotor inertia, have on rotor oscillations.

Before we proceed, let's review the fundamentals of rotor dynamic behavior using a simple one-machine infinite bus system where the machine is represented by a voltage, E', behind its transient reactance, x', as described earlier in Section 10.5.3. The dynamic equation of the rotor, in per unit, given earlier in Table 10.2, can be rewritten using the relationship between rotor angle, δ, and rotor speed, ω_r, as

$$\frac{1}{\omega_b}\frac{d^2\delta}{dt^2} = \frac{1}{2H}\left(T_{mech} + T_{em} - T_{damp}\right) \qquad (10.182)$$

Replacing T_{em} by the expression given in Eq. 10.114 and performing a small perturbation on all variables about an operating point where the rotor angle is at δ_o, we obtain

$$\frac{1}{\omega_b}\frac{d^2\Delta\delta}{dt^2} = \frac{1}{2H}\left\{\Delta T_{mech} - \underbrace{\frac{E'V_\infty}{X}\cos\delta_o}_{K_s}\Delta\delta - D_\omega\Delta\left(\frac{\omega_r-\omega_e}{\omega_b}\right)\right\} \qquad (10.183)$$

where K_s is termed the synchronizing torque coefficient. In the Laplace domain, the above differential equation becomes

$$s^2\Delta\delta + \frac{D_\omega\omega_b}{2H}s\Delta\delta + \frac{K_s\omega_b}{2H}\Delta\delta = \frac{\omega_b}{2H}\Delta T_{mech} \qquad (10.184)$$

Comparing the characteristic equation portion of Eq. 10.184 with that of the normal second-order equation, $s^2 + 2\zeta\omega_n s + \omega_n^2$, with roots $(-\zeta \pm j\sqrt{\zeta^2-1})\omega_n$, the expressions for the undamped or natural frequency, ω_n, and the damping coefficient, ζ, are

$$\omega_n = \sqrt{\frac{K_s\omega_b}{2H}}$$

$$\zeta = \frac{\omega_b}{2\omega_n}\left(\frac{D_\omega}{2H}\right) = \frac{1}{2}\frac{D_\omega}{\sqrt{2H\omega_b K_s}} \qquad (10.185)$$

Note that K_s is inversely proportional to X, the total reactance between E' and V_∞; in other words, the synchronizing torque coefficient will increase with the electrical strength of the network at the generator bus. According to the expressions in Eq. 10.185, a higher K_s raises the value of ω_n, but lowers ζ, whereas a higher H will decrease the value of both ω_n and ζ.

Figure 10.23 shows the overall diagram of SIMULINK simulation, *s2eig*, that is employed by *m2* to determine the linear model about some desired steady-state. It has been constructed using the *tmodel* and *exciter* blocks, after testing them in simpler settings to ensure correctness. Since we will be using the SIMULINK **trim** to obtain the desired operating point, the inputs and outputs must be defined by input and output ports. The output from **trim** can also be used as initial values to get a quick simulation start at the desired operating point.

In Table 10.7 are the two sets of machine parameters that we will be using in this project. Set 1 is the same as that used earlier in Project 1 of this chapter and in Projects 1 and 3 of Chapter 7; it is chosen so that we can compare with the results obtained previously. Since the damper windings are not represented in the transient model, a small value of D_ω between 1 to 3 per units is often used to account for the damping from these omitted windings and the rotor iron. An even higher value of D_ω may be used when windage and friction of the rotor assembly or damping from the network in the case of an equivalent area machine is to be included.

To keep this project to a reasonable length, we will explore only a small number of combinations of changes in system conditions. Summarized in Table 10.8 are the conditions of cases to be conducted. For all five cases, the infinite bus voltage at bus 3 is to be kept at $1 + j0$ pu. At the nominal condition, the generated power from the generators at buses 1 and 2 are $0.8 + j0.6$ per unit on their respective bases. The base case values of the line impedances and load admittance of the network are

$$z_{14} = 0.004 + j0.1 \ pu \qquad\qquad z_{24} = 0.004 + j0.1 \ pu$$

$$\qquad\qquad\qquad\qquad\qquad\qquad\qquad\qquad\qquad\qquad (10.186)$$

$$z_{34} = 0.008 + j0.3 \ pu \qquad\qquad y_{40} = 1.2 - j0.6 \ pu$$

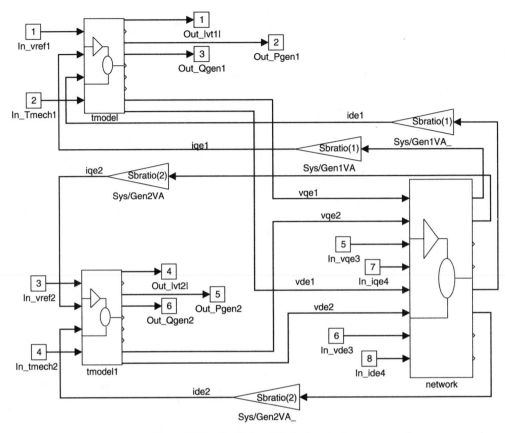

Figure 10.23   Simulation *s2eig* for linearized analysis.

For each case, we will determine the linear model of the system about the given operating point, and from the linear model, the dominant eigenvalues of the system. Following which, we will conduct two simulation runs: The first run is a sequence of small step changes in the externally applied mechanical torque about the operating value of the generator connected to bus 1. The behavior of the simulated rotor oscillations can be cross-checked with the eigenvalues of the linear model. The second run is to simulate a fault at bus 4, representing the fault current drawn by a step change in i_{q4}^e and i_{d4}^e. A fault current of $i_{q4}^e - j i_{d4}^e = -(2 - j2)$ per unit is to be introduced at $t = 5$ seconds for a duration of 0.1 seconds. The disturbances in both of these runs can be applied by programming the signals of the *repeating sequence* sources representing T_{mech1}, i_{q4}^e, and i_{d4}^e.

The given MATLAB M-file, *m2*, can be used to initialize the MATLAB workspace with the system parameters, determine the linear model and its system eigenvalues at some desired operating condition, set up the disturbances for the two simulation runs, and plot certain machine variables after each run.

An examination of the dominant complex eigenvalues and the plots of δ waveforms from the simulation runs will, in most of these cases, show that they are related to the

TABLE 10.7 PARAMETERS OF GENERATOR AND EXCITATION SYSTEM

	Set 1	Set 2		Set 1	Set 2
Rated MVA	920.35	911	Rated kV	18	26
Rated P.F.	0.9	0.9	S.C.R.	0.58	0.64
x_d pu	1.790	2.040	x_q pu	1.660	1.960
x_d' pu	0.355	0.266	x_q' pu	0.570	0.262
x_d'' pu	0.275	0.193	x_q'' pu	0.275	0.191
x_{ls} pu	0.215	0.154	r_s pu	0.0048	0.001
T_{do}' sec	7.9	6.0	T_{qo}' sec	0.41	0.90
T_{do}'' sec	0.032		T_{qo}'' sec	0.055	
H sec	3.77	2.5	D_ω pu	2	2
K_A pu	50	50	T_A sec	0.07	0.06
V_R^{max} pu	1	1	V_R^{min} pu	−1	−1
K_E pu	−0.0465	−0.0393	T_E sec	0.052	0.440
A_{Ex}	0.0012	0.0013	B_{Ex}	1.264	1.156
K_f pu	0.0832	0.07	T_f sec	1.00	1.00

TABLE 10.8 PROJECT 2 CASE STUDIES

Cases	Bus 1 machine excitation type	Bus 2 machine excitation type	Network parameters
1 dc regulated Base case	Set 1 Fixed E'	Set 1 Fixed E'	Base case network
2 ac regulated Base case	Set 1 Ac voltage regulated	Set 1 Ac voltage regulated	Base case network
3 ac regulated Increased network strength	Set 1 Ac voltage regulated	Set 1 Ac voltage regulated	Reduce z_{34} to $z_{34} = 0.002 + j0.1$ pu
4 ac regulated Different machines	Set 1 Ac voltage regulated	Set 2 Ac voltage regulated	Base case network
5 ac regulated Change Set 1's K_A to 200	Set 1 Ac voltage regulated	Set 2 Ac voltage regulated	Base case network

rotor oscillations of the generators, between machines and between machine and infinite bus. A comparison of the dominant eigenvalues for Case 1 with those obtained for the one-machine infinite bus system in Project 1 should review interactions between the two identically connected machines. Differences in results between Case 1 and Case 2 may be attributed to the two excitation methods, namely dc vs. ac voltage regulation.

Using a lower value of z_{34} increases the strength of interconnection between the machines and the infinite bus, as well as the value of the synchronizing torque coefficient,

K_s. According to Eqs. 10.183 and 10.185, the natural frequency of the rotor oscillations with respect to the infinite bus should increase.

The purpose of Case 5 is to explore the effect of raising the gain of the excitation system, which in some situations, can cause the transient behavior of the machine to become less stable.

Presented below are sample results for Case 4. The system eigenvalues for this case with both machines ac voltage regulated are

−8.439e+00	3.556e+01i	−5.325e−01	7.526e+00i
−8.439e+00	−3.556e+01i	−5.325e−01	−7.526e+00i
−2.134e+00	1.224e+01i	−5.084e−01	7.588e−01i
−2.134e+00	−1.224e+01i	−5.084e−01	−7.588e−01i
−8.596e+00	8.207e+00i	−2.481e−01	2.575e−01i
−8.596e+00	−8.207e+00i	−2.481e−01	−2.575e−01i
−6.692e+00	0.000e+00i	−5.506e+00	0.000e+00i

Plots of the response to small step changes of ±0.1 *pu* about the nominal value of T_{mech1} run are shown in Fig. 10.24; those from the fault runs are given in Fig. 10.25. The M-file, *m2*, has the time and output values of the three *repeating sequence sources, Tmech1, iq4e,* and *id4e,* that were used to obtain the results shown.

10.9.3 Project 3: Subsynchronous Resonance

In this project, we will make use of the lumped-mass model of the generator shaft described earlier in this chapter and the synchronously rotating $qd0$ model of the three-phase ac lines described in Section 5.9 to set up the SIMULINK simulation of the first benchmark model for computer simulation of subsynchronous resonance [117]. In the first part of this project, we will make use of the simulations, *s3geig* and *s3g,* to examine the case of the test generator with its shaft torsional system operating directly onto an infinite bus. In the second half of the project, we will make use of another set of simulations, *s3eig* and *s3,* that has the external network connection to examine the phenomenon of subsynchronous resonance introduced by the series compensated line. This project can be made more challenging by having the students create their own MATLAB and SIMULINK files to simulate the second benchmark model given in Reference [118].

Subsynchronous resonance.   Here, subsynchronous resonance refers to the excitation of a torsional frequency of the shaft of a generator by stator current components of subsynchronous frequency of the ac network. The impedance-frequency characteristics of an ac power network is usually quite complex with many poles and zeros. It is also highly variable depending on the components that are connected and their loading. Usually, the first pole is above the fifth harmonic, below which it appears inductive. However, in a series compensated line, where the line equivalent reactance, X_L, is greater than the capacitive reactance, X_C, the natural frequency of the line is less than synchronous, that is

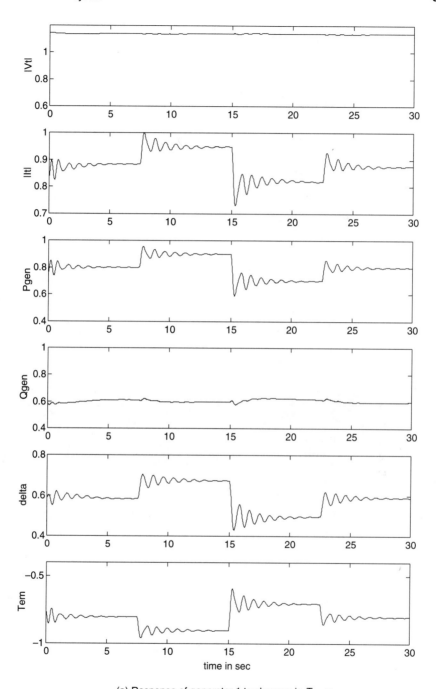

(a) Response of generator 1 to changes in T_{mech1}

Figure 10.24   System response to changes in T_{mech1}.

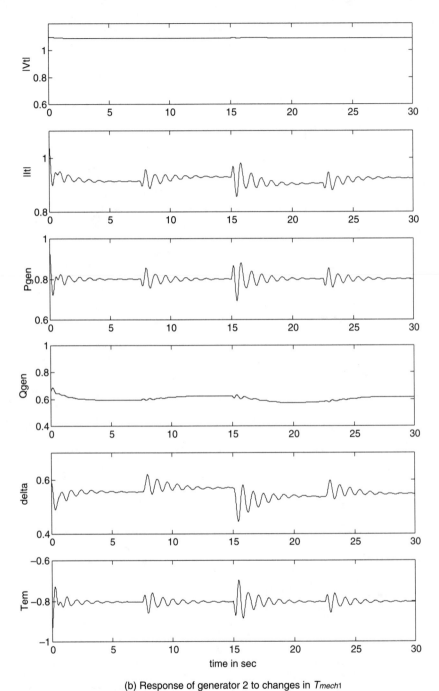

(b) Response of generator 2 to changes in T_{mech1}

Figure 10.24   *(cont.):* System response to changes in T_{mech1}.

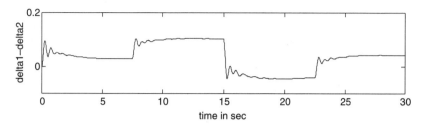

(c) Rotor swing between generators 1 and 2

Figure 10.24   *(cont.):* System response to changes in T_{mech1}.

$$f_n = \frac{1}{2\pi\sqrt{LC}} = f_e\sqrt{\frac{X_C}{X_L}} < f_e \qquad (10.187)$$

When the rotor is rotating at the synchronous speed of f_e, the f_n stator current component will be accompanied by a corresponding $f_n - f_e$ rotor winding current component. The slip corresponding to the slip frequency of $f_n - f_e$ is negative for a f_n that is less than f_e. Thus, with respect to the subsynchronous f_n stator components, the generator is operating as an induction generator and the rotor windings present a negative resistance to the f_n stator current component. Should the negative resistance presented cancel the ac network resistance to the f_n component, resonance between the ac network's subsynchronous f_n component and the complement torsional frequency, $f_n - f_e$ component, will result. From this simple analysis of the situation, we know that the tendency for subsynchronous resonance will be reduced with higher ac network resistance or lower rotor winding resistance, and that the modeling of the phenomenon must correctly portray the dynamics of the ac network, the rotor windings, and the shaft torsional of the generator. Figure 10.26 shows the benchmark system given in Reference [117]. Table 10.9 contains the parameters of the generator model with two q-axis damper windings. A finite value of $r_s = 0.001$ per unit is used here; otherwise, the machine will have unstable stator frequency eigenvalues. The parameters for the lumped-mass model of its rotor in per unit of the generator base are given in Table 10.10.

Fixed terminal voltage case.   For this part of the project, we will be using the simulations *s3geig* and *s3g*; both of these simulations are for a generator that is connected to an infinite bus. Figure 10.27a shows the overall diagram of the simulation, *s3g*, with external inputs to establish the operating point and also to apply a small torque perturbation

TABLE 10.9   PARAMETERS OF GENERATOR WITH TWO Q-AXIS DAMPERS

$x_d = 1.79$	$x_{md} = 1.66$	$x_q = 1.71$	$x_{mq} = 1.58$
$w_b r'_f = 0.53$	$w_b r'_{kd} = 1.54$	$w_b r'_{kq} = 3.1$	$w_b r'_{kq2} = 5.3$
$x'_{lf} = 0.062$	$x'_{lkd} = 0.0055$	$x'_{lkq} = 0.095$	$x'_{lkq2} = 0.326$
$x_{ls} = 0.13$	$r_s = 0.001$	$D_\omega = 0$	

Reference [117] ⓒ 1977 IEEE

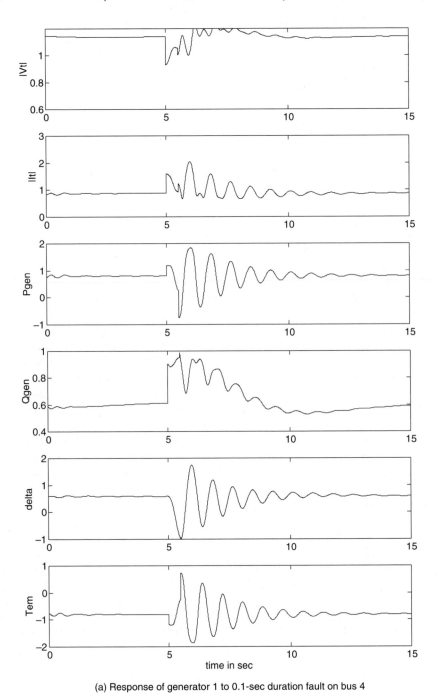

(a) Response of generator 1 to 0.1-sec duration fault on bus 4

Figure 10.25   System response to fault on bus 4.

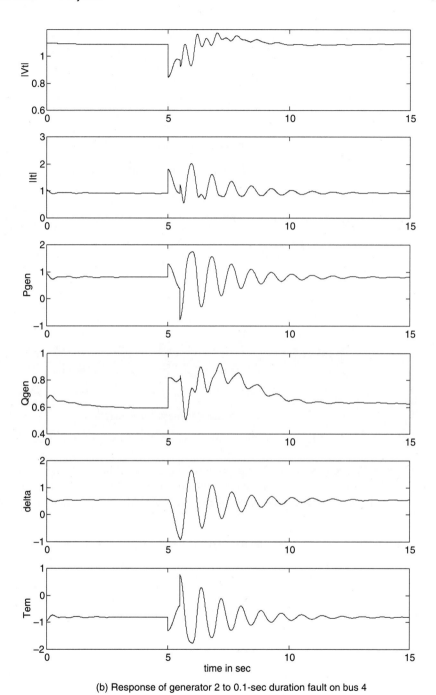

(b) Response of generator 2 to 0.1-sec duration fault on bus 4

Figure 10.25   *(cont.):* System response to fault on bus 4.

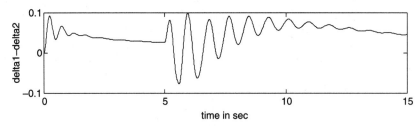

(c) Rotor swing between generators 1 and 2

Figure 10.25   *(cont.):* System response to fault on bus 4.

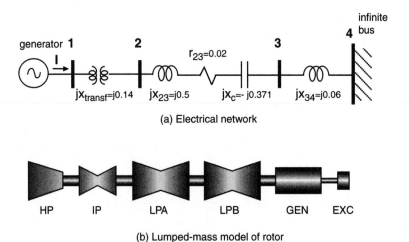

(a) Electrical network

(b) Lumped-mass model of rotor

Figure 10.26   Benchmark model for subsynchronous resonance studies.

about the operating point for the purpose of cross-checking the results from the linearized analysis. Not shown is a similar simulation, *s3geig*, with only the input and output ports, that are being used for linearized analysis. The MATLAB program, *m3g*, uses *s3geig* to

TABLE 10.10   ROTOR SPRING MASS PARAMETERS

Mass	H(sec)	Shaft	K(pu torque/elect. rad)
HP	0.092897		
		HP-IP	7,277
IP	0.155589		
		IP-LPA	13,168
LPA	0.858670		
		LPA-LPB	19,618
LPB	0.884215		
		LPB-GEN	26,713
GEN	0.868495		
		GEN-Exc	1,064
EXC	0.0342165		

Reference [117] © 1977 IEEE

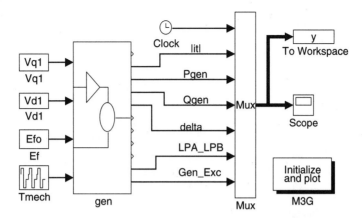

(a) Overall diagram of *s3g* for input torque disturbance

Figure 10.27   Simulation *s3g* of generator connected to an infinite bus.

determine the modal quantities of the torsional system, the linear system equations about an operating point, and the resultant system matrix eigenvalues. *m3g* will also set up the operating condition in *s3g* for the small torque perturbation study.

The simulation details of the generator that has two q-axis damper windings and a torsional shaft system in *s3g* and *s3geig* are the same. The details of their blocks are shown in Figs. 10.27b through 10.27h.

Examine a listing of the MATLAB M-file, *m3g*, and use the M-file to determine the modal quantities of the torsional system, the linear system equation, and eigenvalues of the system. Figure 10.28 shows a plot of the mode shapes of the torsional system using the M-file, *m3g*. From these plots of the mode shapes, we can see that the mode numbering corresponds to the number of sign changes in the mode shape vector or the number of nodes.

Use *m3g* to establish the operating condition where the generator delivers 0.9 per unit power at a power factor of 0.9 lagging to its terminal that is kept at a fixed voltage of one per unit magnitude, and the value of its armature resistance, r_s, is set equal to 0.001 pu.

Rerun the case with r_s set to zero. In each case, take note of the dominant or unstable system eigenvalues and run the simulation, *s3g*, keeping E_f fixed at the steady-state value for the above operating point, but with T_{mech} making small step changes about the operating point value. Step changes in T_{mech} can be applied using a *repeating sequence* source as the input signal for T_{mech}. For example, the time and output sequence of the *repeating sequence* source can be [*0 1 1 4 4 7 7 10*] and [*1 1 1.2 1.2 0.5 0.5 0.8 0.8*]$*T_{mech}$, respectively. Plot the response of the generator's real and reactive output powers, the magnitude of its output current, its rotor angle, and the shaft torques between the LPA and LPB turbines and between the generator and exciter. Compare the response you obtained with those given in Fig. 10.29. Try to identify the major torsional oscillation frequency in the torque or power trace that is associated with the dominant or unstable eigenvalue.

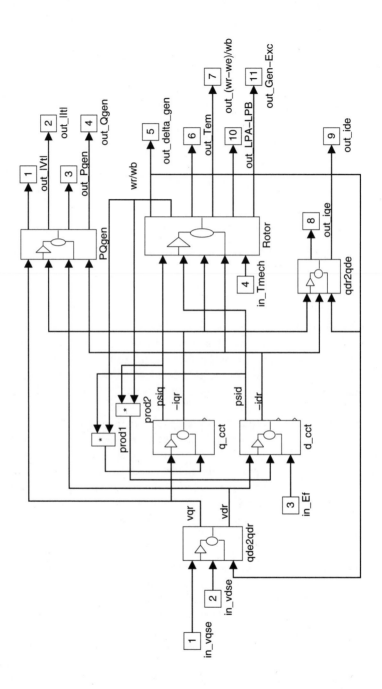

Figure 10.27 (cont.): Simulation *s3g* of generator connected to an infinite bus.

(b) Inside the *gen* block of *s3g*

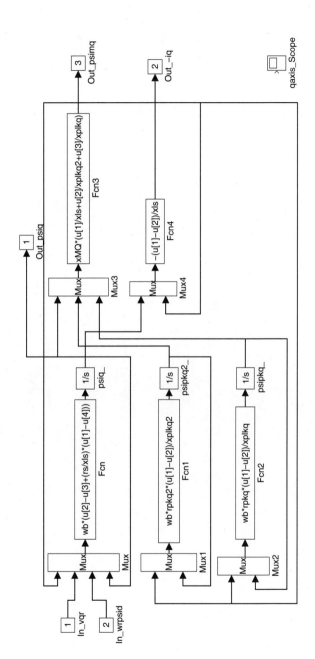

(c) Inside the *q_cct* block of *s3g*

Figure 10.27 (cont.): Simulation *s3g* of generator connected to an infinite bus.

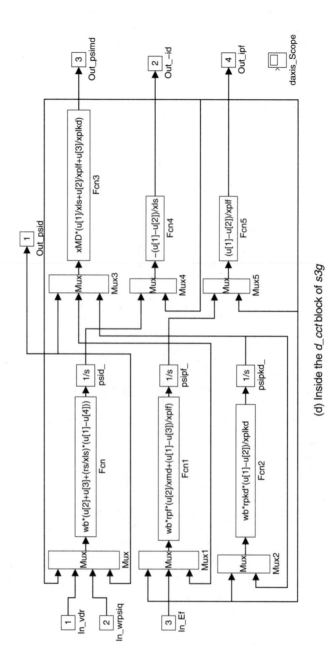

(d) Inside the *d_cct* block of *s3g*

Figure 10.27 (cont.): Simulation *s3g* of generator connected to an infinite bus.

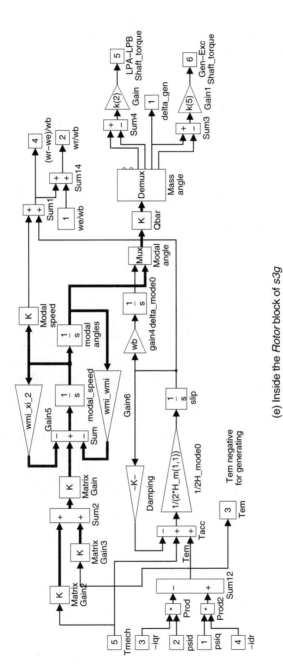

Figure 10.27 (cont.): Simulation $s3g$ of generator connected to an infinite bus.

(e) Inside the *Rotor* block of $s3g$

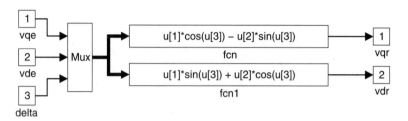

(f) Inside the *qde2qdr* block of *s3g*

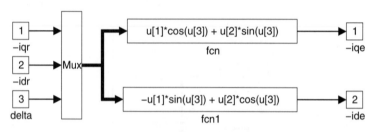

(g) Inside the *qdr2qde* block of *s3g*

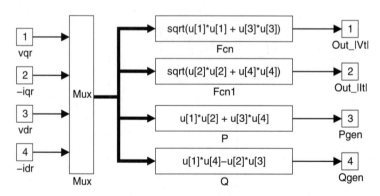

(h) Inside the *PQgen* block of *s3g*

Figure 10.27   *(cont.):* Simulation *s3g* of generator connected to an infinite bus.

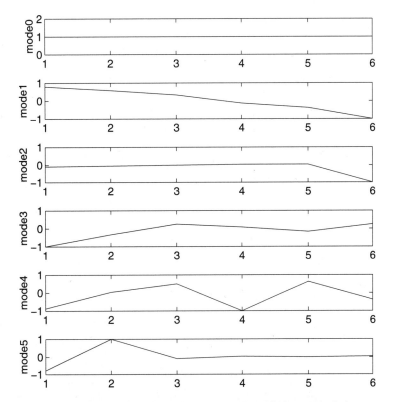

Figure 10.28   Mode shapes of six mass-lumped model of generator shaft.

Generator with series capacitor compensated line.   In the second part of this project, the study is expanded to include the ac network of the benchmark system shown in Fig. 10.26. Figures 10.30a through 10.30c show the simulation given in *s3eig* that may be used for a linearized analysis of the benchmark ssr system of Fig. 10.26. The *gen* module is the same as that used in *s3geig* and *s3g*. The additional components are those of the radial network connecting the generator and infinite bus, represented in the synchronously rotating reference frame.

Assuming that we will be dealing with a balanced network condition where there is no need to model the zero-sequence equation, Eqs. 5.87 and 5.88 for the qd voltage drops across a series connected RL line in the synchronously rotating reference frame and in terms of the positive-sequence parameters may be rewritten as

$$\frac{x_{L1}}{\omega_b}\frac{di_q^e}{dt} = v_{qS}^e - v_{qR}^e - r_1 i_q^e - x_{L1} i_d^e$$

$$\frac{x_{L1}}{\omega_b}\frac{di_d^e}{dt} = v_{dS}^e - v_{dR}^e - r_1 i_d^e + x_{L1} i_q^e$$

(10.188)

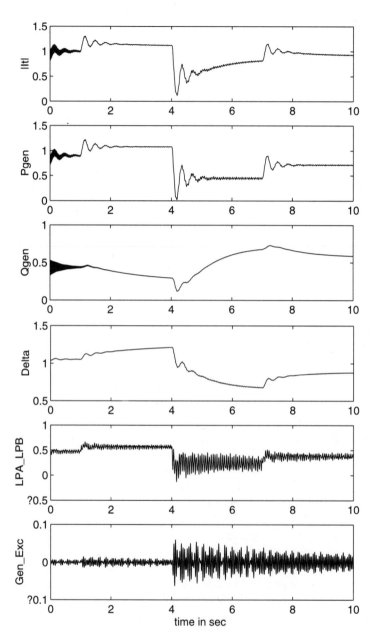

Figure 10.29 Response of generator with torsional system to step changes in input mechanical torque.

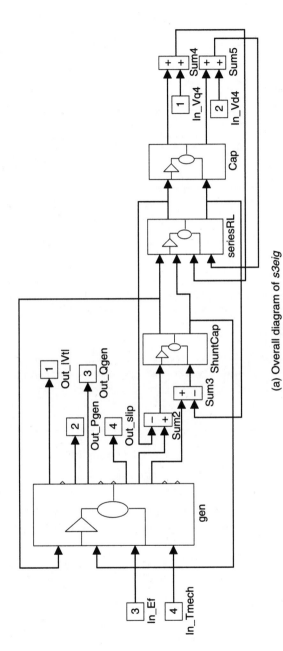

(a) Overall diagram of *s3eig*

Figure 10.30   Simulation *s3eig* for linearized analysis of benchmark ssr system.

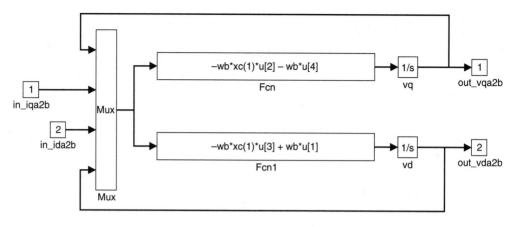

(b) Inside the *Cap* block

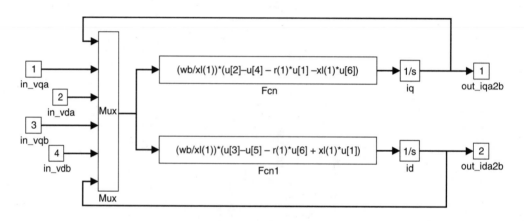

(c) Inside the *seriesRL* block

Figure 10.30   *(cont.)*

where x_{L1} is the positive-sequence inductive reactance of the RL line at the base frequency of ω_b, and the currents, i_q^e and i_d^e, are in the direction from the sending end, S, to the receiving end, R.

Similarly, Eqs. 5.107 and 5.108 for the qd voltage drops across a three-phase set of series capacitors with no mutual capacitances between phases may be rewritten as

$$\frac{d\Delta v_{qCs}^e}{dt} = \omega_b x_{Cs} i_{qCs} - \omega_b \Delta v_{dCs}^e$$
$$\frac{d\Delta v_{dCs}^e}{dt} = \omega_b x_{Cs} i_{dCs} + \omega_b \Delta v_{qCs}^e$$

(10.189)

Small shunt capacitors have been added to each phase of the generator's terminal to develop its terminal voltages at bus 1. The *seriesRL* block represents the positive-sequence equivalent RL circuit of the transformer reactance, the resistance and reactance of the lines between buses 2 and 3, and buses 3 and 4. The intermediate buses 2 and 3 are not represented explicitly in this simulation. If need be, these other buses can also be represented explicitly by introducing small shunt capacitances at each of these buses, as in the case of bus 1. The dynamics of the line's series capacitors are represented in the *Cap* block.

It will be evident later that the benchmark condition for the system of Fig. 10.26 has unstable torsional modes. Left alone, the unstable torsional oscillations will gradually grow in magnitude. However, it can be hastened by an external excitation, such as a fault on any of the network buses. By not explicitly representing the intermediate buses 2 and 3 of Fig. 10.26, that is lumping the resistances and inductances of the generator transformer and lines into one RL representation, we have a simpler simulation. Yet we are able to hasten the unstable behavior of the torsional modes by dropping bus 4's voltage momentarily.

First, we will use the MATLAB M-file, *m3*, to determine the linear model of the simulated benchmark system given in *s3eig* at some desired operating condition, as well as the corresponding system eigenvalues. The M-file, *m3*, sets up the system parameters and estimates initial values to be used as initial guesses for the SIMULINK **trim** function before calling up the **trim** and **linmod** functions. Run the *m3* file once to ensure that the ordering of the state variables in *m3* agrees with how SIMULINK interprets the setup of the simulation file, *s3eig*, in your computer. If not, change the order of the state variables used in *m3* accordingly.

Run the M-file, *m3*, to obtain the linear model and eigenvalues of the benchmark system for the operating condition where the generator delivers 0.9 pu of power to bus 1 at a lagging power factor of 0.9, the voltage magnitude of bus 1 being 1 pu. Knowing that the system condition is unstable, warnings of the ill-conditioned matrix are to be expected. Examine the eigenvalues and relate the frequency of the unstable eigenvalues to the natural frequency of the torsional system.

For the simple electrical radial network given, its natural frequency can be estimated from

$$\omega_n = \sqrt{\frac{1}{C_s L_{total}}} = \omega_b \sqrt{\frac{x_{Cs}}{x_{Ltotal}}} \qquad \text{electrical rad/sec} \qquad (10.190)$$

where x_{Cs} is the per phase capacitive reactance of the series compensation capacitor, and x_{Ltotal} is the total effective inductive reactance to the armature current. For the given system:

$$x_{Ltotal} = x_{transf} + x_{lines} + x'' \qquad (10.191)$$

where x'' may be approximated by the average value of the q- and d-axis sub-transient reactances, that is $x'' = (x_q'' + x_d'')/2$. Thus, for the given per unit values of $x_{transf} = 0.14$, $x_{lines} = 0.50 + 0.06$, $x_{Cs} = 0.371$, and a ω_b of 377 rad/sec, Eq. 10.190 yields a natural frequency of 246.54 rad/sec. The synchronous complement of the network's natural frequency, $\omega_b - \omega_n$, is 130.45 rad/sec or 20.76 Hz, which is close to the computed shaft's natural torsional frequency of 20.2 Hz. The system eigenvalues for the case where $x_{Cs} = 0.371$ are

−4.7962e+00	+8.0356e+03i	−3.2777e+00	+1.2930e+02i
−4.7962e+00	−8.0356e+03i	−3.2777e+00	−1.2930e+02i
−5.0863e+00	+6.6445e+03i	4.0237e−01	+1.2737e+02i
−5.0863e+00	−6.6445e+03i	4.0237e−01	−1.2737e+02i
−4.7160e+00	+6.2364e+02i	3.2319e−02	+1.0002e+02i
−4.7160e+00	−6.2364e+02i	3.2319e−02	−1.0002e+02i
−5.0360e−07	+2.9818e+02i	−3.3023e+01	
−5.0360e−07	−2.9818e+02i	−2.0443e+01	
−2.8469e−02	+2.0280e+02i	−6.4303e−01	+1.0485e+01i
−2.8469e−02	−2.0280e+02i	−6.4303e−01	−1.0485e+01i
−1.3976e−02	+1.6034e+02i	−3.2788e−01	
−1.3976e−02	−1.6034e+02i	−3.9080e+00	

Use Eq. 10.190 to determine the value of x_{Cs} that will create a subsynchronous resonance at another torsional oscillation frequency. Cross-check the value of x_{Cs} with the unstable system eigenvalues obtained from $m3$ with the terminal operating condition of the generator kept the same. With the system near or at resonance, the system matrix will be close to singular; as such, the output of the SIMULINK **trim** function should be closely examined for consistency with the desired operating point.

Finally, as a cross-check, we will perform simulation runs to obtain the dynamic performance of the system about the chosen operating points. Figure 10.31 shows the simulation given in the SIMULINK file, $s3$, with some of the input and output ports reconnected for performing dynamic simulation. Use $s3$ to obtain the system's response to a sudden drop in bus 4 voltage to 50 percent of its predisturbance value for 0.075 second and returning to its predisturbance value after that. The values of E_f and T_{mech} are to remain constant at their predisturbance levels determined by $m3$. Prior to the disturbance, the system is initialized with the operating condition where the generator delivers 0.9 pu power at a power factor of 0.9 lagging to its terminal, and the magnitude of its terminal voltage is 1 pu. With the system properly initialized, the unstable oscillations are small. The disturbance created by a sudden drop in bus 4 voltage merely hastens the growth of the unstable oscillations. Figure 10.32 shows the plot of some variables presented in Reference [117] for the same benchmark system. It is obtained by using a *repeating sequence* signal source with a *time output* vector of [0.2 0.2 0.275 0.275 0.7] and a *value output* vector of [1. 1. 0.5 0.5 1. 1.]∗V_{q4} in $m3$.

10.9.4 Project 4: Power System Stabilizer

Power system stabilizers are a form of supplementary control that is used to provide additional damping to the inter-area oscillations or to stabilize a generator whose voltage regulator gain is such that it may result in negatively damped, machine-to-system oscillations under certain conditions. It has been observed that the damping of these small power oscillations can be improved by feeding back appropriate stabilizing signals to the input of the generator's exciter. Some input signals that have been considered in the literature are slip speed, accelerating power, and frequency. In this project, we will use an established

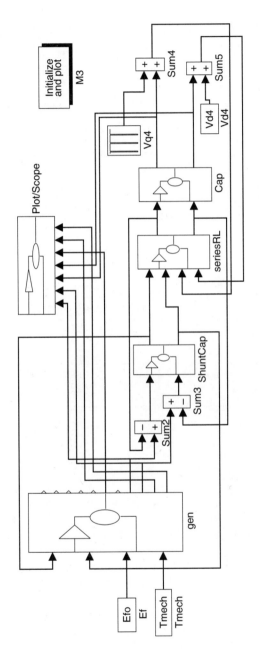

Figure 10.31 Simulation *s3* for response of benchmark system to drop in V_{q4}.

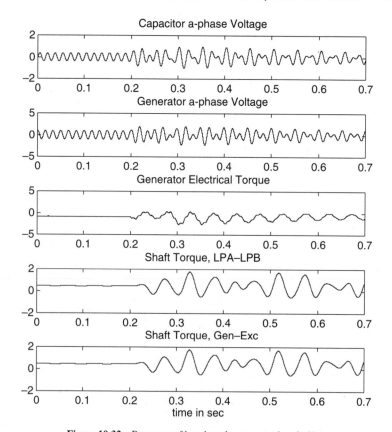

Figure 10.32   Response of benchmark system to drop in V_{q4}.

approach to obtain a preliminary design for a power system stabilizer with slip speed as the feedback signal. The two references that provide a useful background to the modeling and design approach used here are listed in [103] and [104].

For the purpose of this project, we will use the simple system shown in Figure 10.16 of one machine connected to an infinite bus through a series impedance. The series impedance, $r_e + jx_e$, may be regarded as the Thevenin's equivalent impedance of the local load on the generator bus and network as seen from the generator's terminal. Its value may be conveniently adjusted to simulate different electrical strength of the connection. We will model the generator with one field winding by a simplified transient model. With no g field winding in the q-axis, the values of T'_{qo} and E_g in Table 10.2 can be set to zero. The modified equations of the simplified model used here are given in Table 10.11.

TABLE 10.11 SIMPLIFIED TRANSIENT MODEL

Stator winding equations:

$$v_q = -r_s i_q - x'_d i_d + E'_q \qquad pu$$

$$v_d = -r_s i_d + x'_q i_q + E'_d \tag{10.192}$$

$$E_q = E'_q + (x_q - x'_d)i_d$$

Rotor winding equations:

$$T'_{do} \frac{dE'_q}{dt} + E'_q = E_f - (x_d - x'_d)i_d \tag{10.193}$$

$$E'_d = (x_q - x'_q)i_q$$

Torque equation:

$$T_{em} = -\left\{ E'_q i_q + E'_d i_d + (x'_q - x'_d)i_d i_q \right\} = -E_q i_q \qquad pu \tag{10.194}$$

Rotor equation:

$$2H \frac{d\{(\omega_r - \omega_e)/\omega_b\}}{dt} = T_{em(pu)} + T_{mech(pu)} - T_{damp(pu)} \qquad pu \tag{10.195}$$

$$\frac{d\delta_e}{dt} = \omega_r - \omega_e \tag{10.196}$$

As in Project 1 of this chapter, for the small test system of Fig. 10.16 which has a simple external network, it is easier to transform the equations of the simple network to the rotor reference frame of the generator. The stator voltage equations, including the external network impedance, are the same as those given earlier in Eq. 10.173. And, if the phasor voltage of the infinite bus is the reference phasor and the q-axis of the synchronously rotating reference frame, v^e_{di} is zero. The qd components of the infinite bus voltage in the rotor reference frame of the generator may be expressed as

$$v_{qi} = v^e_{qi} \cos\delta$$

$$v_{di} = v^e_{qi} \sin\delta \tag{10.197}$$

where δ is the rotor angle measured with respect to the q-axis of the synchronously rotating reference frame of the infinite bus voltage.

Figure 10.33a shows the overall diagram of the simulation, *s4*, of a synchronous generator with exciter and pss, that is connected by a series RL line to an infinite bus. The *tmodel* block contains the same transient model as that shown in Fig. 10.17. Figure 10.33b shows the SIMULINK simulation of the exciter. The supplementary input, v_{supp}, is the modulation signal from the pss.

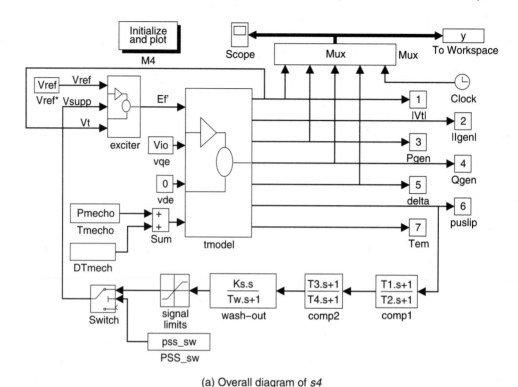

(a) Overall diagram of *s4*

Figure 10.33   Simulation *s4* of generator with exciter and power system stabilizer.

Linearized model of generator and network.   In this section, we begin by deriving a set of linearized equations of the system retaining the identity of variables ΔV, $\Delta E'_q$, ΔE_f, and $\Delta \delta$. The linear model so derived is being used in *m4* to determine certain transfer functions for the preliminary design of the pss.

The usually small stator winding resistance, r_s, will be neglected in this derivation. In per unit, the generated power, P_{em}, may be expressed as $E_q i_q$ or $E'_q i_q + (x_q - x'_d)i_d i_q$. Taking small displacements about the steady-state operating values denoted by an additional subscript o, we will obtain

$$\begin{aligned}
\Delta P_{em} &= E_{qo}\Delta i_q + i_{qo}\Delta E_q \\
&= E_{qo}\Delta i_q + i_{qo}(\Delta v_q + x_q \Delta i_d) \\
&= E_{qo}\Delta i_q + i_{qo}\{\Delta E'_q + (x_q - x'_d)\Delta i_d\}
\end{aligned} \tag{10.198}$$

Taking small displacements of the expression of the generator's terminal voltage magnitude, $V_t^2 = v_q^2 + v_d^2$, and using the relations $E'_q = v_q + x'_d i_d$ and $v_d = x_q i_q$, we obtain

$$2V_t \Delta V_t = 2v_{qo}\Delta v_q + 2v_{do}\Delta v_d$$

$$\Delta V_t = \frac{v_{qo}}{V_t}(\Delta E'_q - x'_d \Delta i_d) + \frac{v_{do}}{V_t}x_q \Delta i_q \tag{10.199}$$

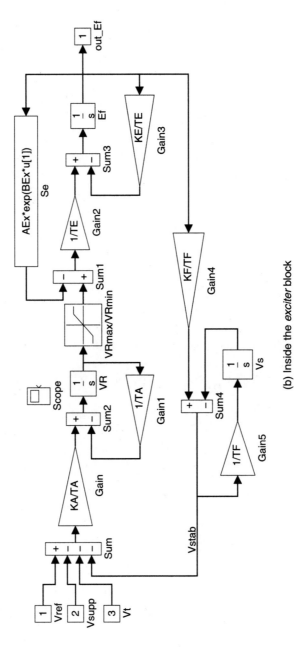

(b) Inside the *exciter* block

Figure 10.33 (cont.): Simulating *s4* of generator with exciter and power system stabilizer.

Taking small displacements of the f rotor winding equation, we obtain

$$T'_{do}\frac{d\Delta E'_q}{dt} + \Delta E'_q = \Delta E_f - (x_d - x'_d)\Delta i_d \tag{10.200}$$

Denoting the infinite bus quantities by an additional subscript, i, the qd voltage equations of the infinite bus in the generator's rotor reference frame are

$$\begin{aligned} v_{qi} &= E'_q - (x'_d + x_e)i_d - r_e i_q \\ v_{di} &= (x_q + x_e)i_q - r_e i_d \end{aligned} \tag{10.201}$$

When expressed in terms of the infinite bus voltage magnitude, V_i, and the rotor angle, δ, we have

$$v_{qi} - jv_{di} = V_i e^{-j\delta} = V_i(\cos\delta - j\sin\delta) \tag{10.202}$$

Solving Eq. 10.201 for i_{qi} and i_{di}, we obtain

$$\begin{aligned} i_q &= \frac{r_e}{Dz}(E'_q - V_i\cos\delta) + \frac{(x_e + x'_d)}{Dz}V_i\sin\delta \\ i_d &= \frac{(x_e + x_q)}{Dz}(E'_q - V_i\cos\delta) - \frac{r_e}{Dz}V_i\sin\delta \end{aligned} \tag{10.203}$$

where $Dz = r_e^2 + (x_e + x_q)(x_e + x'_d)$. Assuming that the infinite bus voltage, V_i, is constant and taking the small displacements of Eq. 10.203, we get

$$\begin{aligned} \Delta i_q &= \frac{r_e}{Dz}\Delta E'_q + \frac{V_i}{Dz}\{r_e\sin\delta_o + (x_e + x'_d)\cos\delta_o)\}\Delta\delta \\ \Delta i_d &= \frac{(x_e + x_q)}{Dz}\Delta E'_q - \frac{V_i}{Dz}\{r_e\cos\delta_o - (x_e + x'_d)\sin\delta_o)\}\Delta\delta \end{aligned} \tag{10.204}$$

Using the expressions in Eq. 10.204 to replace the qd current terms in Eq. 10.198 and regrouping the $\Delta\delta$ and $\Delta E'_q$ terms, Eq. 10.198 can be written as

$$\Delta P_{em} = K_1\Delta\delta + K_2\Delta E'_q \tag{10.205}$$

where

$$\begin{aligned} K_1 &= \frac{E_{qo}V_i}{Dz}\{r_e\sin\delta_o + (x_e + x'_d)\cos\delta_o\} \\ &\quad + \frac{i_{qo}(x_q - x'_d)V_i}{Dz}\{(x_e + x_q)\sin\delta_o - r_e\cos\delta_o\} \end{aligned} \tag{10.206}$$

$$K_2 = \frac{E_{qo}r_e}{Dz} + \frac{i_{qo}}{Dz}\{1 + (x_q - x'_d)(x_e + x_q)\}$$

Similarly, replacing the qd current terms in Eqs. 10.199 and 10.200 and regrouping the $\Delta\delta$ and $\Delta E'_q$ terms, we can express the small displacements equations of the stator terminal voltage and rotor f winding as

$$T'_{do}\frac{d\Delta E'_q}{dt} + \frac{\Delta E'_q}{K_3} = \Delta E_f - K_4\Delta\delta$$

(10.207)

$$\Delta V_t = K_5\Delta\delta + K_6\Delta E'_q$$

where

$$K_3 = \left\{1 + \frac{(x_d - x'_d)(x_e + x_q)}{Dz}\right\}^{-1}$$

$$K_4 = \frac{V_i(x_d - x'_d)}{Dz}\left\{(x_e + x_q)\sin\delta_o - r_e\cos\delta_o\right\}$$

$$K_5 = V_i\frac{v_{do}}{V_t}\frac{x_q}{Dz}\left\{r_e\sin\delta_o + (x_e + x'_d)\cos\delta_o\right\} + V_i\frac{v_{qo}}{V_t}\frac{x'_d}{Dz}\left\{r_e\cos\delta_o - (x_e + x_q)\sin\delta_o\right\}$$

$$K_6 = \frac{v_{qo}}{V_t}\left\{1 - \frac{x'_d(x_e + x_q)}{Dz}\right\} + \frac{v_{do}}{V_t}\frac{x_q r_e}{Dz}$$

(10.208)

From taking the small displacements of the rotor equations, we will obtain

$$\frac{d\Delta(\omega_r/\omega_b)}{dt} = \frac{1}{2H}(\Delta P_{mech} + \Delta P_{em} - D_\omega\Delta(\omega_r/\omega_b))$$

(10.209)

$$\frac{d\Delta\delta}{dt} = \omega_b(\Delta\omega_r/\omega_b)$$

Similarly, taking small displacements of the excitation system's equations, we will obtain

$$\frac{d\Delta E_f}{dt} = \frac{1}{T_E}\Delta V_R$$

$$\frac{d\Delta V'_R}{dt} = \frac{K_A}{T_A}(\Delta V_{ref} - \Delta V_t - \Delta V_{stab} - \Delta V_{supp}) - \frac{\Delta V_R}{T_A}$$

(10.210)

$$\frac{d\Delta V_{stab}}{dt} = \frac{K_f}{T_f T_E}(\Delta V_R - \Delta E_f) - \frac{\Delta V_{stab}}{T_f}$$

Operating point values. If the operating condition were to be specified in terms of the voltage magnitude, $|\tilde{V}_i|$, and the delivered complex power, S_i, at the infinite bus, the q-axis of a synchronously rotating reference frame could then be conveniently defined as in the direction of the voltage, $\tilde{V}_i$. Figure 10.34 shows the phasor diagram and relative orientation between the q-axes of the generator's rotor and the chosen synchronously rotating reference. Knowing $\tilde{V}_i$ and S_i, the steady-state values of the generator variables can be determined from the following equations:

$$\tilde{I} = i^e_q - ji^e_d = \left(\frac{S_i}{\tilde{V}_i}\right)^*$$

$$\tilde{E}_q = |E_q|e^{j\delta} = \tilde{V}_i + \{(r_s + r_e) + j(x_q + x_e)\}\tilde{I}$$

(10.211)

$$\tilde{V}_t = \tilde{V}_i + (r_e + jx_e)\tilde{I}$$

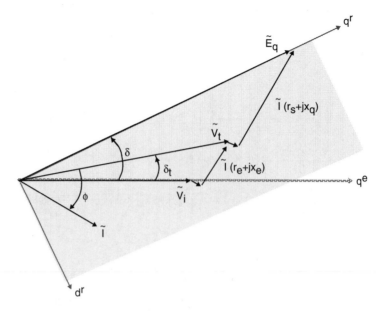

Figure 10.34   Phasor diagram and qd references.

The MATLAB file, *m4*, uses the above equations to compute the steady-state values for the specified operating condition. These steady-state values are used by *m4* to initialize *s4* and to compute the transfer functions.

Small-signal representation of system.   Figure 10.35a shows the small-signal system representation of the generator, excitation system, and power system stabilizer. That of the Type 1 exciter is shown in Fig. 10.35b.

Following the design procedure described in [103] and [104], we will first determine the transfer function, $PSS(s)$, of a pss with slip speed as its input signal. Beginning with the condition that the resulting torque component of ΔP_{em} produced by the pss modulation be $180°$ degrees out of phase with $\Delta(\omega_r - \omega_e)/\omega_b$, as in the case of the damping torque from D_ω, we have

$$\Delta P_{em} = \underbrace{-GEP(s)\,PSS(s)}_{D_{pss}}\,\Delta\left(\frac{\omega_r - \omega_e}{\omega_b}\right) \tag{10.212}$$

$$\frac{\Delta P_{em}}{\Delta\left(\frac{\omega_r - \omega_e}{\omega_b}\right)} = -GEP(s)\,PSS(s) = -D_{pss} \tag{10.213}$$

where ΔP_{em} is the perturbation component of the electromagnetic power that is produced by the pss modulation signal, $GEP(s)$ is the transfer characteristic of the generator and excitation system to the modulation signal, $PSS(s)$ is the desired transfer function of the pss, and D_{pss} is a positive coefficient. The negative sign on the right side of Eq. 10.213 can

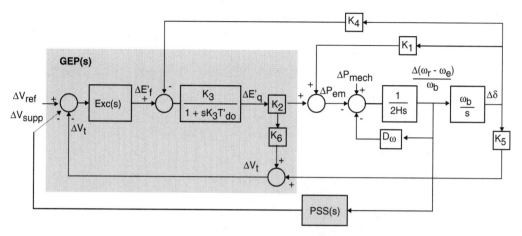

(a) Transfer function blocks of generator with exciter and pss using slip speed

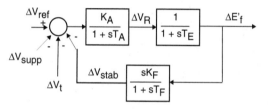

(b) Transfer function of exciter

Figure 10.35   Small-signal representation of system.

be taken care of by inverting the sign of the v_{supp} input to the excitation system, as shown in Fig. 10.33.

An approximate expression for $GEP(s)$ can be determined by ignoring the contributions through K_4 and K_5 to the shaded blocks in Fig. 10.35a, assuming that the speed deviation is small.

$$GEP(s) \approx K_2 \frac{Exc(s)\frac{K_3}{1+sK_3T'_{do}}}{1 + K_6 Exc(s)\frac{K_3}{1+sK_3T'_{do}}}$$
(10.214)

For the excitation system shown in Fig. 10.35b, the transfer function, $Exc(s)$, is

$$Exc(s) = \frac{\Delta E_f}{\Delta V_t} = \frac{\frac{K_A}{1+sT_A}\frac{1}{1+sT_E}}{1 + \frac{K_A}{1+sT_A}\frac{1}{1+sT_E}\frac{sK_f}{1+sT_f}}$$
(10.215)

Figure 10.36 shows the Bode plot of the exciter using the *set*1 parameters given in Table 10.7.

Determining GEP(s) and Exc(s).   As demonstrated in previous projects, we could use **linmod** to determine a linearized model of the system. For this project, however,

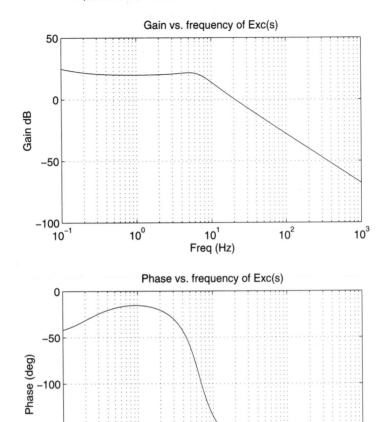

Figure 10.36   Bode plot of Exc(s) using *set* 1 parameters.

the MATLAB M-file, *m4*, uses the expressions given in Eqs. 10.214 and 10.215 instead to obtain these two transfer functions. It first computes the desired operating point values to set up the transfer functions, $GEP(s)$ and $Exc(s)$, and then it uses the $GEP(s)$ to determine the transfer function, $PSS(s)$, using Eq. 10.213. Besides determining the transfer functions, *m4* provides the root-locus and Bode plots of the transfer functions, initializes the SIMULINK simulation, *s4*, and sets it up for a small disturbance study to verify the dynamic response.

Figures 10.36, 10.37, and 10.38 show the Bode plots of $Exc(s)$, $GEP(s)$, and $PSS(s)$ from *m4* for the *set* 1 machine and exciter given in Table 10.7 at the system operating point where $\mathbf{S}_t = 0.8 + j0.6 \, pu$ and $|V_t| = 1.1 \, pu$. The gain of the $PSS(s)$ shown is for a D_{pss} of 6. It can be seen from the phase plot of $GEP(s)$ that it has two poles

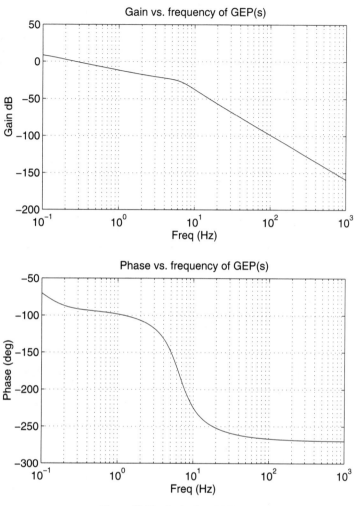

Figure 10.37   Bode plot of GEP(s).

near 6.5 Hz or 40.8 rad/sec. And from the phase plot of $PSS(s)$, we see that the desired phase characteristic of the pss ought to be lagging below 40.8 rad/sec and leading above that angular frequency. The phase compensation can be obtained using a pair of lag-lead and lead-lag networks with zeros centered at 40.8 rad/sec. Thus, picking the initial value of T_1 and T_3 to be equal to 1/40.8 or 0.024, and using roughly a 10:1 ratio, we can choose T_4 of the lag-lead network to be 0.24 and T_2 of the lead-lag network to be 0.002. A sufficiently large T_w of one second for the wash-out network is chosen so as not to disturb the phase compensation of the lead-lag and lag-lead networks from 1 Hz upwards. Too large a value of T_w can result in undesirable variations of the terminal voltage along with speed for a weak or isolated condition. The preliminary value for K_s, or gain of $PSS(s)$, is based on the desired D_{pss} of 6. Too high a gain can result in an unstable pss loop. Finally, a symmetrical

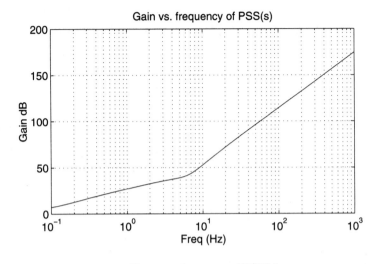

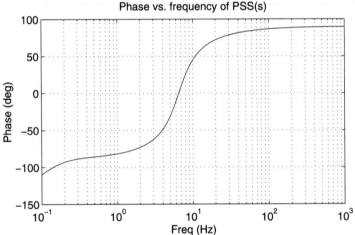

Figure 10.38 Bode plot of PSS(s).

limit of ± 0.1 *pu* is placed on the output of the pss. For the assumed operating condition and based on the above reasoning, we arrive at a preliminary design of the pss with the following parameters:

wash-out network: $K_s = 120$ $T_w = 1$
lead-lag network: $T_1 = 0.024$ $T_2 = 0.002$
lag-lead network: $T_3 = 0.024$ $T_4 = 0.24$

The effects of the K_4 and K_5 connections in Fig. 10.35a can be included by shifting the input of the K_4 connection to the same summing junction for K_5 and PSS, that is

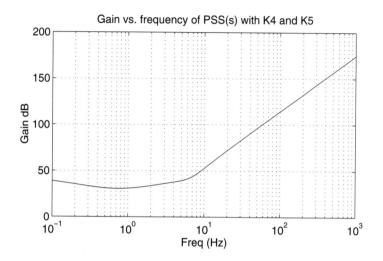

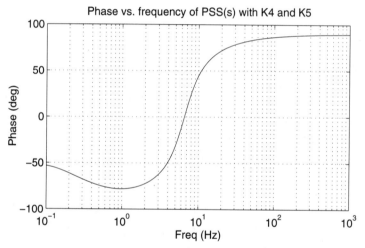

Figure 10.39   Bode plot of PSS(s) with K_4 and K_5 connections.

$$PSS(s) - K_5 \frac{\omega_b}{s} - \frac{K_4}{Exc(s)} \frac{\omega_b}{s} = \frac{-D_{pss}}{GEP(s)} \qquad (10.216)$$

The project M-file, *m4*, also determines the the transfer function of $PSS(s)$ based on Eq. 10.216. Figure 10.39 shows the Bode plot of $PSS(s)$ with the effects of K_4 and K_5 connections for a D_{pss} of 6 and at the same operating point as that used to determine Fig. 10.38. The difference between these two figures over the desired stabilization frequency range centered around the electromechanical oscillation frequency of 40.8 rad/s is not great. Figure 10.40 shows the Bode plot of the pss transfer function with the preliminary design values. For damping purposes, the critical part of these characteristics is that of the phase of $PSS(s)$ about the frequency of the electromechanical oscillations that are to be damped. Figure 10.41 shows the root-locus of the open-loop transfer function for $GEP(s)PSS(s)$

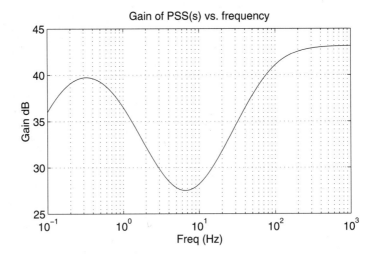

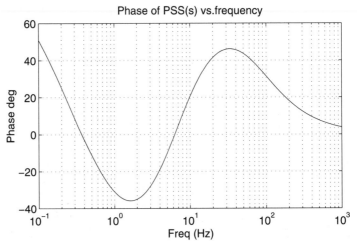

Figure 10.40   Bode plot of PSS(s) with preliminary design values.

over a range of K_s from 10 to 300. The plots in Figs. 10.40 and 10.41 may be obtained using the given MATLAB M-file, *m4comp*.

Simulation studies.   As mentioned earlier, *m4* also establishes the desired starting operating condition for the simulation, *s4*. As given, *m4* uses the data file, *set1*, which contains the parameters of the 9375 kVA synchronous generator and exciter used in Project 1, and will prompt the user to provide the impedance value of the RL line connecting the generator to the infinite bus, $r_e + jx_e$, and the voltage and delivered complex power, $\tilde{\mathbf{V}}_i$ and $\mathbf{S}_i$, at the infinite bus for it to establish the operating condition. In the design mode, *m4* will also prompt the user for the desired value of D_{pss}.

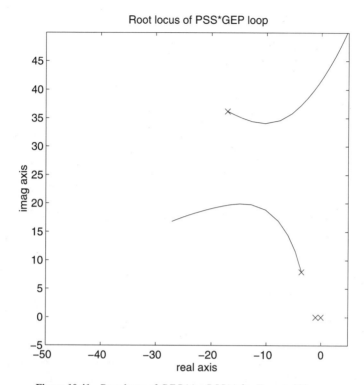

Figure 10.41   Root-locus of $GEP(s) * PSS(s)$ for K_s up to 300.

In the simulation mode, *m4* is programmed to set array values in the *repeating sequence* source, *DTmech*, to introduce a small step torque perturbation, *DT*, of ± 0.1 *pu* about the chosen operating point in a fixed sequence, using the *time* values of [*0 7.5 7.5 15 15 22.5 22.5 30*] and *output* values of [*0 0 DT DT -DT -DT 0 0*]. As given, it will also set the *PSS_sw* flag in *s4*, first for a run without the pss, and later for run with pss, storing the values of $|V_t|$, δ, P_{gen}, and Q_{gen} of the first run for plotting alongside those from the second run. It will plot the values of $|V_t|$, δ, P_{gen}, and Q_{gen} for each chosen operating condition. Figure 10.42 shows a plot of these variables for the system condition where $r_e + jx_e = 0.027 + j0.1$ *pu*, $\tilde{\mathbf{V}}_i = 1 + j0$ *pu*, and $\mathbf{S}_i = 0.8 + j0.6$ *pu*.

Use *m4* to obtain a preliminary design of the pss with slip speed as the input signal using a $r_e + jx_e$ of $0.013 + j0.05$ *pu*. Obtain the Bode plots of the transfer functions $GEP(s)$ and $PSS(s)$.

Conduct simulation runs for the various system conditions listed in Table 10.12 using either the given pss design or your own preliminary design. Use *m4* to conduct the two runs for each system condition: first without the stabilizing action of the pss, followed by another on the same system condition but with pss reconnected. Comment on the effects that the pss has on the dynamic response, and on the observed changes with the power factor of the generator and the strength of the network connection.

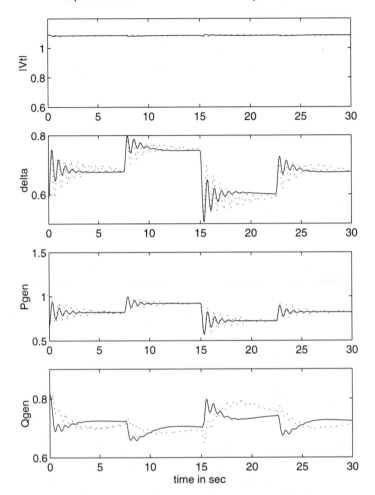

Figure 10.42   System response: dotted curve for without pss and solid for with pss
$r_e + jx_e = 0.027 + j0.1 \ pu$, $\tilde{\mathbf{V}}_i = 1 + j0 \ pu$, and $\mathbf{S}_i = 0.8 + j0.6 \ pu$.

Using the same operating condition of a $r_e + jx_e = 0.027 + j0.1 \ pu$, $\tilde{\mathbf{V}}_i = 1.0 + j0$,
and $\mathbf{S}_i = 0.8 + j0.6$ that was used to obtain Fig. 10.41, raise the value of K_s of the pss until
instability becomes apparent in your simulation. Note the value of K_s and the nature of the
instability and try to relate these with the information you can extract from the root-locus
plot.

10.9.5 Project 5: Self-controlled Permanent Magnet Motor Drive

Many synchronous motor drives are of the current-fed type. In the higher power ranges it
is with line-commutated inverters, and in the lower power ranges it is with pulse-width
modulated inverters. In the low power range, permanent magnet motors are becoming

TABLE 10.12   OPERATING CONDITIONS

$r_e + jx_e = 0.027 + j0.1\ pu$	$r_e + jx_e = 0.013 + j0.05\ pu$
$\tilde{\mathbf{V}}_i = 1.0 + j0$ $\mathbf{S}_i = 0.8 + j0.6$	$\tilde{\mathbf{V}}_i = 1.0 + j0$ $\mathbf{S}_i = 0.8 + j0.6$
$\tilde{\mathbf{V}}_i = 1.0 + j0$ $\mathbf{S}_i = 0 + j0$	$\tilde{\mathbf{V}}_i = 1.0 + j0$ $\mathbf{S}_i = 0 + j0$
$\tilde{\mathbf{V}}_i = 1.0 + j0$ $\mathbf{S}_i = 0.8 - j0.6$	$\tilde{\mathbf{V}}_i = 1.0 + j0$ $\mathbf{S}_i = 0.8 - j0.6$

increasingly popular. With a permanent magnet motor, however, the magnet excitation is not controllable. For operations higher than base speed, the control strategy used will have to provide the necessary demagnetization if the motor's terminal voltage is to be kept within the rated voltage of the inverter.

In a brushless dc motor drive where the magnets of the motor are surface-mounted and the magnet field is quite independent of the stator field, the stator current is kept in phase with the internal voltage by turning on the stator winding currents at right angles to the poles of the magnet. The signals for turning on the phase currents in such a scheme can be obtained from Hall-effect devices that sense the crossing of the magnet poles at points orthogonal to each of the stator winding axes. The field in the airgap from embedded magnets are more dependent on the armature reaction field of the stator currents and has a spatial distribution along the airgap that is sinusoidal rather than trapezoidal. With a standard distributed winding stator, the output torque will be smooth if the stator currents are sinusoidal. Continuous vector control of sinusoidally varying stator currents can be produced by a pulse-width modulated inverter with continuous rotor position feedback from a resolver. For some applications, self-control using an internal power factor angle that is adjusted to maintain greater linearity between torque and stator current magnitude may be desirable.

The objectives of this project are to examine the basis for a self-controlled permanent magnet drive with torque control via armature current control, and the implementation of a simulation of the drive to verify the performance of the control scheme.

TABLE 10.13   STEADY-STATE EQUATIONS OF PERMANENT MAGNET MOTOR IN PER UNIT

$qd\,0$ voltage equations:

$$v_q = r_s i_q + \frac{\omega_e}{\omega_b}\psi_d \qquad pu$$

$$v_d = r_s i_d - \frac{\omega_e}{\omega_b}\psi_q$$

$$V_s = \sqrt{v_q^2 + v_d^2}$$

$$I_s = \sqrt{i_q^2 + i_d^2}$$

(10.217)

Flux linkages:

$$\psi_q = x_q i_q \qquad pu$$

$$\psi_d = x_d i_d + \underbrace{x_{md} i'_m}_{E_m} \qquad (10.218)$$

where E_m, the stator excitation voltage from the magnets, is sometimes denoted by ψ'_m or $\omega_e \lambda'_m$.

Electromagnetic torque:

$$T_{em} = (\psi_d i_q - \psi_q i_d) \qquad pu$$

$$= (x_d - x_q) i_d i_q + E_m i_q \qquad (10.219)$$

or, with stator resistance neglected,

$$T_{em} = -\left\{ \frac{V_s E_m}{x_d} \sin\delta + V_s^2 \left(\frac{1}{x_q} - \frac{1}{x_d} \right) \sin 2\delta \right\} \qquad pu \qquad (10.220)$$

We will begin with the steady-state equations for a permanent magnet motor given in Table 10.13. By imposing the condition that the output torque be varying linearly with the stator current, I_s, we can reduce Eq. 10.219 into a nonlinear equation with one unknown, namely $\sin\delta$. Solving for $\sin\delta$, we can then use it to determine the current and voltage components in the qd rotor reference frame, the internal and terminal power factor angles, and also the current components in the synchronous qd reference frame, that is i_q^e and i_d^e. Using the above approach in the MATLAB M-file, $m5$, we can determine the steady-state curves of many of the variables of interest over the range of output torque from rated motoring to rated generating. Figure 10.43 shows these curves for a 70-hp, four-pole permanent magnet motor described in [120].

The above plots were obtained for the condition that the stator current magnitude, I_s, varies linearly with the output torque, T_{em}. It is evident from these plots that the relationship between I_q^e or I_q and output torque is almost linear. Torque control, therefore, can be accomplished through either I_q^e or I_q, provided that the corresponding d-axis component, I_d^e or I_d, is also controlled accordingly in the manner shown in these plots. Note that the internal and external power factor angles, γ and ϕ, are equal to one-half the power angle, δ, that is to say the phasor of the stator current that is varying with torque is along the direction of the bisector between the phasors, V_s and E'_m. Figure 10.44 shows the phasor diagrams

TABLE 10.14   PARAMETERS OF PERMANENT MAGNET MOTOR

Rated power = 70 hp	*Base speed*, ω_b = 710.48 $elect.rad/sec$
Rated rms stator phase voltage = 58.5 V	*Magnet's rms phase voltage*, E_m = 40.2 V
r_s = 0.00443 Ω	x_{ls} = 0.0189 Ω
x_{md} = 0.0785 Ω	x_{mq} = 0.1747 Ω
J_{rotor} = 0.292* $Nmsec^2$	D_ω = 1*

Reference [120] ⓒ 1988 IEEE, modified or added value denoted by *.

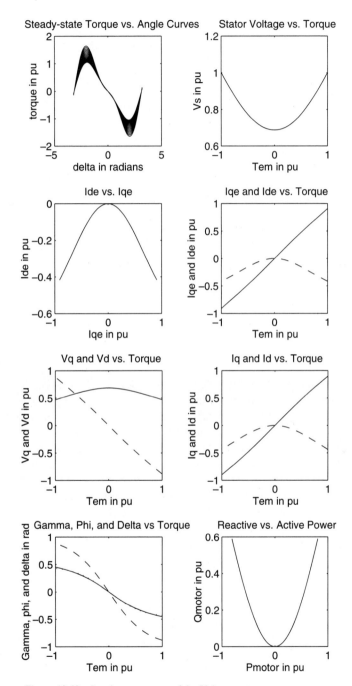

Figure 10.43   Steady-state curves of the 70-hp permanent magnet motor.

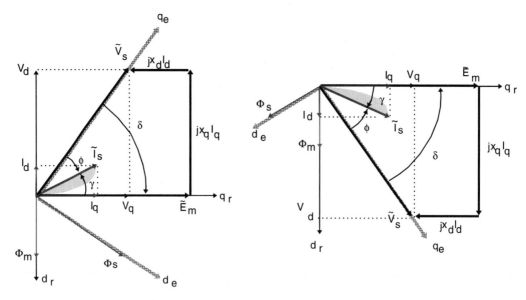

Figure 10.44   Phasor diagram for leading and lagging power factor.

for both a motoring and generating condition, where $\tilde{\mathbf{I}}_s$ is along the bisector of $\tilde{\mathbf{V}}_s$ and $\tilde{\mathbf{E}}_m$. As the output torque changes, the phasor $\tilde{\mathbf{I}}_s$ is kept within the area shown shaded in each diagram.

Figure 10.45a shows the overall diagram of the SIMULINK simulation, *s5*, for a self-controlled permanent magnet drive using coordinated values of i_q and i_d to control the output torque. In this simulation, the signal for the torque command is taken from a *repeating sequence* source. A rate limit is placed on the reference torque going into the input of the torque controller. An estimated value of the output torque from the *Feedback* block closes the outer torque loop. The inner current loops of i_q and i_d are also closed loops. Figure 10.45b shows the inside of the *Feedback* block in which the values of the qd current components are determined from the measured values of stator phase currents and rotor position. The simulation inside of the *pm_motor* block is very much like that given earlier in Figs. 7.21b through 7.21f, except that the blocks associated with the damper windings in the *q_cct* and *d_cct* blocks are deleted in this case. Figures 10.45c and 10.45d show the inside of the modified *q_cct* and *d_cct* blocks, respectively. The inside of the *qdr2abc1* block is shown in Fig. 10.45e.

The coordinated reference values for i_d^* and V_s^* are generated by separate function generators: i_d^* from the function generator, *Id-Iq*, and V_s^* from the function generator, *Vs-Tem*. The expressions for these function generators are determined in *m5* using the polynomial curve-fitting MATLAB function, **polyfit**, on the steady-state data of i_d vs. i_q and of V_s vs. T_{em} over the full range.

Figure 10.46 shows sample results of the simulation using the torque command sequence as programmed in *m5*. It shows the machine operating under no external load, motoring up from standstill under a torque command. The value and time arrays of the

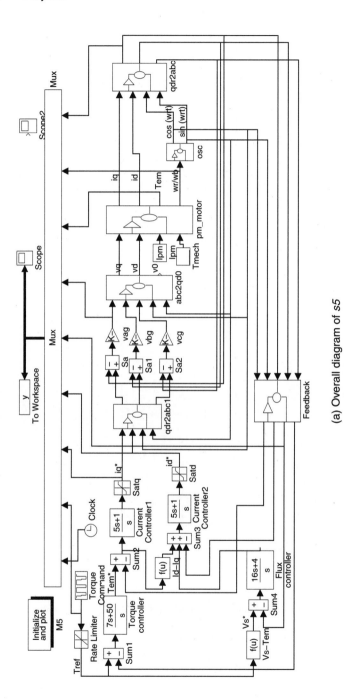

(a) Overall diagram of *s5*

Figure 10.45 Simulation *s5* of a self-controlled permanent magnet motor drive.

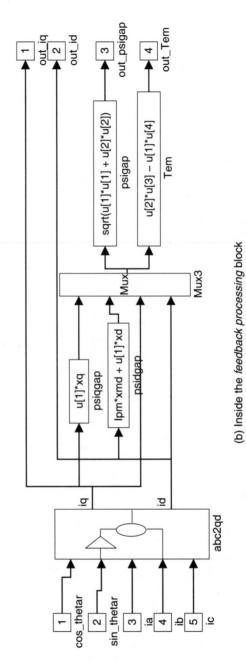

Figure 10.45 Simulation s5 of a self-controlled permanent magnet motor drive.

(b) Inside the *feedback processing* block

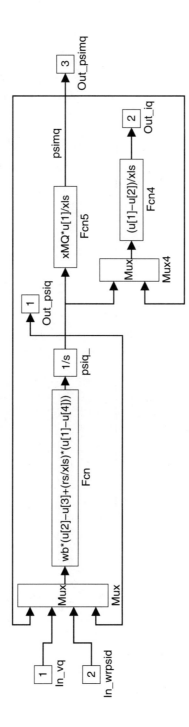

(c) Inside the *q_cct* block

Figure 10.45 Simulation *s5* of a self-controlled permanent magnet motor drive.

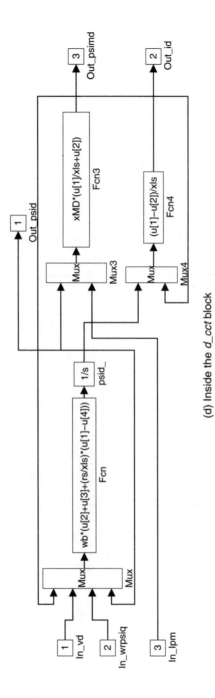

Figure 10.45 Simulation *s5* of a self-controlled permanent magnet motor drive.

(d) Inside the *d_cct* block

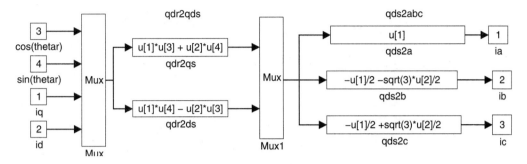

(e) Inside the *qdr2abc1* block

Figure 10.45   *(cont.):* Simulation *s5* of a self-controlled permanent magnet motor drive.

repeating source used to provid the torque command signal are

$$tref\_time = [0\ 1\ 1\ 1.2\ 1.2\ 2.2\ 2.2\ tstop]$$

$$tref\_value = [1\ 1\ 0\ 0\ -1\ -1\ 1\ 1]\ \%\ \text{negative for motoring}$$

The MATLAB file, *m5*, is programmed to load the machine parameters into the MATLAB workspace, determine and plot the steady-state curves shown in Fig. 10.43, initialize the SIMULINK file, *s5*, and prompt the user about repeating a run before plotting the simulated results. Use it to conduct a simulation run as programmed and cross-check your results with those given in Fig. 10.46. Repeat the simulation with the rotor inertia reduced by a factor of 2 and note any difference in response.

Next, modify a copy of *s5* to simulate the same motor operating under self-control with an outer speed control loop instead of the torque control given. Adjust, if necessary, the values of the proportional-integral controllers to obtain good speed regulation.

Modify the simulation further to simulate a speed-controlled drive that operates above base speed by changing over from a pwm current regulated mode to a six-step voltage fed mode as described in [120].

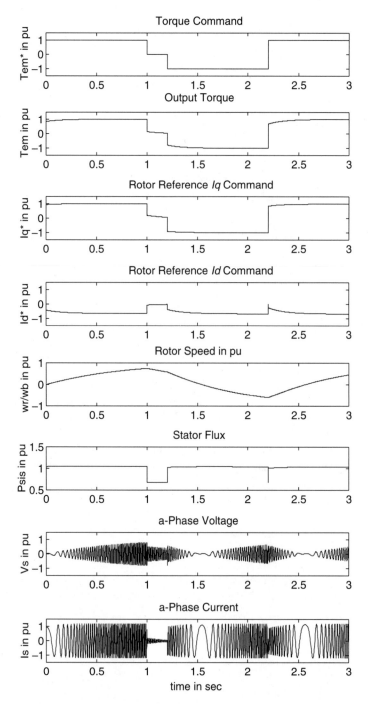

Figure 10.46   Sample results showing drive response to torque command.

Appendix A

Numerical Methods

A.1 INTRODUCTION

As the convenience of powerful computers becomes more affordable, the use of numerical methods for solving problems that cannot be easily solved analytically becomes more popular. The choice of mathematical model and numerical solution often are interrelated in that simplifications of the model may be deliberately made to adapt to specific methods of solution. Often a balance is struck between acceptable accuracy for the application at hand and the expediency in solution.

It is worthwhile for someone using digital simulation to have a little insight to some popular numerical integration methods-insight which may help him/her pick a suitable algorithm for the problem that he/she is simulating. Besides the behavior of the solution and nature of the model, the main factors to consider when picking a numerical method are compatible accuracy, speed of solution, and reliability. The solution may have a mix of fast and slow transient components. Numerical accuracy and stability considerations usually dictate using a time step that is many times smaller than the smallest time constant of the dominant eigenvalues at each stage. Thus, the time steps needed to capture the fast transient components must be small. But, as these fast components die away, larger time steps, limited only by stability consideration, should be used to reduce the overall computational effort.

In practice, errors in digital computation are unavoidable because the number representation in a computer is discrete or granular. The two sources of computational errors are from roundoff and truncation. Roundoff error is the difference between an answer obtained with rounded arithmetic and with exact arithmetic, using the same method.

$$y_{machine} + roundoff\ error = y_{true} \tag{A.1}$$

Rounding is inevitable when real numbers are to be represented by machine numbers having a finite number of bits, as in the situation of taking the quotient of two N-bits numbers whose result could require more than 2N bits. The closeness of the spacing is a measure of the precision or the capacity for accuracy.

Truncation errors are also introduced when using truncated series, such as in polynomial approximations, and from terminating an iteration before convergence. They are the difference between the answers obtained with exact arithmetic using a numerical method and the true answer.

$$y_{true} + truncation\ error = y_{exact} \tag{A.2}$$

Thus

$$y_{machine} - y_{exact} = y_{machine} - y_{true} + y_{true} - y_{exact}$$
$$\leq |truncation\ error| + |round\ off| \tag{A.3}$$

In numerical analysis, different measures of errors are defined.

$$(absolute)\ error \triangleq approximate\ value - true\ value \tag{A.4}$$

Since the relative size of the error is more helpful, we also have

$$relative\ error \triangleq \frac{error}{true\ value} \tag{A.5}$$

For most applications, it is sufficient to determine the approximate value to within some prescribed error tolerance of the exact value. Thus, the so-called acceptable solution will be no more than the specified tolerance from the exact solution or

$$|computed\ solution - exact\ solution| \leq error\ tolerance \tag{A.6}$$

The mathematical models of power components for transient studies are often described by a mix of algebraic and ordinary or partial differential equations. For some situations, the mathematical model can consist of a mix of differential and algebraic equations. In the case of ordinary differential equations, the independent variable is usually time. Partial differential equations, on the other hand, will have two or more independent variables, such as time and position in the network.

Let's examine the numerical process of integrating the ordinary differential equation (ODE) given by

$$y' = f(t, y) \tag{A.7}$$

where y is the dependent variable and t is the independent variable. To completely specify the solution in addition to the ODE, we need an auxiliary condition. For transient studies, the auxiliary condition is the value of all the dependent variables at some initial

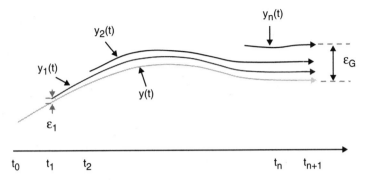

Figure A.1   Propagation of local errors in subsequent steps.

time, often referred simply to as the initial values. Such a problem is called an initial-value problem(IVP). With a higher-order ODE, a more general boundary-value problem may have values of dependent variables specified at different values of the independent variables.

Typically, the solution of an initial-value ODE problem is determined by using an integration formula to construct the values of y at finite time instants, $t_1 < t_2 < \ldots t_{n-1} < t_n < t_{n+1}$, called the mesh points, beginning with the given initial point, (t_o, y_o). The difference between successive time instants, such as $h_n = t_{n+1} - t_n$, is called the step size. In going from one time instant to the next, the process uses some discrete approximation of the function $f(t, y)$. The approximation of $f(t, y)$ by a finite series in the integration introduces some local truncation, or discretization, error (LTE) at each time instant. However, the total or global error accumulated from start to finish is not just the sum of the local errors.

Along the way, the local errors propagate, in that errors in the preceding steps are compounded by some growth factors as shown in Fig. A.1. The global error over the entire time interval may be expressed as

$$\epsilon_G^n = \sum_{i=1}^{n} G_i \epsilon_i \tag{A.8}$$

where G_i is the growth factor for the local error at the ith time step and ϵ_i is the local error between the two neighboring solution trajectories, $y_i(t_i) - y_{i-1}(t_i)$.

The propagation of errors will depend on the nature of the local errors and stability of the numerical method and system model. An integration scheme is considered stable when its global error remains within bounds, that is

$$y_n - y(t_n) = \epsilon_G^n \qquad \text{remains bounded as } n \to \infty \tag{A.9}$$

Integration methods usually have finite regions of stability, where the size of the stable region is a function of the integration step size. A method with a large stable region is considered to be robust. In a variable step size ODE solver, estimates of the local truncation error are made at each step so that the next step size can be adjusted accordingly. If the local truncation error is larger than some specified local error tolerance, the next step size will be reduced to achieve smaller local error. On the other hand, if the estimate is smaller than some fraction of the specified tolerance, the next step size may be increased to save computing time. Since a change in step size requires some overhead computations that are greater in multi-step methods than in single-step methods, the fraction is usually set much smaller than one in multi-step methods.

Overall, the error of concern is the global error which depends on the size of the local errors, their growth factors, and the number of steps. While the first is under control of the user, the last two depend on the nature of the problem.

Many of the digital methods of solving initial-value ordinary differential equations can be classified as one- or multi-step methods. Essentially, a one-step method uses information from a single point, y_n, to predict the value y_{n+1} at the next point in time, whereas a multi-step method will make use of curvature information from several past points, that is $y_{(n+1)-i}, i = 1, 2 \ldots K$, to predict the next value.

A.2 EULER METHODS

Let's begin by considering the first-order ODE given by

$$y' = f(t, y) \tag{A.10}$$

with an initial condition of $y(t_0)$ at $t = t_0$, abbreviated from this point on as y_0.

Starting with the given initial value at time t_0, the task of a numerical method is to determine the approximates of the solution $y(t)$, namely $y_1, y_2, \ldots$ at the times $t_1, t_2, \ldots$ and so on untill the final time. Integrating Eq. A.10 forward from the point y_n at t_n to the next instant at t_{n+1}, we have

$$y_{n+1} = y_n + \int_{t_n}^{t_{n+1}} y'(t, y) dt \tag{A.11}$$

Note that the integrand in Eq. A.11 is a function of y and that the value of y_{n+1} is as yet unknown, as opposed to the situation when evaluating an integral where the dependence of the integrand in terms of the independent variable is known ahead of time. The one-step methods use a prediction formula of the form;

$$y_{n+1} = y_n + \phi h_n \tag{A.12}$$

Comparing the second term on the right-hand side of Eqs. A.11 and A.12, we obtain

$$\phi = \frac{1}{h_n} \int_{t_n}^{t_{n+1}} y'(t, y) dt \tag{A.13}$$

Thus, the increment function, ϕ, may be interpreted as the average of slope of y over the time step h_n defined by $t_{n+1} - t_n$.

The simplest of the one-step methods is the *explicit*, or *forward*, Euler method. It uses the derivative of y at t_n as the increment function. A geometrical interpretation of the approach is shown in Fig. A.2a. The approximate value at t_{n+1} is given by

$$y_{n+1} - y_n = y'_n \underbrace{(t_{n+1} - t_n)}_{h_n} \tag{A.14}$$

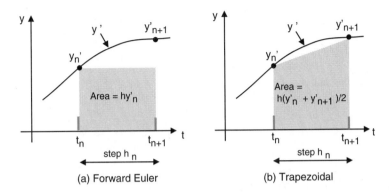

Figure A.2   Forward Euler step.

or

$$y_{n+1} = y_n + h_n y'_n \qquad (A.15)$$

The above equation yields an approximate of y_{n+1} by a linear extrapolation of step size h_n from the point (t_n, y_n) along a slope determined at that same point, that is y'_n or $f(t_n, y_n)$. The method is self-starting. Repeated application of Eq. A.15, beginning from the given initial value sequentially, will produce the approximates $y_1, y_2, \ldots, y_{n+1}$.

To establish an error estimate for the above Euler method, let's begin with the Taylor series expansion of the function at the point (t_n, y_n), that is

$$y(t_{n+1}) = y(t_n) + y'(t_n)h_n + \frac{y''(t_n)h_n^2}{2!} + \frac{y'''(t_n)h_n^3}{3!} + \ldots + \frac{y^{(m)}(t_n)h^m}{m!} + R_m \qquad (A.16)$$

The time derivatives of y are denoted by $y', y'', y''', \ldots, y^{(m)}$. R_m denotes the remainder of all other terms from $m+1$ to infinity; it is given by

$$R_m = \frac{y^{(m+1)}(\xi)}{(m+1)!} h^{m+1} \qquad \text{where } t_n < \xi < t_{n+1} \qquad (A.17)$$

R_m provides an estimate of the local truncation error. It is also used in criteria for determining the step size or order of the approximating formula in many integration schemes. When Eq. A.15 is compared to the above full Taylor-series expansion about y at t_n, it is evident that the remainder is of order (h^2), that is $O(h^2)$. Since the number of steps over a fixed integration interval is inversely proportional to h, the accumulated truncation error will be one order less in h, that is $O(h)$. The order of the method is the same as the order of the accumulated truncation error, thus the Euler method is a first-order method.

A lower value of accumulated truncation error can be achieved using a smaller step size. But, when the order of the method is too low, the increase in roundoff error from the additional computations using significantly more steps may offset the benefit gained from reducing the truncation errors.

The stability of the Euler method may be examined using the simple case of $y' = -\lambda y$, $y(t_o) = y_0$, which has the analytical solution of $y(t) = y_0 e^{-\lambda(t-t_0)}$. Assuming a uniform set of mesh points and introducing Θ as the extrapolation function from t_n to t_{n+1}, we have

$$y_{n+1} + \epsilon_{n+1} = \Theta(y_n + \epsilon_n) \tag{A.18}$$

From the analytical solution, we obtain $y_{n+1} = y_n e^{-\lambda h}$. The accumulated truncation error at $(t_{n+1}, \epsilon_{n+1})$ can therefore be expressed as

$$\epsilon_{n+1} = \underbrace{[\Theta - e^{-\lambda h}]y_n}_{present\ step} + \underbrace{\Theta\epsilon_n}_{previous\ steps} \tag{A.19}$$

The method is called *absolute stable* if $|\Theta| \le 1$ and *relative stable* if $|\Theta| \le e^{-\lambda h}$. Specifically, the forward Euler formula yields

$$y_{n+1} = y_n - \lambda h y_n \tag{A.20}$$

Hence the extrapolation function for the forward Euler is

$$\Theta = 1 - \lambda h \tag{A.21}$$

For the method to be absolute stable, $|1 - \lambda h| \le 1$, which for positive λ is equivalent to $h \le 2/\lambda$. Consequently, the step size of the forward Euler method will have to be small when used on systems with a large λ or small time constant.

A larger stability region is obtained with an *implicit*, or *backward*, Euler method, whose equation can be derived from the following truncated Taylor series expanded backwards from the point y_{n+1}:

$$y_n = y_{n+1} + (-h_n)y'_{n+1} \tag{A.22}$$

or

$$y_{n+1} = y_n + h_n f(t_{n+1}, y_{n+1}) \tag{A.23}$$

The backward Euler method is implicit in that the above formula does not yield y_{n+1} explicitly in terms of past values; instead, the unknown y_{n+1} which also appears on the right-hand side of Eq. A.23 must be solved iteratively. It can be readily shown that the *implicit* formula has a $\Theta = 1/(1 - \lambda h)$. The *implicit* Euler method remains absolutely stable for large step sizes, even though its accuracy may be poor.

A.3 TRAPEZOIDAL RULE

As evident from Fig.A.2b, a closer approximation of the elemental area under the curve $y' = f(t, y)$ over the interval h_n can be obtained using the trapezoid rule, that is

$$elemental\ area\ A_n = y_{n+1} - y_n = \frac{h_n}{2}(y'(t_n, y_n) + y'(t_{n+1}, y_{n+1})) \tag{A.24}$$

Since y_{n+1} appears in the $y'(t_{n+1}, y_{n+1})$ term on the right side of the equation, the rule yields an implicit relation from which the unknown y_{n+1} must be solved for iteratively.

The Taylor-series expansion of the derivative y' at t_n is

$$y'_{n+1} = y'_n + y''_n h_n + \frac{y'''_n h_n^2}{2!} + \dots,$$ (A.25)

Using Eq. A.25 to eliminate the $y''_n(t)$ term in Eq. A.16, the latter becomes

$$y_{n+1} - y_n = \frac{y'_n + y'_{n+1}}{2} h_n - \frac{1}{12} y'''_n(t) h_n^3 + \dots, \quad t_n < t < t_{n+1}$$ (A.26)

Thus, the local truncation error is

$$\epsilon = -\frac{1}{12} y'''_n(t) h_n^3$$ (A.27)

The accumulated truncation error is of the order $O(h^2)$, hence, the method is second-order. The method is implicit and self-starting. The region of stability of the trapezoidal method covers the entire left half of the complex z plane, and it is A-stable. Although its global error stays bounded, the result can contain spurious numerical [147].

A.4 RUNGE-KUTTA METHODS

The RK methods achieve a higher order of accuracy of a Taylor series without having to calculate the higher derivatives of $f(t, y)$ explicitly. These methods make use of midpoint quadratures. The general form of the prediction formula of an mth RK method is

$$y_{n+1} = y_n + (a_1 k_1 + a_2 k_2 + \dots + a_m k_m) h_n$$ (A.28)

Clearly, when Eq. A.28 is compared to the form given in Eq. A.12, the increment function of the mth order RK method is

$$\phi(t_n, y_n, h_n) = (a_1 k_1 + a_2 k_2 + \dots + a_m k_m)$$ (A.29)

The increment function of the mth order RK method may be interpreted as the weighted mean of the slopes at several points between t_n and t_{n+1}. The k's are defined by the following recurrence relationships, which facilitate the computation of their values:

$$\begin{aligned}
k_1 &= f(t_n, y_n) \\
k_2 &= f(t_n + p_1 h_n, y_n + q_{1,1} k_1) \\
&\vdots = \vdots \\
k_m &= f(t_n + p_{m-1} h_n, y_n + q_{m-1,1} k_1 + \dots + q_{m-1,m-1} k_{m-1})
\end{aligned}$$ (A.30)

For the integration formula to be at least of order one,

$$\sum_{j=1}^{m} a_j = 1$$ (A.31)

The parameter's a's, p's, and q's can be determined from relationships that are obtained by equating coefficients of powers of h in the expansion of Eq. A.28 to those of the Taylor series expansion of the solution given in Eq. A.16.

To obtain the relationships, let's consider the following Taylor series expansion of a function with two variables about the point (t_n, y_n), that is

$$f(t_n + \Delta t, y_n + \Delta y)$$

$$= f(t_n, y_n) + \frac{\partial f(t_n, y_n)}{\partial t} \Delta t + \frac{\partial f(t_n, y_n)}{\partial y} \Delta y$$

$$+ \frac{1}{2!} \left[\frac{\partial^2 f(t_n, y_n)}{\partial t^2} (\Delta t)^2 + 2 \frac{\partial^2 f(t_n, y_n)}{\partial t \partial y} \Delta t \Delta y + \frac{\partial^2 f(t_n, y_n)}{\partial y^2} (\Delta y)^2 \right] + \dots \quad \text{(A.32)}$$

$$= f(t_n, y_n) + \left(\Delta t \frac{\partial}{\partial t} + \Delta y \frac{\partial}{\partial y} \right) f(t_n, y_n)$$

$$+ \frac{1}{2!} \left(\Delta t \frac{\partial}{\partial t} + \Delta y \frac{\partial}{\partial y} \right)^2 f(t_n, y_n) + \frac{1}{3!} \left(\Delta t \frac{\partial}{\partial t} + \Delta y \frac{\partial}{\partial y} \right)^3 f(t_n, y_n) + \dots$$

Using Eq. A.32 to replace the functional terms on the right-hand side of Eq. A.31, we obtain for the case of $m = 4$:

$$k_1 = f(t_n, y_n)$$

$$k_2 = f(t_n, y_n) + \left(p_1 h_n \frac{\partial}{\partial t} + q_{1,1} k_1 \frac{\partial}{\partial y} \right) f(t_n, y_n)$$

$$+ \frac{1}{2!} \left(p_1 h_n \frac{\partial}{\partial t} + q_{1,1} k_1 \frac{\partial}{\partial y} \right)^2 f(t_n, y_n)$$

$$+ \frac{1}{3!} \left(p_1 h_n \frac{\partial}{\partial t} + q_{1,1} k_1 \frac{\partial}{\partial y} \right)^3 f(t_n, y_n) + \dots$$

$$k_3 = f(t_n, y_n) + \left(p_2 h_n \frac{\partial}{\partial t} + (q_{2,1} k_1 + q_{2,2} k_2) \frac{\partial}{\partial y} \right) f(t_n, y_n)$$

$$+ \frac{1}{2!} \left(p_2 h_n \frac{\partial}{\partial t} + (q_{2,1} k_1 + q_{2,2} k_2) \frac{\partial}{\partial y} \right)^2 f(t_n, y_n) \quad \text{(A.33)}$$

$$+ \frac{1}{3!} \left(p_2 h_n \frac{\partial}{\partial t} + (q_{2,1} k_1 + q_{2,2} k_2) \frac{\partial}{\partial y} \right)^3 f(t_n, y_n) + \dots$$

$$k_4 = f(t_n, y_n) + \left(p_3 h_n \frac{\partial}{\partial t} + (q_{3,1} k_1 + q_{3,2} k_2 + q_{3,3} k_3) \frac{\partial}{\partial y} \right) f(t_n, y_n)$$

$$+ \frac{1}{2!} \left(p_3 h_n \frac{\partial}{\partial t} + (q_{3,1} k_1 + q_{3,2} k_2 + q_{3,3} k_3) \frac{\partial}{\partial y} \right)^2 f(t_n, y_n)$$

$$+ \frac{1}{3!} \left(p_3 h_n \frac{\partial}{\partial t} + (q_{3,1} k_1 + q_{3,2} k_2 + q_{3,3} k_3) \frac{\partial}{\partial y} \right)^3 f(t_n, y_n) + \dots$$

With $y' = f(t, y(t))$, the next two higher time derivatives of y at (t_n, y_n) are given by

$$y''(t_n) = \frac{\partial f}{\partial t} + \frac{\partial f}{\partial y} y'$$

$$y'''(t_n) = \frac{\partial^2 f}{\partial t^2} + 2f\frac{\partial^2 f}{\partial t \partial y} + f^2\frac{\partial^2 f}{\partial y^2} + \frac{\partial f}{\partial y}\left(\frac{\partial f}{\partial t} + f\frac{\partial f}{\partial y}\right) \tag{A.34}$$

Substituting these expressions of the higher derivatives into the Taylor series expansion of the function given in Eq. A.16, we obtain

$$y_{n+1} = y_n + fh_n + \left(\frac{\partial f}{\partial t} + \frac{\partial f}{\partial y} y'\right)\frac{h_n^2}{2!}$$

$$+ \left(\frac{\partial^2 f}{\partial t^2} + 2f\frac{\partial^2 f}{\partial t \partial y} + f^2\frac{\partial^2 f}{\partial y^2} + \frac{\partial f}{\partial y}\left(\frac{\partial f}{\partial t} + f\frac{\partial f}{\partial y}\right)\right)\frac{h_n^3}{3!} + \ldots \tag{A.35}$$

The functions and their derivatives in Eq. A.35 are to be evaluated at the point (t_n, y_n).

For orders up to the fourth, the order of the RK method corresponds to the number of terms of the Taylor series used. For example, by equating the coefficients of the partial derivative terms in Eq. A.35 to the corresponding terms in Eq. A.28 after the k's have been replaced by the expressions given in Eq. A.33, we obtain the following relationships of the coefficients for a fourth-order RK method:

$$p_1 = q_{1,1}$$
$$p_2 = q_{2,1} + q_{2,2}$$
$$p_3 = q_{3,1} + q_{3,2} + q_{3,3}$$
$$a_1 + a_2 + a_3 + a_4 = 1$$
$$a_2 p_1 + a_3 p_2 + a_4 p_3 = \frac{1}{2}$$
$$a_2 p_1^2 + a_3 p_2^2 + a_4 p_3^2 = \frac{1}{3}$$
$$a_2 p_1^3 + a_3 p_2^3 + a_4 p_3^3 = \frac{1}{4} \tag{A.36}$$
$$a_3 p_1 q_{2,2} + a_4(p_1 q_{3,2} + p_2 q_{3,3}) = \frac{1}{6}$$
$$a_3 p_1^2 q_{2,2} + a_4(p_1^2 q_{3,2} + p_2^2 q_{3,3}) = \frac{1}{12}$$
$$a_3 p_1 p_2 q_{2,2} + a_4 p_3(p_1 q_{3,2} + p_2 q_{3,3}) = \frac{1}{8}$$
$$a_4 p_1 q_{2,2} q_{3,3} = \frac{1}{24}$$

The above set of eleven equations has 13 unknowns. Since the number of unknowns is greater than the number of constraint equations, some of the unknowns can be chosen to achieve any of the following:

- To reduce storage requirements(e.g., Gill).
- To minimize the bound on the local truncation error(e.g., Ralston).
- To maximize the order of the scheme(e.g., King).
- To increase the region of stability (e.g., Lawson).

To obtain Runge's coefficient, we set $a_2 = a_3 = 1/3$ and solve for the rest.

$$a_1 = \tfrac{1}{6} \quad p_1 = \tfrac{1}{2} \quad q_{1,1} = \tfrac{1}{2}$$

$$a_2 = \tfrac{1}{3} \quad p_2 = \tfrac{1}{2} \quad q_{2,1} = 0 \quad q_{2,2} = \tfrac{1}{2}$$

$$a_3 = \tfrac{1}{3} \quad p_3 = 1 \quad q_{3,1} = 0 \quad q_{3,2} = 0 \quad q_{3,3} = 1$$

$$a_4 = \tfrac{1}{6}$$

The fourth-order RK formula with Runge's coefficients is given by

$$y_{n+1} = y_n + \frac{h_n}{6}(k_1 + 2k_2 + 2k_3 + k_4) \tag{A.37}$$

The k's in Eq. A.37 are determined by evaluating the function value as follows:

$$k_1 = f(t_n, y_n)$$

$$k_2 = f\left(t_n + \frac{h_n}{2}, y_n + \frac{k_1}{2}h_n\right)$$

$$k_3 = f\left(t_n + \frac{h_n}{2}, y_n + \frac{k_2}{2}h_n\right) \tag{A.38}$$

$$k_4 = f(t_n + h_n, y_n + k_3 h_n)$$

On the other hand, the Kutta's coefficients are obtained by setting $p_1 = 1/3$, $p_2 = 2/3$, and solving for the rest:

$$a_1 = \tfrac{1}{8} \quad p_1 = \tfrac{1}{3} \quad q_{1,1} = \tfrac{1}{3}$$

$$a_2 = \tfrac{3}{8} \quad p_2 = \tfrac{2}{3} \quad q_{2,1} = -\tfrac{1}{3} \quad q_{2,2} = 1$$

$$a_3 = \tfrac{3}{8} \quad p_3 = 1 \quad q_{3,1} = 1 \quad q_{3,2} = -1 \quad q_{3,3} = 1$$

$$a_4 = \tfrac{1}{8}$$

The fourth-order RK formula with Kutta's coefficients is given by

$$y_{n+1} = y_n + \frac{h_n}{8}(k_1 + 3k_2 + 3k_3 + k_4) \tag{A.39}$$

Its k's are computed using

$$k_1 = f(t_n, y_n)$$

$$k_2 = f\left(t_n + \frac{h_n}{3}, y_n + \frac{k_1}{3}h_n\right)$$

$$k_3 = f\left(t_n + \frac{2h_n}{3}, y_n + (k_2 - k_1)\frac{h_n}{3}\right)$$

$$k_4 = f(t_n + h_n, y_n + (k_1 - k_2 + k_3)h_n). \tag{A.40}$$

The local truncation error of the above fourth-order RK methods is of order h_n^5. The accumulated truncation error of these methods is $O(h^4)$; hence, they are fourth-order methods.

It can also be shown that the expressions for a third-order RK method are

$$y_{n+1} = y_n + \frac{1}{6}(k_1 + 4k_2 + k_3)h_n \tag{A.41}$$

where the k's are determined from

$$k_1 = f(t_n, y_n)$$

$$k_2 = f\left(t_n + \frac{h_n}{2}, y_n + \frac{h_n k_1}{2}\right)$$

$$k_3 = f(t_n + h_n, y_n - h_n k_1 + 2h_n k_2) \tag{A.42}$$

The local truncation error of the third-order RK method is of order h_n^4. Note that the derivative of the function is independent of the function, that is $y' = f(t)$, and the expression of the third-order RK method reduces to that of the Simpson's Rule.

Even though the third- and especially the fourth-order RK methods are very popular, the second-order RK methods are seldom of interest because they do not produce much better accuracy than the simpler modified Euler method. Nevertheless, it may be of interest to know that certain well-known methods may be considered as second-order RK methods. The expressions for a second-order RK method can be written as

$$y_{n+1} = y_n + (a_1 k_1 + a_2 k_2)h_n \tag{A.43}$$

where

$$k_1 = f(t_n, y_n)$$

$$k_2 = f(t_n + p_1 h_n, y_n + q_{1,1} k_1 h_n) \tag{A.44}$$

Equating coefficients with a truncated Taylor series up to the second-order derivative, we obtain the following constraint equations:

$$a_1 + a_2 = 1$$

$$a_2 p_1 = \frac{1}{2}$$

$$a_2 q_{1,1} = \frac{1}{2} \tag{A.45}$$

Since there are four unknowns with three constraints, we can arbitrarily assign a value to one of the unknowns. The three most common cases are:

Heun Method: $a_1 = a_2 = 1/2$, $p_1 = q_{1,1} = 1$

$$y_{n+1} = y_n + \frac{h_n}{2}(k_1 + k_2) \tag{A.46}$$

where

$$k_1 = f(t_n, y_n)$$
$$k_2 = f(t_n + h_n, y_n + k_1 h_n) \tag{A.47}$$

Using such a second-order RK is equivalent to using the Heun method with a single correction.

Polygon Method: $a_1 = 0, a_2 = 1$, $p_1 = q_{1,1} = 1/2$

$$y_{n+1} = y_n + k_2 h_n \tag{A.48}$$

where

$$k_1 = f(t_n, y_n)$$
$$k_2 = f\left(t_n + \frac{h_n}{2}, y_n + \frac{k_1 h_n}{2}\right) \tag{A.49}$$

Ralston Method: $a_1 = 1/3, a_2 = 2/3$, $p_1 = q_{1,1} = 3/4$

$$y_{n+1} = y_n + \left(\frac{k_1}{3} + \frac{2k_2}{3}\right)h_n \tag{A.50}$$

where

$$k_1 = f(t_n, y_n)$$
$$k_2 = f\left(t_n + \frac{3h_n}{4}, y_n + \frac{3k_1 h_n}{4}\right) \tag{A.51}$$

Since all three second-order RK methods match the accuracy of the Taylor series expansion up to the second order, their local truncation errors are of order h_n^3. However, the a's, p's and q's in Ralston's method correspond to the smallest upper bound on the local truncation error.

In summary, the RK methods are self-starting, even though they rely on information from more than one point-that is because these points are generated in the current step, not past points. The explicit RK methods have certain drawbacks. One is that they require several function evaluations at every integration step. For instance the fourth-order RK method requires four function evaluations per integration step. Another drawback is that their regions of stability are not large; as such, they are not suitable for handling stiff systems. These drawbacks can be offset with implicit or semi-implicit RK-type schemes at the expense of additional effort required for solving the resulting nonlinear equations.

Popular examples of the implicit RK-type are the first-order backward Euler and the second-order trapezoidal methods. For higher-order implicit or semi-implicit RK schemes, see [121, 122]. Finally, the basic formulation does not provide an estimate of the truncation error at each step. Such estimates, though, can be obtained using extra function evaluations per step.

A.5 LINEAR MULTI-STEP METHODS

Linear multi-step methods make use of information about the solution from more than one past value. The solution or its derivative is replaced by an analytic function that can be readily handled. There is a variety of ways to express the interpolating formula, the choice of which will depend on the type of values available, the preferred class of the function, and the criterion of goodness of fit. The interpolating formula can be determined using the method of undetermined coefficients, numerical differentiation, or numerical integration.

Newton's divided difference interpolating polynomial is widely used for approximating derivatives, integrals, and the solution of IVP. Let's consider $P_k(t)$, a kth-order polynomial that fits through the $k+1$ function values at $t_0, t_1, \ldots, t_k$, of the following form:

$$P_k(t) = a_0 + a_1(t - t_0) + a_2(t - t_0)(t - t_1) + \ldots + a_k(t - t_0)(t - t_1) \cdots (t - t_{k-1}) \quad \text{(A.52)}$$

The a's can be determined by equating the above expressions to the function values at the various time instants. For the case of $k = 1$, $a_0 = f(t_0)$ is obtained by equating $P_1(t)$ to $f(t)$ at t_0. Next, equating $P_1(t)$ to $f(t)$ at t_1 yields

$$P_1(t_1) = f(t_0) + a_1(t_1 - t_0) = f(t_1) \quad \text{(A.53)}$$

from which we obtain

$$a_1 = \frac{f(t_1) - f(t_0)}{t_1 - t_0} \quad \text{(A.54)}$$

The interpolating formula for $k = 1$ gives the classical linear interpolating expression:

$$P(t) = f(t_0) + \frac{f(t_1) - f(t_0)}{t_1 - t_0}(t - t_0) \qquad t_1 > t > t_0 \quad \text{(A.55)}$$

For compactness, we will use the divided difference notation: The *zero*th divided difference of the function with respect to t_k is simply the value of the function at t_k, that is $f[t_k] = f(t_k)$. The first finite divided difference is an approximation of the first derivative between two points; thus, the first finite difference of the function with respect to t_i and t_{i-1} is defined as

$$f[t_i, t_{i-1}] = \frac{f(t_i) - f(t_{i-1})}{t_i - t_{i-1}} \quad \text{(A.56)}$$

With this notation, it can be shown that the kth-order Newton's polynomial that fits over $k+1$ points, can be written as

$$P_k(t) = f(t_0) + f[t_1, t_0](t - t_0) + f[t_2, t_1, t_0](t - t_0)(t - t_1) + \ldots$$

$$+ f[t_k, t_{k-1}, \cdots, t_1, t_0](t - t_0)(t - t_1) \cdots (t - t_{k-1})$$

$$\text{(A.57)}$$

Analogous to the Taylor series remainder R_m of Eq. A.17, the truncation error of the kth-order Newton's divided difference polynomial is

$$R_k = f[t, t_k, t_{k-1}, \cdots, t_1, t_0](t - t_0)(t - t_1) \cdots (t - t_k) \qquad \text{(A.58)}$$

The above expression of R_k can be used to estimate the local error introduced by a finite kth-order polynomial when an extra point, say at $t = t_{k+1}$, becomes available. Interpolating polynomials based on finite differences is preferred because changing the order of the approximation can be accomplished simply by adding or dropping terms, and because the truncation error can be estimated.

Interpolation is simpler when dealing with data that are evenly spaced. As background to our discussion of two popular multi-step methods, let's review the relationships between finite differences and derivatives, assuming that the mesh points are equally spaced by a time interval, h.

Differences and derivatives of a function are closely related. In the case of forward differences, the relations are as follows:

$$\Delta f(t) = f(t + h) - f(t) = h f'(\xi) \qquad t < \xi < t + h$$

$$\Delta^2 f(t) = \Delta f(t + h) - \Delta f(t) = f(t + 2h) - 2f(t + h) + f(t)$$
$$= h^2 f''(\xi) \qquad t < \xi < t + 2h$$

$$\Delta^3 f(t) = \Delta^2 f(t + h) - \Delta^2 f(t)$$
$$= f(t + 2h) - 3f(t + 2h) + 3f(t + h) - f(t)$$
$$= h^3 f'''(\xi) \qquad t < \xi < t + 3h \qquad \text{(A.59)}$$

$$\vdots \qquad \vdots$$

$$\Delta^m f(t) = \Delta^{m-1} f(t + h) - \Delta^{m-1} f(t)$$

$$= f(t + mh) - \binom{k}{1} f(t + (m - 1)h)$$

$$+ \binom{k}{2} f(t + (m - 2)h) - \ldots + (-1)^m f(t)$$

$$= h^m f^m(\xi) \qquad t < \xi < t + mh$$

Similar relations for backward differences are

TABLE A.1   DIFFERENCE TABLE ABOUT THE FUNCTION VALUES AT TIME t

Time	Function	Δf	$\Delta^2 f$	$\Delta^3 f$	$\Delta^4 f$
$t-4h$	$f(t-4h)$				
		Δf_{-4}			
$t-3h$	$f(t-3h)$		$\Delta^2 f_{-4}$		
		Δf_{-3}		$\Delta^3 f_{-4}$	
$t-2h$	$f(t-2h)$		$\Delta^2 f_{-3}$		$\Delta^4 f_{-4}$
		Δf_{-2}		$\Delta^3 f_{-3}$	
$t-h$	$f(t-h)$		$\Delta^2 f_{-2}$		$\Delta^4 f_{-3}$
		Δf_{-1}		$\Delta^3 f_{-2}$	
t	$f(t)$		$\Delta^2 f_{-1}$		$\Delta^4 f_{-2}$
		Δf_0		$\Delta^3 f_{-1}$	
$t+h$	$f(t+h)$		$\Delta^2 f_0$		$\Delta^4 f_{-1}$
		Δf_1		$\Delta^3 f_0$	
$t+2h$	$f(t+2h)$		$\Delta^2 f_1$		$\Delta^4 f_0$
		Δf_2		$\Delta^3 f_1$	
$t+3h$	$f(t+3h)$		$\Delta^2 f_2$		
		Δf_3			
$t+4h$	$f(t+4h)$				

$$\Delta f(t-h) = f(t) - f(t-h) = hf'(\xi) \qquad t-h < \xi < t$$

$$\Delta^2 f(t-2h) = \Delta f(t-h) - \Delta f(t-2h) = f(t) - 2f(t-h) + f(t-2h)$$
$$= h^2 f''(\xi) \qquad t-2h < \xi < t$$

$$\Delta^3 f(t-3h) = \Delta^2 f(t-2h) - \Delta^2 f(t-3h)$$
$$= f(t) - 3f(t-h) + 3f(t-2h) - f(t-3h)$$
$$= h^3 f'''(\xi) \qquad t-3h < \xi < t \tag{A.60}$$

$$\vdots \qquad \vdots$$

$$\Delta^m f(t-mh) = \Delta^{m-1} f(t-(m-1)h) - \Delta^{m-1} f(t-mh)$$

$$= f(t) - \binom{k}{1} f(t-h) + \binom{k}{2} f(t-2h) - \ldots + (-1)^m f(t-mh)$$

$$= h^m f^m(\xi) \qquad t-mh < \xi < t$$

Table A.1 gives the difference table about the function value at t.

Substituting $s = (t - t_0)/h$ or $t - t_k = t - (t_0 + kh) = h(s-k) \quad k = 0, 1, \ldots, (k-1)$, the coefficient of the kth difference term in the Newton's interpolating polynomial of Eq. A.58 can be expressed as

$$\frac{(t-t_0)(t-t_1)\cdots(t-t_{k-1})}{h^k k!} = \frac{h^k(s)(s-1)(s-2)\cdots(s-(k-1))}{h^k k!} = \binom{s}{k} \quad\text{(A.61)}$$

Equation A.58, rewritten in terms of the forward differences, becomes

$$P_k(s) = \underbrace{\underbrace{f(t_0)}_{P_0} + \binom{s}{1}\Delta f_0 + \binom{s}{2}\Delta^2 f_0 + \ldots + \binom{s}{k}\Delta^k f_0}_{P_k} \quad\text{(A.62)}$$

Successive degree polynomials of the sequence of polynomials, $P_0, P_1, P_2, \ldots, P_k$, will interpolate through one more point of the set $t_0, t_1, t_2, \ldots, t_{k-1}$ than the previous polynomial. In fact, the polynomials can be generated using the recursion formula:

$$P_m(s) = P_{m-1}(s) + \binom{s}{m}\Delta^m f_0 \qquad m = 1, 2, \ldots \quad\text{(A.63)}$$

The other advantage of using the interpolating polynomial of Eq. A.62 is that its truncation error can be determined from

$$LTE = \binom{s}{k+1} h^{k+1} f^{k+1}(t) \qquad t_0 < t < t_k \quad\text{(A.64)}$$

It is evident from Eq. A.62 and Table A.1 that a certain minimum number of points will have to be stored for a given degree of interpolating polynomial. For example, a third-degree interpolating polynomial using forward divided differences from $f(t = t_0)$ will require the divided differences up to $\Delta^3 f_0$, which as shown in Table A.1, will in turn require the function values $f(t)$, $f(t+h)$, $f(t+2h)$, and $f(t+3h)$.

Instead of constructing an interpolating formula based on forward values, we can construct one using past values, if past values of the function are known at $t = t_0, t_0 - h, \ldots, t_0 - kh$. The formula for the Newton's backward-difference polynomial of degree k, which interpolates these $(k+1)$th points, is

$$P_k(s) = \underbrace{\underbrace{f(t_0)}_{P_0} + \binom{s}{1}\Delta f_{-1} + \binom{s+1}{2}\Delta^2 f_{-2} + \ldots + \binom{s+k-1}{k}\Delta^k f_{-k}}_{P_k} \quad\text{(A.65)}$$

In this case, s being negative, the identity:

$$\binom{s+k-1}{k} = (-1)^k \binom{-s}{k} \quad\text{(A.66)}$$

can be used to rewrite the above equation as

$$P_k(s) = f(t_0) + (-1)\binom{-s}{1}\Delta f_{-1} + (-1)^2\binom{-s}{2}\Delta^2 f_{-2} + \ldots + (-1)^k\binom{-s}{k}\Delta^k f_{-k} \quad\text{(A.67)}$$

The recursion formula for generating higher-degree polynomials is

$$P_m(s) = P_{m-1}(s) + (-1)^m \binom{-s}{m} \Delta^m f_{-m} \qquad m = 1, 2, \ldots \qquad (A.68)$$

and the truncation error of Eq. A.67 can be determined from

$$LTE = (-1)^{k+1} \binom{-s}{k+1} h^{k+1} f^{k+1}(t) \qquad t_0 < t < t_k \qquad (A.69)$$

The integration formula of many linear multi-step methods may be expressed in the form:

$$\begin{aligned} y_{n+1} &= \alpha_{m-1} y_n + \alpha_{m-2} y_{n-1} + \ldots + \alpha_1 y_{n-m} + \alpha_0 y_{n-m+1} \\ &+ h(\beta_m f(t_{n+1}, y_{n+1}) + \beta_{m-1} f(t_n, y_n) + \ldots \\ &+ \beta_1 f(t_{n-m}, y_{n-m}) + \beta_0 f(t_{n-m+1}, y_{n-m+1}) \end{aligned} \qquad (A.70)$$

where m is an integer greater than 1 and $y_0, y_1, \ldots, y_{m-1}$ are the required starting values. The initial set of starting values can be generated with an RK method whose order is at least as high as that of m. The right-hand side is a linear combination of the derivatives $f(t_{n+i}, y_{n+i}), i = 0, 1, \ldots, m$. The coefficients, $\alpha_i, \beta_i, i = 1, 2, \ldots, m$, are real. The method is *explicit* or *open* when $\beta_m = 0$, since y_{n+1} is then explicitly given in terms of previously determined values. But, when $\beta_m \neq 0$, the method is called *implicit* or *closed*.

A simple example of a multi-step method is the modified Heun's method which uses the slope at y_n as a predictor, that is

$$y_{n+1}^0 = y_{n-1} + 2hy'(t_n, y_n) \qquad (A.71)$$

Comparing the coefficients of $y's$ with those in Eq. A.71, we can see that the above predictor equation has $m = 2, \alpha_0 = 1$, and $\beta_1 = 2$. The predicted value is corrected by a corrector that uses the trapezoidal rule given by

$$y_{n+1}^\nu = y_n + h \frac{y'(t_n, y_n) + y'(t_{n+1}, y_{n+1}^{\nu-1})}{2} \qquad \text{for } \nu = 1, 2, \ldots, M \qquad (A.72)$$

The corrector, which refines the value of y_{n+1}^ν, is applied iteratively until $\nu = M$ or until the stopping criterion, such as that given below, is satisfied.

$$\left| \frac{y_{n+1}^\nu - y_{n+1}^{\nu-1}}{y_{n+1}^\nu} \right| \leq \text{specified error tolerance} \qquad (A.73)$$

In terms of the parameters given in Eq. A.71, the corrector has $m = 2, \alpha_1 = 1, \beta_1 = 1/2$, and $\beta_2 = 1/2$.

A.6 ADAMS METHODS

The Adams-Bashforth formulas are *open* or *explicit* integration formulas, They can be obtained by substituting the derivative, y', on the right-hand side of Eq. A.11 with a polynomial interpolation that fits a certain number of previous points, including that at t_n. For example, the formula for an m-step Adams-Bashforth formula can be obtained as follows using a $(m-1)$th-order Newton's backward-difference formula to approximate the derivative in Eq. A.11:

$$y_{n+1} = y_n + \int_{t_n}^{t_{n+1}} y' \, dt$$

$$= y_n + \int_{t_n}^{t_{n+1}} f(t, y(t)) \, dt \tag{A.74}$$

$$\approx y_n + \int_{t_n}^{t_{n+1}} P_{m-1}(t) \, dt$$

Substituting $t = t_n + sh$, $dt = h \, ds$, the integral term becomes

$$\int_{t_n}^{t_{n+1}} f(t, y(t)) \, dt = \int_{t_n}^{t_{n+1}} \sum_{k=0}^{m-1} (-1)^k \binom{-s}{k} \Delta_{-k} f(t_n, y_n) \, dt$$

$$+ \int_{t_n}^{t_{n+1}} \frac{f^{(m)}(\xi_n, y(\xi_n))}{m!} (t - t_n)(t - t_{n-1}) \cdots (t - t_{n+1-m}) \, dt$$

$$= \sum_{k=0}^{m-1} \Delta_{-k} f(t_n, y_n) h (-1)^k \int_0^1 \binom{-s}{k} \, ds \tag{A.75}$$

$$+ \frac{h^{m+1}}{m!} \int_0^1 s(s+1) \cdots (+m-1) f^{(m)}(\xi_n, y(\xi_n)) \, ds$$

The values of the integral $(-1)^k \int_0^1 \binom{-s}{k} \, ds$ for $k = 0, 1, 2, \ldots, m-1$ can be easily evaluated. For instance, the value for $k = 3$ is

$$(-1)^3 \int_0^1 \binom{-s}{3} \, ds = - \int_0^1 \frac{-s(-s-1)(-s-2)}{3!} \, ds$$

$$= \frac{1}{6} \int_0^1 (s^3 + 3s^2 + 2s) \, ds = \frac{3}{8} \tag{A.76}$$

Substituting the above expression for the integral and also the values of the coefficients of the first few terms in the summation series, Eq. A.75 becomes

$$y_{n+1} = y_n + h[f(t_n, y_n) + \frac{1}{2} \Delta_{-1} f(t_n, y_n) + \frac{5}{12} \Delta_{-2} f(t_n, y_n)$$

$$+ \frac{9}{24} \Delta_{-3} f(t_n, y_n) + \frac{251}{720} \Delta_{-4} f(t_n, y_n) + \frac{475}{1440} \Delta_{-5} f(t_n, y_n) + \ldots] + \tag{A.77}$$

$$+ h^{m+1} f^{(m)}(\xi_n, y(\xi_n))(-1)^m \int_0^1 \binom{-s}{m} \, ds$$

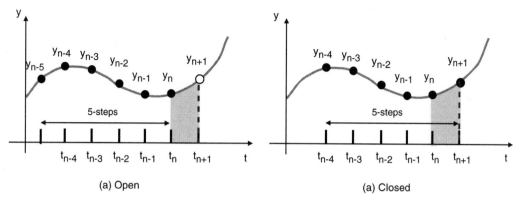

Figure A.3   Range of the sixth order Adams *open* and *closed* formulas.

Using the relationship in Table A.1 to express the backward differences in terms of the function at t_n and before, and ignoring the remainder term, the formula of an m-step and mth-order Adams-Bashforth formula can be expressed as

$$y_{n+1} = y_n + h \sum_{k=0}^{m-1} \beta_k f(t_{n-k}, y_{n-k}) \qquad (A.78)$$

If the derivative y' on the right-hand side of Eq. A.11 is substituted with a polynomial interpolation that fits a number of previous points and the point at t_{n+1}, we obtain the following *closed* or *implicit* integration formula known as the Adams-Moulton formula:

$$y_{n+1} = y_n + h[f(t_n, y_n) - \frac{1}{2}\Delta_{-1}f(t_n, y_n) - \frac{1}{12}\Delta_{-2}f(t_n, y_n)$$

$$- \frac{1}{24}\Delta_{-3}f(t_n, y_n) - \frac{19}{720}\Delta_{-4}f(t_n, y_n) - \frac{27}{1440}\Delta_{-5}f(t_n, y_n) + \ldots] \qquad (A.79)$$

$$+ h^{m+1} f^{(m)}(\xi_n, y(\xi_n))(-1)^m \int_0^1 \binom{-s}{m} ds$$

The formula of an $(m-1)$step and mth-order Adams-Moulton formula can be expressed in the form

$$y_{n+1} = y_n + h \sum_{k=0}^{m-1} \beta_k f(t_{n+1-k}, y_{n+1-k}) \qquad (A.80)$$

The essential difference between the *open* and *closed* Adams integration formula is in the range as illustrated in Fig A.3 for the case of $m = 6$.

The β coefficients and accumulated truncation errors of the lower-order Adams-Bashforth and Adams-Moulton formulas are given in Tables A.2 and A.3. Note that the truncation error of an $(m-1)$-step Adams-Moulton implicit formula is of the same order as that of a m-step Adams-Bashforth explicit formula. In general, the coefficients of the *implicit* formula are smaller than those of the *explicit* formula and are of the opposite sign.

TABLE A.2   COEFFICIENTS AND TRUNCATION ERRORS OF ADAMS-BASHFORTH FORMULAS

Steps	β_0	β_1	β_2	β_3	β_4	β_5	Accumulated truncation error
1	1						$\frac{1}{2}hf'(\xi)$
2	$\frac{3}{2}$	$-\frac{1}{2}$					$\frac{5}{12}h^2 f''(\xi)$
3	$\frac{23}{12}$	$-\frac{16}{12}$	$\frac{5}{12}$				$\frac{9}{24}h^3 f'''(\xi)$
4	$\frac{55}{24}$	$-\frac{59}{24}$	$\frac{37}{24}$	$-\frac{9}{24}$			$\frac{251}{720}h^4 f^{(4)}(\xi)$
5	$\frac{1901}{720}$	$-\frac{2774}{720}$	$\frac{2616}{720}$	$-\frac{1274}{720}$	$\frac{251}{720}$		$\frac{475}{1440}h^5 f^{(5)}(\xi)$
6	$\frac{4277}{720}$	$-\frac{7923}{720}$	$\frac{9982}{720}$	$-\frac{7298}{720}$	$\frac{2877}{720}$	$-\frac{475}{720}$	$\frac{19,087}{60,480}h^6 f^{(6)}(\xi)$

TABLE A.3   COEFFICIENTS AND TRUNCATION ERRORS OF ADAMS-MOULTON FORMULAS

Steps	β_0	β_1	β_2	β_3	β_4	β_5	Accumulated truncation error
1	$\frac{1}{2}$	$\frac{1}{2}$					$-\frac{1}{12}h^2 f''(\xi)$
2	$\frac{5}{12}$	$\frac{8}{12}$	$-\frac{1}{12}$				$-\frac{1}{24}h^3 f'''(\xi)$
3	$\frac{9}{24}$	$\frac{19}{24}$	$-\frac{5}{24}$	$\frac{1}{24}$			$-\frac{19}{720}h^4 f^{(4)}(\xi)$
4	$\frac{251}{720}$	$\frac{646}{720}$	$-\frac{264}{720}$	$\frac{106}{720}$	$-\frac{19}{720}$		$-\frac{27}{1440}h^5 f^{(5)}(\xi)$
5	$\frac{475}{1440}$	$\frac{1427}{1440}$	$-\frac{798}{1440}$	$\frac{482}{1440}$	$-\frac{173}{1440}$	$\frac{27}{1440}$	$-\frac{863}{60,480}h^6 f^{(6)}(\xi)$

In practice, the implicit formula is used as a corrector to improve the prediction made by the explicit formula.

The well-known Adams-Bashforth-Moulton predictor-corrector integration scheme makes use of the Adams-Bashforth formulas as predictors and the Adams-Moulton formulas as correctors. The Adams-Moulton formulas are used as correctors because they have smaller truncation errors and are of opposite sign when compared to the Adams-Bashforth formulas of the corresponding order. This scheme is recommended when the expected solution is smooth and is to be of high precision, and when the function evaluation is time-consuming. Typically, a predictor-corrector scheme consists of the following steps:

P   Extrapolate from a number of past values to predict the value of the point at the new time.

E   Use the latest available value at the new time to evaluate the derivative of the function at the new time.

C   Use a corrector formula to recompute the value at the new time.

There are variations to these basic steps from scheme to scheme. When the last two steps are repeated n times, the scheme is said to have a $P(EC)^n$ mode. Another scheme is the $PECE$ mode, in which an evaluation is made after the correction. A practical algorithm will have the ability to change the order of the Adams formula to maximize the benefits attainable from the extrapolation and also to adjust the step size to satisfy prescribed criteria on local truncation error, accuracy, and stability. To do these things, an algorithm needs to have a bookkeeping method of previous points, an efficient interpolation method for generating the necessary in-between points when halving the step size, and an effective criterion for changing order or step size so as to keep the overhead involved with changing back and forth to a minimum. Changes in order and step size in multi-step methods are more costly than in one-step methods because new equally spaced values have to be generated to effect the change.

A.7 GEAR'S BACKWARD DIFFERENCE FORMULAS

Stiffness in numerical problems is often associated with mathematical models which have time constants distributed over a wide range, as in a system with a slow-acting mechanical component coupled to electrical circuits having fast exponentially decaying transients. However, numerical stiffness depends not only on system characteristics, but also on the initial conditions, desired accuracy, and duration of integration [128]. Hence, it cannot be easily determined in advance. When it comes to handling numerically stiff equations, *implicit* integration algorithms are better than *explicit* ones [124]. And of the *implicit* algorithms, the backward difference formulas (BDFs) are supposedly reliable for dealing with stiff systems. The BDFs are excellent for solving stiff systems with eigenvalues not close to the imaginary axis [128, 129, 132, 144]. Lightly damped systems with eigenvalues close to the imaginary axis are better handled by *implicit/semi-implicit* RK methods.

Reference [130] describes a variable-step, variable-order Gear's algorithm using Newton's divided difference. A kth-order Gear's integration formula replaces the first derivative, y', by $y' = \beta y_{n+1} + S$, where S is a linear combination of the past values, $y_n, y_{n-1}, y_{n-2}, \ldots, y_{n-k}$.

The algorithm to determine β, S, and y_{n+1}^p for the kth-order formula consists of the following steps:

1. Calculate the divided differences, $y[t_n, t_{n-1}, \ldots, t_{n-j}]$, for $j = 1, 2, 3, \ldots, k$. The first-order difference is defined as

$$y[t_n, t_{n-1}] = \frac{y(t_n) - y(t_{n-1})}{t_n - t_{n-1}}$$

The second-order difference is defined as

$$y[t_n, t_{n-1}, t_{n-2}] = \frac{y[t_n, t_{n-1}] - y[t_{n-1}, t_{n-2}]}{t_n - t_{n-2}}$$

The kth-order difference is defined as

$$y[t_n, t_{n-1}, \ldots, t_{n-k}] = \frac{y[t_n, t_{n-1}, \ldots, t_{n-k-1}] - y[t_{n-1}, t_{n-2}, \ldots, t_{n-k}]}{t_n - t_{n-k}} \tag{A.81}$$

2. Calculate α_j, β_j, and L_j using

$$L_j = t_{n+1} - t_{n-j+1} \qquad\qquad j = 1, 2, \ldots, k$$

$$\alpha_1 = L_1 \quad \alpha_j = \alpha_{j-1} L_j \qquad j = 2, 3, 4, \ldots, k$$

$$\beta_1 = \frac{1}{L_1}, \quad \beta_j = \beta_{j-1} + \frac{1}{L_j} \quad j = 2, 3, 4, \ldots, k \qquad (A.82)$$

3. Calculate S and y_{n+1}^P using

$$y_{n+1}^{P0} = y(t_n), \qquad y_{n+1}^{Pj} = y_{n+1}^{Pj-1} + \alpha_j y[t_n, t_{n-1}, \ldots, t_{n-j}] \qquad j = 1, 2, 3, \ldots, k$$

$$S_1 = \frac{y_{n+1}^{P0}}{L_1}, \qquad S_j = S_{j-1} + \frac{y_{n+1}^{Pj-1}}{L_j} \qquad j = 2, 3, 4, \ldots, k-1 \qquad (A.83)$$

With Gear's formula, the mathematical model of the form, $f(\mathbf{y}, \mathbf{y}', t) = 0$, upon substituting for the first derivatives, yields an algebraic equation of the form $g(\mathbf{y}_{n+1}, t_{n+1}) = 0$. For stiff initial-value problems, better convergence of the solution of the resulting nonlinear equation, $g(y, t) = 0$, is obtained using a Newton's scheme. The predicted value, $\mathbf{y}_{n+1}^P$, from the above algorithm can be used as a good initial guess to start the Newton's scheme. As we will see in the next section, the same predicted value also provides an estimate of the local truncation error.

A.8 SOLVING THE RESULTING ALGEBRAIC EQUATIONS

When the mathematical model consists of differential and algebraic equations(DAE), there will be additional non-state variables associated with the algebraic constraint equations, that is y is a vector of unknown state and non-state variables at the next time step. For the beginning time step, either default zero initial-values or user-supplied initial-values can be used to start the Newton iteration. Since the convergence of the Newton iterations improves significantly with initial estimates that are close to the solution, every effort should be made to obtain good initial estimates. Thereafter, the predicted values, $\mathbf{y}_{n+1}^P$ from Eq. A.83, serve as the initial values of the state variables at the beginning of each new round of Newton iterations. With DAE systems or where non-states variables are associated with switching functions, previous values of non-state variables should also be stored for interpolations. Predicted values of the non-state variables can be extrapolated from the stored values and used as initial values for the Newton iterations.

The $(\nu + 1)$th update of the iteration of the Newton's scheme is computed using

$$\mathbf{y}^{\nu+1} = \mathbf{y}^\nu + \Delta \mathbf{y}^\nu = \mathbf{y}^\nu - \mathbf{J}^{-1} g(\mathbf{y}^\nu) \qquad (A.84)$$

where ν is the iteration count of the Newton's scheme and the Jacobian, J, is $\partial g(y)/\partial y$. Functions with exponential characteristics can sometimes cause the Newton algorithm to diverge when there is no limit placed on the correction, Δy.

The elements of the Jacobian matrix can be determined by making infinitesimal perturbations of the corresponding variables in g; for instance, the element, $J(i, j)$, can be computed using

$$\frac{\partial g_i}{\partial y_j} \approx \frac{g_i(y(1),\dots,y(j)+\sigma_j,\dots,y(m)) - g_i(y(1),\dots,y(j),\dots,y(m))}{\sigma_j} \quad \text{(A.85)}$$

where m is the dimension of the unknown vector, y. The value of σ_j is critical to avoiding numerical cancellation and to achieving fast convergence with the Newton algorithm. Stoer and Bulirsch [133] suggested choosing a σ_j that is about half a machine precision of the function itself.

In each new iteration of y^ν, a new correction, Δy^ν, must be solved for using $J\Delta y^\nu = -g(y^\nu)$. Since the solution of the equation, $J\Delta y^\nu = -g(y^\nu)$, must be repeated with more than one function value on the right-hand side, an LU decomposition approach will have the advantage over a Gauss elimination in operations count.

There are several methods of LU decomposition. The Crout algorithm decomposes J in a lower triangular L and an upper triangular U matrix with unit diagonal elements in the following way:

1. Begin with

$$L_{j1} = J_{j1} \qquad\qquad \text{for } j = 1,2,\dots,m$$

$$U_{1j} = \frac{J_{1j}}{L_{11}} \qquad\qquad \text{for } j = 2,3,\dots,m \tag{A.86}$$

2. For $i = 2,3,\dots,m$

a. first step through

$$L_{ij} = J_{ij} - \sum_{k=1}^{i-1} L_{ik}U_{kj} \qquad\qquad \text{for } j = i, i+1,\dots,m$$

$$U_{ji} = \frac{1}{L_{ii}}\left(J_{ji} - \sum_{k=1}^{i-1} L_{jk}U_{ki} \right) \qquad \text{for } j = i+1, i+2,\dots,m \tag{A.87}$$

b. then process the remaining $(i-1)$ by $(i-1)$ submatrix:

$$J_{ij} = \frac{J_{ij}}{J_{ii}} \qquad\qquad \text{for } j = i+1, i+2,\dots,m$$

$$J_{jk} = J_{jk} - J_{ji}J_{ik} \qquad\qquad \begin{cases} \text{for } j = i+1, i+2,\dots,m \\ \text{for } k = i+1, i+2,\dots,m \end{cases} \tag{A.88}$$

Clearly, the pivot, J_{ii}, should not be zero. Since every element of J is used only once in the decomposition and there is no need to provide storage for the unit diagonal elements of U, the remaining elements of U and the elements of L can all be stored in the same location occupied by J as the decomposition progresses.

With J decomposed into LU, the solution of $J\Delta y^\nu = -g(y^\nu)$ is obtained using two steps:

1. In the forward substitution step, the solution, $\mathbf{y}^{tmp}$, of the equation, $\mathbf{L}\mathbf{y}^{tmp} = -\mathbf{g}$, is computed as follows:

$$y_1^{tmp} = \frac{-g_1}{L_{11}}$$

$$y_k^{tmp} = \frac{1}{L_{kk}}\left(-g_k - \sum_{i=1}^{k-1} L_{ki} y_i^{tmp}\right) \qquad \text{for } k = 2, 3, \ldots, m$$

(A.89)

2. The second step is a backward substitution step, where the desired correction vector, $\Delta\mathbf{y}^v$, of the Newton iteration is calculated as follows using the $\mathbf{y}^{tmp}$ determined from the forward substitution step:

$$\Delta y_m = y_m^{tmp}$$

$$\Delta y_k = y_k^{tmp} - \sum_{i=k+1}^{m} U_{k,i} \Delta y_i \qquad \text{for } k = m-1, m-2, \ldots, 1$$

(A.90)

Successive updates are made until the value of $\mathbf{y}_{n+1}$ converges to within some specified tolerance. The convergence criterion may be of the form:

$$|\Delta\mathbf{y}(n)| = \epsilon_{rel}|\mathbf{y}(n)| + \epsilon_{abs}$$

(A.91)

where ϵ_{rel} and ϵ_{abs} are the relative and absolute error tolerances to be specified by the user.

Savings in computation associated with the evaluation and LU decomposition of the Jacobian can be considerable if the same Jacobian can be reused for several time steps, that is if the Newton's iterations continue to converge satisfactorily. The rate of convergence of the Newton's iterative scheme may be gauged from

$$ROC = \frac{\mathbf{y}^{v+1} - \mathbf{y}^v}{\mathbf{y}^v - \mathbf{y}^{v-1}}$$

(A.92)

Petzold has used the following scheme in DASSL [127] to speed up the Newton convergence:

$$\mathbf{y}^{v+1} = \mathbf{y}^v + c\Delta\mathbf{y}^v$$

(A.93)

where elements of vector $\mathbf{c}$ associated with β terms are calculated from

$$c_{\beta_i} = \frac{2}{1 + \beta_i^{curr}/\beta_i^{old}}$$

(A.94)

β_i^{curr} is the current value of β_i and β_i^{old} is the value of β_i when the Jacobian matrix was last evaluated.

It has also been observed that with stiff systems [143], the values of β_i are usually very large compared to other terms in the system equations; as a result, entries with β_i are usually chosen as pivots. Another scheme to extend the use of the same Jacobian matrix for several more time steps with minor computation is to partially update the LU decomposition by selectively modifying only those pivots associated with β as follows:

$$pivot_i^{new} = \frac{\beta_i^{curr}}{\beta_i^{old}} pivot_i^{old} \tag{A.95}$$

Pivots not associated with β_i will not be updated.

Some packages allow the user to specify the limits that are to be imposed on the corrections of some or all variables.

A.8.1 Pivoting

To avoid a division by zero in Eq. A.89 during decomposition, the pivots should be non-zero. When the pivot element is not zero but comparatively small, row interchanges may be desirable to get a larger pivot so that the roundoff errors can be reduced. Crout's algorithm with partial pivoting can be used for the LU decomposition [134]. In addition, choosing the biggest entries in the rows for row pivoting (or columns for column pivoting) of the submatrices will limit the growth of values in the decomposition process. Even though large values do not necessarily imply a large backward error in the solution [135], limiting growth in values usually results in better numerical stability.

There are several strategies for reordering equations to minimize the number of fill-ins. Nagel [124] and Duff, et al. [134] have, however, reported that none seems to perform better than the Markowitz strategy. The strategy of Markowitz [136] is to select the pivot from the entry with the lowest Markowitz count, defined as

$$(r_k - 1) \, (c_k - 1) \tag{A.96}$$

where r_k and c_k are the numbers of non-zero entries in the row and column, respectively, of the kth candidate for pivot. The Markowitz strategy with threshold pivoting offers a compromise between sparsity and numerical stability [134, 143]. In this pivoting scheme, entries with values larger than some fraction of the maximum value in the same row are considered candidates for the pivots. Thus

$$J_{pv,k} \geq \mu \, max_j \, |J_{i,j}| \tag{A.97}$$

where $J_{pv,k}$ is the kth candidate for the pivot of the ith row, and μ is a user specified constant between 0 and 1. Partial pivoting is obtained with a μ of 1. Duff, et al. [134], suggest a μ of 0.1 from their extensive testing.

The Markowitz strategy requires information on the number of non-zero entries in the rows and columns. In the order of the increasing number of non-zero entries, the row and column linked lists are scanned alternately. A limit on the number of rows and columns to be scanned can be specified by the user. When two or more candidates for pivoting have the same minimum Markowitz count, the entry with the largest value is usually chosen, as suggested by Osterby and Zlatev [137].

A.9 STEP SIZE CONTROL

For purposes of adjusting the step size of integration in Gear's BDF, the current local truncation error, LTE_{curr} [129], may be computed from

$$LTE_{curr} = \frac{|y_{n+1} - y^p_{n+1}|}{\beta L_{K+1}} \tag{A.98}$$

It is compared with the allowable local truncation error, LTE_{allow} [129, 130, 132], that is determined from

$$LTE_{allow} = \epsilon_\tau \, h_{allow} \tag{A.99}$$

The truncation error per unit time, ϵ_τ, may be specified by the user or calculated from the ratio of the specified global truncation error to the total integration interval. With uniform time steps, the local truncation error of an mth-order integration formula is proportional to h^{m+1}, or

$$\frac{LTE_{allow}}{LTE_{curr}} = \left(\frac{h_{allow}}{h_{curr}}\right)^{m+1} \tag{A.100}$$

Substituting Eq. A.99 into Eq. A.100, the next allowable step size can be determined from

$$h_{allow} = h_{curr}\left(\frac{\epsilon_\tau h_{curr}}{LTE_{curr}}\right)^{\frac{1}{m}} \tag{A.101}$$

When there is more than one state variables, the next allowable step size is determined by the smallest h_{allow} value.

The use of Eq. A.99 in conjunction with Eq. A.101 can sometimes result in very small step size when handling numerically stiff systems or when the duration of the simulation is long. Alternatively, as in IVPAG from IMSL[125], ACSL[123], ODEPACK[126], or DASSL[127], LTE_{allow} may be calculated from

$$LTE_{allow} = \epsilon_{rel}|y_{n+1}| + \epsilon_{abs} \tag{A.102}$$

where ϵ_{rel} and ϵ_{abs} are the relative and absolute error tolerances specified by the user, respectively. Allowing the user to specify these two error tolerances on the variables can avoid the need to scale the variables, even when their dynamic ranges are orders of magnitude apart. With LTE_{allow} calculated from Eq. A.102, the step size will be influenced more by the local dynamics as opposed to the global influence with Eq. A.99. Such responsiveness will make the maintenance of accuracy for both fast and slow transients easier.

In some packages, the size of the first time step after a discontinuity or at the beginning of the simulation is arbitrarily set to some tiny fraction of the desired duration, because reliable information needed to apply Eq. A.98 is lacking. Such an approach will be inefficient when there are many discontinuities. On the other hand, allowing a larger time step could risk stepping over the next discontinuity that is close. Clearly, some LTE control to adjust the initial step size is desirable. Immediately after a discontinuity, the orders of the integration and extrapolating polynomials should be reset to one, and the value of y^p_{i+1} in Eq. A.98 should be calculated from the forward Euler formula instead, that is

$$y^p_{n+1} = y_n + L_1 y'_n \tag{A.103}$$

At each subsequent step, the aim is for as few Newton's iterations as possible by picking values of h_{allow} small enough so that the predicted values of state variables are mostly within the error tolerance range of the corresponding corrector values. For example, our experimentation with TARDIS[143] on electrical drive systems led to the following relation between h_{allow} and ϵ_{NR}:

$$
h_{allow} = \left(.75 \frac{\epsilon_{NR}}{|y_{n+1} - y^p_{n+1}|} \right)^{\frac{.7}{K}+1}
\tag{A.104}
$$

where ϵ_{NR} was the convergence tolerance specified by the user for the Newton's algorithm. The .75 factor was used to reduce the ratio of the expected error in the Newton's algorithm to the correction of the predictors at the current step. The .7 factor in the exponent was an adjustment to the fact that Eqs. A.100 and A.102 are just rough estimates of the relationship between local truncation error and step size.

A.10 CHANGING ORDER OF INTEGRATION

When dealing with a system that has a smooth response, a higher-order integration formula generally will permit larger time steps, which translates to fewer steps for a given duration; however, each step will require slightly more overhead computation than that of a lower-order scheme. The order of integration is adjusted to yield the largest allowable h_{allow} that is obtained by using values of y^p_{n+1}, β, and L_K of three different orders: the current order, one order lower, and one order higher. Some hysteresis is introduced in these comparisons to avoid frequent changes of the order of integration. For example, the order will be reduced only if

$$
\frac{h_{allow(K-1)}}{1.1} > h_{allow(K)} > \frac{h_{allow(K+1)}}{1.21}
\tag{A.105}
$$

Conversely, the order will be raised only if

$$
\frac{h_{allow(K-1)}}{1.095} < h_{allow(K)} < \frac{h_{allow(K+1)}}{1.2}
\tag{A.106}
$$

A.11 HANDLING OF CONDITIONAL DISCRETE EVENTS

Scheduled and conditional discrete events are to be handled differently. With a scheduled event, the instant is known in advance. Integration can be carried right up to the scheduled instant, adjusting the size of the latest time step for the scheduled instant to be one of the meshpoints. With a conditional event, however the instant will not be known in advance and is detected only when it has been passed. In the switching function techniques, the actual instant can be determined from an interpolation between the previous and current values by locating the zero-crossing of the corresponding switching function that characterizes that event. For example, the switching function of a diode turning off and on can be that describing its current and forward voltage, respectively. Other suggested techniques of marking singularities are the perturbed polynomial [145] and rational function [146].

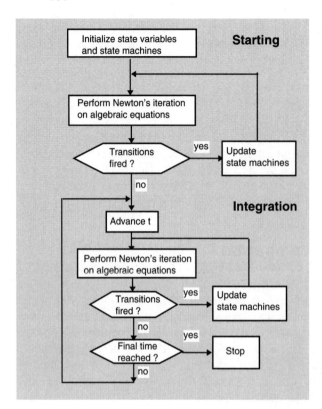

Figure A.4   Updating state transitions during the startup and integration phases.

The use of simple IF-ELSE statements to handle discrete events within the Newton's iteration loop will cause changes in parameters and/or equations from one Newton's iteration to the next. Such changes, if drastic, can disrupt the convergence. In [143], state machines are introduced in the model of a component capable of generating discrete events. The state of the state machine changes when any of the transitions connected to the current state are satisfied (or fired on). When there is more than one fired transition, the one with highest priority assigned by the user will be chosen by virtue of its order in the list of state machines. The program codes for state machines are kept separate from those of the system equations. The states of the state machines are updated together in the order of their assigned priority, only after the system equations and variables have been updated.

Figure A.4 shows a simple flowchart of how the state transitions are handled. At the start of a fresh run, the initial states of the state machines are determined in conjunction with the initial iterations of the state and non-state variables of the system equations, until a consistent set of conditions is obtained, following which, the integration proceeds a step forward. At the end of a step, the program determines whether any state machine has been fired. If none is detected, another step forward in integration can proceed. If there are fired transitions, the states of the corresponding state machines will be changed, the new system equations reflecting these changes in states will be solved, and the consistency of the new solution with the states of the state machines will be checked. The update and solution cycle will be repeated until the status of the state machines shows no further changes, that is, until the values of the variables and the states of the state machines are consistent with each other.

A change in state of the state machines may be accompanied by changes in the input excitation and/or parameters of the system. If the change causes an abrupt change in the system response, the interpolating polynomials approximated from values at previous time steps will not be applicable. The order of these polynomials and the integration must be reset to 1. If the order is not reset, it is likely that the next integration step will fail repeatedly and the step size eventually reduced to an undesirably small value. Similarly, with an abrupt change in the system parameters, the Jacobian matrix used in the Newton's method must also be updated.

A.11.1 Locating Discontinuities

Gear [138] uses the step size control to locate discontinuities. Ellison [139] and Birta, et al. [140] use the values of switching functions and the derivatives of their interpolating polynomials to locate discontinuities in conjunction with a Runge-Kutta integration algorithm. Some integration routines also have a root-finding capability [131, 141] that will locate the zeros of user-specified switching functions from the changes in the sign of their values.

Switching functions can be expressed in a manner such that a transition of its value from positive to negative marks the occurrence of some discrete event. Values of switching functions with active transitions are stored in terms of divided differences for interpolation purposes. The order of interpolation is equal to or less than the current integration order. When the order of integration is reset, the order of interpolating polynomials is also reset. Negative-going zero-crossings can be located by a routine modified from Brent's [142]. Depending on the nature of the switching function, Brent's routine uses one of the following three methods: linear interpolation, inverse quadratic interpolation, and binary search. Locating the zero-crossing points with interpolating polynomials is not always reliable because false zero-crossings can occur, usually a little ahead of the real zero-crossings. The located crossings will have to be checked by re-integrating from the previous point to the located instant of zero-crossing using very small time steps.

For resolving multiple transitions, a "band of certainty", as suggested by Birta, et al. [140], of width depending on the resolution desired, can be placed on the negative side. Short-lived discrete events are characterized by the value of their switching functions dipping negative briefly within a time step. They can be detected by the change in the sign of the slopes of the switching function at the two cross-over points. When the signs of the slopes at the cross-over points are different, the minimum, or turning, point of the switching function can be estimated from the corresponding interpolating polynomial.

Appendix B

CD-ROM

B.1 FILE FORMAT

The CD-ROM that accompanies this book contains the MATLAB and SIMULINK files of exercises and projects described in the book

Files for Macintosh machines are placed in a separate folder from that containing files for the UNIX and windows machines. In view of the pending release of MATLAB 5/SIMULINK 2 to replace the current version of MATLAB 4/SIMULINK 1, files for the new version are placed in the MATLAB5 folder and those for the current version are placed in MATLAB4. Within each of these two folders are sub-folders, C2 through C10, containing exercise and project files mentioned in Chapters 2 through 10 of the text.

In general, filenames beginning with the letter m are MATLAB M-files and those beginning with s are SIMULINK files. Many of the files will run on both versions, but some need minor changes here and there. For convenience, I have included the files for both MATLAB 4/SIMULINK 1 and MATLAB 5/SIMULINK 2. Those for SIMULINK version 1 will have an .m filename extension, whereas those for SIMULINK version 2 have an .mdl filename extension. For example, the MATLAB M-files for Project No. 3, m3.m say, and the corresponding SIMULINK files are

s3.m for SIMULINK version 1,
or S3.mdl for SIMULINK version 2.

The given files have been tested on MATLAB Version 4.2c and SIMULINK Version 1.0a, and on the pre-release MATLAB Version 5 and SIMULINK Version 2.

Also included on the CD-ROM are the following utility files:

fftplot.m - for computing and plotting the fft of a variable. The times and values of the variables are stored in separate row arrays.

sizeplot.m - for resizing the figure window

These utility files should be placed in a directory that is on the MATLAB's search path, for example, in the MATLAB/bin directory.

Depending on the display resolution used, the SIMULINK window may or may not display the given SIMULINK simulations in full. The viewing area can be increased by setting the display resolution higher, at the expense of smaller lettering. The simulation diagrams given should fit within an open screen when using a display resolution of 800×600 pixels or higher.

```
C*************************************************************
CONTENTS OF FOLDERS

CHAPTER 2 (C2 Folder) INTRODUCTION TO MATLAB/SIMULINK

Exercise 1: Variable Frequency Oscillator
SIMULINK file s1.mdl
MATLAB m-file m1.m

Exercise 2: Parallel RLC Circuit
SIMULINK file s2.mdl
MATLAB m-file m2.m
MATLAB m-file m2init.m(1st half of m2.m)
MATLAB m-file m2plot.m(2nd half of m2.m)

Exercise 3: ac Energization of a RL Circuit
SIMULINK file s3.mdl
MATLAB m-file m3.m

Exercise 4: Series RLC Resonant Circuit
SIMULINK file s4.mdl
MATLAB m-file m4.m

***************

CHAPTER 3 (C3 Folder) BASICS ON MAGNETICS AND LINE MODELING

Project 1: Line Parameters and Circuit Models:

   (a) Relative Accuracy of Circuit Models for
Different Line Lengths
   (b) Real and Reactive Power Transfer
SIMULINK file none
MATLAB M-file m1.m
```

Project 2: Switching Transients in Single-phase Line

SIMULINK file s2.mdl
MATLAB M-file m2.m

****

CHAPTER 4 (C4 Folder) TRANSFORMERS

Project 1: Short-circuit and RL Load Terminations

SIMULINK file s1a.mdl (linear transformer)
SIMULINK file s1b.mdl (piece-wise linear saturation)
SIMULINK file s1c.mdl (lookup table saturation)
MATLAB M-file m1.m
MATLAB M-file fftplot.m (fftplot.m should be
          placed in a directory on MATLAB's search path)

Project 2: Open-circuit Termination, In-rush Current,
and dc Bias Core Saturation

SIMULINK file s1c.mdl (lookup table saturation)
MATLAB M-file m1.m

  Use the given fftplot.m to obtain the discrete Fourier transform.

Project 3: Autotransformer Connection

SIMULINK file none
MATLAB M-file none

Project 4: Delta-Wye Transformer:Voltage and Current Ratios
Zero-sequence Current

SIMULINK file s4.mdl
MATLAB M-file m4.m

It also contains the following M-files for translating an
open-circuit curve to an instantaneous magnetization
curve and comparison study:

SIMULINK file smag.mdl
maginit.m (translates open-circuit curve and
sets up SIMULINK smag.m)
magplt.m (plot results from smag.m)

****

CHAPTER 5 (C5 Folder) BASICS OF ELECTRIC MACHINES AND TRANSFORMATION

Project 1: qd0 Transformation of Network Components

SIMULINK file none
MATLAB M-file none

Project 2: Space Vectors

SIMULINK file s2.mdl
MATLAB M-file m2.m (also used by masked block)

Project 3: Sinusoidal and Complex Quantities in qd0

SIMULINK file s3.mdl
MATLAB M-file m3.m (also used by masked block)

****

CHAPTER 6 (C6 Folder) THREE-PHASE INDUCTION MACHINES

Project 1: Operating Characteristics

SIMULINK file s1.mdl
MATLAB M-file m1.m to initialize simulation and plot results
MATLAB M-file p1hp.m (1 hp induction motor)

Project 2: Starting Methods

SIMULINK file (modify a copy of s1.mdl)
MATLAB M-file (modify a copy of m1.m)
MATLAB M-file p1hp.m (1-hp, three-phase induction motor)

Project 3: Open-circuit Conditions

SIMULINK file (modify a copy of s1.mdl)e
MATLAB M-file (modify a copy of m1.m)
MATLAB M-file p1hp.m (1-hp, three-phase induction motor)

Project 4: Linearized Analysis

SIMULINK file s4.mdl
MATLAB M-file m4.m to initialize, determine linear
 model, transfer function, and plot root-locus
MATLAB M-file p20hp.m (20-hp induction motor)

MATLAB M-file m4ustp.m calculates the unit step
 response of the transfer function set up in the
 MATLAB workspace by running m4.m, using the
 Control System Toolbox function step.

SIMULINK file s4stp.m for step response
MATLAB M-file m4stp.m to initialize s4step.m

Project 5: Some Non-zero v_{sg} Conditions

SIMULINK M-files s5a.mdl and s5b.mdl
MATLAB M-file m5.m to initialize simulation and plot results
MATLAB M-file p1hp.m (1-hp, three-phase induction motor)

Project 6: Single-phase Induction Motor

SIMULINK file s6.mdl
MATLAB M-file m6.m to initialize simulation and plot results
MATLAB M-file psph.m (1/4-hp, single-phase induction motor)

****

CHAPTER 7 (C7 Folder) SYNCHRONOUS MACHINES

Project 1: Operating Characteristics and Parameter Variations

SIMULINK file s1.mdl
MATLAB M-file m1.m to initialize simulation and plot results
MATLAB M-file set1.m (Set 1 synchronous generator parameters)

Project 2: Terminal Faults on a Synchronous Generator

SIMULINK file (modify a copy of s1.mdl)
MATLAB M-file m1.m
MATLAB M-file set1.m (Set 1 synchronous generator parameters)

Project 3: Linearized Analysis of a Synchronous Generator

SIMULINK files s3.mdl and s3eig.mdl
MATLAB M-file m3.m to initialize simulation,
        determine linear model, transfer functions, and plot
  root-locus and results.
MATLAB M-file set1.m (Set 1 synchronous generator parameters)

Project 4: Permanent Magnet Synchronous Motor

SIMULINK file s4.mdl
MATLAB M-file m4.m to initialize, calculate operating
 values, and plot results.

Project 5: Simulation of Synchronous Machine Model with Unequal
        Stator and Rotor Mutual and Coupling of Rotor Circuits

SIMULINK files s5.mdl and s1.mdl
MATLAB M-files m5.m and m1.m to initialize, calculate operating
 values, and plot results.
MATLAB M-file set3a.m (Set 3 generator parameters for s5.m)
MATLAB M-file set3b.m (Set 3 generator parameters for s5.m)
MATLAB M-file set3c.m (Set 3 generator parameters for s1.m)

Project 6: Six-phase Synchronous Machine

None

```
***************
```

CHAPTER 8 (C8 Folder) dc MACHINES

Project 1: Startup and Loading of a Shunt dc Generator

SIMULINK file s1.mdl
MATLAB M-file m1.m to initialize simulation
  and plot magnetization curve.

Project 2: Resistance Starting of a dc Shunt Motor

SIMULINK file s2.mdl
MATLAB M-file m2.m to initialize simulation
        and plot results.

Project 3: Methods of Braking

SIMULINK files s3a.mdl and s3b.mdl
MATLAB M-files m3a.m and m3b.m to initialize simulation
  and plot results.
m3a.m and s3a.mdl are for plugging and dynamic braking.
m3b.m and s3b.mdl are for regenerative braking.

Project 4: Universal Motor

SIMULINK file s4.mdl
MATLAB M-file m4.m to initialize simulation and plot results.

Project 5: Series dc Machine Hoist

SIMULINK files s5a.mdl and s5b.mdl
MATLAB M-file m5.m to initialize simulation
  and plot results.

```
***************
```

CHAPTER 9 (C9 Folder) CONTROL OF INDUCTION MACHINES

Project 1: Closed-loop Speed Using Scalar Control

   (a) Open loop system
SIMULINK file s1o.mdl
MATLAB M-file m1o.m to initialize s1o.m
  and plot magnetization curve.
MATLAB data file p20hp.m

   (b) Closed loop system
SIMULINK file s1c.mdl
MATLAB M-file m1c.m to initialize S1CLOSED.m
  and plot magnetization curve.
MATLAB M-file p20hp.m to put parameters

of the 20-hp motor in MATLAB workspace.

Project 2: Six-step Inverter/Induction Motor Drive

  (a) Open loop system
SIMULINK file s2O.mdl
MATLAB M-file m1O.m to initialize s2o.m
  and plot magnetization curve.
MATLAB data file p20hp.m

  (b) Closed loop system
SIMULINK file s2c.mdl
MATLAB M-file m1c.m to initialize s2c.m
  and plot magnetization curve.
MATLAB M-file p20hp.m to put parameters
  of the 20-hp motor in MATLAB workspace.

Project 3: Field-oriented Control

SIMULINK file s3.mdl
MATLAB M-file m3.m to initialize simulation
  and plot results.
MATLAB M-file p20hp.m to put parameters
  of the 20-hp motor in MATLAB workspace.

****

CHAPTER 10 (C10 Folder) SYNCHRONOUS MACHINES IN POWER SYSTEMS
                        AND DRIVES

Project 1: Transient Models

  (a) Linearized Analysis of a Synchronous Generator
SIMULINK file s1eig.mdl
MATLAB M-file m1.m to initialize s1eig.m and compute
  its steady-state and linearized model.
MATLAB M-file set1.m to put parameters
  of synchronous generator set 1 in MATLAB workspace.

  (b) Operating Characteristics and Parameter Sensitivity
Studies
SIMULINK file s1.mdl
MATLAB M-file m1.m to initialize s1.mdl
  and plot results of simulation.
MATLAB M-file set1.m to put parameters
  of synchronous generator set 1 in MATLAB workspace.

Project 2: Multi-machines System

SIMULINK files s2.mdl and s2eig.mdl:
  s2.mdl for simulation and s2eig.mdl for linearized model
MATLAB M-file m2.m to initialize conditions, determine
  linearized model, and plot results of simulation.

MATLAB M-files set1.m and set2.m to put parameters
   of synchronous generators, set 1 and set 2,
   in MATLAB workspace.

Project 3: Subsynchronous Resonance

   (a)   Fixed Terminal Voltage Case
SIMULINK files s3g.mdl and s3eig.mdl:
   s3g.mdl for simulation and s3geig.mdl for linearized model
MATLAB M-file m3g.m to initialize conditions, determine
   linearized model and eigenvalues, and plot mode shapes.
MATLAB M-file i3essr.m to put parameters
   of synchronous generator in the IEEE subsynchronous
   benchmark system in MATLAB workspace.

   (b)   Generator with Series Capacitor Compensated Line
SIMULINK files s3.mdl and s3eig.mdl:
   s3.mdl for simulation and s3eig.mdl for linearized model
MATLAB M-file m3.m to initialize conditions, determine
   linearized model, and plot results.
MATLAB M-file i3essr.m to put parameters
   of the IEEE subsynchronous benchmark system in
   MATLAB workspace.

Project 4: Power System Stabilizer

SIMULINK file s4.mdl for simulation.
MATLAB M-file m4.m to determine linearized model
   of synchronous generator and exciter,
   GEP(s) and Exc(s), and produce Bode plots.
MATLAB M-file m4comp.m to check preliminary design
   of pss and provide Bode and root-locus plots.
MATLAB M-file set1.m to put parameters
   of synchronous generator set 1 in MATLAB workspace.

Project 5: Self-controlled Permanent Magnet Motor Drive

SIMULINK file s5.mdl for simulation
MATLAB M-file m5.m to initialize drive system parameters
   and simulation conditions, plot steady-state curves,
   and plot results of simulation. It uses the MATLAB
   function files m5torqi.m and m5torqv.m.

**** END OF FILE ****

Bibliography

[1] Mitchell, Edward E. (1982), "Advanced Continuous Simulation Language (ACSL): An Update," IMACS World Congress on System Simulation and Scientific Computation, Montreal, Canada, Vol. 1, August 1982, pp. 462-464.

[2] Crosbie, R. E. and Hay, J. L.(1986), "Description and Processing of Discontinuities with the ESL Simulation Language," Proceedings of the Conference on Continuous System Simulation Languages, Francois E. C. (ed.), Society for Computer Simulation, San Diego, CA, 1986, pp. 30-35.

[3] Ummel, B. R.(1986), "Simplified Modeling of Discontinuous Phenomena Using EASY5 Switch states," Proceedings of the 1986 Summer Computer Simulation Conference, Roy Crosbie and Paul Luker (eds.), July 1986, pp. 99-104.

[4] Nagel, Laurence W. (1975), "SPICE2: A Computer Program to Simulate Semiconductor Circuits," Memorandum No. UCB/ERL M520, Electronics Research Laboratory, College of Engineering, University of California, Berkeley, CA, May 9, 1975.

[5] Halin, H.J. and Benz, H. (1982), "Continuous-System Simulation with PSCSP a New Simulation Program Based Upon Semi-Analysis Methods," IMACS World Congress on System Simulation and Scientific Computation, Montreal, Canada, Vol. 1, August 1982, pp. 358-360.

[6] Electromagnetic Transients Program (EMTP) Application Guide (1986), EPRI Report No:EL-4650, Project 2149-1, Westinghouse Electric Corp., Pittsburgh, PA.

[7] Rajagopalan, V. (1987), Computer-Aided Analysis of Power Electronic Systems, Dekker, 1987.

[8] Smith, D. W., Majdecki, S. A., and Johnson, D. (1987), "Interactive Control of Analog System Simulation," VLSI Systems Design, July 1987, pp. 46-54.

[9] MATH/LIBRARY FORTRAN Subroutines for Mathematical Applications, IMSL Inc., U.S.A., 1987, pp. 640-651.

[10] Hindmarsh, A. C. (1982), "Toward a Systematized Collection of ODE Solvers," IMACS World Congress on System Simulation and Scientific Computation, Montreal, Canada, Vol. 1, August 1982, pp. 427-429.

[11] Petzold, L. (1982), "A Description of DASSL: A Differential/Algebraic System Solver," IMACS World Congress on System Simulation and Scientific Computation, Montreal, Canada, Vol. 1, August 1982, pp. 430-432.

[12] Hanselman, D. and Littelfield. B. (1996), Mastering MATLAB, Prentice Hall, Englewood Cliffs, NJ.

[13] The MathWorks Inc. (1996), The Student Edition of SIMULINK, User's Guide, Prentice Hall, Englewood Cliffs, NJ, 1996, ISBN 0-13-452435-7.

[14] Parker, R. J. (1990), Advances in Permanent Magnetism, John Wiley & Sons, New York.

[15] Slemon, G. R. (1992), Electric Machines and Drives, Addison-Wesley Publishing Co., Inc., Reading, MA.

[16] Gumaste, A. V. and Slemon, G. R. (1981), "Steady-State Analysis of a Permanent Magnet Synchronous Motor Drive with Voltage-Source Inverter," IEEE Trans. on Industry Applications, Vol. IA-17, No. 2, March/April 1981, pp. 143-151.

[17] Pillay, P. and Krishnan, R. (1989), "Modeling, Simulation, and Analysis of Permanent Magnet Motor Drives, Part I: The Permanent Magnet Synchronous Motor Drive," IEEE Trans. on Industry Applications, Vol. IA-25, No. 2, March/April 1989, pp. 265-273.

[18] Hammond, P. and Sykulski J.K. (1994), Engineering Electromagnetism, Physical Processes and Computation, Oxford University Press, New York.

[19] Weedy, B. M. (1972), Electric Power Systems, second edition, John Wiley & Sons, New York.

[20] Grainger, J.J. and Stevenson, Jr. W. D. (1994), Power System Analysis, McGraw-Hill Publishing Co., New York.

[21] Triezenberg, D. M. (1978), Electric Power Systems, Class notes, Purdue University, West Lafayette, IN.

[22] Bergen, A.R. (1986), Power System Analysis, Prentice Hall, Englewood Cliffs, NJ.

[23] Central Station Engineers (1989), Electrical Transmission and Distribution Reference Book, Westinghouse Electric Corp., Pittsburgh, PA.

[24] Members of the Staff of the Department of Electrical Engineering, MIT (1950), Magnetic Circuits and Transformers, John Wiley & Sons, New York.

[25] Cherry, E. C. (1949), "The Duality Between Interlinked Electric and Magnetic Circuits and the Formation of Transformer Equivalent Circuits," Proc. Physical Society, Vol. 62B, 1949, pp. 101-111.

[26] Fitzgerald, A. E., Kingsley, Jr. C., and Umans S. D. (1990), Electric Machinery, Fifth Ed., McGraw-Hill Publishing Co., New York.

[27] Trutt, F. C., Erdelyi, E. A., and Hopkins, R. E. (1968), "Representation of the Magnetization Characteristic of DC Machines for Computer Use," IEEE Trans. on Power Apparatus and Systems, Vol. 87, No. 3, 1968, pp. 665-669.

[28] Widger, G. F. T. (1969), "Representation of Magnetization Curves over Extensive Range by Rational-Fraction Approximations," Proc. IEE, Vol. 116, 1969, pp. 156-160

[29] Macfadyen, W.K.,Simpson, R.R.S, Slater, R. D., and Wood, W. S. (1973), "Representation of Magnetization Curves by Exponential Series," Proc. IEE, Vol. 121, 1973, pp. 992-994.

[30] Prusty, S. and Rao, M.V.S. (1980), "A Direct Piecewise Linearized Approach to Convert RMS Saturation Characteristic to Instantaneous Saturation Curve," IEEE Trans. on Magnetics, Vol.16, No.1, 1975, pp. 156-160.

[31] Dommel, H. W. (1985), "Transformer Representations in the EMTP," Electromagnetic Transients Program Reference Manual, Bonneville Power Administration, Portland, OR.

[32] Langsdorf, A. S. (1955), Theory of Alternating-Current Machinery, McGraw-Hill Publishing Co., New York.

[33] Fortescue, C. L. (1918), "Method of Symmetrical Co-ordinates Applied to the Solution of Polyphase Networks," A.I.E.E. Transactions, Vol. 37, 1918, pp. 1027-1115.

[34] Park, R. H. (1929), "Two-Reaction Theory of Synchronous Machines Generalized Method of Analysis - Part I," A.I.E.E. Transactions, Vol. 48, 1929, pp. 716-727.

[35] Clarke, E. (1943), Circuit Analysis of Power Systems - Vol. I, Symmetrical and Related Components, John Wiley & Sons, New York.

[36] Langsdorf, A. S. (1955), Theory of Alternating-Current Machinery, McGraw-Hill Publishing Co., New York.

[37] Alger, P.L. (1970), Induction Machines, Their Behavior and Uses, Second Edition, Gordon and Breach Science Publishers, New York.

[38] Say, M.G. (1976), Alternating Current Machines, Pitman Publishing, London.

[39] Slemon, G. R. and Straughen, A. (1980), Electric Machines, Addison-Wesley Publishing Co., Reading, MA.

[40] McPherson, G. (1981), An Introduction to Electrical Machines and Transformers, John Wiley & Sons, New York.

[41] Stanley, H.C. (1938), "An Analysis of the Induction Motor," AIEE Trans., Vol. 57 (Supplement), 1938, pp. 751-755.

[42] Krause, P. C. and Thomas, C. H. (1965), "Simulation of Symmetrical Induction Machinery," IEEE Trans. Power Apparatus and Systems, Vol. 84, No.11, 1965, pp. 1038-1053.

[43] Krause, P. C. (1965), "Simulation of Unsymmetrical 2-Phase Induction Machines," IEEE Trans. Power Apparatus and Systems, Vol. 84, No. 11, 1965, pp. 1025-1037.

[44] Rogers, G. J. (1965), "Linearized Analysis of Induction Motor Transients," Proc. IEE, Vol. 112, 1965, pp. 1917-1926.

[45] Grace, A., Laub, A. J., Little, J. N., and Thompson, C. (1990), Control System Toolbox User's Guide, The Mathworks, Inc., 1990, Natick, MA.

[46] Novotny, D. W. and Wouterse, J. H. (1976), "Induction Machine Transfer Functions and Dynamic Response by Means of Complex Time Variables," IEEE Trans. Power Apparatus and Systems, Vol. 95, No. 4, 1976, pp. 1325-1335.

[47] Roger, G. J., Di Manno, J., and Alden, R. T. H. (1984), "An Aggregate Induction Motor Model for Industrial Plants," IEEE Trans. Power Apparatus and Systems, Vol. 103, No. 4, 1984, pp. 683-690.

[48] Willis, J. R., Brock, G. J. and Edmonds, J. S. (1989), "Derivation of Induction Motor Models from Standstill Frequency Response Tests," IEEE Trans. on Energy Conversion, Vol. 4, No. 4, 1989, pp. 608-615.

[49] Adkins, B. (1957), The General Theory of Electrical Machines, Chapman & Hall Ltd., London.

[50] Slemon, G. R. and Straughen, A. (1980), Electric Machines, Addison-Wesley Publishing Co., Reading, MA.

[51] Anderson, P.M. and Fouad, A. A. (1977), Power System Control and Stability, The Iowa State University Press, Ames, IA.

[52] McPherson, G. (1981), An Introduction to Electrical Machines and Transformers, John Wiley & Sons, New York.

[53] Krause, P. C., Wasynczuk, O., and Sudhoff, S. D. (1995), Analysis of Electric Machinery, 1995, IEEE Press, New York.

[54] Wright, S. H. (1931), "Determination of Synchronous Machine Constants by Test - Reactances, Resistances and Time Constants," A.I.E.E. Trans., Vol. 50, 1931, pp. 1331-1351.

[55] IEEE Standard 1110, "Guide for Synchronous Generator Modeling Practices in Stability Analysis," November 1991.

[56] IEEE Joint Working Group Paper, "Supplementary Definitions and Associated Test Methods for Obtaining Parameters for Synchronous Machine Stability Study Constants," Paper No: F 79 647-9, IEEE PES Summer Power Meeting, 1979, Vancouver.

[57] Kundur, P. (1994), Power System Stability and Control, McGraw-Hill Publishing Co., New York.

[58] Canay, I. M. (1969), "Causes of Discrepancies on Calculation of Rotor Quantities and Exact Equivalent Diagram of Synchronous Machines," IEEE Trans. on Power Apparatus and Systems, Vol. PAS 88, July 1969, pp. 1114-1120.

[59] Takeda, Y. and Adkins, B. (1974), "Determination of Synchronous Machine Parameters Allowing for Unequal Mutual Inductances," Proc. IEE, Vol. 121, No. 12, December 1974, pp. 1501-1504.

[60] Canay, I. M. (1983), "Determination of Model Parameters of Synchronous Machines," Proc. IEE, Vol. 130, No. 20, March 1983, pp. 86-94.

[61] Hannett, L. N. (1988), "Confirmation of Test Methods for Synchronous Machine Dynamic Performance Models," Final Report EPRI EL-5736, August 1988.

[62] Canay, I. M. (1993), "Determination of the Model Parameters of Machines from the Reactance Operators $x_d(p)$, $x_q(p)$, (Evaluation of Standstill Frequency Response Test)," IEEE Trans. on Energy Conversion, Vol. 8, No. 2, June 1993, pp. 272-279.

[63] Canay, I. M. (1993), "Modelling of Alternating-Current Machines Having Multiple Rotor Circuits," IEEE Trans. on Energy Conversion, Vol. 8, No. 2, June 1993, pp. 280-296.

[64] Alger, P. L., Ivanes, M., and Poloujadoff, M. (1974), "Equivalent Circuits for Double Squirrel-Cage Induction Motors," Paper C 74 222-6 Winter Power Meeting, January 1974.

[65] Minnich, S. H. (1986), "Small Signals, Large Signals, and Saturation in Generator Modeling," IEEE Trans. on Energy Conversion, Vol. 1, No. 1, March 1986, pp. 94-102.

[66] Tahan, S. A. and Kamwa, I. (1995), "A Two-Factor Saturation Model for Synchronous Machines with Multiple Rotor Circuits," IEEE Trans. on Energy Conversion, Vol. 10, No. 4, December 1995, pp. 609-616.

[67] Rahman, M. A. and Little, T. A. (1984), "Dynamic Performance Analysis of Permanent Magnet Synchronous Magnet Motors," IEEE Trans. on Power Apparatus and Systems, Vol. 103, No. 6, June 1984, pp. 1277-1282.

[68] Pillay, P. and Krishnan, R. (1989), "Modeling, Simulation and Analysis of Permanent Magnet Motor Drives, Part I: The Permanent Magnet Synchronous Motor Drive, Part II: The Brushless DC Motor Drive," IEEE Trans. on Industry Applications, Vol. 25, No. 2, March/April 1989, pp. 265-273 and pp. 274-279.

[69] Sebastian, T. and Slemon, G. R. (1989), "Transient Modeling and Performance of Variable Speed Permanent Magnet Motors," IEEE Trans. on Industry Applications, Vol. 25, No. 1, Jan. 1989, pp. 101-106.

[70] Fuchs, E. F. and Rosenberg, L. T. (1974), "Analysis of an Alternator with Two Displaced Stator Windings," IEEE Trans. on Power Apparatus and Systems, Vol. 93, No. 6, Nov/Dec. 1974, pp. 1776-1786.

[71] Kataoka, T., Watanabe, E. H. and Kitano J. (1981), "Dynamic Control of a Current Source Inverter/Double-Wound Synchronous Machine System for AC Power Supply," IEEE Trans. on Industry Applications, Vol. 17, No. 3, 1981, pp. 314-320.

[72] Schiferl, R. F. and Ong, C. M. (1983), "Six Phase Synchronous Machine with AC and DC Stator Connections, Part I: Equivalent Circuit Representation and Steady-State Analysis and Part II: Harmonic Studies and a Proposed Uninterruptible Power Supply Scheme," IEEE Trans. on Power Apparatus and Systems, Vol. 102, No. 8, August 1983, pp. 2685-2693 and pp. 2694-2701.

[73] Shackshaft, G. and Henser, P. B. (1979), "Model of Generator Saturation for Use in Power System Studies," IEEE Proc. C, Vol. 126, No. 8, 1979, pp. 759-763.

[74] Say, M. G. and Taylor, E. O. (1980), Direct Current Machines, Pitman Publishing Ltd., London.

[75] Landsdorf, A. S. (1959), Principles of Direct-Current Machines, McGraw-Hill Publishing Co., New York.

[76] Siskind, C. S. (1949), Direct-Current Armature Windings, Theory and Practice, McGraw-Hill Publishing Co., New York.

[77] Dubey, G. K. (1989), Power Semiconductor Controlled Drives, Prentice Hall, Englewood Cliffs, NJ.

[78] Leonhard, W. (1985), Control of Electrical Drives, Springer-Verlag, New York.

[79] Murphy, J. M. D. and Turnbull, F. G. (1988), Power Electronic Control of AC Motors, Pergamon Press, Oxford, England.

[80] Dubey, G. K. (1989), Power Semiconductor Controlled Drives, Prentice Hall, Englewood Cliffs, NJ.

[81] Bose, B. K. (1986), Power Electronics and AC Drives, Prentice Hall, Englewood Cliffs, NJ.

[82] Thorborg, K. (1986), "Staircase PWM, An Uncomplicated and Efficient Modulation Technique for AC Motor Drives," IEEE Power Electronics Specialist Conference Record, 1986, pp. 593-602.

[83] Klaes, N. R. and Ong, C. M. (1987), "Implementation and Study of a Digital Staircase Pulse Width Modulator," IEEE Power Electronics Specialist Conference Record, 1987, pp. 128-135.

[84] Patel, H. S. and Hoft, R. G. (1973), "Generalized Techniques of Harmonic Elimination and Voltage Control, Part I - Harmonic Elimination," IEEE Trans. on Industry Applications, Vol. IA-10, May/June 1973, pp. 310-317.

[85] Patel, H. S. and Hoft, R. G. (1973), "Generalized Techniques of Harmonic Elimination and Voltage Control, Part II - Voltage Control Techniques," IEEE Trans. on Industry Applications, Vol. IA-10, Sept/Oct. 1973, pp. 666-673.

[86] Buja, G. and Indri, G. (1975), "Improvement of Pulse Width Modulation Techniques," Archive fur Elektrotechnik, Vol. 57, 1975, pp. 281-289.

[87] Bowes, S. R. and Bullough, R. (1986), "PWM Switching Strategies for Current-Fed Inverter Drives," Proc. IEE, Vol. 131, Pt. B, No. 5, Sept. 1984, pp. 195-202.

[88] van der Broeck, H. W., Skudelny, H. C., and Stanke, G. V. (1988), "Analysis and Realization of a Pulse-width Modulator Based on Voltage Space Vector," IEEE Trans. on Industry Applications, Vol. IA-24, Jan/Feb. 1988, pp. 142-150.

[89] Kliman, G. B. and Plunkett, A. B. (1979), "Development of a Modulation Strategy for a PWM Inverter Drive," IEEE Trans. on Industry Applications, Vol. IA-15, Jan/Feb. 1979, pp. 72-79.

[90] Pollmann, A. J. (1986), "Software Pulse Width Modulation for μP Control of AC Drives," IEEE Trans. on Industry Applications, Vol. IA-22, July/Aug. 1986, pp. 691-696.

[91] Zubek, J., Abbondanti, A., and Nordby, C. J. (1975), "Pulse-width Modulated Inverter Motor Drives With Improved Modulation," IEEE Trans. on Industry Applications, Vol. IA-11, Nov/Dec. 1975, pp. 695-703.

[92] Abbondanti, A. (1977), "Method of Flux Control in Induction Motors Driven by Variable Frequency, Variable Voltage Supplies," Proc. of IEEE/IAS International Semiconductor Power Conference, 1977, pp. 177-184.

[93] Blaschke, F. (1972), "The Principle of Field Orientation as Applied to the New Transvektor Closed-Loop Control System for Rotating-Field Machines," Siemens Review, Vol. 34, May 1972, pp. 217-220.

[94] Gabriel, R., Leonhard, W., and Norby, C. J. (1980), "Field-Oriented Control of a Standard AC Motor Using Microprocessors," IEEE Trans. on Industry Applications, Vol. IA-16, Mar/April 1980, pp. 186-192.

[95] Novotny, D. W. and Lorenz, R. D. (1985), "Introduction to Field Orientation and High Performance AC Drives," IEEE IAS Tutorial Course, Oct. 1985, Toronto.

[96] Novotny, D. W. and Lipo, T. A. (1996), Vector Control and Dynamics of AC Drives, Oxford University Press Inc., New York.

[97] Bayer, K. H. and Blaschke, F. (1977), "Stability Problems with the Control of Induction Machines Using the Method Field Orientation," Proc. Second IFAC Symposium on Control in Power Electronics and Electrical Drives, 1977, pp. 483-492.

[98] Garces, L. J. (1980), "Parameter Adaption for Speed-Controlled Static AC Drive With a Squirrel-Cage Induction Motor," IEEE Trans. on Industry Applications, Vol. IA-16, Mar/April 1980, pp. 173-178.

[99] Rowan, T. M. and Kerkman, R. J. (1985), "A New Synchronous Current Regulator and an Analysis of Current Regulated PWM Inverters," Conf. Record of the 1985 IEEE-IAS Annual Meeting, pp. 487-495.

[100] Matsuo, T. and Lipo, T. (1985), "A Rotor Parameter Identification Scheme for Vector-Controlled Induction Motor Drives," IEEE Trans. on Industry Applications, Vol. IA-21, May/June 1985, pp. 624-632.

[101] Concordia, C. (1951), Synchronous Machines, Theory and Performance, John Wiley & Sons, New York.

[102] Riaz, M. (1974), "Hybrid-Parmater Models of Synchronous Machines," IEEE Trans. on Power Apparatus and Systems, Vol. 93, No. 3, 1974, pp. 849-858.

[103] DeMello, F. P. and Concordia, C. (1969), "Concepts of Synchronous Machine Stability as Affected by Excitation Control," IEEE Trans. on Power Apparatus and Systems, Vol. 88, No. 4, 1969, pp. 316-329.

[104] Larsen, E. V. and Swann, D. A. (1981), "Applying Power System Stabilizers; Part I: General Concepts; Part II: Performance Objectives and Tuning Concepts; Part III: Practical Considerations," IEEE Trans. on Power Apparatus and Systems, Vol. 100, No. 6, 1981, pp. 3017-3046.

[105] Walker, D. N., Bowler, C. E. J., Jackson, R. L., and Hodges, D. A. (1975), "Results of Subsynchronous Resonance Test at Mohave," IEEE Trans. on Power Apparatus and Systems, Vol. 94, No. 5, 1975, pp. 1878-1889.

[106] IEEE Subsynchronous Resonance Working Group (1980), "Proposed Terms and Definitions for Subsynchronous Oscillations," IEEE Trans. on Power Apparatus and Systems, Vol. 99, No. 2, 1980, pp. 506-511.

[107] Bahrman, M., Larsen, E. V., Piwko, R. J., and Patel, H. S. (1980), "Experience with HVDC-Turbine Generator Torsional Interaction at Square Butte," IEEE Trans. on Power Apparatus and Systems, Vol. 99, No. 3, 1980, pp. 966-975.

[108] IEEE Committee Report (1968), "Computer Representation of Excitation Systems," IEEE Trans. on Power Apparatus and Systems, Vol. 87, No. 5, 1968, pp. 1460-1464.

[109] IEEE Committee Report (1969), "Proposed Excitation System Definitions for Synchronous Machines," IEEE Trans. on Power Apparatus and Systems, Vol. 88, No. 8, 1969, pp. 1248-1258.

[110] IEEE Committee Report (1973), "Excitation System Dynamic Characteristics," IEEE Trans. on Power Apparatus and Systems, Vol. 92, No. 1, 1973, pp. 64-75.

[111] IEEE Committee Report (1981), "Excitation System, Model for Power System Stability Studies," IEEE Trans. on Power Apparatus and Systems, Vol. 100, No. 2, 1981, pp. 494-509.

[112] IEEE Recommended Practice for Excitation System Models for Power System Stability Studies, IEEE Standard 421.5-1992.

[113] IEEE Committee Report (1973), "Dynamic Models for Steam and Hydro Turbines in Power System Studies," IEEE Trans. on Power Apparatus and Systems, Vol. 92, No. 6, 1973, pp. 1904-1915.

[114] IEEE Working Group Report (1991), "Dynamic Models for Fossil Fueled Steam Units in Power System Studies," IEEE Trans. on Power Systems, Vol. 6, No. 2, 1991, pp. 753-761.

[115] IEEE Working Group Report (1992), "Hydraulic Turbine and Turbine Control Models for System Dynamic Studies," IEEE Trans. on Power Systems, Vol. 7, No. 1, 1992, pp. 167-179.

[116] Bowler, C. E. J. (1976), "Understanding Subsynchronous Resonance in Analysis and Control of Subsynchronous Resonance," IEEE PES Special Publication 76 CH 1066-0-PWR, pp. 66-73.

[117] IEEE SSR Working Group Report (1977), "First IEEE Benchmark Model for Computer Simulation of Subsynchronous Resonance," IEEE Trans. on Power Apparatus and Systems, Vol. 96, No. 5, 1977, pp. 1565-1572.

[118] IEEE SSR Working Group Report (1985), "Second Benchmark Model for Computer Simulation of Subsynchronous Resonance," IEEE Trans. on Power Apparatus and Systems, Vol. 104, No. 5, 1985, pp. 1057-1066.

[119] Bayer, K-H., Waldmann, H., and Weibelzahl, M. (1972), "Field-Oriented Closed-Loop Control of A Synchronous Machine With the New Transvektor Control System," Siemens Review XXXIX, No. 3, 1972, pp. 220-223.

[120] Bose, B. K. (1988), "A High-Performance Inverter-Fed Drive System of an Interior Permanent Magnet Synchronous Machine," IEEE Trans. on Industry Applications, Vol. 24, No. 6, Nov/Dec. 1988, pp. 987-997.

[121] Rosenbrock, H. H. (1963), "Some General Implicit Processes for the Numerical Solution of Differential Equations," Computer Journal 5, pp. 329-330.

[122] Butcher, J. C. (1964), "Implicit Runge Kutta Processes," Mathematics of Computation 18, pp. 50-64.

[123] Mitchell, E. E. (1982), "Advanced Continuous Simulation Language (ACSL),: an update," IMACS World Congress on System Simulation and Scientific Computation, Montreal, Canada, Vol. 1, Amsterdam: North-Holland, pp. 462-464.

[124] Nagel, L. W. (1975), "SPICE2: A Computer Program to Simulate Semiconductor Circuits," Memorandum No. UCB/ERL M520, Electronics Research Laboratory, College of Engineering, University of California, Berkeley, CA.

[125] MATH/LIBRARY FORTRAN Subroutines for Mathematical Applications, IMSL Inc., U. S. A., 1987, pp. 640-651.

[126] Hindmarsh, A. C. (1982), "Toward a Systematized Collection of ODE Solvers," IMACS World Congress on System Simulation and Scientific Computation, Montreal, Canada, Vol. 1, Amsterdam: North-Holland, pp. 427-429.

[127] Petzold, L. (1982), "A Description of DASSL: A Differential/Algebraic System Solver," IMACS World Congress on System Simulation and Scientific Computation, Montreal, Canada, Vol. 1, Amsterdam: North-Holland, pp. 430-432.

[128] Enright, W. H. (1985), "The Comparison of Numerical Methods for Stiff ODEs," Stiff Computation, Richard C. Aiken (ed.), Oxford University Press, New York, pp. 175-180.

[129] Vlach, J. and Singhal, K. (1983), Computer Methods for Circuit Analysis and Design, Van Nostrand Reinhold Company, New York, pp. 364-394.

[130] Van Bokhoven, W. M. G. (1975), "Linear Implicit Differentiation Formulas of Variable Step and Order," IEEE Trans. Circuits and Sys., Vol. CAS-22, No. 2, pp. 109-115.

[131] Petzold, L. and Hindmarsh, A. C., LSODAR Code from ODEPACK.

[132] Brayton, R. K., Gustavson, F. G., and Hachtel, G. D. (1972), "New Efficient Algorithm for Solving Differential-Algebraic Systems Using Implicit Backward Differentiation Formulas," Proc. IEEE, Vol. 60, No. 1, pp. 98-108.

[133] Stoer, J. and Bulirsch, R. (1980), Introduction to Numerical Analysis, Springer-Verlag, New York, p. 267.

[134] Duff, I. S., Erisman, A. M., and Reid, J. K. (1989), Direct Methods for Sparse Matrices, Oxford UP, Oxford.

[135] Higham, N. and Higham, D. J. (1989), "Large Growth Factors in Gaussian Elimination with Pivoting," SIAM J. Matrix Anal. Appl., Vol. 10, No. 2, pp. 155-164.

[136] Markowitz, H. (1957), "The Elimination Form of the Inverse and its Application to Linear Programming," Management Science, Journal of the Institute of Management Sciences, Vol. 3, No. 3, pp. 255-269.

[137] Osterby, O. and Zlatev, Z. (1983), Direct Methods for Sparse Matrices, Lecture Notes in Computer Science, Springer-Verlag, Berlin, pp. 48-49.

[138] Gear, C. W. (1984), "Efficient Step Size Control for Output and Discontinuity," Trans. of the Society for Computer Simulation, Vol. 1, No. 1, pp. 27-31.

[139] Ellison, D. (1981), "Efficient Automatic Integration of Ordinary Differential Equations with Discontinuities," Math. and Comp. in Simulation 23, pp. 12-20.

[140] Birta, L. G., Eren, T. I., and Kettenis, D. L. (1985), "A Robust Procedure for Discontinuity Handling in Continuous System Simulation," Trans. of the Society for Computer Simulation, Vol. 2, No. 3, 1985, pp. 189-205.

[141] Hindmarsh, A. C. (1985), "The ODEPACK Solvers," Stiff Computation, Richard C. Aiken (ed.), Oxford University Press, New York, 1985, pp. 167-174.

[142] Brent, R. P. (1973), Algorithms for Minimization without Derivatives, Prentice Hall, Englewood Cliffs, NJ, pp. 47-60, 187-191.

[143] Suwanwisoot, W. (1989), TARDIS: A Numerical Simulation Package for Drive Systems, Ph. D. thesis, Purdue University, West Lafayette, IN, December 1989.

[144] Hindmarsh, A. C. (1974), "GEAR: ODE System Solver," Revision 3, Report No. UCID-30001, Lawrence Livermore Lab., University of California, Livermore, CA, 1974.

[145] Lambert, J. D. and Shaw, B. (1965), "On the Numerical Solution of $y' = f(x, y)$, by a Class of Formulae Based on Rational Approximations," Mathematics of Computation 19, pp. 456-462.

[146] Luke, Y. L., Fair, W., and Wimp, J. (1975), "Predictor Corrector Formulas Based on Rational Interpolants," Journal on Computers and Mathematics with Applications 1, pp. 3-12.

[147] Marti, J. R. and Lin, J. (1989), "Suppression of Numerical Oscillations in the EMTP," IEEE Trans. on Power Systems, Vol. 4, No. 2, 1989, pp. 739-747.

[148] The Mathworks Inc. (1997), Student Editions of MATLAB 5 and SIMULINK 2, Prentice Hall, Englewood Cliffs, NJ; MATLAB 5: ISBN 0-13-272477-4 (for Windows), ISBN 0-13-272485-5 (for Macintosh); SIMULINK 2: ISBN 0-13-659673-8 (for Windows), ISBN 0-13-659681-9 (for Macintosh).

Index

A

absolute error, 570
Adams methods, 585
airgap, 126, 132
   flux density, 132
Ampere's circuit law, 33
average flux density, 128

B

backward difference, 582
base frequency, 91, 178
Bewley lattice diagram, 68
BH characteristics, 32
block modulation, 430
brushless dc, 502, 559

C

changing order of integration, 595
characteristic equation, 495
circuit breaker, 80
circuit breaker, closing resistance, 80
circuit breaker, opening resistance, 80
closed or implicit integration formula, 587
closure domains, 31
co-energy, 38, 138
coercive force, H_c 33
coercivity, 33
computing tools, 4
   analog simulator, 4
   digital computer, 5
   network analyzer, 4
conditional discrete events, 595
constant flux linkage theorem, 293
Crout's algorithm, 593
crystal anisotropy, 31
current source, 515
current-fed inverter, 428

D

damping coefficient, 313, 494, 497, 519
dc machine, 351

truncation error, 570
turns per pole pair, 135

V

variable frequency oscillator, 15
voltage-fed inverter, 428

W

wash-out, 553
winding
    armature, 123
    back pitch, 354
    coil sides, 353
    commutator, 351, 364
    concentrated coil mmf, 131
    concentric, 122
    distributed, 123
    distributed winding mmf, 134
    distribution factor, 129
    excitation, 123
    field, 123
    flux per pole, 127
    front pitch, 355

full-pitch, 129
fundamental mmf, 134
induced voltage, 128
inductance, 136
lap-wound, 353
parallel path, 354
pitch factor, 129, 130
plexity, 354
progressive, 354
resultant mmf, 138
retrogressive, 354
rotating mmf, 137
short-pitch, 130
slot angle, 129
slot pitch, 353
turns, N_{eff}, 135
turns, N_{ph}, 134
turns, N_{sine}, 135
two-layer arrangement, 132, 355
wave-wound, 355
winding factor, 129

Z

zero-sequence, 54, 120, 141

LICENSE AGREEMENT AND LIMITED WARRANTY

READ THE FOLLOWING TERMS AND CONDITIONS CAREFULLY BEFORE OPENING THIS SOFTWARE PACKAGE. THIS LEGAL DOCUMENT IS AN AGREEMENT BETWEEN YOU AND PRENTICE-HALL, INC. (THE "COMPANY"). BY OPENING THIS SEALED SOFTWARE PACKAGE, YOU ARE AGREEING TO BE BOUND BY THESE TERMS AND CONDITIONS. IF YOU DO NOT AGREE WITH THESE TERMS AND CONDITIONS, DO NOT OPEN THE SOFTWARE PACKAGE. PROMPTLY RETURN THE UNOPENED SOFTWARE PACKAGE AND ALL ACCOMPANYING ITEMS TO THE PLACE YOU OBTAINED THEM FOR A FULL REFUND OF ANY SUMS YOU HAVE PAID.

1. **GRANT OF LICENSE:** In consideration of your purchase of this book, and your agreement to abide by the terms and conditions of this Agreement, the Company grants to you a nonexclusive right to use and display the copy of the enclosed software program (hereinafter the "SOFTWARE") on a single computer (i.e., with a single CPU) at a single location so long as you comply with the terms of this Agreement. The Company reserves all rights not expressly granted to you under this Agreement.

2. **OWNERSHIP OF SOFTWARE:** You own only the magnetic or physical media (the enclosed SOFTWARE) on which the SOFTWARE is recorded or fixed, but the Company and the software developers retain all the rights, title, and ownership to the SOFTWARE recorded on the original SOFTWARE copy(ies) and all subsequent copies of the SOFTWARE, regardless of the form or media on which the original or other copies may exist. This license is not a sale of the original SOFTWARE or any copy to you.

3. **COPY RESTRICTIONS:** This SOFTWARE and the accompanying printed materials and user manual (the "Documentation") are the subject of copyright. You may not copy the Documentation or the SOFTWARE, except that you may make a single copy of the SOFTWARE for backup or archival purposes only. You may be held legally responsible for any copying or copyright infringement which is caused or encouraged by your failure to abide by the terms of this restriction.

4. **USE RESTRICTIONS:** You may not network the SOFTWARE or otherwise use it on more than one computer or computer terminal at the same time. You may physically transfer the SOFTWARE from one computer to another provided that the SOFTWARE is used on only one computer at a time. You may not distribute copies of the SOFTWARE or Documentation to others. You may not reverse engineer, disassemble, decompile, modify, adapt, translate, or create derivative works based on the SOFTWARE or the Documentation without the prior written consent of the Company.

5. **TRANSFER RESTRICTIONS:** The enclosed SOFTWARE is licensed only to you and may not be transferred to any one else without the prior written consent of the Company. Any unauthorized transfer of the SOFTWARE shall result in the immediate termination of this Agreement.

6. **TERMINATION:** This license is effective until terminated. This license will terminate automatically without notice from the Company and become null and void if you fail to comply with any provisions or limitations of this license. Upon termination, you shall destroy the Documentation and all copies of the SOFTWARE. All provisions of this Agreement as to warranties, limitation of liability, remedies or damages, and our ownership rights shall survive termination.

7. **MISCELLANEOUS:** This Agreement shall be construed in accordance with the laws of the United States of America and the State of New York and shall benefit the Company, its affiliates, and assignees.

8. **LIMITED WARRANTY AND DISCLAIMER OF WARRANTY:** The Company warrants that the SOFTWARE, when properly used in accordance with the Documentation, will operate in substantial conformity with the description of the SOFTWARE set forth in the Documentation. The Company does not warrant that the SOFTWARE will meet your requirements or that the oper-

ation of the SOFTWARE will be uninterrupted or error-free. The Company warrants that the media on which the SOFTWARE is delivered shall be free from defects in materials and workmanship under normal use for a period of thirty (30) days from the date of your purchase. Your only remedy and the Company's only obligation under these limited warranties is, at the Company's option, return of the warranted item for a refund of any amounts paid by you or replacement of the item. Any replacement of SOFTWARE or media under the warranties shall not extend the original warranty period. The limited warranty set forth above shall not apply to any SOFTWARE which the Company determines in good faith has been subject to misuse, neglect, improper installation, repair, alteration, or damage by you. EXCEPT FOR THE EXPRESSED WARRANTIES SET FORTH ABOVE, THE COMPANY DISCLAIMS ALL WARRANTIES, EXPRESS OR IMPLIED, INCLUDING WITHOUT LIMITATION, THE IMPLIED WARRANTIES OF MERCHANTABILITY AND FITNESS FOR A PARTICULAR PURPOSE. EXCEPT FOR THE EXPRESS WARRANTY SET FORTH ABOVE, THE COMPANY DOES NOT WARRANT, GUARANTEE, OR MAKE ANY REPRESENTATION REGARDING THE USE OR THE RESULTS OF THE USE OF THE SOFTWARE IN TERMS OF ITS CORRECTNESS, ACCURACY, RELIABILITY, CURRENTNESS, OR OTHERWISE.

IN NO EVENT, SHALL THE COMPANY OR ITS EMPLOYEES, AGENTS, SUPPLIERS, OR CONTRACTORS BE LIABLE FOR ANY INCIDENTAL, INDIRECT, SPECIAL, OR CONSEQUENTIAL DAMAGES ARISING OUT OF OR IN CONNECTION WITH THE LICENSE GRANTED UNDER THIS AGREEMENT, OR FOR LOSS OF USE, LOSS OF DATA, LOSS OF INCOME OR PROFIT, OR OTHER LOSSES, SUSTAINED AS A RESULT OF INJURY TO ANY PERSON, OR LOSS OF OR DAMAGE TO PROPERTY, OR CLAIMS OF THIRD PARTIES, EVEN IF THE COMPANY OR AN AUTHORIZED REPRESENTATIVE OF THE COMPANY HAS BEEN ADVISED OF THE POSSIBILITY OF SUCH DAMAGES. IN NO EVENT SHALL LIABILITY OF THE COMPANY FOR DAMAGES WITH RESPECT TO THE SOFTWARE EXCEED THE AMOUNTS ACTUALLY PAID BY YOU, IF ANY, FOR THE SOFTWARE.
SOME JURISDICTIONS DO NOT ALLOW THE LIMITATION OF IMPLIED WARRANTIES OR LIABILITY FOR INCIDENTAL, INDIRECT, SPECIAL, OR CONSEQUENTIAL DAMAGES, SO THE ABOVE LIMITATIONS MAY NOT ALWAYS APPLY. THE WARRANTIES IN THIS AGREEMENT GIVE YOU SPECIFIC LEGAL RIGHTS AND YOU MAY ALSO HAVE OTHER RIGHTS WHICH VARY IN ACCORDANCE WITH LOCAL LAW.

ACKNOWLEDGMENT

YOU ACKNOWLEDGE THAT YOU HAVE READ THIS AGREEMENT, UNDERSTAND IT, AND AGREE TO BE BOUND BY ITS TERMS AND CONDITIONS. YOU ALSO AGREE THAT THIS AGREEMENT IS THE COMPLETE AND EXCLUSIVE STATEMENT OF THE AGREEMENT BETWEEN YOU AND THE COMPANY AND SUPERSEDES ALL PROPOSALS OR PRIOR AGREEMENTS, ORAL, OR WRITTEN, AND ANY OTHER COMMUNICATIONS BETWEEN YOU AND THE COMPANY OR ANY REPRESENTATIVE OF THE COMPANY RELATING TO THE SUBJECT MATTER OF THIS AGREEMENT.

Should you have any questions concerning this Agreement or if you wish to contact the Company for any reason, please contact in writing at the address below.

Robin Short
Prentice Hall PTR
One Lake Street
Upper Saddle River, New Jersey 07458